J. Frank Adams was one of the world's leading topologists. He solved a number of celebrated problems in algebraic topology, a subject in which he initiated many of the most active areas of research. He wrote a large number of papers during the period 1955–1988, and they are characterised by elegant writing and depth of thought. Few of them have been superseded by later work.

This selection brings together all his major research contributions. They are organised by subject matter rather than in strict chronological order. This first volume contains papers on: the cobar construction, the Adams spectral sequence, higher order cohomology operations, and the Hopf invariant one problem; applications of K-theory; generalised homology and cohomology theories. The second volume is mainly concerned with Adams' contributions to: characteristic classes and calculations in K-theory; modules over the Steenrod algebra and their Ext groups; finite H-spaces and compact Lie groups; maps between classifying spaces of compact groups.

THE SELECTED WORKS OF J. FRANK ADAMS

THE SELECTED WORKS OF J. FRANK ADAMS VOLUME I

Edited by

J. P. MAY
Department of Mathematics, University of Chicago

and

C. B. THOMAS
Department of Pure Mathematics and Mathematical Statistics, University of Cambridge

CAMBRIDGE
UNIVERSITY PRESS

CAMBRIDGE UNIVERSITY PRESS
Cambridge, New York, Melbourne, Madrid, Cape Town, Singapore, São Paulo, Delhi

Cambridge University Press
The Edinburgh Building, Cambridge CB2 8RU, UK

Published in the United States of America by Cambridge University Press, New York

www.cambridge.org
Information on this title: www.cambridge.org/9780521110679

First published 1992
This digitally printed version 2009

A catalogue record for this publication is available from the British Library

ISBN 978-0-521-41063-2 hardback
ISBN 978-0-521-11067-9 paperback

Contents

Contents of Volume II

Introduction

Frank Adams was a great mathematician with a fine expository style. These two volumes contain the bulk of his numerous papers, grouped according to subject, and roughly chronologically within groups. We chose this organisation because of Adams' practice of returning periodically to certain subjects dear to his heart, re-examining them in the light of intervening research. We have added no editorial material since we prefer to let Adams' own introductions to his papers speak for themselves. We have also made no attempt to note errors since Adams was a scrupulously careful author. Several of his papers have published corrections; these we have included.

Adams' final bibliography contains 82 published items. These volumes contain 52 of them. The omitted items fall into five categories. There are five early papers (to 1957), an appendix to a paper of another author, and a published letter which we feel are probably not of lasting mathematical interest. There are four announcements of results which were published elsewhere. There are three biographical items. There are fifteen primarily expository articles, most of which were published in conference proceedings; we have chosen to include four of these, which we feel are of more lasting interest than the others.

The five remaining omitted items are books. The first of these, from 1964, is *Stable Homotopy Theory, Volume 3* in the Springer Lecture Notes in Mathematics. It contains material that has been superseded mathematically by results published elsewhere. Still, it has many delightful passages, and comparison of it with later work shows how rapidly algebraic topology in general, and Adams' work in particular, was progressing in those days. A second, *Algebraic Topology: a student's guide* consists primarily of selected reprints. The remaining three are highly recommended reading. Two of them, *Lectures on Lie Groups* and *Stable Homotopy and Generalized Homology* are being kept in print by the University of Chicago Press and remain among the best references on their subjects. The third, *Infinite Loop Spaces* is Study 90 of the Annals of Mathematics, published by Princeton University Press. It is vintage Adams, beautifully and humorously written, and it contains capsule summaries of various topics in algebraic topology other than the one described by the title.

The second volume ends with two expository articles which perhaps were not intended for publication. They well illustrate Adams' interest in the work of young algebraic topologists and his success in illuminating the most recent developments in his subject.

Biographical data

J. Frank Adams was born on 5 November 1930, in Woolwich, London, and died on 7 January 1989. He was married in 1953 and had one son and three daughters.

The main dates in his education were Bedford School until 1948, followed by military service in the Royal Engineers (1948–9) and attendance at Trinity College, Cambridge (1949–55), obtaining first class honours in Part II of the mathematical tripos (1951) and a distinction in Part III a year later. He was awarded his B.A. in 1952, his Ph.D. in 1955, his M.A. in 1956 and the higher degree of Sc.D. in 1982.

Following the completion of his Ph.D., he was appointed a Junior Lecturer at Oxford (1955–6). and he held a Research Fellowship at Trinity College, Cambridge, until 1958. During this time he also made his first visit to the University of Chicago (Summer 1957) and held a Commonwealth Fund Fellowship at the Institute of Advanced Study in Princeton (1957–8).

On returning to England he became an Assistant Lecturer at Cambridge, combining this with the Directorship of Studies in Mathematics at Trinity Hall. He was again a visiting member of the Institute at Princeton in the fall of 1961.

In 1962 Adams moved to the University of Manchester first as a Reader and then as Fielden Professor of Pure Mathematics (1964–71). In 1970 he returned to Cambridge as Lowndean Professor of Astronomy and Geometry, and he was also active as a Fellow of Trinity College.

His professional affiliations included the Association of University Teachers, the American Mathematical Society (since 1957) and the London Mathematical Society (since 1958). He was also elected a Fellow of the Royal Society in 1964 and a Foreign Associate of the National Academy of Sciences of the United States in 1985.

His frequent trips to the United States included ten lengthy visits to the University of Chicago, the last in 1985. Adams was in great demand as a speaker – his more important lectures included addresses to the International Congress of Mathematicians in Stockholm (1962) and in Moscow (1966), the Herman Weyl Lectures at the Institute in Princeton (1975) and an address to the American Mathematical Society as Bicentennial Exchange Lecturer (1976).

Over many years he contributed to the *Mathematical Reviews*, and in addition acted as referee for numerous other journals. He served as Editor, Member of the Editorial Board, or Editorial Advisor for three journals of the London Mathematical Society, the *Annals of Mathematics*, *Inventiones Mathematicae*, the *Journal of Pure and Applied Algebra*, and the *Mathematical Proceedings of the Cambridge Philosophical Society*. He also served on subcommittees to choose speakers in algebraic topology for the International Congress (once as Chairman), on the committee to choose Fields Medallists, and on the Consultative (programme) committee of the ICM. He gave similar organisational help to other conferences.

As a Fellow of the Royal Society he served on the mathematical sectional committee (two terms, one as chairman) and on the Council. He was also a member of the Council of the London Mathematical Society, and of the mathematics committee of the Science and Engineering Research Council (two terms).

Adams' honours include the Junior Berwick Prize (1963), the Senior Whitehead Prize (1974) and the Sylvester Medal (1982). Besides the Sc.D. from Cambridge he was also made a Doctor (h.c.) by the Universität Heidelberg in 1986.

Acknowledgements

The following papers appear in these volumes with grateful thanks to the original publishers:

On the chain algebra of a loop space, *Comment. Math. Helv.* **30** (1956) 305–330

On the structure and applications of the Steenrod algebra, *Comment. Math. Helv.* **32** (1958) 180–214

Reproduced here by kind permission of Birkäuser-Verlag.

Lectures on generalised cohomology, Lecture Notes in Mathematics 99 (1969) 1–138

Maps between classifying spaces, *Invent. Math.* **35** (1976) 1–41

Maps between classifying spaces II, *Invent. Math.* **49** (1978) 1–65

2-Tori in E_8, *Math. Ann.* **278** (1978) 29–39

Reproduced here by kind permission of Springer-Verlag.

The sphere, considered as an H-space mod p, *Quart. J. Math.* **12** (1961) 52–60

K-theory and the Hopf invariant, *Quart. J. Math.* **17** (1966) 31–38

Primitive elements in the K-theory of BSU, *Quart. J. Math.* **27** (1976) 253–262

Reproduced here by kind permission of Oxford University Press.

Atomic spaces and spectra, *J. Edinburgh. Math. Soc.* **32** (1989) 473–481

Reproduced here by kind permission of the Edinburgh Mathematical Society.

Finite H-spaces and Lie groups, *J. Pure Appl. Algebra* **19** (1980) 1–8

Reproduced here by kind permission of Elsevier Science Publishers.

Operations of the nth kind in K-theory and what we don't know about RP^∞, in *New developments in topology,* London Math. Soc. Lecture Note Series 11 (1974) 1–9

Maps between classifying spaces, III, London Math. Soc. Lecture Note Series 86 (1983) 136–153

On formulae of Thom and Wu, *Proc. London Math. Soc.* **11** (1961) 741–752

Hopf algebras of cooperations for real and complex K-theory, *Proc. London Math. Soc.* **23** (1971) 385–408

The Hurewicz homomorphism for MU and BP, *J. London Math. Soc.* **5** (1972) 539–545

Reproduced here by kind permission of the London Mathematical Society.

Maps between p-completed classifying spaces

Reproduced by kind permission of the Royal Society of Edinburgh from *Proc. Roy. Soc. Edinburgh*, **112**A, 231–235.

On matrices whose real linear combinations are nonsingular, *Proc. Amer. Math. Soc.* **16** (1965) 318–322 & Correction, *ibid.* **17** (1966) 193–222

Algebraic topology in the last decade, *Proc. Sympos. Pure Math.* **22** (1971) 1–22
Graeme Segal's Burnside ring conjecture, *Bull. Amer. Math. Soc.* **6** (1982) 201–210
The fundamental representations of E_8, *Contemp. Math.* **3** (1985) 1–10
Reproduced here by kind permission of the American Mathematical Society.

H-spaces with few cells, *Topology* **1** (1962) 67–72
On the groups $J(X)$—I, *Topology* **2** (1963) 181–195
On the groups $J(X)$—II, *Topology* **3** (1965) 137–171
On the groups $J(X)$—III, *Topology* **3** (1965) 193–222
On the groups $J(X)$—IV, *Topology* **5** (1966) 21–71 & Correction, *ibid.* **7** (1968) 331
Modules over the Steenrod algebra, *Topology* **10** (1971) 271–282
A variant of E. H. Brown's representability theorem, *Topology* **10** (1971) 185–198
The Segal conjecture for elementary abelian *p*-groups, *Topology* **24** (1985) 435–460
A generalization of the Atiyah–Segal completion theorem, *Topology* **27** (1988) 1–6
A generalization of the Segal conjecture, *Topology* **27** (1988) 7–21
Reproduced here by kind permission of Pergamon Press.

On the non-existence of elements of Hopf invariant one, *Ann. of Math.* **72** (1960) 20–104
Vector fields on spheres, *Ann. of Math.* **75** (1962) 603–632
Finite *H*-spaces and algebras over the Steenrod algebra, *Ann. of Math.* **111** (1980) 95–143 & Correction, *ibid.* **113** (1981) 621–622
Reproduced here by kind permission of The Annals of Mathematics.

Chern characters revisited, *Illinois J. Math.* **17** (1973) 333–336 & Addendum, *ibid.* **20** (1976) 372
Stable operations on complex *K*-theory, *Illinois J. Math.* **21** (1977) 826–829
Reproduced here by kind permission of the Illinois Journal of Mathematics.

An example in homotopy theory, *Proc. Cambridge Philos. Soc.* **53** (1957) 922–923
A finiteness theorem in homological algebra, *Proc. Cambridge Philos. Soc.* **57** (1961) 52–60
On Chern characters and the structure of the unitary group, *Proc. Cambridge Philos. Soc.* **57** (1961) 189–199
An example in homotopy theory, *Proc. Cambridge Philos. Soc.* **60** (1964) 699–700
On complex Stiefel manifolds, *Proc. Cambridge Philos. Soc.* **61** (1965) 81–103
A periodicity theorem, *Proc. Cambridge Philos. Soc.* **62** (1966) 365–377
The Kahn–Priddy theorem, *Proc. Cambridge Philos. Soc.* **73** (1973) 43–55
Sub-Hopf-algebras of the Steenrod algebra, *Proc. Cambridge Philos. Soc.* **76** (1974) 45–52
Operations on *K*-theory of torsion-free spaces, *Math. Proc. Cambridge Philos. Soc.* **79** (1976) 483–491
Uniqueness of *BSO*, *Math. Proc. Cambridge Philos. Soc.* **80** (1976) 475–509
Calculation of Lin's Ext groups, *Math. Proc. Cambridge Philos. Soc.* **87** (1980) 459–469
Reproduced here by kind permission of the Cambridge Philosophical Society.

Apart from vol. 30, fasc. 4 (1956)
COMMENTARII MATHEMATICI HELVETICI

On the chain algebra of a loop space

by J. F. ADAMS and P. J. HILTON

1. Introduction

An important concept in homotopy theory is that of the loop space of a given space. Given a CW-complex K, James has described in [4] a reduced product complex K_∞ which has the singular homotopy type of the space of loops on the suspension of K; and Toda has also introduced a standard path space (in [9]), performing essentially the same function[1]). In this paper, we consider the loop space of a CW-complex K which need not be a suspension but such that K^1 is a single point, the base-point[2]). We do not construct a combinatorial equivalent of ΩK, the loop space, but instead obtain a chain-equivalent of the cubical chain group of ΩK. Our method lends itself readily to the computation of the homology groups of ΩK.

There is a fibre-space (LK, p, K), where LK is the space of paths on K terminating in the base-point and p associates with every path its initial point. Then ΩK is the fibre. We will in fact construct a system of chain groups and maps equivalent to that given by the fibre-space.

In this paper we adopt J. C. Moore's definition of a path in a space X. In this definition a path is a pair (f, r) where r is a non-negative real number and f is a map of the closed interval $[0, r]$ into X. Paths (f, r), (g, s) such that $f(r) = g(0)$ are added by the rule $(f, r) + (g, s) = (h, r + s)$, where

$$h(t) = f(t) , \qquad 0 \leqslant t \leqslant r ,$$

$$h(t) = g(t - r) , \qquad r \leqslant t \leqslant r + s .$$

Let X^I be the space of maps of the unit interval I into X and let R be the set of non-negative real numbers with its usual topology. A function

1) We understand that J. C. Milnor has described a construction replacing the space of loops on a suitably restricted complex by an equivalent topological group.

2) This restriction could be avoided at the cost of an increase in complication in the proofs of our results (and a small modification in some statements). However, the restriction is not so serious in practice, since, for any CW-complex K, the universal cover of K is of the homotopy type of a CW-complex of the given kind.

$h : EX \to X^I \times R$, where EX is the set of paths on X, is given by $h(f, r) = (f', r)$ where $f'(t) = f(rt)$, $0 \leqslant t \leqslant 1$. Then EX is topologized by requiring h to be a homeomorphism onto its image. Let

$$\varrho_t : X^I \times R \to X^I \times R$$

be the deformation given by $\varrho_t(f, r) = (f, r(1-t) + t)$. Let LX, ΩX be the subsets of EX consisting of paths (f, r) such that $f(r) = x_*$, $f(0) = f(r) = x_*$ respectively, where x_* is the base-point in X. Then LX, ΩX are topologized as subsets of EX. Let $L'(X)$, $\Omega'(X)$ be the subspaces of X^I corresponding to LX, ΩX in the classical definition. Then $\varrho_1 h(LX) = L'(X) \times 1$, $\varrho_1 h(\Omega X) = \Omega'(X) \times 1$ and ϱ_t respects the subspaces $h(LX)$, $h(\Omega X)$. This shows that $LX \simeq L'(X)$, which is contractible, and $\Omega X \simeq \Omega'(X)$. Moreover a homotopy equivalence

$$(LX, \Omega X) \simeq (L'(X), \Omega'(X))$$

is given by $g(f, r) = f'$ where $f'(t) = f(rt)$.

The advantage of Moore's definition is that the pairing of LX and ΩX to LX, by composition of paths, is associative and ΩX possesses a unit. The chain groups $C_*(LX)$, $C_*(\Omega X)$ inherit these properties and the algebraical analogue we construct when X is a CW-complex will reproduce the multiplicative features of the chain groups of the fibre-space. In particular, we define in section 2 the notion of a chain algebra [3]) $A(K)$ which describes the additive and multiplicative structure of $C_*(\Omega K)$.

In section 2 we state and prove the main theorem. In section 3 we prove that our constructions behave properly with respect to maps (not necessarily cellular) of CW-complexes. In section 4 we consider the problem of the relation of $A(K_1 \times K_2)$ to $A(K_1)$ and $A(K_2)$. A generalization of Samelson's result (see [8]) on the relation between Whitehead and Pontryagin products is obtained by considering products of arbitrarily many spheres. We also study a product whose role in homotopy groups is closely related to that of the torsion product in homology groups and obtain an analogue of Samelson's result for this product.

It should be noted that the mapping $\Psi : \Omega(X_1 \times X_2) \to \Omega X_1 \times \Omega X_2$, given by $\Psi l = (p_1 l, p_2 l)$ where $p_i : X_1 \times X_2 \to X_i$, $i = 1, 2$, is the projection, is not a homeomorphism in Moore's definition. However it follows from the commutativity of the diagram

---

[3]) This will differ from a DGA-algebra over the integers, in the sense of Cartan ([2]), in not requiring that multiplication be anti-commutative.

$$\begin{array}{ccc} \Omega(X_1 \times X_2) & \xrightarrow{\Psi} & \Omega X_1 \times \Omega X_2 \\ \downarrow g & & \downarrow g_1 \times g_2 \\ \Omega'(X_1 \times X_2) & \xrightarrow{\Psi'} & \Omega'(X_1) \times \Omega'(X_2) \end{array}$$

that Ψ is a homotopy equivalence.

2. Chain-algebras and the main theorem

Let A be a differential graded free abelian group, $A = \sum_n A^n$ such that $A^n = 0$, $n < 0$, and $dA^n \subseteq A^{n-1}$. Then A will be called a chain algebra if a product is defined in A such that

(i) A is a ring with unit element;
(ii) $A^p A^q \subseteq A^{p+q}$;
(iii) $d(xy) = (dx)y + (-1)^p x(dy), \quad x \in A^p$.

We write 1 for the unit element; condition (ii) implies that $1 \in A^0$. A function φ from the chain algebra A to the chain algebra A' will be called a map if it is a chain mapping and a ring homomorphism [4]). An augmentation $\alpha : A \to A$ is a map whose image is the ring generated by 1. A map φ of augmented chain algebras is required to commute with α. Henceforth it will be understood that a chain algebra is augmented. The homology group $H_*(A)$ is an augmented graded ring with unit element and a map $\varphi : A \to A'$ induces a homomorphism

$$\varphi_* : H_*(A) \to H_*(A') .$$

Let $Q(\Omega K)$ be the group generated by the singular cubes of ΩK. Then the multiplication in ΩK induces a ring structure in $Q(\Omega K)$ in the usual way. Moreover the subgroup $D(\Omega K)$ generated by the degenerate singular cubes of ΩK (with respect to any co-ordinate) is an ideal in $Q(\Omega K)$. Let $C_*(\Omega K)$ be the quotient ring $Q(\Omega K)/D(\Omega K)$. Then $C_*(\Omega K)$ is a chain algebra with respect to the boundary operator induced by that in $Q(\Omega K)$; the unit element is the 0-cube at the unit element of ΩK and $C_*(\Omega K)$ is augmented by requiring α to be 1 on every 0-cube. The homology ring of $C_*(\Omega K)$ is the (singular) Pontryagin homology ring of ΩK. Our object is to use the structure of K as a CW-complex to construct a chain algebra A and a map $\theta : A \to C_*(\Omega K)$ such that θ_* is an isomorph-

---

[4]) We require a ring-homomorphism to have the property $\varphi(1) = 1$.

ism. With this end in view we write A' for $C_*(\Omega K)$. We recall that K is being restricted to having one 0-cell (the base-point) and no 1-cells.

Let $\{e_i^n\}$, $n = 0, 2, 3, \ldots$, $i \in$ indexing set T_n, be the cells of K and to each e_i^n except the vertex choose a generator $a_i = a_i^{n-1}$ of dimension $(n - 1)$. Let $A = A(K)$ be the ring with unit element freely generated by the elements a_i, and augmented by $\alpha(1) = 1$, $\alpha(a_i) = 0$, all i. Then A, provided with a suitable differential, will turn out to be the appropriate chain algebra.

Let LK be the space of paths on K terminating at the base-point and let $p : LK \to K$ associate with every path its initial point. Then $(LK, p; K)$ is a fibre-space with ΩK as fibre. Let $C_*(LK)$ be the group generated by the non-degenerate singular cubes of LK whose vertices lie in ΩK. Then $C_*(LK)$, given a graduation, differential and augmentation in the usual way, is the singular chain group of LK, which is, of course, acyclic. The pairing $LK \times \Omega K \to LK$, given by composition of paths, induces a pairing $(C_*(LK) \times C_*(\Omega K) \to C_*(LK)$ which is associative with a unit [in $C_*(\Omega K)$]. $C_*(LK)$ contains $C_*(\Omega K)$ and the pairing, restricted to $C_*(\Omega K) \times C_*(\Omega K)$, induces the ring structure in $C_*(\Omega K)$.

Let $C_*(K)$ be the singular chain group of K generated by the non-degenerate cubes of K all of whose vertices are at the base-point. Then the projection $p : LK \to K$ induces a chain mapping[5]) $p : C_*(LK) \to C_*(K)$. We proceed to construct a system of chain groups and maps equivalent to that given by the fibre-space.

To this end, we introduce a free graded abelian group $B = B(K)$, freely generated by elements $b_i = b_i^n$ in $(1 - 1)$ dimension-preserving correspondence with the cells of K. The element b^0 will be written 1. B is augmented by $\alpha(1) = 1$, $\alpha(b_i^n) = 0$, $n > 0$. Then B is intended to play the role of $C_*(K)$; the latter will therefore be called B'. Define $C = C(K)$ as the tensor product $B \otimes A$, graded and augmented by the usual rules. Then A, B may be embedded in C by identifying y with $1 \otimes y$, x with $x \otimes 1$, $y \in A$, $x \in B$. There is a pairing $C \times A \to C$ given by $(x \otimes y, y') \to x \otimes yy'$; restricted to $A \times A$, this pairing induces the multiplication in A. It is clearly legitimate to write a typical generator of C as xy; this will be done when convenient. A projection[6]) $\pi : C \to B$ is given by $\pi(xy) = \alpha(y)x$. Since C is to play the role of $C_*(LK)$, the latter will be called C'. We may now state the main theorem.

---

[5]) Where no confusion will arise, we will use the same symbol for a map and the induced chain mapping.

[6]) We may regard the augmentation of an element in A, B or C as an ordinary integer.

Theorem 2.1. *Differentials* $d : C, A \to C, A$, $\bar{d} : B \to B$, *and chain maps* $\theta : C, A \to C', A'$, $\bar{\theta} : B \to B'$ *may be defined such that*

(i) A *is a chain algebra with respect to* $d \mid A$;

(ii) $\theta \mid A$ *is a map of chain algebras and* θ *is product-preserving*[7]);

(iii) $\bar{\theta}\pi = p\theta$, $\pi d = \bar{d}\pi$;

(iv) $\theta_* : H_*(A) \cong H_*(A') = H_*(\Omega K)$.
$\bar{\theta}_* : H_*(B) \cong H_*(B') = H_*(K)$.
$\theta_* : H_*(C) \cong H_*(C') = H_*(LK)$.

Notice that since π maps C onto B, $\bar{d}$ and $\bar{\theta}$ are determined by d and θ.

The differential d and the map θ will be defined inductively on the sections of K. Let K^n be the n-section of K and let ${}^nA$, ${}^nB$, ${}^nC$ be $A(K^n)$, $B(K^n)$, $C(K^n)$ respectively; we regard them as embedded in A, B, C. Similarly we define ${}^nA'$, ${}^nB'$, ${}^nC'$ and embed them in A', B', C'.

For $n = 1$, define $d(1) = 0$, $\theta(1) = 1$; the theorem is trivially verified. Suppose now that d and θ have been determined on ${}^nC$, ${}^nA$ so that the theorem is verified. We proceed to determine d and θ on ${}^{n+1}C$, ${}^{n+1}A$. To determine d on ${}^{n+1}A$ it is sufficient to determine it on the generators. On the generators of dimension $<n$ we determine it by the embeddings ${}^nA \subseteq {}^{n+1}A$, ${}^nA' \subseteq {}^{n+1}A'$. Let a be a generator of dimension n, corresponding to a cell e^{n+1} in K^{n+1}. Let $f : E^{n+1}, S^n \to K^{n+1}, K^n$ be the characteristic map for this cell, inducing $f' : LS^n, \Omega S^n \to LK^n, \Omega K^n$, $f'' : LE^{n+1}, \Omega E^{n+1} \to LK^{n+1}, \Omega K^{n+1}$; and let $\beta \in H_{n-1}(\Omega S^n)$, with $\alpha(\beta) = 0$ if $n = 1$, be such that the suspension of β generates[8]) $H_n(S^n)$. Choose an $(n-1)$ cycle z in ${}^nA$ such that $\theta_*\{z\} = f'_*\beta$ – this is possible by the inductive hypothesis – and define $da = z$. Then $d^2 = 0$ on all cells and, hence, by the product rule, d^2 is zero on ${}^{n+1}A$. If $n = 1$, we must take $da = 0$, since $\alpha(\beta) = 0$, so that α is obviously an augmentation of A with respect to the differential being defined on A.

We next define a retraction $s : {}^{n+1}C \to {}^{n+1}C$, raising dimension by 1, by

$$(R1)\ s(1) = 0, \quad sa_i^{r-1} = b_i^r, \quad sb_i^r = 0, \quad r > 1 ,$$
$$(R2)\ s(xy) = (sx)y + (\alpha x)sy, \quad x \in {}^{n+1}C, \quad y \in {}^{n+1}A$$

and extend the differential to a differential d on ${}^{n+1}C$ by defining[9])

$$(D1)\ db_i^r = (1 - sd)a_i^{r-1}, \quad r > 1 ,$$
$$(D2)\ d(xy) = (dx)y + (-1)^p x dy, \quad x \in {}^{n+1}C^p, \quad y \in {}^{n+1}A .$$

[7]) In the sense of the pairings $C \times A \to C$, $C' \times A' \to C'$.

[8]) If $n > 1$, β generates $H_{n-1}(\Omega S^n)$.

[9]) Notice that the chain group B has the differential $\bar{d}$. B is only embedded in C as a subgroup.

Then s is clearly consistent with the two distributive laws; it is also consistent with the associative law of multiplication since

$$s(x(yz)) = (sx)(yz) + (\alpha x)s(yz) = (sx)(yz) + \alpha(x)(sy)z + \alpha(x)\alpha(y)sz\,,$$

while

$$s((xy)z) = (s(xy))z + \alpha(xy)sz = (sx)yz + \alpha(x)(sy)z + \alpha(x)\alpha(y)sz\,.$$

Similarly d is consistent with the two distributive laws and the associative law of multiplication.

We now prove

Lemma 2.1. *For* $x \in {}^{n+1}C$, $(ds+sd)x = (1-\alpha)x$.

If $x = 1$, this is trivial. Thus it holds for $x \in {}^{n+1}C^0$. Now let $x = a$, a generator of ${}^{n+1}A$ with $sa = b$. Then $\alpha(a) = 0$ and $(ds+sd)(a) = db + sda = a$ by $(D1)$. Next let $x = b$, a generator of ${}^{n+1}B$ with $sa = b$, then $\alpha(b) = 0$ and

$$(ds+sd)b = sdb = s(1-sd)a = sa - s^2da\,.$$

Now by $(R1)$ s^2 is zero on the generators of ${}^{n+1}B$ and of ${}^{n+1}A$; thus by $(R2)$ s^2 is zero on ${}^{n+1}C$. It follows that $(ds+sd)b = sa = b$, so that the lemma is verified on the generators of ${}^{n+1}B$ and of ${}^{n+1}A$.

Now suppose that $x \in {}^{n+1}C^p$, $y \in {}^{n+1}A$ and the lemma is verified for x and y. Then, using $(R2)$ and $(D2)$ we have

$$\begin{aligned}(ds+sd)(xy) &= d((sx)y + (\alpha x)sy) + s((dx)y + (-1)^p xdy)\\ &= (dsx)y + (-1)^{p+1}(sx)(dy) + (\alpha x)dsy\\ &\quad + (sdx)y + (-1)^p(sx)(dy) + (-1)^p(\alpha x)sdy\\ &= (dsx+sdx)y + (\alpha x)(dsy + (-1)^p sdy)\,.\end{aligned}$$

Now if $p > 0$, $\alpha x = 0$ and $(ds+sd)(xy) = xy = (1-\alpha)(xy)$. If $p = 0$, then

$$(ds+sd)(xy) = xy - \alpha x\cdot y + \alpha x(y - \alpha y) = xy - \alpha x\cdot\alpha y = (1-x)xy\,.$$

Thus the lemma is completely established.

Lemma 2.2. *d is a differential on ${}^{n+1}C$.*

The only assertion to be proved is that $d^2 = 0$. This certainly holds on ${}^{n+1}A$ and so, in the light of $(D2)$ it is sufficient to verify it on a generator of ${}^{n+1}B$. Let b be a generator with $sa = b$. Then $d^2b = d(1-sd)a = (d - dsd)a = (1-ds)da$. Now $(ds+sd)da = (1-\alpha)da$. Thus $dsda = da$ since $d^2a = 0$, $\alpha da = 0$. This implies $d^2b = 0$ and hence the lemma.

Lemma 2.3. *${}^{n+1}C$ is acyclic.*

For, by lemma 2.1, s is a chain-homotopy between α and the identity.

Lemma 2.4. *The kernel of π, restricted to ${}^{n+1}C$, is stable under d.*

For an arbitrary element of ${}^{n+1}C$ is expressible as $x_0 \otimes 1 + \sum_{i>0} x_i \otimes y_i$, where $x_i \in {}^{n+1}B$ and $y_i \in {}^{n+1}A^{n_i}$, $n_i > 0$. The π-image of this is x_0, so that the kernel of π, restricted to ${}^{n+1}C$, consists of elements of the form

$$\sum_{i>0} x_i \otimes y_i \;, \qquad \text{or} \qquad \sum_{i>0} x_i y_i \;.$$

The set of such expressions is obviously stable under d since $d({}^{n+1}A^1) = 0$.

It follows that d induces a differential $\bar{d}$ on ${}^{n+1}B$; it is given by

$$\bar{d}b = -\pi s d a \;.$$

Notice also that the definitions of s and d respect the embedding of ${}^{n}C$, ${}^{n}A$ in ${}^{n+1}C$, ${}^{n+1}A$.

We next define θ; we recall that θ is to be a product-preserving map ${}^{n+1}C$, ${}^{n+1}A \to C_*(LK^{n+1})$, $C_*(\Omega K^{n+1})$. It is sufficient to define θ on the generators of ${}^{n+1}B$, ${}^{n+1}A$ and, as above, we determine it on the generators of ${}^{n+1}B$ of dimension $< n+1$ and on those of ${}^{n+1}A$ of dimension $< n$ by means of the embeddings ${}^{n}C$, ${}^{n}A \subseteq {}^{n+1}C$, ${}^{n+1}A$; $C_*(LK^n)$, $C_*(\Omega K^n) \subseteq C_*(LK^{n+1})$, $C_*(\Omega K^{n+1})$. We conserve the notation of this section and let $i : LS^n, \Omega S^n \to LE^{n+1}, \Omega E^{n+1}$, $j : LK^n, \ \Omega K^n \to LK^{n+1}$, ΩK^{n+1} be injections; then $jf' = f''i$ and $\theta = j\theta$ on ${}^{n}C$. Let ζ be a cycle in the class β and let $i\zeta = d\eta$, $\eta \in C_n(\Omega E^{n+1})$. Now $\theta z - f'\zeta = dx'$, $x' \in C_n(\Omega K^n)$. We define[10] $\theta a = jx' + f''\eta$. Then

$$d\theta a = djx' + df''\eta = j\theta z - jf'\zeta + f''i\zeta = j\theta z = \theta da \;.$$

Now let b, as before, be the generator of B corresponding to e^{n+1} (and hence to a above). Since LS^n is acyclic, $\zeta = d\xi$, $\xi \in C_n(LS^n)$. Moreover, $p\xi$ is an n-cycle of S^n whose class generates $H_n(S^n)$ – by the definition of β. Since LE^{n+1} is acyclic and $i\xi - \eta$ is a cycle of LE^{n+1}, it follows that $i\xi - \eta = d\varkappa$, $\varkappa \in C_{n+1}(LE^{n+1})$. Moreover $p\varkappa$ is an $(n+1)$-relative cycle of $E^{n+1} \bmod S^n$ whose class generates $H_{n+1}(E^{n+1}, S^n)$ – in fact, under $d \colon H_{n+1}(E^{n+1}, S^n) \to H_n(S^n)$, we have $d\{p\varkappa\} = \{p\xi\} = S\beta$. We now proceed to define θb. We have

$$d(f'\xi - \theta s z + x') = f'\zeta - \theta z + \theta z - f'\zeta = 0 \;,$$

since $\alpha z = 0$, $dz = 0$. Thus $f'\xi - \theta s z + x'$ is a cycle in LK^n and so

---

[10]) If $n = 1$, then $x' = 0$ and $\theta a = f''\eta$.

$f'\xi - \theta sz + x' = dx''$, $x'' \in C_{n+1}(LK^n)$. We define[11] $\theta b = jx'' - f''\varkappa$. Then $\theta db = \theta(1 - sd)a = jx' + f''\eta - \theta sz$, and $d\theta b = djx'' - df''\varkappa = jf'\xi - \theta sz + jx' - f''i\xi + f''\eta$, so that $\theta db = d\theta b$. Thus θ is defined on ${}^{n+1}C$.

We next show that a map $\bar{\theta} : {}^{n+1}B \to {}^{n+1}B'$ is defined by $\bar{\theta}\pi = p\theta$; it is sufficient to show that $\bar{\theta}$ is single-valued. As above, let $\Sigma x_i y_i$ be a typical element of the kernel of π, $x_i \in {}^{n+1}B$, $y_i \in {}^{n+1}A^{n_i}$, $n_i > 0$. Then $\theta(x_i y_i) = \theta x_i \theta y_i$; but $\theta y_i \in C_{n_i}(\Omega K^{n+1})$ so that $p\theta(x_i y_i)$ is a sum of degenerate cubes and so is zero in $C_*(K)$. Thus $p\theta$ is zero on the kernel of π so that $\bar{\theta}$ is single-valued.

The inductive definition of d and θ will be established when we have shown that

$$\theta_* : H_*({}^{n+1}A) \cong H_*(\Omega K^{n+1}) \tag{2.1}$$

$$\bar{\theta}_* : H_*({}^{n+1}B) \cong H_*(K^{n+1}) \tag{2.2}$$

$$\theta_* : H_*({}^{n+1}C) \cong H_*(LK^{n+1}) \tag{2.3}$$

(2.3) is trivial since ${}^{n+1}C$, LK^{n+1} are acyclic and $\theta(1) = 1$. To prove (2.2), observe that $\bar{\theta}b = p\theta b = pjx'' - pf''\varkappa = pjx'' - fp\varkappa$. Thus $\bar{\theta}b$ is a relative cycle of K^{n+1} mod K^n whose class generates

$$H_{n+1}(K^n \cup e^{n+1}, K^n)\ .$$

Thus $\bar{\theta}_* : H_{n+1}({}^{n+1}B, {}^{n}B) \cong H_{n+1}(K^{n+1}, K^n)$ and (2.2) follows from the inductive hypothesis and the 5-lemma.

To establish (2.1) we introduce a filtration into ${}^{n+1}C$. Then θ will be a filtration-preserving map from ${}^{n+1}C$ to $C_*(LK^{n+1})$, filtered by the Serre filtration, and we will be able to apply a theorem due to J. C. Moore (see [6]) which asserts that, since the first terms of the spectral sequence are well behaved[12]), and since the map induces isomorphisms of the homology groups of the fibre-spaces and of the bases, it must therefore induce isomorphisms of the homology groups of the fibres. To avoid an undue proliferation of superscripts and subscripts, we will permit ourselves in this part of the argument to write A, B, C for ${}^{n+1}A$, ${}^{n+1}B$, ${}^{n+1}C$.

We filter C by putting $C_p = \sum_{q \leqslant p} B^q \otimes A$; equivalently if $x \in B^p$, $y \in A$, then $w(xy) = p$. Moreover if b is a q-dimensional generator of B and $y \in A$ then $d(by) = (db)y + (-1)^q b dy = ay - (sz)y + (-1)^q b dy$

---

[11]) If $n = 1$, then $x'' = 0$ and $\theta b = -f''\varkappa$. Note that, in defining θa, θb, we have used ζ, η, ξ, $\varkappa$ for fixed chains of standard spaces and x', x'' depend on f.

[12]) We make the notion of 'good behaviour' precise in our application below.

and so clearly belongs to C_q. Thus $dC_p \subseteq C_p$ and C is a differential filtered group. Also $\theta(by) = \theta b \cdot \theta y$ and $\theta y \in C(\Omega K^{n+1})$. Thus $p\theta(by)$ is a sum of cubes only depending on their first q co-ordinates. It follows that $\theta(by) \in C'_q$, so that θ respects filtration. Let $E_r^{p,q}$, $E_r'^{p,q}$ be the terms of the spectral sequences associated with C, C' so that θ induces $\theta_* : E_r^{p,q} \to E_r'^{p,q}$.

Define $\psi : B^p \otimes A^q \to E_0^{p,q}$ by $\psi(x \otimes y) = \{xy\}$. Then ψ is an isomorphism and $\psi d_F = d_0 \psi$ where $d_F(x \otimes y) = (-1)^p x \otimes dy$. Thus the induced map $\psi_* : B^p \otimes H_q(A) \to E_1^{p,q}$ is an isomorphism. Define $d_B : B^p \otimes H_q(A) \to B^{p-1} \otimes H_q(A)$ by $d_B(x \otimes \{y\}) = \bar{d}x \otimes \{y\}$. We will show that $\psi_* d_B = d_1 \psi_*$.

Now $d_1 \psi_*(x \otimes \{y\}) = d_1\{xy\} = \{(dx)y\}$, while $\psi_* d_B(x \otimes \{y\}) = \psi_*(\bar{d}x \otimes \{y\}) = \{(\bar{d}x)y\}$. Suppose $x \in B^p$; then $dx = x_0 + \sum_{i>0} x_i y_i$, $y_i \in A^{n_i}$, $x_i \in B^{p-1-n_i}$, where $n_i > 0$ if $i > 0$, and $\bar{d}x = x_0$. Thus $(\bar{d}x)y - (dx)y = \sum_{i>0} x_i y_i y \in C_{p-2}$, whence $\{(dx)y\} = \{(\bar{d}x)y\}$. It follows that ψ_* induces an isomorphism $\psi_{**} : H_p(B\,; H_q(A)) \cong E_2^{p,q}$.

Let φ be the map $E_0'^{p,q} \to B'^p \otimes A'^q$ introduced by Serre. Then since K is simply-connected we know that φ induces isomorphisms

$$\varphi_* : E_1'^{p,q} \cong B'^p \otimes H_q(A')\ , \qquad \varphi_{**} : E_2'^{p,q} \cong H_p(B'\,; H_q(A'))\ .$$

Consider the diagram

$$\begin{array}{ccc} B^p \otimes H_q(A) & \xrightarrow{\ \Theta\ } & B'^p \otimes H_q(A') \\ \Big\downarrow \psi_* & & \Big\uparrow \varphi_* \\ E_1^{p,q} & \xrightarrow{\quad \theta_* \quad} & E_1'^{p,q} \end{array}$$

where $\Theta(y \otimes \{x\}) = \bar{\theta} y \otimes \{\theta x\}$. Then $\Theta = \varphi_* \theta_* \psi_*$. For

$$\theta_* \psi_*(y \otimes \{x\}) = \theta_*\{yx\} = \{\theta y x\}\ .$$

Now if u is a p-cube of LK^{n+1}, v a q-cube of ΩK^{n+1}, then $\varphi(uv) = pu \otimes v$. Thus $\varphi\theta(yx) = \varphi(\theta y \cdot \theta x) = p\theta y \otimes \theta x = \bar{\theta} y \otimes \theta x$ and so $\varphi_*\{\theta y x\} = \bar{\theta} y \otimes \{\theta x\} = \Theta(y \otimes \{x\})$.

We have now verified the conditions of validity of Moore's theorem [13]). The proof of this theorem sets up and filters the chain mapping-cylinder of $\theta : C \to C'$. It then follows from the diagram above that the first terms of the spectral sequence of this filtration also are properly related to the appropriate tensor products, and then an inductive argument

[13]) Théorème B, p. 3–04, of [6]. The fact that ψ_* goes in the opposite direction in the statement of the theorem is, of course, of no consequence.

shows that the spectral sequence is trivial. This leads immediately to the conclusion that

$$\theta_* : H_q(A) \simeq H_q(A') \ .$$

The proof of Theorem 2.1 is now practically complete. We have shown that differentials d, $\bar{d}$ and maps θ, $\bar{\theta}$ may be defined verifying (i), (ii) and (iii) and such that

$$\theta_* : H_*({}^nA) \simeq H_*(\Omega K^n)$$
$$\bar{\theta}_* : H_*({}^nB) \simeq H_*(K^n)$$
$$\theta_* : H_*({}^nC) \simeq H_*(LK^n)$$

for all n. It follows immediately that $\bar{\theta}_* : H_*(B) \simeq H_*(K)$. Since the retraction s may be defined over all C, it follows that C is acyclic so that $\theta_* : H_*(C) \simeq H_*(LK)$. We again apply the spectral sequence argument to deduce that $\theta_* : H_*(A) \simeq H_*(\Omega K)$ and the proof is complete.

Corollary 2.1. *If K is a subcomplex of K^* and if d, θ are given on $C(K)$, $A(K)$ then d^*, θ^* may be chosen so that $d^* \mid C(K) = id$, $\theta^* \mid C(K) = j\theta$, where $i : C(K) \to C(K^*)$, $j : C_*(LK) \to C_*(LK^*)$ are injections.*

Corollary 2.2. *Let K be the union of subcomplexes K_i with a single common point, the single 0-cell of each K_i. Then $A(K)$ may be chosen as the free product of the $A(K_i)$, and θ may be given by $\theta b_i = \theta_i b_i$, $\theta a_i = \theta_i a_i$ where $\theta_i : C(K_i) \to C_*(LK_i)$.*

These two corollaries follow immediately from the definitions of d and θ. By a free product of chain-algebras A_i we understand the chain algebra which is, qua algebra, the free product of the algebras A_i and whose differential is given by

$$d(a_{i_1} \ldots a_{i_k}) = \sum_{q=1}^{k} (-1)^{r_q} a_{i_1} \ldots (da_{i_q}) \ldots a_{i_k} \ , \qquad a_{i_q} \in A_{i_q}^{n_q} \ ,$$

where $r_q = \sum_{s=1}^{q-1} n_s$.

In the light of theorem 2.1. corollary 2.2 may be regarded as a generalization of the theorem due to Bott and Samelson (see [17]) when K is a wedge of spheres.

Before stating the next corollary, which is in the nature of an example, we draw attention to the fact that the map $\bar{\theta} : B \to C_*(K)$ reverses orientation, in the sense that the generator b^n corresponds to the negative of the class of the oriented n-cell e^n in $H_n(K^n, K^{n-1})$.

Corollary 2.3. *Let* $K = S^m \cup e^{m+1}$, $m \geqslant 2$, *where* e^{m+1} *is attached by a map of degree* r. *Then* $A(K)$ *is the chain algebra generated by* a, a', *with* $\dim a = m-1$, $\dim a' = m$ *and* $da' = -ra$.

For certainly $da' = ka$, for some integer k. Now let b, b' be the generators of B. Then since the attaching map is of degree r, we have $db' = rb$. Thus $\pi db' = rb$; but $\pi db' = \pi(a' - sda') = \pi(a' - ksa) = \pi(a' - kb) = -kb$, whence[14] $k = -r$. We note that the differential in C is given by $db = a$, $db' = a' + rb$.

Corollary 2.4. Let $K = S^m \times S^n$, $m, n \geqslant 2$. Then we may take for $A(K)$ the chain algebra (a_1, a_2, a) with $\dim a_1 = m-1$, $\dim a_2 = n-1$, $\dim a = m+n-1$ and

$$da = \varepsilon(a_1 a_2 - (-1)^{(m-1)(n-1)} a_2 a_1), \qquad \varepsilon = \pm 1 .$$

Let $K_0 = S^m \vee S^n$. Then $A(K_0) = (a_1, a_2)$ and θa_1 belongs to a generator g_1 of $H_{m-1}(\Omega S^m)$, θa_2 belongs to a generator g_2 of $H_{n-1}(\Omega S^n)$. Now e^{m+n} is attached to K_0 by a map, f, in the class $[\iota_m, \iota_n]$ and, by Samelson's theorem (see [8], $f'_* \beta = \varepsilon(g_1 g_2 - (-1)^{(m-1)(n-1)} g_2 g_1)$. It follows therefore that we may choose $da = \varepsilon(a_1 a_2 - (-1)^{(m-1)(n-1)} a_2 a_1)$. We note that the differential in C is given by $db_1 = a_1$, $db_2 = a_2$, $db = (1 - sd)a = a - \varepsilon(b_1 a_2 - (-1)^{(m-1)(n-1)} b_2 a_1)$. We note also that θa is a relative cycle in the class generating $H_{m+n-1}(\Omega(S^m \times S^n), \Omega(S^m \vee S^n))$.

For further discussion of product complexes, see section 4.

3. Induced maps of chain-algebras

Let $f : K_1 \to K_2$ be a map[15]) of CW-complexes, inducing $f' : LK_1, \Omega K_1 \to LK_2, \Omega K_2$. Our main object in this section is to realize the induced homology homomorphism f'_* by an appropriate $\varphi_* : H_*(A(K_1)) \to H_*(A(K_2))$, induced by a map $\varphi : C(K_1), A(K_1) \to C(K_2), A(K_2)$. Although $A(K)$ is not uniquely determined by K, we may then think of the passage from the category of CW-complexes and maps to that of chain algebras and maps given by $(K, f) \to (A(K), \varphi)$ as a (multi-valued) covariant functor. We will prove

---

[14]) The minus sign can be avoided by replacing $(R1)$, $(D1)$ by $sa^{r-1} = (-1)^r b^r$, $db^r = (-1)^r (1 - sd) a^{r-1}$. Of course, to compute $H_*(\Omega K)$ one may take a chain algebra generated by a, a' with $da' = ra$.

[15]) Recall that all complexes considered in this paper have one 0-cell and no 1-cells. A map is required to send 0-cell to 0-cell.

Theorem 3.1. *There are chain-maps* $\varphi : C(K_1), A(K_1) \to C(K_2), A(K_2)$, $\bar{\varphi} : B(K_1) \to B(K_2)$ *such that*

(*i*) φ *is product-preserving and* $\varphi s = s\varphi$;

(*ii*) $\bar{\varphi}\pi_1 = \pi_2\varphi$;

(*iii*) *the diagrams*

$$\begin{array}{ccc} C(K_1), A(K_1) & \xrightarrow{\theta_1} & C'(K_1), A'(K_1) \\ \downarrow \varphi & & \downarrow f' \\ C(K_2), A(K_2) & \xrightarrow{\theta_2} & C'(K_2), A'(K_2) \end{array} \qquad \begin{array}{ccc} B(K_1) & \xrightarrow{\bar{\theta}_1} & B'(K_1) \\ \downarrow \bar{\varphi} & & \downarrow f \\ B(K_2) & \xrightarrow{\bar{\theta}_2} & B'(K_2) \end{array}$$

are commutative to within chain homotopy.

We will define φ and a chain homotopy $\psi : C(K_1), A(K_1) \to C'(K_2), A'(K_2)$, such that $d\psi + \psi d = f'\theta_1 - \theta_2\varphi$, inductively on the sections of K_1. Define $\varphi(1) = 1$, $\psi(1) = 0$. Suppose φ, ψ defined on $C(K_1^n)$, and let a be the generator of $A(K_1)$ corresponding to the cell e^{n-1} in K_1. Then φ, ψ are defined on da and

$$\theta_2\varphi da = f'\theta_1 da - d\psi da = d(f'\theta_1 a - \psi da) .$$

Since θ_{2*} is $(1-1)$, there is an element $g_2 \in A(K_2)$ with $dg_2 = \varphi da$. Now $f'\theta_1 a - \psi da - \theta_2 g_2$ is a cycle in $A'(K_2)$; since θ_{2*} is onto, there is a cycle z_2 in $A(K_2)$ and an element g_2' in $A'(K_2)$ such that

$$\theta_2 z_2 + dg_2' = f'\theta_1 a - \psi da - \theta_2 g_2 .$$

We put $\varphi a = g_2 + z_2$, $\psi a = g_2'$. Then $d\varphi a = dg_2 = \varphi da$ and

$$f'\theta_1 a - \theta_2\varphi a = f'\theta_1 a - \theta_2 g_2 - \theta_2 z_2 = dg_2' + \psi da = (d\psi + \psi d)a ,$$

as required. Extend φ to a map of $A(K_1^{n+1})$ into $A(K_2)$; direct computation shows that ψ is extended to $A(K_1^{n+1})$ by the formula $\psi(xy) = (\psi x)(f'\theta_1 y) + (-1)^p(\theta_2\varphi x)(\psi y)$, for $x \in {}^{n+1}A^p$, $y \in {}^{n+1}A$, $A = A(K_1)$. Extend φ to $C(K_1^{n+1})$ by putting $\varphi b = s\varphi a$. Then $s\varphi = \varphi s$ on the generators of $A(K_1^{n+1})$, $B(K_1^{n+1})$ and hence on the whole of $C(K_1^{n+1})$. Certainly ψb may be defined since $C'(K_2)$ is acyclic and ψ is extended to the whole of $C(K_1^{n+1})$ by the same formula as above, where now $x \in {}^{n+1}C^p$ and $y \in {}^{n+1}A$. The inductive definitions of φ and ψ are complete.

Now $\bar{\varphi}$ is defined by (ii), provided we can show that $\pi_2\varphi$ is zero on the kernel of π_1. A typical element of the kernel is $\Sigma x_i y_i$, $y_i \in A^{n_i}(K_1)$, $n_i > 0$. Then $\varphi(x_i y_i) = \varphi(x_i)\varphi(y_i)$, $\varphi(y_i) \in A^{n_i}(K_2)$, so that

$$\pi_2(\varphi(x_i)\varphi(y_i)) = 0 .$$

Thus $\overline{\varphi}$ is defined. A chain homotopy $\overline{\psi} : B(K_1) \to B'(K_2)$ such that $\overline{d}\overline{\psi} + \overline{\psi}\overline{d} = \overline{f}\overline{\theta}_1 - \overline{\theta}_2\overline{\varphi}$ is defined by $\overline{\psi}\pi_1 = p_2\psi$, provided $p_2\psi$ is zero on the kernel of π_1. Now if $y_i \in A^{n_i}(K_1)$, $n_i > 0$, then

$$f'\theta_1 y_i \in A'^{n_i}(K_2) \ , \qquad \psi y_i \in A'^{n_i+1}(K_2) \ .$$

It follows from the product formula for ψ that $p_2\psi(x_i y_i) = 0$ in $B'(K_2)$, $x_i \in C^{p_i}(K_1)$, so that $\overline{\psi}$ is defined as required. This completes the proof of the theorem.

We make some remarks about this theorem. First we note that if φ' is any other suitable chain map $C(K_1)$, $A(K_1) \to C(K_2)$, $A(K_2)$, then $\theta_2\varphi' \simeq f'\theta_1 \simeq \theta_2\varphi$; since θ_2 is a chain equivalence, it follows that any two choices of φ are chain homotopic. Similarly any two choices of $\overline{\varphi}$ are chain homotopic. Let us write $\varphi(f)$ for φ; then we see that if $f : K_1 \to K_2$, $g : K_2 \to K_3$ are maps we may choose $\varphi(gf)$ to be $\varphi(g)\varphi(f)$. We also note the trivial fact that if f is an injection and if d, θ have been chosen on K_2 consistently with their values on K_1, then $\varphi, \overline{\varphi}$ may be taken as injections. Finally, we remark that if f is a map $K_1, L_1 \to K_2, L_2$ where L_i is a subcomplex of K_i, $i = 1, 2$, then φ, ψ may be chosen so that

$$\varphi(C(L_1), A(L_1)) \subseteq C(L_2), A(L_2), \qquad \psi(C(L_1), A(L_1)) \subseteq C'(L_2), A'(L_2).$$

Now let $f_0 : L_1 \to L_2$ be a map of CW-complexes and let $K_i = L_i \cup e_i^{n+1}$, where $g_i : E^{n+1}, S^n \to K_i, L_i$ is a characteristic map for e_i^{n+1}, $i = 1, 2$. Suppose $f_0 g_1 \mid S^n \simeq g_2 \mid S^n$. Then we may extend f_0 to a map $f : K_1 \to K_2$ with $fg_1 \simeq g_2$. Now $A(K_i)$ is formed from $A(L_i)$ by adjoining a new generator a_i. We prove

Theorem 3.2. *If we*[16]) *have chosen d and θ on K_1 and L_2 and φ on L_1, then we may choose d and θ on a_2 and φ, ψ on a_1 so that $\varphi a_1 = a_2$.*

We first choose da_2. Adopting the notation of the previous section, we have only to choose da_2 so that $\theta_2 da_2 \sim g_2'\zeta$. Now

$$\theta_2\varphi da_1 = f'\theta_1 da_1 - d\psi da_1 = f'g_1'\zeta + f'dx' - d\psi da_1 \ .$$

Now $f'g_1' \simeq g_2'$; there is a chain homotopy $\omega : C_*(LE^{n+1})$, $C_*(\Omega E^{n+1})$, $C_*(LS^n)$, $C_*(\Omega S^n) \to C'(K_2)$, $A'(K_2)$, $C'(L_2)$, $A'(L_2)$ with[17])

$$f'g_1' - g_2' = d\omega + \omega d \ , \qquad f'g_1'' - g_2'' = d\omega + \omega d \ .$$

It follows that $\theta_2\varphi da_1 = g_2'\zeta + d(\omega\zeta + f'x' - \psi da_1)$. We may, and

[16]) We will say that d and θ are chosen on K if they are chosen on $A(K)$.

[17]) In the argument which follows it is cumbersome and unnecessary always to distinguish g_1', g_2' from g_1'', g_2''; however, we simply copy the notation of theorem 2.1.

do, choose $da_2 = \varphi da_1$. Then we may take $x'_2 = \omega\zeta + f'x'_1 - \psi da_1$ and $\theta_2 a_2 = j_2 x'_2 + g''_2\eta$. Now

$$f'\theta_1 a_1 - \psi da_1 - \theta_2 a_2 = f'j_1 x'_1 + f'g''_1\eta - \psi da_1 - j_2\omega\zeta - j_2 f'x'_1 + j_2\psi da_1 - g''_2\eta = f'g''_1\eta - g''_2\eta - j_2\omega\zeta = d\omega\eta + \omega d\eta - j_2\omega\zeta = d\omega\eta + \omega i\zeta - j_2\omega\zeta = d\omega\eta\ .$$

Thus we may choose $\varphi a_1 = a_2$, $\psi a_1 = \omega\eta$, and the theorem is proved.

Suppose f_0 is a homotopy equivalence. Then $\varphi_* : H_*(A(L_1)) \cong H_*(A(L_2))$. Also f is a homotopy equivalence so that $\varphi_* : H_*(A(K_1)) \cong H_*(A(K_2))$. This is the topological analogue of the following purely algebraic theorem.

Theorem 3.3. *Let $\varphi : A \to A'$ be a map of chain algebras inducing an isomorphism $\varphi_* : H_*(A) \to H_*(A')$. Let $\overline{A}$ be defined by adjoining a generator a to A and let $\overline{A}'$ be defined by adjoining a generator a' to A' of the same dimension, n, as a. Let $\varphi da = da'$. Then the map $\overline{\varphi} : \overline{A} \to \overline{A}'$ given by $\overline{\varphi} \mid A = \varphi$, $\varphi a = a'$, induces an isomorphism $\overline{\varphi}_* : H_*(\overline{A}) \cong H_*(\overline{A}')$.*

Filter $\overline{A}$ by the rule $\omega(x_0 a x_1 a \ldots a x_p) = p$, $x_i \in A$ and filter $\overline{A}'$ similarly. Let the associated groups of the spectral sequence be $E_r^{p,q}$, $E_r'^{p,q}$. Then $E_r'^{p,q} = E_r'^{p,q} = 0$ if $q < pn - p$. Now $\overline{\varphi}$ is filtration-preserving and $d\overline{A}_p \subseteq \overline{A}_p$, $d\overline{A}'_p \subseteq \overline{A}'_p$, where $(\overline{A}_p)$, $(\overline{A}'_p)$ are the filtering subgroups. Thus $\overline{\varphi}$ is a map of differential filtered groups.

Let $A^{(p)}$ be the tensor product of p copies of A and define $A'^{(p)}$ similarly. Then φ induces $\tilde{\varphi} : A^{(p)} \to A'^{(p)}$ which is a chain equivalence since φ is a chain equivalence. Let $A^{(p)q}$, $q \geqslant pn - p$, be the homogeneous component of $A^{(p)}$ of dimension $p + q - pn$ and let $\psi : A^{(p)q} \to E_0^{p,q}$ be defined by $\psi(x_0 \otimes \ldots \otimes x_p) = (-1)^\sigma x_0 a x_1 \ldots a x_p$, where $x_i \in A^{n_i}$ and $\sigma = n \sum_{i=0}^{p} i n_i$. Then ψ is an isomorphism and $\psi d = d_0 \psi$ so that ψ induces $\psi_* : H_{p+q-pn}(A^{(p)}) \cong E_1^{p,q}$. Similarly $\psi' : A'^{(p)q} \to E_0'^{p,q}$ induces $\psi'_* : H_{p+q-pn}(A'^{(p)}) \cong E_1'^{p,q}$. Also $\psi'\tilde{\varphi} = \overline{\varphi}\psi : A^{(p)q} \to E_0'^{p,q}$, so that $\overline{\varphi}$ induces $\overline{\varphi}_* : E_1^{p,q} \cong E_1'^{p,q}$. It follows that $\overline{\varphi}$ induces $\overline{\varphi}_* : E_\infty^{p,q} \cong E_\infty'^{p,q}$ and hence $\overline{\varphi}_* : H_*(\overline{A}) \cong H_*(\overline{A}')$.

The spectral sequence $E_r^{p,q}$ seems the appropriate tool for studying the effect on $H_*(\Omega K)$ of adding a cell to K, since $\overline{A}_0 = A$.

Our next result is in the nature of an example.

Theorem 3.4. *Let $K = S^n \cup e^{2n}$, $n \geqslant 2$, and suppose $A(K)$ is the chain algebra generated by a_1, a_2 with $da_2 = pa_1^2$. Then p is the Hopf invariant of the attaching map for e^{2n}.*

Let K', K'' be copies of K and let $K' \times K''$ be decomposed into cells in the obvious way. We write a_1, a_2 for the generators of $A(K)$ corresponding to the cells e^n, e^{2n} of K and $a_1', a_2', a_1'', a_2'', a_{11}, a_{12}, a_{21}, a_{22}$ for the generators of $A(K' \times K'')$ corresponding to the cells e'^n, e'^{2n}, e''^n, e''^{2n}, $e'^n \times e''^n$, $e'^n \times e''^{2n}$, $e'^{2n} \times e''^n$, $e'^{2n} \times e''^{2n}$ of $K' \times K''$. Let $f: K \to K' \times K''$ be the diagonal map. Suppose d, θ chosen on $A(K' \times K'')$ consistent with the embedding of $K' \vee K''$ in $K' \times K''$. Let $\varphi: A(K) \to A(K' \times K'')$, $\bar{\varphi}: B(K) \to B(K' \times K'')$ be associated with f, let the cells of $B(K)$, $B(K' \times K'')$ be symbolized similarly to the generators of $A(K)$, $A(K' \times K'')$ and let the Hopf invariant of the attaching map be q. Then

$$\bar{\varphi} b_1 = b_1' + b_1'', \quad \bar{\varphi} b_2 = b_2' + q b_{11} + b_2'' \,.$$

A dimensionality argument[18]) shows that $\varphi a_1 = \varrho a_1' + \sigma a_1''$, $\varphi a_2 = \lambda a_2' + \mu a_{11} + \nu a_2''$. Applying s and comparing with the formulae for $\bar{\varphi}$, we find $\varrho = \sigma = \lambda = \nu = 1$, $\mu = q$. Now $d\varphi a_2 = \varphi d a_2 = p \varphi a_1^2 = p a_1'^2 + p a_1' a_1'' + p a_1'' a_1' + p a_1''^2$, while $d(a_2' + q a_{11} + a_2'') = p a_1'^2 + (-1)^n q a_1' a_1'' + q a_1'' a_1' + p a_1''^2$ (we orient the cell a_{11}, or a in corollary 2.4, so that $\varepsilon = (-1)^m$). Comparing coefficients, $p = (-1)^n q$, $p = q$. This proves the theorem and also shows that the Hopf invariant is zero if n is odd.

If e^{2n} is attached by a map of Hopf invariant 1, then $H_*(\Omega K) \cong H_*(A)$ where A is the chain algebra generated by $a_1 = a_1^{n-1}$, $a_2 = a_2^{2n-1}$ with $d a_2 = a_1^2$. It may be of interest to compute the ring $H_*(A)$. We prove

Theorem 3.5. *$H_{r(3n-2)}(A) = Z_\infty$, generated by $\{g\}^r$, $g = a_1 a_2 - (-1)^{n-1} a_2 a_1$, $H_{r(3n-2)+n-1}(A) = Z_\infty$, generated by $\{g^r a_1\}$, $H_m(A) = 0$, for other values of m. Moreover $\{a_1 g\} = (-1)^n \{g a_1\}$.*

We remark first that in the topological case n is even so that $g = a_1 a_2 + a_2 a_1$ and $\{a_1 g\} = \{g a_1\}$. Thus the theorem asserts that $H_*(A)$ is a commutative ring in this case, isomorphic with the tensor product of an exterior ring generated by $\{a_1\}$ and a polynomial ring generated by $\{g\}$. We now prove the theorem.

Consider the exact sequence $0 \to A_{k-n+1} \overset{i}{\to} A_k \overset{j}{\to} A_{k-2n+1} \to 0$, where $ix = x a_1$, $j(x a_1 + y a_2) = y$, $x \in A_{k-n+1}$, $y \in A_{k-2n+1}$. This induces the exact homology sequence

---

[18]) This argument only holds if $n > 2$; if $n = 2$, the expression for φa_2 could, a priori, contain terms in a'^3 and a''^3. We may either eliminate this possibility by considering projections $K' \times K'' \to K'$, $K' \times K'' \to K''$, (whereby we may also deduce $\lambda = \nu = 1$) – or leave these terms in the expression until they are annihilated in the passage from φ to $\bar{\varphi}$.

$$\cdots \to H_{k-2n+2}(A) \overset{d_*}{\to} H_{k-n+1}(A) \overset{i_*}{\to} H_k(A) \overset{j_*}{\to} H_{k-2n+1}(A) \overset{d_*}{\to} H_{k-n}(A) \to \cdots,$$

where $i_*\{x\} = \{xa_1\}$, $j_*\{xa_1 + ya_2\} = \{y\}$, and $d_*\{y\} = (-1)^\sigma\{ya_1\}$, $\sigma = \dim y$. Now the homology groups of A are certainly as stated in dimensions $<3n-2$. Suppose inductively that they are as stated in dimensions $<(r+1)(3n-2)$, $r \geqslant 0$. The part of the homology sequence beginning $H_{(r+1)(3n-2)+2n-2}(A) \overset{i_*}{\to} \cdots$ and ending... $\overset{d_*}{\to} H_{r(3n-2)+2n-2}(A)$ will be trivial except for

$$j_* : H_{(r+1)(3n-2)} \cong H_{r(3n-2)+n-1},$$
$$i_* : H_{(r+1)(3n-2)} \cong H_{(r+1)(3n-2)+n=1},$$
$$d_* : H_{(r+1)(3n-2)} \cong H_{(r+1)(3n-2)+n-1},$$

Thus $H_{(r+1)(3n-2)} = Z_\infty$ generated by $\{g^r a_1 a_2 + x a_1\}$, x being chosen arbitrarily, subject only to the condition that $g^r a_1 a_2 + x a_1$ be a cycle; since g is a cycle and g^{r+1} is of this form, it follows that $H_{(r+1)(3n-2)}$ is generated by $\{g\}^{r+1}$. It then follows that $H_{(r+1)(3n-2)+n-1} = Z_\infty$, generated by $\{g^{r+1}a_1\}$ and that H_* is zero in all other dimensions $<(r+2)(3n-2)$.

We complete the proof of the theorem by observing that $d(a_2^2) = a_1^2 a_2 - a_2 a_1^2 = a_1(a_1 a_2 - (-1)^{n-1} a_2 a_1) - (-1)^n (a_1 a_2 - (-1)^{n-1} a_2 a_1) a_1$.

4. Product complexes.

The main object of this section is to obtain a chain equivalence from $A(K_1 \times K_2)$ to $A(K_1) \otimes A(K_2)$. We first provide a universal example for the chain algebra of a product complex.

Theorem 4.1. *Let $E_1 = e^0 \cup e^p \cup e^{p+1}$ be a $(p+1)$-element decomposed in the usual way into the cells e^0, e^p, e^{p+1} and let E_2 be a $(q+1)$-element similarly decomposed, $p, q \geqslant 2$. Then $A(E_1 \times E_2)$ is freely generated by elements*[19] *$a_1, a_2, c_1, c_2, b, e, \bar{e}, t$, corresponding to the cells e^p, e^q, e^{p+1}, e^{q+1}, $e^p \times e^q$, $e^{p+1} \times e^q$, $e^p \times e^{q+1}$, $e^{p+1} \times e^{q+1}$ and d, θ may be chosen on $E_1 \times E_2$ to give $dc_1 = -a_1$, $dc_2 = -a_2$,*

$$db = (-1)^p (a_1 a_2 - (-1)^{(p-1)(q-1)} a_2 a_1)$$
$$de = -b + (-1)^{p+1}(c_1 a_2 - (-1)^{p(q-1)} a_2 c_1),$$
$$d\bar{e} = (-1)^{p-1} b + (-1)^p (a_1 c_2 - (-1)^{(p-1)q} c_2 a_1),$$
$$dt = (-1)^p e - \bar{e} + (-1)^{p+1}(c_1 c_2 - (-1)^{pq} c_2 c_1).$$

The first two boundary formulae are given by corollaries 2.2 and 2.3.

[19]) The notations used for generators of $A(E_1 \times E_2)$ are chosen for their convenience in studying product complexes and are not related to previous notation.

The formula for db is given by corollary 2.4, the orientation being chosen so that, under the map $\varphi : A(E^{p+q}) \to A(S^p \times S^q)$ induced by the characteristic map E^{p+q}, $S^{p+q-1} \to S^p \times S^q$, $S^p \vee S^q$, the cell corresponding to S^{p+q-1} is mapped precisely by the Samelson formula (cf. Theorem 3.4).

Now consider $A(E_1 \times S_2^q) = \{a_1, a_2, c_1, b, e\}$. Since the injection $i_2 : S_2^q \to E_1 \times S_2^q$ and the projection $p_2 : E_1 \times S_2^q \to S_2^q$ are homotopy equivalences such that $p_2 i_2 = 1$, it follows readily that[20]) we may choose $\varphi_2 = \varphi(p_2)$ such that $\varphi_2 a_2 = a_2$, $\varphi_2 a = 0$ for $a = a_1, c_1$, or b and φ_{2*} is an isomorphism. Now, if z is the element proposed for de, then z is a cycle and $\varphi_2 z = 0$. Thus z is a boundary; it follows that, for an arbitrary choice of de, there exist an integer k and an element $x \in \{a_1, a_2, c_1, b\}$ such that $d(ke + x) = z$ or $k(de) + dx = z$. It may be seen by inspection that no such equation can subsist in $\{a_1, a_2, c_1, b\}$ unless $k = \pm 1$. Thus, if e is suitably oriented, z is a proper choice for de. The orientation of e is chosen to give the correct boundary formula in $B(E_1 \times S_2^q)$, when the cells of $E_1 \times S_2^q$ are given the product orientation.

A similar argument establishes the formula of $d\bar{e}$; the orientation of $\bar{e}$ is chosen by the same considerations.

Finally the element z' proposed for dt is a cycle and therefore a boundary; it follows that, for an arbitrary choice of dt, there exist an integer k' and an element $x' \in \{a_1, a_2, c_1, c_2, b, e, \bar{e}\}$ such that $k'(dt) + dx' = z'$. It may be seen by inspection of $\{a_1, a_2, c_1, c_2, b, e, \bar{e}\}$ that this implies $k' = \pm 1$, so that z' is a proper choice for dt if t is suitably oriented; we choose the orientation for t as for e and $\bar{e}$ and the theorem is proved.

Now let $K_1 \times K_2$ be the topological product of two CW-complexes with its usual cellular decomposition[21]). Let

$$j : A(K_1 \times K_2) \to A(K_1) \otimes A(K_2)$$

be the ring homomorphism given by

$$\begin{aligned} ja &= a \otimes 1, \quad a \in A(K_1), \\ ja &= 1 \otimes a, \quad a \in A(K_2), \\ ja &= 0, \quad \text{for any other generator } a \text{ of } A(K_1 \times K_2). \end{aligned}$$

Let $\varphi_i : A(K_1 \times K_2) \to A(K_i)$, $i = 1, 2$, be the ring homomorphism given by

$$\begin{aligned} \varphi_i a &= a, \quad a \in A(K_i), \\ \varphi_i a &= 0, \quad \text{for any other generator } a \text{ of } A(K_1 \times K_2). \end{aligned}$$

[20]) Clearly a suitable φ for the projection $S_1^p \times S_2^q \to S_2^q$ is given by $\varphi(a_1) = 0$, $\varphi(a_2) = a_2$, $\varphi(b) = 0$, provided ∂b has been appropriately chosen.

[21]) We are not disturbed by the fact that $K_1 \times K_2$ need not be a CW-complex; theorem 2.1 holds for products of CW-complexes.

Let $\varrho : C_* X \times C_* Y \to C_*(X \times Y)$ be the standard chain equivalence of cubical homology theory.

The main theorem of this section is as follows.

Theorem 4.2. *We may choose d and θ on $K_1 \times K_2$ so that j is a chain mapping; with this choice φ_i is a φ-map*[22]) *associated with the projection $p_i : K_1 \times K_2 \to K_i$, and the diagram*

$$\begin{array}{ccc}
A(K_1 \times K_2) & \xrightarrow{\;j\;} & A(K_1) \otimes A(K_2) \\
\downarrow{\scriptstyle \theta} & & \\
A'(K_1 \times K_2) & & \Big\downarrow{\scriptstyle \theta_1 \otimes \theta_2} \\
\downarrow{\scriptstyle \Psi} & & \\
C_*(\Omega K_1 \times \Omega K_2) & \xrightarrow{\;\varrho\;} & A'(K_1) \otimes A'(K_2)
\end{array}$$

is homotopy-commutative and leads to a commutative diagram of isomorphisms of homology rings.

Suppose that d and θ have been chosen so that j is a chain mapping. Let $i: A(K_1) \otimes A(K_2) \to A(K_1 \times K_2)$ be the chain mapping of chain groups given by $i(x \otimes y) = xy$, $x \in A(K_1)$, $y \in A(K_2)$. Then $ji = 1$. Let ΩK_1, ΩK_2 be embedded in $\Omega(K_1 \times K_2)$ and let

$$\eta : \Omega K_1 \times \Omega K_2 \to \Omega(K_1 \times K_2)$$

be the map given by $\eta(l_1, l_2) = l_1 l_2$ (composition of loops). Then[23]) $\theta i = \eta\varrho(\theta_1 \otimes \theta_2)$. We next show that η is a homotopy inverse of Ψ. Since Ψ is a homotopy equivalence it is sufficient to show that $\Psi\eta \simeq 1$. Now $\Psi\eta(l_1, l_2) = (l_1 \omega_s, \omega_r l_2)$ where l_1 is a loop of 'duration' r, l_2 is a loop of 'duration' s and ω_r, ω_s are constant loops of duration r, s. Thus a homotopy of the identity to $\Psi\eta$ is given by $h_t(l_1, l_2) = (l_1 \omega_{st}, \omega_{rt} l_2)$. Then $\Psi\theta i \simeq \varrho(\theta_1 \otimes \theta_2)$. Since Ψ, θ, ϱ and $\theta_1 \otimes \theta_2$ are chain equivalences it follows that i is a chain equivalence. Since $ji = 1$, it follows that j is a chain inverse of i so that $\Psi\theta \simeq \varrho(\theta_1 \otimes \theta_2) j$ and j_* is an isomorphism.

We still assume that j is a chain mapping and next prove that if $p_1' : A'(K_1 \times K_2) \to A'(K_1)$ is induced by p_1, then $p_1'\theta \simeq \theta_1 \varphi_1$. Let $p_1'' : C_*(\Omega K_1 \times \Omega K_2) \to A'(K_1)$ be induced by the projection

$$\Omega K_1 \times \Omega K_2 \to \Omega K_1 \;;$$

22) In the sense of theorem 3.1.

23) We always suppose d and θ chosen consistently with the embedding

$$K_1 \vee K_2 \subseteq K_1 \times K_2 \;.$$

let $\tilde{p}_1 : A(K_1) \otimes A(K_2) \to A(K_1)$ be the map given by $\tilde{p}_1(x \otimes 1) = x$, $\tilde{p}_1(1 \otimes y) = 0$ and let $\tilde{p}'_1 : A'(K_1) \otimes A'(K_2) \to A'(K_1)$ be defined similarly. Then the relations

$$p''_1\Psi = p'_1 , \quad \theta_1 \tilde{p}_1 = \tilde{p}'_1(\theta_1 \otimes \theta_2) , \quad \tilde{p}'_1 = p''_1 \varrho , \quad \varphi_1 = \tilde{p}_1 j ,$$

are obvious.

We have proved that $\theta \simeq \eta\varrho(\theta_1 \otimes \theta_2)j$ and $p'_1\eta \simeq p''_1$ since $\Psi\eta \simeq 1$. Thus

$$p'_1\theta \simeq p'_1\eta\varrho(\theta_1 \otimes \theta_2)j \simeq p''_1\varrho(\theta_1 \otimes \theta_2)j = \tilde{p}'_1(\theta_1 \otimes \theta_2)j = \theta_1\tilde{p}_1 j = \theta_1\varphi_1 .$$

Thus φ_1 is a suitable choice for $\varphi(p_1)$ and a similar argument shows that φ_2 is a suitable choice for $\varphi(p_2)$.

It remains to show that d and θ may be chosen so that j is a chain mapping. We observe first that j is a chain mapping on $A(K_1 \vee K_2)$, embedded in $A(K_1 \times K_2)$, and second that j is a chain mapping on the universal example $A(E_1 \times E_2)$.

We now prove that d and θ may be chosen on $E_1 \times K_2$, $E_1 = E_1^{p+1}$, $p \geqslant 2$, so that j is a chain mapping. The argument proceeds by induction on the sections of K_2. It is trivial for $E_1 \times K_2^0$ and follows easily for $E_1 \times K_2^2$ from theorem 4.1. Suppose inductively that d and θ have been chosen on $E_1 \times K_2^q$, $q \geqslant 2$, so that j is a chain mapping and let e be a $(q+1)$-cell attached to K_2^q, the characteristic map being $f : E_2^{q+1}$, $S_2^q \to K_2^q \cup e, K_2^q$ Let $\varphi_2 : A(E_2^{q+1}) \to A(K_2^q \cup e)$ be associated with f. We proceed to define a map $\varphi : A(E_1 \times S_2^q) \to A(E_1 \times K_2^q)$. In the notation of theorem 4.1, we put $\varphi(a_1) = a_1$, $\varphi(a_2) = \varphi_2(a_2)$, $\varphi(c_1) = c_1$. Then, so far as φ is defined, the diagram

$$\begin{array}{ccc} A(E_1 \times S_2^q) & \xrightarrow{j} & A(E_1) \otimes A(S_2^q) \\ \downarrow \varphi & & \downarrow 1 \otimes \varphi_2 \\ A(E_1 \times K_2^q) & \xrightarrow{j} & A(E_1) \otimes A(K_2^q) \end{array} \tag{4.3}$$

is commutative.

Consider the element $b \in A(E_1 \times S_2^q)$. Then $j\varphi db = 0$; since j is a chain equivalence onto $A(E_1) \otimes A(K_2^q)$, it follows that the kernel of j is acyclic, so that there exists an element $x \in A(E_1 \times K_2^q)$ with $dx = \varphi db$ and $jx = 0$. Define $\varphi b = x$. Then $\varphi d = d\varphi$ on b and the commutativity of (4.3) is preserved. Then $j\varphi de = 0$ and the same argument shows that there exists an element $y \in A(E_1 \times K_2^q)$ with $dy = \varphi de$, $jy = 0$; we take $\varphi e = y$. Thus we have defined a map φ making (4.3) a commutative diagram.

We now assert that φ is associated with the map

$$1\times f : E_1\times S_2^q \to E_1\times K_2^q \ .$$

To establish this, we consider the diagram

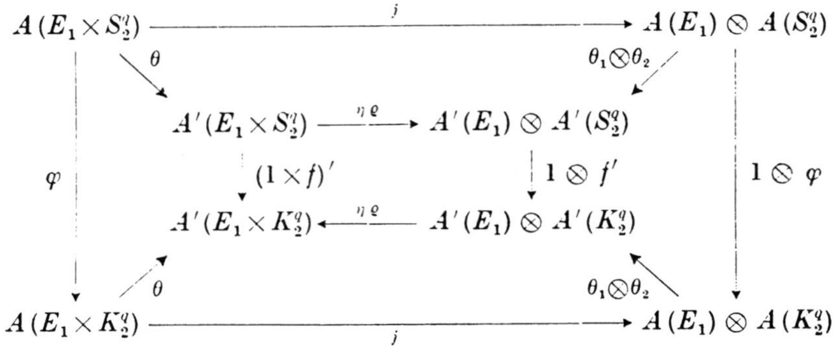

We wish to show that $\theta\varphi \simeq (1\times f)'\theta$; but this follows from the commutativity properties of the diagram. Now $E_1\times E_2$ is obtained from $E_1\times S_2^q$ by attaching cells e^{q+1}, $e^p\times e^{q+1}$, $e^{p+1}\times e^{q+1}$ and each of these cells is mapped homeomorphically onto a cell of $E_1\times(K_2^q \smile e)$. The generators of $A(E_1\times E_2)$ corresponding to these cells are c_2, $\bar{e}$, t; let the generators of $A(E_1\times(K_2^q\smile e)$ corresponding to the cells $(1\times f)(e^{q+1})$, $(1\times f)(e^p\times e^{q+1})$, $(1\times f)(e^{p+1}\times e^{q+1})$ be called c_2, $\bar{e}^*$, t^*. Then by theorem 3.2, we may define d, θ on $\bar{e}^*$, t^* and extend φ to a map associated with $1\times f : E_1\times E_2 \to E_1\times(K_2^q\smile e)$ by putting $\varphi\bar{e} = \bar{e}^*$, $\varphi t = t^*$; but then $jd\bar{e}^* = jd\varphi\bar{e} = j\varphi d\bar{e} = (1\otimes\varphi_2)jde = 0$ and $j\bar{e}^* = 0$ so that j is a chain mapping on $\bar{e}^*$; and $jt^*=0$, $jdt^*=jd\varphi t = j\varphi dt = (1\otimes\varphi_2)jdt$ (since, by definition, $j\varphi\bar{e} = (1\otimes\varphi_2)j\bar{e}\ (=0)) = 0$, so that j is a chain mapping on t^* and hence on the whole of $A(E_1\times(K_2^q\smile e))$. We proceed in this way over all the $(q+1)$-cells of K_2 and so define d and θ on $A(E_1\times K_2^{q+1})$ so that j is a chain mapping. This establishes the induction and hence the result when $K_1 = E_1$.

Finally we consider the general case, and proceed by induction over the sections of K_1. It is an immediate consequence of the argument above that we may choose d and θ on $K_1^2\times K_2$ so that j is a chain mapping. Suppose inductively that d and θ have been chosen on $K_1^p\times K_2$ so that j is a chain mapping and let e be a $(p+1)$-cell attached to K_1^p, the characteristic map being $f : E_1^{p+1}, S_1^p \to K_1^p\smile e, K_1^p$. Let

$$\varphi_1 : A(E_1) \to A(K_1^p\smile e)$$

be associated with f. We assert that a map

$$\varphi : A(S_1^p \times K_2) \to A(K_1^p \times K_2)$$

may be defined so that the diagram

$$\begin{array}{ccc} A(S_1^p \times K_2) & \xrightarrow{j} & A(S_1^p) \otimes A(K_2) \\ \varphi \downarrow & & \downarrow \varphi_1 \otimes 1 \\ A(K_1^p \times K_2) & \xrightarrow{j} & A(K_1^p) \otimes A(K_2) \end{array}$$

is commutative[24]). We may define $\varphi a_1 = \varphi_1 a_1$, $a_1 \in A(S_1^p)$, $\varphi a_2 = a_2$, $a_2 \in A(K_2)$. We then define φa where a is a generator corresponding to a cell $e^p \times e^{n+1}$ in $S_1^p \times K_2$ inductively with respect to n. For if φ is defined on $A(S_1^p \times K_2^n \cup (S_1^p \vee K_2))$ so that $j\varphi = (\varphi_1 \otimes 1)j$ and if the generator a corresponds to a cell $e^p \times e^{n+1}$, then $j\varphi da = (\varphi_1 \otimes 1)jda = 0$ and so, as previously, there exists an element $x \in A(K_1^p \times K_2)$ such that $dx = \varphi da$ and $jx = 0$; we put $\varphi a = x$ and then $j\varphi a = (\varphi_1 \otimes 1)ja = 0$. This establishes that such a map φ may be defined. Arguing from a diagram analogous to (4.4) shows that φ is associated with the map $f \times 1 : S_1^p \times K_2 \to K_1^p \times K_2$.

Let e^n be an arbitrary cell of K_2, let a be the generator of $A(E_1 \times K_2)$ corresponding to $e^{p+1} \times e^n$ and let a^* be the generator of $A((K_1^p \cup e) \times K_2)$, $e = e^{p+1}$, corresponding to $e \times e^n$. Again applying theorem 3.2, we deduce that d, θ may be chosen on $A((K_1^p \cup e) \times K_2)$ so that the map φ may be extended to a map associated with $f \times 1 : E_1 \times K_2 \to (K_1^p \cup e) \times K_2$ by defining $\varphi a = a^*$ for all e^n in K_2. Then we still have $j\varphi = (\varphi_1 \otimes 1)j$. It remains to show that $jda^* = 0$ for all a^*; but $jda^* = jd\varphi a = j\varphi da = (\varphi_1 \otimes 1)jda = 0$, since $ja = 0$ and j is a chain mapping on $A(E_1 \times K_2)$. We proceed in this way over all the $(p+1)$-cells of K_1 and so define d and θ on $A(K_1^{p+1} \times K_2)$ so that j is a chain mapping. This establishes the induction and completes the proof of the theorem.

Corollary 4.1. *Let $\varphi_i : A(K_i) \to A(L_i)$ be associated with maps $f_i : K_i \to L_i$, $i = 1, 2$, and let d, θ be chosen on $K_1 \times K_2$, $L_1 \times L_2$ so that j is a chain mapping. Then we may choose a map*

$$\varphi : A(K_1 \times K_2) \to A(L_1 \times L_2)$$

so that $j\varphi = (\varphi_1 \otimes \varphi_2)j$ and any such φ is associated with the product map $f_1 \times f_2$.

---

[24]) We suppose $A(E_1 \times K_2)$ furnished with suitable d, θ to make j a map.

We establish the existence of such a map φ by an inductive argument analogous to that following diagram (4.3) and the required property of φ by an argument based on a diagram analogous to (4.4).

Corollary 4.2. *If* $L_i \subseteq K_i$, $i = 1, 2$, *and if* d, θ *have been chosen on* $L_1 \times L_2$ *so that* j *is a chain mapping, then* d, θ *may be extended to* $K_1 \times K_2$ *so that* j *remains a chain mapping.*

For this is essentially the procedure in the last part of the proof of theorem 4.2.

Now let $K = S_1 \times \cdots \times S_t$, where S_i is an n_i-sphere, $n_i \geqslant 2$, $i = 1, \ldots, t$. Then K may be decomposed into cells in the usual way: for each non-empty subset D of $\{1, 2, \ldots, t\}$, let e_D be the cell $\prod_{i \in D} e_i$, and let a_D be the corresponding generator of $A(K)$. We prove

Theorem 4.3. *d and θ may be chosen on K so that*

$$da_D = \sum_{A,B} (-1)^{\varepsilon(A,B)} a_A a_B$$

where the sum extends over all partitions of D into non-empty subsets A, B and

$$\varepsilon(A, B) = \sum_{a \in A} n_a + \sum_{\substack{a \in A,\, b \in B \\ a > b}} n_a n_b .$$

We prove this by induction on t; it is trivial if $t = 1$ and reduces to the Samelson formula if $t = 2$. Suppose the theorem established for products of $t - 1$ spheres, $t \geqslant 3$, and consider K. We propose to choose d and θ on K so that $j : A(K) \to A(S_1 \times \cdots \times S_{t-1}) \otimes A(S_t)$ is a map. For any a_D, $D \neq \{1, 2, \ldots, t\}$, choose the proposed formula for da_D; the inductive hypothesis tells us this is possible, and we observe (by direct computation) that $jda_D = dja_D$. Now let $D = \{1, 2, \ldots, t\}$; by corollary 4.2, there exists a choice for the boundary of a_D, say $d'a_D$, such that j remains a map and therefore a chain equivalence. Now we observe (by direct computation) that $x = \sum_{A,B} (-1)^{\varepsilon(A,B)} a_A a_B$ is a cycle and $jx = 0$. If follows that x is a boundary, so that $x = dy + kd'a_D$, where $y \in A(K - e_D)$ and k is an integer. Now (arguing as in theorem 4.1) we observe that $a_A a_B$, for example, cannot appear in the boundary of an element of $A(K - e_D)$ with non-zero coefficient. Thus it must appear in the boundary of a_D and we must have $k = \pm 1$. Thus, reorienting e_D if necessary, we have proved that x is a legitimate choice for da_D. We observe, of course, that j remains a map with this choice. In fact, partitioning the ordered array $\{1, 2, \ldots, t\}$ in any way we please as $A_1, \ldots, A_s$, where A_i is the array $\{n_i, n_i+1, \ldots, n_{i+1}-1\}$, $n_1 = 1$, $n_{s+1} = t+1$, we find that

$$j : A(K) \to A(K_1) \otimes \cdots \otimes A(K_s)$$

is a map, where the definitions of j is an obvious extension of that for $s = 2$ and $K_i = \prod_{r \in A_i} S_r$.

Theorem 4.3 constitutes a generalization of Samelson's formula; it is consistent with the formula contained in remark (i) on p. 5 of [7].

J. C. Moore considers in [5] spaces with a single non-vanishing homology group, in dimension p, say. If this group is finitely generated, then an appropriate space is a wedge of subspaces X_i, where each X_i is a p-sphere or a p-sphere with a $(p+1)$-cell attached by a map of non-zero degree. We study here the Pontryagin ring of the loop space of a Moore space. The method is exemplified by the case when the Moore space Z is the wedge of two such subspaces X_i, but we will generalize the problem slightly by allowing $Z = X_1 \vee X_2$, where X_1 is a p-sphere or a p-sphere with a $(p+1)$-cell attached by a map of non-zero degree and X_2 is a q-sphere or a q-sphere with a $(q+1)$-cell attached by a map of non-zero degree. We take p, $q \geqslant 2$. We will also consider $H_*(\Omega P)$, where $P = X_1 \times X_2$. We first observe that, for quite arbitrary spaces X_1, X_2, $H_*(\Omega Z)$ and $H_*(\Omega P)$ contain $H_*(\Omega X_1) + H_*(\Omega X_2)$ as a direct summand; we will use the congruence symbol to indicate that we are computing modulo this subgroup.

We prove

Theorem 4.4. *Let $P = X_1 \times X_2$, where $X_1 = S^p \cup e^{p+1}$, e^{p+1} being attached by a map of degree $m \neq 0$, and $X_2 = S^q \cup e^{q+1}$, e^{q+1} being attached by a map of degree $n \neq 0$, $p, q \geqslant 2$. Then $A(P)$ is generated by a_1, a_2, c_1, c_2, b, e, $\bar{e}$, t, corresponding to the cells e^p, e^{p+1}, e^q, e^{q+1}, $e^p \times e^q$, $e^{p+1} \times e^q$, $e^p \times e^{q+1}$, $e^{p+1} \times e^{q+1}$ and d, θ may be chosen on P to give*
$dc_1 = -ma_1$, $dc_2 = -na_2$, $db = (-1)^p(a_1a_2 - (-1)^{(p-1)(q-1)}a_2a_1)$,
$de = -mb + (-1)^{p+1}(c_1a_2 - (-1)^{p(q-1)}a_2c_1)$,
$d\bar{e} = (-1)^{p-1}nb + (-1)^p(a_1c_2 - (-1)^{(p-1)q}c_2a_1)$,
$dt = (-1)^p ne - m\bar{e} + (-1)^{p+1}(c_1c_2 - (-1)^{pq}c_2c_1)$.

Consider $E_1 \times E_2$ and use the same symbols for the generators of $A(E_1 \times E_2)$. Let $f_i : E_i \to X_i$ be characteristic maps, $i = 1, 2$. Then we may take $\varphi_1 a_1 = ma_1$, $\varphi_1 c_1 = c_1$, $\varphi_2 a_2 = na_2$, $\varphi_2 c_2 = c_2$. We will define d on $A(P)$ so that j is a chain mapping, and we will also define an appropriate $\varphi : A(E_1 \times E_2) \to A(P)$, in accordance with corollary 4.1.

The formula for db is already established. The formula proposed for de is a cycle of $A(K_1 \times S_2^q)$ in the kernel of j and hence a boundary; arguing as in theorem 4.1, we see that it is a legitimate choice for de; similarly we justify the formula for $d\bar{e}$. Then it follows, from corollary 4.1, that we may take $\varphi(b) = mnb$, $\varphi(e) = ne$, $\varphi(\bar{e}) = m\bar{e}$. By theorem 3.2 we may now take $\varphi(t) = t$, getting the given formula for dt.

From theorem 4.4 we may calculate $H_*(Z)$ and $H_*(P)$. In particular we consider the injection $H_{p+q-1}(\Omega Z) \to H_{p+q-1}(\Omega P)$. Let h be the $g{\cdot}c{\cdot}d$ of m, n, so that $m = hm'$, $n = hn'$. We will restrict attention to the case $p, q \geqslant 3$, though, by complicating the argument, it would be possible to include the cases $p = 2$ or $q = 2$ (or both). With this restriction we have $H_{p+q-1}(\Omega Z) \equiv Z_h + Z_h$, with generators

$$\{\xi\} = \{m'a_1c_2 + (-1)^p n'c_1a_2\}, \qquad \{\eta\} = \{n'a_2c_1 + (-1)^q m'c_2a_1\} .$$

On the other hand,

$$H_{p+q-1}(\Omega P) \equiv \mathrm{Tor}\,(H_{p-1}(\Omega X_1), H_{q-1}(\Omega X_2)) = Z_h ,$$

generated by $\{\xi\}$ or $\{\eta\}$. In fact, we see that

$$\xi - (-1)^{pq}\eta = d((-1)^p n'e - m'\bar{e}) .$$

It follows that the injection $H_{p+q-1}(\Omega Z) \to H_{p+q-1}(\Omega P)$ is onto $H_{p+q-1}(\Omega P)$ with kernel[25]) $\{\xi - (-1)^{pq}\eta\}$.

Consider the diagram

$$\begin{array}{ccc}
\pi_{p+q+1}(P, Z) & \xrightarrow{d} & \pi_{p+q}(Z) \\
\downarrow \omega_1 & & \downarrow \omega_2 \\
\pi_{p+q}(\Omega P, \Omega Z) & \xrightarrow{d'} & \pi_{p+q-1}(\Omega Z) \\
\downarrow h_1 & & \downarrow h_2 \\
H_{p+q}(\Omega P, \Omega Z) & \xrightarrow{d''} & H_{p+q-1}(\Omega Z)
\end{array}$$

where ω_i are the usual isomorphisms and h_i are Hurewicz homomorphisms, $i = 1, 2$. Then each square is commutative or anti-commutative and h_1 is onto $H_{p+q}(\Omega P, \Omega Z)$. Moreover d'' maps $H_{p+q}(\Omega P, \Omega Z)$

---

25) We permit ourselves here and subsequently to identify $A(Z)$ with $C_*(\Omega Z)$, and thus to omit the maps θ, θ_*.

onto Z_h, generated by $\{\xi - (-1)^{pq}\eta\}$. The group

$$\pi_{p+q}(\Omega P, \Omega Z) \cong \pi_{p+q+1}(P, Z)$$

was computed in [3]; we have

$$\begin{aligned} \pi_{p+q}(\Omega P, \Omega Z) &= Z_h , & &\text{if} \quad h \text{ is } odd , \\ &= Z_{2h} , & &\text{if} \quad h = 4k , \\ &= Z_h + Z_2 , & &\text{if} \quad h = 4k + 2 . \end{aligned}$$

If h is even the Z_2 subgroup (direct factor if $h = 4k + 2$) is certainly annihilated by $h_2 d'$. Thus there is an element, ϱ, in $\pi_{p+q-1}(\Omega Z)$ which is mapped by h_2 to $\{\xi - (-1)^{pq}\eta\}$; it follows from arguments in [3] that the image of $(d'\omega_1)^{-1}\varrho$ in the Hurewicz homomorphism

$$\pi_{p+q+1}(P, Z) \to H_{p+q+1}(P, Z)$$

generates the latter group which is isomorphic to Tor $(H_p(X_1), H_q(X_2))$.

For further simplicity we now take $m = n$; we leave the slight modifications in the general case to the reader. Let S be any path-connected space and let $\alpha \in \pi_p(S)$, $\beta \in \pi_q(S)$ be elements whose order divides m. Then we may map Z to S by a map g which, restricted to S^p, represents α and, restricted to S^q, represents β. Let $f : S^{p+q} \to Z$ represent $\omega_2^{-1}\varrho$. Then $gf : S^{p+q} \to S$ represents an element $\{\alpha, \beta\} \in \pi_{p+q}(S)$ which is determined modulo the subgroup generated by elements $[\alpha, \varkappa]$, $[\lambda, \beta]$, $\varkappa \in \pi_{q+1}(S)$, $\lambda \in \pi_{p+1}(S)$. Let $u \in A_{p-1}(S)$, $v \in A_{q-1}(S)$ be cycles such that[26] $\{u\} = h_2\omega_2\alpha$, $\{v\} = h_2\omega_2\beta$, and let $-mu = du'$, $-mv = dv'$. Then $uv' + (-1)^p u'v - (-1)^{pq}(vu' + (-1)^q v'u)$ is a $(p + q - 1)$-cycle of $A(S)$ whose homology class is determined modulo the ideal generated by $\{u\}$ and $\{v\}$. We call the element of

$$H_{p+q-1}(\Omega S)/(\{u\}, \{v\})$$

so determined $|\alpha, \beta|$. Since $h_2\omega_2[\alpha, \pi_{q+1}(S)]$ lies in the ideal generated by $\{u\}$ and $h_2\omega_2\ [\pi_{p+1}(S), \beta]$ lies in the ideal generated by $\{v\}$, we may discuss unambiguously the element $h_2\omega_2\ \{\alpha, \beta\}$ in the quotient ring $H_{p+q-1}(\Omega S)/(\{u\}, \{v\})$. It follows by naturality that

$$h_2\omega_2\{\alpha, \beta\} = |\alpha, \beta| ,$$

the space Z being a universal example for the construction $\{\alpha, \beta\}$.

---

[26]) We use ω_2, h_2 for the maps $\pi_r(Y) \to \pi_{r-1}(\Omega Y)$, $\pi_{r-1}(\Omega Y) \to H_{r-1}(\Omega Y)$ for any r and any space Y. See also the previous footnote.

A direct computation shows that

$$(-1)^{pr} \| \alpha, \beta |, \gamma | + (-1)^{qp} \| \beta, \gamma |, \alpha | + (-1)^{rq} \| \gamma, \alpha |, \beta | = 0 ,$$

where $\gamma \in \pi_r(S)$, $r \geqslant 3$, $m\gamma = 0$ and the calculation is made in $H_{p+q+r-1}(\Omega S)$ modulo the ideal generated by $\{u\}$, $\{v\}$, and $\{w\} = h_2 \omega_2 \gamma$.

Note added in proof. W. S. Massey [Annals of Mathematics 62 (1955) p. 327] has raised (as problem 18) the question of homotopy operations of higher kinds. It is clear that the product $\{\alpha, \beta\}$ introduced above is an operation of the sort indicated.

REFERENCES

[1] *R. Bott and H. Samelson,* On the Pontryagin product in spaces of paths, Comment. Math. Helv. 27 (1953) 320–337.

[2] *H. Cartan,* Sur les groupes d'Eilenberg-Maclane I, Proc. Nat. Acad. Sci. U. S. A. 40 (1954) 467–471.

[3] *P. J. Hilton,* On the homotopy groups of the union of spaces, Comment. Math. Helv. 29 (1955) 59–92.

[4] *I. M. James,* On reduced product spaces, Ann. Math. 62 (1955) 170–197.

[5] *J. C. Moore,* On the homotopy groups of spaces with a single non-vanishing homology group, Ann. Math. 59 (1954) 549–557.

[6] *J. C. Moore's* isomorphism theorem for spectral sequences, Seminaire H. Cartan, E. N. S. 1954–1955, Exposé 3.

[7] *M. Nakaoka and H. Toda,* On Jacobi identity for Whitehead products, J. Inst. Polytech., Osaka City Univ., Ser. A (1954) 1–13.

[8] *H. Samelson,* On the relation between the Whitehead and Pontryagin product, Amer. J. Math. 75 (1953) 744–752.

[9] *H. Toda,* On standard path spaces and homotopy theory I, Proc. Japan Acad. 29 (1953) 299–304.

(Received October 20, 1955)

DRUCK: ART. INSTITUT ORELL FÜSSLI AG., ZÜRICH

On the Cobar Construction

par M. J. F. Adams (Cambridge, England)

1. Introduction

P. J. Hilton [1] and the present author [1] have given means of computing $H_*(\Omega K)$ when K is a (suitably restricted) CW-complex. This method leaves two things to be desired. First, it passes from K, which is a CW-complex, to A, which is a chain complex; and thus it cannot be iterated. Secondly, the boundary operator d in A is not explicitly given.

In the present paper, I give an algebraic construction, which passes from one chain complex (with suitable extra structure) to another chain complex. This construction may be called the cobar [2] construction, since it is dual to the bar construction of Eilenberg and MacLane [3] (cf. also [2]). Each construction is analogous to a fibration with acyclic total space. The bar construction passes from a chain complex A, analogous to a fibre, to a chain complex B(A), analogous to a base-space. The cobar construction passes from a chain complex C, analogous to a base-space, to a chain complex F(C), analogous to a fibre.

The cobar construction permits one to calculate the homology of loop-spaces. Since the boundary d in F(C) is explicitly given, it answers both the points made above.

2. Definition of the Construction

Let Λ be a fixed principal ideal domain of coefficients. We shall make the following assumptions about C and Δ.

(1) C is a Λ-free graded chain complex with $C_0 = \Lambda$, $C_1 = 0$.

[1] I am indebted to P. J. Hilton for comments on the present paper; and also to H. Cartan, S. Mac Lane, W. S. Massey and J. C. Moore for their comments on and interest in the subject.

[2] This name is due to H. Cartan, who has (independently) considered the same construction.

Thus C has a natural augmentation. Further, C C is defined and is a Λ-free graded augmented chain complex, with the usual boundary.

(2) $\Delta : C \to C \otimes C$ is a chain map.

We now define $\Delta_{p,q}$ to be the component of Δ on $C_p \otimes C_q$.

(3) For $x \in C_p$,

$$\Delta_{p,0}(x) = x \otimes 1 \ , \ \Delta_{0,p}(x) = 1 \otimes x \ .$$

(4) Δ is associative, in the sense that (if I is the identity map)

$$(\Delta \otimes I)\,\Delta = (I \otimes \Delta)\,\Delta \ .$$

We may now define

$$\overline{C} = \frac{C}{\Lambda} = \sum_{n \geq 2} C_n \, , \quad (\overline{C})^r = \overline{C} \otimes \overline{C} \otimes \ldots \otimes \overline{C}$$

(r factors) and

$$F(C) = \Lambda + \sum_{r \geq 1} (\overline{C})^r \, .$$

The canonical isomorphism

$$(\overline{C})^r \otimes (\overline{C})^s \to (\overline{C})^{r+s}$$

induces a product

$$\mu : F(C) \otimes F(C) \to F(C) \ .$$

We grade F(C) by setting $\deg^F C_p = p - 1$ and making μ preserve degree. We give F(C) a boundary d^F by the formulae

$$d^F \mu = \mu d^F \, , \ d^F(c) = - dc + \sum_{2 \leq p \leq n-2} (-1)^p \Delta_{p,n-p} c \quad (c \in C_n) \, .$$

It may easily be verified that

$$(d^F)^2 = 0 \ .$$

Algebraically, F(C) is a free associative algebra. We may therefore define a boundary d and contracting homotopy s for the "total space" $F(C) \otimes C$ by the formulae R1, R2, D1, D2 given in [1] § 2. The "projected" boundary $\overline{d}$ on C coincides with d.

Structures such as C (³) form a category. A map $f : C \to C'$ is a Λ-linear function wich preserves augmentation and grading and commutes with d and Δ. Such a map f induces a map $F(f) : F(C) \to F(C')$ which commutes with d^F and μ.

(³) J. Cartier has called such structures "co-algebras".

3. Main Theorem

Let B, b be a 1-connected space with base-point. Let CS(B) be the simplicial singular chain complex of B. Let $\sigma^{n,1}$ be the 1-skeleton of σ^n; let $CS^1(B)$ be the subcomplex of CS(B) generated by maps

$$f : \sigma^n, \sigma^{n,1} \to B, b.$$

Let

$$C_n(B) = \frac{CS_n^1(B)}{CS_n^1(b)} \quad (n > 0), \; C_0(B) = \Lambda.$$

We obtain a chain complex C(B) with

$$C_0 = \Lambda, \quad C_1 = 0.$$

With the standard (Cech-Alexander) diagonal map Δ, it satisfies the axioms (4). Let $\Omega(B)$ be the space of loops ω, of variable length r, in B, based at b.

Theorem. — *There is a natural isomorphism*

$$H_*(F(CB)) \cong H_*(\Omega B) \; (^5).$$

It is easily proved (by the method of [3], p. 84) that in this theorem C(B) may be replaced by a smaller equivalent chain complex for purposes of computation.

Proof. — Let P(B) be the singular polytope of B [4]. Let $P^1(B)$, $P^1(b)$ correspond to $CS^1(B)$, $CS^1(b)$. Define

$$Q(B) = \frac{P^1(B)}{P^1(b)}$$

by identifying $P^1(b)$ to a point p. Then

$$H_*(\Omega QB) \cong H_*(\Omega B);$$

I shall show how to apply the methods of [1] to prove

$$H_*(F(CB)) \cong H_*(\Omega QB).$$

The method is based on the (essentially explicit) geometric construction of a map from I^{n-1} to $L(\sigma^n)$. Let I^n be the standard n-cube, σ^n the standard n-simplex (with vertices v_i). Let $L_{i,j}(\sigma^n)$ be the space of paths ω in σ^n, of variable length r, with

$$\omega(0) = v_i, \quad \omega(r) = v_j$$

(4) I gather from W. S. Massey that the idea of dualising the bar construction by using the cup-product instead of the Pontrjagin product is not new, but has not previously been successful.

(5) By convention, I shall cease to insert brackets when they cease to add clarity.

Let

$$\lambda_i^\varepsilon : I^{n-1} \to I^n, \quad d_i : \sigma^{n-1} \to \sigma^n$$

be the standard injections of faces. Let

$$f_i : \sigma^i \to \sigma^n, \quad l_i : \sigma^{n-i} \to \sigma^n$$

be the standard injections of first and last faces, so that

$$\Delta_{i,n-i} f = ff_i \otimes f l_i \quad \text{for } f : \sigma_n \to B .$$

These maps induce

$$L(f_i) : L_{0,i}(\sigma^i) \to L_{0,i}(\sigma^n), \; L(l_i) : L_{0,n-i}(\sigma^{n-i}) \to L_{i,n}(\sigma^n) ;$$

similarly for $L(d_i)$. There is a pairing

$$L_{0,i}(\sigma^n) \times L_{i,n}(\sigma_n) \to L_{0,n}(\sigma^n) .$$

When such a pairing $X \times Y \to Z$ exists, and

$$f : I^p \to X, \quad g : I^q \to Y$$

are maps, we write

$$f \cdot g : I^{p+q} \to Z$$

for the product map.

We shall construct maps

$$\Theta_n : I^{n-1} \to L_{0,n}(\sigma^n)$$

to satisfy the following conditions.

(1) $\qquad \Theta_1(I^0) = (\omega, r) \in L_{0,1}(\sigma^1),$

where $r=1$ and

$$\omega : [0, 1] \to \sigma^1$$

is given by

$$\omega(x) = (1-x, x) .$$

(2) $\qquad \Theta_n \lambda_i^0 = [L(f_i)\Theta_i] \cdot [L(l_i)\Theta_{n-i}] .$

(3) $\qquad \Theta_n \lambda_i^1 = L(d_i)\Theta_{n-1} .$

The construction is by induction over n. For $n=1$, Θ_1 is defined by (1). Suppose Θ_r defined to satisfy (1), (2), (3) for $r<n$. Then the conditions (2), (3) define Θ_n consistently on $\dot{I}^{n-1}$. Since $L_{0,n}(\sigma^n)$ is contractible, we may extend the definition of Θ_n over I^{n-1}. Θ_n is thus defined for all n.

By normalising the paths to length 1, we may interpret Θ_n as a map

$$\Theta_n' : I \times I^{n-1} \to \sigma^n .$$

In fact,

$$\Theta_n' : I^n, \dot{I}^n \to \sigma^n, \dot{\sigma}^n$$

has degree $+1$ (6). This is easily seen by induction over n, using the explicit form of the boundary $\Theta_n' | \dot{I}^n$.

If $f : \sigma^n \longrightarrow B$ is a map, and σ_f^n is the corresponding cell in $P(B)$, let $c_f : \sigma^n \longrightarrow P(B)$ be the characteristic map for σ_f^n. Let $q : P^1(B) \rightarrow Q(B)$ be the identification map. The composite $L(q c_f) \Theta_n : I^{n-1} \rightarrow \Omega QB$ maps certain points to paths stationary at p, but not of zero length. We shall deform these paths to the path 1 of zero length at p. Formally, for each map $f : \sigma^n, \sigma^{n,1} \rightarrow B, b$ we shall construct a homotopy.

$$\chi_t(f) : I^{n-1} \rightarrow \Omega QB$$

to satisfy the following conditions

(4) $\chi_0(f) = L(q c_f)\Theta_n$.

(5) $\chi_t(f) I^{n-1} \subset \Omega(q c_f(\sigma_n))$.

(6) If $f(\sigma^n) = b$ then $\chi_1(f)(I^{n-1}) = 1$.

(7) $\chi_t(f)\lambda_i^0 = \chi_t(f f_i) \cdot \chi_t(f l_i)$.

(8) $\chi_t(f)\lambda_i\lambda_i^1 = \chi_t(f d_i)$.

The construction is by induction over n. Suppose $\chi_t(f)$ defined to satisfy (4) to (8) for $f : \sigma^r, \sigma^{r,1} \rightarrow B, b$ with $r < n$ (this assumption being vacuous if $n = 1$). Consider

$$f : \sigma^n, \sigma^{n,1} \rightarrow B, b .$$

Possibly $f(\sigma^n) = b$; in this case, $\chi_t(f)$ is consistently defined on $\dot{I} \times I^{n-1} \cup I \times \dot{I}^{n-1}$ and lies in $\Omega(p)$. Since $\Omega(p)$ is contractible, the definition may be extended over $I \times I^{n-1}$. Possibly $f(\sigma^n) \neq b$. In this case, $\chi_t(f)$ is consistently defined on $0 \times I^{n-1} \cup I \times \dot{I}^{n-1}$ and lies in $\Omega(q c_f(\sigma^n))$. Again, the definition may be extended over $I \times I^{n-1}$. $\chi_t(f)$ is thus defined for all f.

By normalising the paths to lenght 1, we may interpret $\chi_1(f)$ as a map

$$\chi_1'(f) : I \times I^{n-1} \rightarrow QB .$$

In fact, $$\chi_1'(f) : I^n, \dot{I}^n \rightarrow q c_f(\sigma^n), q c_f(\dot{\sigma}^n)$$

is degenerate if $f(\sigma^n) = b$, and has degree $+1$ *if* $f(\sigma^n) \neq b$.

(6) This assumes some conventions on orientation; I use the orientations of homology theory. I assume that the orientations (generators of $H_n(E^n, \dot{E}^n)$) of the cube and simplex are given by their identity maps in cubical and simplicial singular homology. I assume that a natural equivalence between cubical and simplicial singular homology is so set up that the map

$$x \longrightarrow (1 - x, x) : I^1 \longrightarrow \sigma^1$$

has degree $+1$.

Let $CU(\Omega QB)$ be the cubical chain complex of ΩQB (normalised, as in [1]). We define a function

$$\varphi : (\overline{C}(B))^r \leftarrow CU(\Omega QB)$$

by setting

$$\varphi(f_1 \otimes f_2 \otimes \ldots \otimes f_r) = \chi_1(f_1) \cdot \chi_2(f_2)\sigma \cdot \ldots \cdot \chi_1(f_r),$$

and

$$\varphi : \Lambda \rightarrow CU_0(\Omega QB) \text{ by } \varphi(1) = 1 .$$

Thus we obtain a function

$$\varphi : F(CB) \rightarrow CU(\Omega QB) .$$

By construction, we have

$$d\varphi(f) = \sum_{1 \leq i \leq n-1} (-1)^i (\chi_1(f)\lambda_i^0 - \chi_1(f)\lambda_i^1$$

$$= -\sum_{0 \leq i \leq n} (-1)^i \chi_1(fd_i) + \sum_{2 \leq i \leq n-2} (-1)^i \chi_1(ff_i) \cdot \chi_1(fl_i)$$

(for $n \geqslant 3$; for $n = 2$ it is zero)

$$= \varphi(-df + \sum_{2 \leq p \leq n-2} (-1)^p \Delta_{p,n-p} f)$$

$$= \varphi d^F f .$$

Moreover, both in $F(CB)$ and $CU(\Omega QB)$ we have the product formula $d\mu = \mu d$. It follows that φ is a chain map.

Let $L(QB)$ consist of paths ω, of variable length r, in QB, with $\omega(0) = p$. We shall extend φ to a map from

$$F(CB) \otimes CB \text{ to } CU(LQB) .$$

We shall satisfy the conditions

(9) $d\varphi = \varphi d$.

(10) $\mu(\varphi \otimes \varphi) = \varphi\mu$.

(11) $\varphi(1 \otimes f) \subset CU_n(L(qc_f\sigma^n))$.

(12) If $f(\sigma^n) \neq b$, $p_* \varphi(1 \otimes f)$ is of degree $+1$ in

$$Z_n(qc_f\sigma^n,\ qc_f\dot{\sigma}^n) .$$

The construction is by induction over n. Suppose φ so defined on the generators of dimension $r < n$; the assumption is vacuous for $n = 2$. Let $f : \sigma^n, \sigma^{n,1} \rightarrow B, b$ be a map with $f(\sigma^n) \neq b$. From $\chi_1(f)$, define

$$u(f) : I \times I^{n-1} \rightarrow L(qc_f\sigma^n)$$

using the standard contraction of $L(qc_f\sigma^n)$, so that

$$du(f) - \chi_1(f) \in CU_{n-1} L(qc_f\dot{\sigma}^n) .$$

By the above, $p_* u(f)$ is of degree $+1$ in

$$Z_n(q\, c_f\, \sigma^n,\ q\, c_f\, \dot\sigma^n), \quad \text{and} \quad \varphi d(1 \otimes f) - du(f)$$

is a cycle in $Z_{n-1}\, L(q\, c_f \dot\sigma^n)$. $L(q\, c_f \dot\sigma^n)$ is acyclic; thus

$$\varphi d(1 \otimes f) - du(f) = d\nu(f)$$

with

$$\nu(f) \in CU_n(L(q\, c_f \dot\sigma^n)).$$

We may set

$$\varphi(1 \otimes f) = u(f) + \nu(f),$$

satisfying (9), (11), (12), and define φ on products by (10), so satisfying (9).

The map φ is thus constructed. The remainder of the proof consists of an appeal to Moore's Theorem ([2] Exp. 3); this follows precisely the method given in [1]. The isomorphism of homology of the "bases" C(B), CU(QB) follows from (12). We thus reach the conclusion that

$$\varphi_* : H_*(FC(B)) \cong H_*(\Omega Q B).$$

I shall omit the proof that φ is natural.

Trinity College, Cambridge.

Bibliography

[1] ADAMS, J. F., and HILTON, *On the Chain Algebra of a Loop Space* (*Commentarii Mathematici Helvetici*, to appear).

[2] CARTAN, H., *Séminaire H. Cartan de l'École Normale Supérieure*, 1954-55.

[3] EILENBERG, S. and MAC LANE, *On the Groups* H (π, n) I (*Annals of Mathematics* 58, 1953, pp. 55-106).

[4] GIEVER, J. B., *On the Equivalence of Two Singular Homology Theories* (*Annals of Mathematics*, 51. 1950, pp. 178-191).

Offprint of Commentarii Mathematici Helvetici, vol. 32, fasc. 3, 1958

On the Structure and Applications of the STEENROD Algebra

by J. F. ADAMS

§ 1. Introduction

This paper contains a proof of the following theorem in homotopy-theory.

Theorem 1.1. *If* $\Pi_{2n-1}(S^n)$ *and* $\Pi_{4n-1}(S^{2n})$ *both contain elements of* HOPF *invariant one, then* $n \leqslant 4$.

This theorem, of course, is only significant if n is of the form 2^m. We note that it yields an independent proof of the following theorem of H. TODA [5]; there is no element of HOPF invariant one in $\Pi_{31}(S^{16})$.

The author conjectures that this theorem can be improved; it is included mainly to motivate and illustrate the methods here introduced.

These methods depend on a certain spectral sequence. It leads, roughly speaking, from the cohomology of the (mod p) STEENROD algebra[1]) to the p-components of the stable homotopy groups of spheres. This spectral sequence may by regarded, on the one hand, as an extension of ADEM's method of studying homotopy groups by considering cohomology operations of the second and higher kinds. On the other hand, it may be regarded as a reformulation of the method of killing homotopy groups.

Theorem 1.1 follows from a superficial study of this spectral sequence. It requires, however, some knowledge (though very little) of the cohomology of the STEENROD algebra. Our methods for studying the cohomology of the STEENROD algebra depend on a thorough knowledge of the structure of the STEENROD algebra. This is obtained by classical methods (cf. [2])[1a]).

[1]) The (mod p) STEENROD algebra, where p is a prime, has as generators the symbols β_p, P_p^k if $p > 2$, Sq^k if $p = 2$. The relations are those which are universally satisfied by the BOCKSTEIN boundary β_p and the STEENROD operations P_p^k or Sq^k in the cohomology of topological spaces. See [2].

The cohomology of the STEENROD algebra is defined below.

[1a]) Note added in proof. I learn that J. W. MILNOR has made an elegant study of the structure of the STEENROD algebra, which overlaps in content with § 5.

§ 2. Summary of Results and Methods

We will next summarise these methods in more detail. We will consider the spectral sequence, and its multiplicative structure; we will give some data on the cohomology of the STEENROD algebra, and show how to deduce Theorem 1.1. We begin with the spectral sequence.

Let X be a space, let $S^n X$ be the iterated suspension of X, and let $\Pi_m^S(X)$ be the stable or S-homotopy group $\operatorname{Dir Lim} \Pi_{m+n}(S^n X)$. Let p be a prime, and let K^m be the subgroup of $\Pi_m^S(X)$ which consists of elements whose order is finite and prime to p. Let $H_*(X)$ be the (augmented) singular homology group of X; we suppose $H_t(X)$ finitely generated for each t. We make the convention that when we write "cohomology", it means "cohomology with Z_p coefficients", and when we write "$H^*(X)$" it means "$H^*(X; Z_p)$", the (augmented) singular cohomology group of X with Z_p coefficients. The group $H^*(X)$, then, has the structure of a left module over A, the (mod p) STEENROD algebra. We give Z_p the trivial A-module structure. That is, the unit in A acts as a unit, while $a(Z_p) = 0$ if $a \in A_q$ with $q > 0$. (Here the grading q of the STEENROD algebra $A = \sum_q A_q$ is defined by $\deg \beta_p = 1$, $\deg P_p^k = 2k(p-1)$, $\deg Sq^k = k$.) The group $\operatorname{Ext}_A^*(H^*(\mathrm{X}), Z_p)$ is now defined[2]). It is bigraded; the grading s is the grading of Ext_A^s, while the grading t arises from the grading of $H^t(X)$ and that of the STEENROD algebra A[3]). With these notations, we have the following theorem.

Theorem 2.1. *There is a spectral sequence, with terms $E_r^{s,t} = E_r^{s,t}(X)$ which are zero if $s < 0$ or if $t < s$, and with differentials*

$$d_r: E_r^{s,t} \to E_r^{s+r,t+r-1}$$

satisfying the following conditions.

(i) *There is a canonical isomorphism*

$$E_2^{s,t} \cong \operatorname{Ext}_A^{s,t}(H^*(X), Z_p) .$$

(ii) *There is a canonical isomorphism*

$$E_{r+1}^{s,t} \cong H^{s,t}(E_r; d_r) .$$

(iii) *There is a canonical monomorphism from $E_R^{s,t}$ to $E_r^{s,t}$ for $s < r < R \leqslant \infty$.*

(iv) *If (using (iii)) we regard $E_r^{s,t}$ as a subgroup of $E_{s+1}^{s,t}$ for $s < r \leqslant \infty$, we have*

$$E_\infty^{s,t} = \bigcap_{s<r<\infty} E_r^{s,t} .$$

---

2) See [4].

3) See § 3.

(v) *There exist groups* $B^{s,t}$ *such that*

$$B^{s,t} \subset B^{s-1,t-1} \subset \cdots \subset B^{0,t-s}, \quad B^{0,m} = \Pi_m^S(X)$$

and

$$E_\infty^{s,t} \cong B^{s,t}/B^{s+1,t+1} .$$

(vi) $\bigcap_{t-s=m} B^{s,t} = K^m$.

This is the spectral sequence referred to in § 1. The convergence statements (iv) and (vi) are not needed for the proof of Theorem 1.1.

If we take $X = S^0$, then $\Pi_m^S(X)$ becomes the stable group of the m-stem, and the term E_2 becomes $\mathrm{Ext}_A^*(Z_p, Z_p)$. We shall write this $H^*(A)$, and refer to it as the cohomology of the STEENROD algebra.

We turn next to the products in this spectral sequence.

Theorem 2.2. *If* $X = S^0$, *then it is possible to define products*

$$E_r^{s,t} \otimes E_r^{s',t'} \to E_r^{s+s',t+t'}$$

(in the spectral sequence of Theorem 2.1) with the following properties.

(i) *The products are associative, and anticommutative for the degree* $t - s$.

(ii) *The product* $E_2^{s,t} \otimes E_2^{s',t'} \to E_2^{s+s',t+t'}$ *coincides, except for a sign* $(-1)^{ts'}$, *with the cup-product*[4])

$$H^{s,t}(A) \otimes H^{s',t'}(A) \to H^{s+s',t+t'}(A) .$$

(iii) $d_r(uv) = (d_r u)v + (-1)^{(t-s)} u(d_r v)$.

(iv) *The products commute with the isomorphisms* $E_{r+1}^{s,t} \cong H^{s,t}(E_r; d_r)$ *and with the monomorphisms from* $E_R^{s,t}$ *to* $E_r^{s,t}$ *(if* $s < r < R \leqslant \infty$*).*

(v) *The products in* E_∞ *may be obtained by passing to quotients from the composition product*

$$\Pi_m^S(S^0) \otimes \Pi_{m'}^S(S^0) \to \Pi_{m+m'}^S(S^0) .$$

We offer next some remarks on the interpretation of these theorems. We should explain that it is possible to define a filtration F_s of $\Pi_m^S(S^0)$ by considering cohomology operations of higher kinds. We consider only those operations which act on cohomology with Z_p coefficients. Let $\alpha: S^{n+m} \to S^n$ be a map. Form a complex $K = S^n \cup E^{n+m+1}$ by using α as an attaching map. Then $H^n(K) = Z_p$ and $H^{n+m+1}(K) = Z_p$, at least so long as α induces the zero map of cohomology. Suppose that the following condition holds: if Φ is any non-trivial stable cohomology operation of the rth kind, with $r < s$, and of degree $(m+1)$, then $\Phi: H^n(K) \to H^{n+m+1}(K)$ is defined and zero. Then we set $\alpha \in F_s$; this defines $F_s \subset \Pi_m^S(S^0)$. It is a subgroup, and $F_s \supset F_{s+1}$. We

[4]) This cup-product will be defined in § 4.

complete the definition by setting $\alpha \in F_1$ if α induces the zero map of cohomology.

The author supposes that this filtration coincides with that given by Theorem 2.1 (in case $X = S^0$); that is, $F_s = B_{s,s+m}$. However, he has not tried to prove this proposition, which is not material to this paper.

We have next to compare the classical method of killing homotopy groups with the method of calculation provided by Theorem 2.1. It is clear that both rely on the information contained in CARTAN's calculation of $H^*(\Pi; n)$. However, from accounts of the classical method, one obtains the impression that it enables one to calculate a great deal, but that one cannot guarantee in advance exactly how much. With the formalism of Theorem 2.1 the situation is more clear; we can effectively compute the term E_2 (to any finite dimension); we cannot, at present, give a convenient method for effective computation of the differentials d_r, or of the group extensions involved.

In the case $X = S^0$, it is possible to obtain information about the group extensions involved in $\Pi_m^S(S^0)$ from Theorem 2.2; for this theorem will in particular inform us about the composite of an element in $\Pi_m^S(S^0)$ and the element of degree p in $\Pi_0^S(S^0)$.

The author's interest is particularly attracted to the phenomena which arise because the differentials d_r may not be zero; it will appear that Theorem 1.1 is a case in point.

It is clear that in order to study the spectral sequence of Theorems 2.1, 2.2 we shall need some information at least about the term E_2. Although $H^*(A)$ is defined by means of resolutions, to study it in this way seems unrewarding. We therefore employ the spectral sequence which relates the cohomology rings $H^*(\Lambda)$, $H^*(\Gamma)$ and $H^*(\Omega)$ of an algebra Γ, a normal subalgebra Λ and the corresponding quotient algebra Ω[5]). To make use of this spectral sequence, we prove results on the structure of the STEENROD algebra. Some of these concern a descending sequence of subalgebras A^r of the STEENROD algebra $A = A^1$. These are such that

(i) $\bigcap_r A^r$ has the unit as a base.

(ii) A^r is a normal subalgebra of A^s if $r > s$; and A^s is then a free module over A^r.

(iii) The quotient $A^r//A^{r+1}$ is an algebra whose cohomology is known.

Such results enable one to apply the method of calculation indicated. They form a large part of the technical labour of this paper.

We will show next how Theorem 1.1 can be deduced (using Theorems 2.1 and 2.2) from a very superficial knowledge of $H^*(A)$ (in the case $p = 2$). We

[5]) See [4], p. 349.

write $H^s(A) = \sum_t H^{s,t}(A)$, and catalogue various facts we suppose known. First, we suppose it known that $H^0(A)$ has as a base the element 1 in dimension $t = 0$, and that $H^1(A)$ has as a base elements h_m in dimension $t = 2^m$, $m = 0, 1, 2, \ldots$ [6]).

We further suppose known the following lemma.

Lemma 2.3. *If there is an element of* HOPF *invariant one in* $\Pi_{2n-1}(S^n)$, *whose S-class is* h'', *then* $n = 2^m$, *and the class* h' *of* h'' *in* $E_\infty^{1,n}$ *passes by the canonical monomorphism to the class* h_m *in* $E_2^{1,n}$.

Conversely, if h_m *lies in the image of* $E_\infty^{1,2^m}$, *then there is an element of* HOPF *invariant one in* $\Pi_{2n-1}(S^n)$ *for* $n = 2^m$.

This lemma will follow as soon as the spectral sequence is set up. In case $m = 0$, the lemma is still true with a suitable interpretation; we may take (and now define) a "map of HOPF invariant one in $\Pi_1(S^1)$" to be the map of degree two [7]).

Lastly, we will assume the following theorem on $H^*(A)$ (in case $p = 2$).

Theorem 2.4. *The products* $h_i h_j$ *in* $H^2(A)$ *are subject to the following relation only:* –

$$h_i h_{i+1} = 0.$$

The products $h_i h_j h_k$ *in* $H^3(A)$ *are subject to the following three relations only:* –

$$h_i h_{i+1} h_j = 0\ , \quad (h_i)^2 h_{i+2} = (h_{i+1})^3\ , \quad h_i (h_{i+2})^2 = 0\ .$$

We will now deduce Theorem 1.1. Let us suppose (for a contradiction) that h''_m, h''_{m+1} are S-classes containing maps of HOPF invariant one, and that $m \geqslant 3$. Consider the element $h_0(h_m)^2$ in $E_2^{3,2^{m+1}+1}$. By Theorem 2.4 it is non-zero. It is a cycle for d_2 by Theorem 2.2. It is not a boundary for d_2, because $d_2 E_2^{1,2^{m+1}}$ is generated by $d_2 h_{m+1}$, and this is zero because h_{m+1} is in the image of $E_\infty^{1,2^{m+1}}$ (Lemma 2.3). Therefore $h_0(h_m)^2$ yields a non-zero element in $E_3^{3,2^{m+1}+1}$. This implies that, in $\Pi^S_{2^{m+1}-2}$, the element $h''_0(h''_m)^2$ is non-zero, that is, $2(h''_m)^2$ is non-zero. But since composition in stable homotopy groups is anti-commutative and the dimension of h''_m is odd, we have $2(h''_m)^2 = 0$. This contradiction proves Theorem 1.1.

We note as a corollary of the proof, that the differential $d^2 \colon E_2^{1,16} \to E_2^{3,17}$ maps h_4 to $h_0(h_3)^2$, and is thus non-zero. This remark may be paraphrased as follows.

Corollary 2.5. *If* Sq^{16} *is considered as a cohomology operation of the second kind, it has a non-trivial decomposition.*

---

[6]) These are related to the elements Sq^{2^m} in A, and will be defined at the beginning of § 6.

[7]) When we use this map as an attaching map, Sq^1 is non-zero in the resulting complex.

The decomposition asserted is of the form

$$Sq^{16}u = \sum_i a_i \Phi_i(u) \pmod{Q} \quad (\text{if } u \in K).$$

Here $a_i \in \sum_{1\leq j\leq 15} A_j$, Φ_i is of the second kind, and for the space X concerned we have

$$H^*(X) \supset K \supset \bigcap_{\substack{a\in A_j\\ 1\leq j\leq 15}} \operatorname{Ker} a\,, \quad H^*(X) \supset Q \subset \sum_{\substack{a\in A_j\\ 1\leq j\leq 15}} \operatorname{Im} a\,.$$

Such a decomposition evidently shows that it is impossible to form a complex $S^n \cup E^{n+16}$ in which Sq^{16} is non-zero. It would be interesting to know whether such decompositions can be proved directly by Adem's method, or by any other method.

Theorem 2.4 has the following obvious corollary.

Corollary 2.6. *If h''_i, h''_j, h''_k are S-classes of dimensions 2^i-1, 2^j-1, 2^k-1 containing elements of Hopf invariant one, then the S-classes $h''_i h''_j$ and $h''_i h''_j h''_k$ are non-zero except perhaps in the following cases (where h''_0 is to be interpreted as the class 2ι of dimension 0).*

$$h''_i h''_{i+1}\,, \quad h''_i h''_{i+1} h''_j\,, \quad h''_i (h''_{i+2})^2\,, \quad h''_0 (h''_j)^2\,.$$

The case which concerns $(h''_i)^2$ is due to Adem [1].

This concludes our summary of results and methods.

§ 3. The Spectral Sequence

In this section we prove Theorem 2.1 by constructing the spectral sequence. We do this, roughly speaking, by taking the homotopy exact couple of a sequence $Y_0 \supset Y_1 \supset Y_2 \ldots$ of spaces. These are such that Y_0 is equivalent to an iterated suspension $S^n X_*$ and $\sum_s H^*(Y_s, Y_{s+1})$ (with the cohomology boundary) is an A-free resolution of $H^*(X)$. Actually we only obtain this property for a finite number of dimensions at one time, as we have to keep to a stable range. We shall therefore consider a finite sequence $Y_0 \supset Y_1 \supset \cdots \supset Y_k$ of spaces which have the required properties in a finite range of dimensions, specified by a parameter l. By increasing k and l we obtain increasing portions of the spectral sequence. The reader will lose little (except the details needed for rigour) if he replaces formulae containing k and l by suitable phrases containing the words "sufficiently large". We conclude by proving the convergence of the spectral sequence.

We proceed to give the details. Let X be a space and C a chain complex of left A-modules, with augmentation onto $H^*(X)$: –

$$\sum_t H^t(X) \xleftarrow{\varepsilon} \sum_t C_{0,t} \xleftarrow{d} \sum_t C_{1,t} \leftarrow \cdots \leftarrow \sum_t C_{s,t} \leftarrow \cdots$$

Here, for example, C may be an acyclic resolution of $H^*(X)$ by free left A-modules: it is understood that a free bigraded module has a base whose elements are bihomogeneous. The algebra A is still the (mod p) Steenrod algebra. The second grading t is to be preserved by d and ε and to become the topological dimension in $H^t(X)$; the operations of A on C satisfy

$$A_q \cdot C_{s,t} \subset C_{s,t+q} \, .$$

We suppose that $C_{s,t} = 0$ if $t < s$ and that each $C_{s,t}$ is finitely generated; it is always possible to find resolutions satisfying these conditions (recall that $H^t(X)$ is finitely generated).

Suppose given also integers k, l. By a *realisation* of the resolution C, we understand an integer n ($n \geqslant l + 1$) and a sequence $Y_0 \supset Y_1 \supset \cdots \supset Y_k$ of CW-complexes and subcomplexes with the following properties.

(1) Y_0 and $S^n X$ are of the same singular homotopy type. (This induces isomorphisms[8]) i: $H^t(X) \cong_A H^{n+t}(Y_0)$).

Y_s is $(n-1)$-connected (for $0 \leqslant s \leqslant k$); $\Pi_r(Y_s, Y_{s+1})$ is finite and p-primary (for all r, $0 \leqslant s < k$).

(2) There are isomorphisms φ: $C_{s,t} \cong_A H^{n+t-s}(Y_s, Y_{s+1})$ for $0 \leqslant s < k$, $t \leqslant l$.

(3) The following diagrams are commutative (for $t \leqslant l$ and for $s + 1 < k$, $t \leqslant l$ respectively).

$$\begin{array}{ccc} C_{s,t} & \xleftarrow{\quad d \quad} & C_{s+1,t} \\ \downarrow \varphi & & \downarrow \varphi \\ H^{n+t-s}(Y_s, Y_{s+1}) & \xleftarrow{(-1)^n d} & H^{n+t-s-1}(Y_{s+1}, Y_{s+2}) \end{array} \qquad \begin{array}{ccc} H^t(X) & \xleftarrow{\varepsilon} & C_{0,t} \\ \downarrow i & & \downarrow \varphi \\ H^{n+t}(Y_0) & \longleftarrow & H^{n+t}(Y_0, Y_1) \end{array}$$

We note that a realisation for some l is also a realisation for any less l. Similarly, from a realisation we can obtain realisations with less k, by ignoring some subspaces Y_s, or with greater n, by suspension. This is the reason for the sign $(-1)^n d$; it is inserted so that the diagram is preserved on suspension. (The suspension isomorphism is defined using a coboundary map, and it anticommutes with other coboundary maps.)

We know that resolutions of $H^*(X)$ exist; it is necessary to prove that realisations of them exist.

---

[8]) The symbol $\cong_A$ indicates an isomorphism commuting with the operations from A.

Lemma 3.1. *Let C be an acyclic resolution of $H^*(X)$ by free left A-modules (as above); let k, l be integers. Then X, C, k, l have a realisation.*

Proof. Suppose given X, C, k and l, as above. We take $n = l + 1$. Let $Z_{s,t}$ be the subgroup of cycles in $C_{s,t}$. Let $\mathrm{Hom}_A(C_{s,t}, Z_p)$ stand by convention for the "component" of $\mathrm{Hom}_A(C, Z_p)$ in dimension (s, t), that is, the image of $\mathrm{Hom}_A(\sum_u C_{s,u}, Z_p)$ in $\mathrm{Hom}(C_{s,t}, Z_p)$. Suppose, as an inductive hypothesis, that we have defined a space F_s such that

$$H^{n+t-s-1}(F_s) \cong_A Z_{s,t} \quad \text{for} \quad t<n \ .$$

Take a space B_{s+1} with

$$\Pi_{n+t-s-1}(B_{s+1}) \cong \mathrm{Hom}_A(C_{s+1,t}, Z_p) \ , \qquad k^{n+t-s}(B_{s+1}) = 0 \ .$$

Since $\sum_t C_{s+1,t}$ is a free left A-module and $C_{s+1,t}$ is finitely generated, we deduce that

$$H^{n+t-s-1}(B_{s+1}) \cong_A C_{s+1,t} \quad (\text{for } t - s - 1 < n) \ .$$

Take also a CW-complex F'_s of the same singular homotopy type as F_s, and take a (singular) equivalence[9] $F'_s \to F_s$. We may now take a map $f_{s+1}: F'_s \to B_{s+1}$ such that the following diagram is commutative (for $t<n$).

$$\begin{array}{ccc} H^{n+t-s-1}(F'_s) & \xleftarrow[f^*_{s+1}]{} & H^{n+t-s-1}(B_{s+1}) \\ \downarrow \cong_A & & \downarrow \cong_A \\ Z_{s,t} & \xleftarrow{(-1)^n d} & C_{s+1,t} \end{array}$$

Factor the map f_{s+1} through an equivalence and a fibration; we obtain a fibre-space $F_{s+1} \to E_{s+1} \to B_{s+1}$ and a (singular) equivalence $e_s: E_{s+1} \to F_s$. The spectral sequence of the fibre-space reduces to an exact sequence in the low dimensions, and we easily show that

$$H^{n+t-s-2}(F_{s+1}) \cong_A Z_{s+1,t} \quad \text{for} \quad t<n \ .$$

This induction is started (with $s = -1$) by interpreting F_{-1} as $S^n X$ and $(-1)^n d: C_{0,t} \to Z_{-1,t}$ as $\varepsilon: C_{0,t} \to H^t(X)$. We use it to define pairs E_s, F_s for $s \leqslant k - 1$, with singular equivalences $e_s: E_s \to F_{s-1}$.

Let Y'_s be the total mapping cylinder of the maps e_v for $v \geqslant s$; it is obtained from

$$F_{s-1} \cup \bigcup_{s \leqslant v \leqslant k-1} I \times E_v$$

by identifying $0 \times x$ with $1 \times e_v(x)$ if $x \in E_v$ for $v > s$, and $0 \times x$ with $e_s(x)$ if $x \in E_s$. We have embeddings $Y'_0 \supset Y'_1 \supset \cdots \supset Y'_k$.

[9]) A singular equivalence is a map inducing isomorphisms of all homotopy groups.

It is clear that Y'_0 is equivalent to $S^n X$, and that $\Pi_r(Y'_s, Y'_{s+1})$ is finite and p-primary. Applying the HUREWICZ isomorphism mod p to F_s, we see that Y'_s is $(n-1)$-connected. With a routine use of spectral theory, we see that

$$H^{n+t-s}(Y'_s, Y'_{s+1}) \cong_A C_{s,t} \quad (\text{for } t-s<n) .$$

We have also two commutative diagrams, namely those set out in the definition of a realisation as condition (3). These follow on setting up the appropriate inclusive diagrams, and hold for $t<n$, $t-s<n$ respectively.

It remains only to replace these spaces Y'_s by CW-complexes Y_s. The proof of Lemma 3.1 is complete.

It is clear that a resolution admits more than one realisation. Therefore, if we construct anything from a realisation, we must prove a uniqueness theorem. Such a theorem will follow by standard naturality arguments from two naturality lemmas, which we give as lemmas 3.4, 3.5. We preface them with some remarks on realisations.

Let C be an acyclic left A-complex, and let $\{Y_s\}$ be a realisation of it; we have the following lemma.

Lemma 3.2. (i) *For* $s<k$, $t \leqslant l$ *we have the following commutative diagram, in which the columns are isomorphisms.*

$$\begin{array}{ccc} H^{n+t-s}(Y_s, Y_{s+1}) & \xleftarrow{d} & H^{n+t-s-1}(Y_{s+1}) \\ \downarrow \cong_A & & \downarrow \cong_A \\ C_{s,t} & \longleftarrow & Z_{s,t} \end{array}$$

ii) *For* $s<k$, $t \leqslant l$ *the map* $H^{n+t-s}(Y_s) \to H^{n+t-s}(Y_{s+1})$ *is zero.*

This lemma is proved by a trivial induction over s, using the exact cohomology sequence of the pair (Y_s, Y_{s+1}).

Now let C be a complex of free left A-modules, and let $\{Y_s\}$ be a realisation of it; we have the following lemma.

Lemma 3.3. (i) *For complexes* W *of dimension* $<n+l$, *the compression of a map* $f: W \to Y_s$ *into* Y_{s+1} *is equivalent to its compression to a point in* Y_s/Y_{s+1}.

(ii) *If* $t<l$, *then*

$$\Pi_{n+t-s}(Y_s, Y_{s+1}) \cong \Pi_{n+t-s}(Y_s/Y_{s+1}) \cong \mathrm{Hom}_A(C_{s,t}, Z_p)$$

$$k^{n+t-s+1}(Y_s/Y_{s+1}) = 0 .$$

Proof. The first part depends only on the data that Y_s and Y_{s+1} are $(n-1)$-connected and (by the definition of a realisation) $n \geqslant l+1$. As for the second, the projection

$$\Pi_r(Y_s, Y_{s+1}) \to \Pi_r(Y_s/Y_{s+1})$$

is isomorphic for $r \leqslant 2n - 2$, hence for $r < n + l$. For a prime p' distinct from p the p'-component of $\Pi_r(Y_s/Y_{s+1})$ is zero, by inspecting homology. Since $\sum_t C_{s,t}$ is A-free, there is a map from Y_s/Y_{s+1} to

$$\times_t K(\mathrm{Hom}_A(C_{s,t}, Z_p), n + t - s)$$

which induces isomorphisms of cohomology in dimensions $\leqslant n + l - s$.

Next suppose that we have the following data.

(1) $f: X \to Z$ is a map.

(2) C, D are left A-complexes, with augmentations onto $H^*(X)$, $H^*(Z)$; C is acyclic and D is A-free.

(3) $\{Y_s\}$, $\{W_s\}$ are realisations of C, D.

We will suppose that these realisations have the same n, k and l; this will be sufficient for our purposes, by remarks above. Then we have the following lemma.

Lemma 3.4. *There is a map $g: Y_0 \to W_0$ equivalent to $S^n f$ with*

$$g(Y_s^{n+l-s}) \subset W_s^{n+l-s} \quad (\text{for } s \leqslant k) .$$

We postpone the proof of this lemma until we have stated Lemma 3.5.

Next suppose that we have the following data.

(1) $\{Y_s\}$, $\{W_s\}$ are realisations, as above.

(2) g_0, g_1 are homotopic maps, with

$$g_\varepsilon(Y_s^{n+l-s}) \subset W_s^{n+l-s} \quad (\text{for } s \leqslant k,\ \varepsilon = 0, 1) .$$

Then we have the following lemma.

Lemma 3.5. *There is a homotopy $h: g_0 \sim g_1$ with*

$$h(I \times Y_s^{n+l-s}) \subset W_{s-1}^{n+l-s+1} \quad (0 < s \leqslant k) .$$

It is clear that the map g constructed by Lemma 3.4 will yield a map $g^*: D_{s,t} \to C_{s,t}$ of resolutions (at least for $s < k$, $t < l - 1$). Similarly, the homotopy constructed by Lemma 3.5 will yield a homotopy $h^*: D_{s-1,t} \to C_{s,t}$ between two such maps.

Proof of Lemma 3.4. Let us assume that $f: X \to Z$, C, D, $\{Y_s\}$ and $\{W_s\}$ are as given in the data. There is some map $g: Y_0 \to W_0$ equivalent to $S^n f: S^n X \to S^n Z$; we have to examine the obstruction to compressing it so that Y_s^{n+l-s} maps into W_s^{n+l-s}. Suppose we have compressed it so that

$$g(Y_u^{n+l-u}) \subset W_u^{n+l-u} \quad \text{for} \quad u \leqslant s.$$

By Lemma 3.3 we can deform $g | Y^{n+l-s-1}_{s+1}$ through W_s into W_{s+1} if (and only if) the map

$$H^{n+t-s}(W_s/W_{s+1}) \to H^{n+t-s}(Y_{s+1})$$

is zero (for $t<l$). But this map can be factored through

$$H^{n+t-s}(Y_s) \to H^{n+t-s}(Y_{s+1}) ,$$

which is zero by Lemma 3.2. This completes the proof of Lemma 3.4 by induction over s.

The proof of Lemma 3.5 is analogous to that of Lemma 3.4; the obstruction is the composite map

$$\begin{aligned} H^{n+t-s+1}(W_{s-1}/W_s) &\to H^{n+t-s+1}(I \times Y_s, \dot{I} \times Y_s) \\ &\to H^{n+t-s+1}(I \times Y_{s+1}, \dot{I} \times Y_{s+1}) . \end{aligned}$$

We next proceed to obtain the spectral sequence of Theorem 2.1. As indicated above, this is that determined by the homotopy exact couple of the complexes Y_s. Each particular group or homomorphism in it may be obtained from some realisation with finite k and l. Corresponding terms obtained from different realisations may be identified, using homomorphisms constructed with the use of Lemma 3.4.

The details are as follows. Let X be a space, C an acyclic A-free resolution of $H^*(X)$, and let $\{Y_s\}$ be a realisation of C with $k \geqslant s+r$, $l>r+t$. Let $G_r^{s,t}$, $D_r^{s,t}$ be the images by i, d of $\Pi_{n+t-s}(Y_s, Y_{s+r})$, $\Pi_{n+t-s+1}(Y_{s-r+1}, Y_s)$ in $\Pi_{n+t-s}(Y_s, Y_{s+1})$. (If $s-r+1<0$, Y_{s-r+1} is to be interpreted as Y_0.) Then we may define $E_r^{s,t} = G_r^{s,t}/D_r^{s,t}$.

Similarly, let $\{Y_s\}$ be a realisation with $k \geqslant s+1$, $l>t+1$. Let $G_\infty^{s,t}$, $D_\infty^{s,t}$ be the images of $\Pi_{n+t-s}(Y_s)$, $\Pi_{n+t-s+1}(Y_0, Y_s)$ in $\Pi_{n+t-s}(Y_s, Y_{s+1})$. Then we may define $E_\infty^{s,t} = G_\infty^{s,t}/D_\infty^{s,t}$.

The map d_r: $E_r^{s,t} \to E_r^{s+r,t+r-1}$ is obtained (if both groups are defined) by passing to the quotient from the homotopy boundary $(-1)^n d$. (The sign is introduced so that d_r is preserved on suspension.)

If we consider only the terms which can be obtained from a single realisation, the formal properties of a spectral sequence are easily verified. In particular, we have $E_{r+1}^{s,t} \cong H^{s,t}(E_r; d_r)$; there is a canonical monomorphism $E_R^{s,t} \to E_r^{s,t}$ for $s<r<R \leqslant \infty$; and $E_\infty^{s,t} \cong B^{s,t}/B^{s+1,t+1}$, where $B^{s,t}$ is the image of $\Pi_{n+t-s}(Y_s)$ in $\Pi_{n+t-s}(Y_0)$. Thus $E_\infty^{s,t}$, for $t-s=m$, is a quotient obtained by filtering $B^{0,m} = \Pi_m^S(X)$. We may also identify $E_2^{s,t}$. In fact, if $k \geqslant s+2$, $l>t+2$ we have

$$E_1^{s,t} = \Pi_{n+t-s}(Y_s, Y_{s+1}) \cong \mathrm{Hom}_A(C_{s,t}, Z_p)$$

and similarly for $(s+1)$. The map $d_1\colon\ E_1^{s,t} \to E_1^{s+1,t}$ is obtained by transposition from $d\colon\ C_{s+1,t} \to C_{s,t}$, as one sees by using the pairing of Π_* and H^*. Thus

$$E_2^{s,t} \cong \mathrm{Ext}^{s,t}(H^*(X), Z_p)\,.$$

The next point we should consider is the identification of corresponding terms obtained from different realisations. This follows by standard methods from Lemmas 3.4, 3.5, and is omitted.

It remains only to prove the convergence of the spectral sequence, by proving

$$\text{(vi)}\quad \bigcap_{t-s=m} B^{s,t} = K^m\,, \qquad \text{(iv)}\quad \bigcap_{r>s} E_r^{s,t} = E_\infty^{s,t}\,.$$

The inclusions

$$\bigcap_{t-s=m} B^{s,t} \supset K^m\,, \qquad \bigcap_{r>s} E_r^{s,t} \supset E_\infty^{s,t}$$

are elementary; we have to prove the opposite inclusions.

We begin with (vi), in the case when $H_t(X)$ is finite and p-primary for each t, so that the same is true of $\Pi_{m'}^S(X)$. Given m, we will construct a realisation $\{Y_s\}$ and an integer u such that $\Pi_{n+m'}(Y_u) = 0$ for $m' \leqslant m$. The corresponding complex C will be A-free but not necessarily acyclic. The construction is by induction, as for Lemma 3.1. Suppose constructed a space F_s with finite p-primary homotopy groups; let $\Pi_{n+m'}(F_s)$ $(= G$ say$)$ be the first that is not zero. Let F_s' be an equivalent complex, and let

$$f_s\colon\ F_s' \to K(G/pG,\, n+m')$$

be a map inducing the projection $G \to G/pG$ of homotopy groups. Factor f_s through an equivalence and a fibration; let the fibre be F_{s+1}. The induction is started with $F_{-1} = S^n X$. If we form a mapping-cylinder and take equivalent complexes, as in the proof of Lemma 3.1, we obtain a realisation $\{Y_s\}$ with the required properties.

This construction gives the integer u; for if integers f_i are taken so that $p^{f_i}\Pi_i^S(X) = 0$ $(0 \leqslant i \leqslant m)$ it is sufficient to take $u = \sum_0^m f_i$. Thus u depends only on X. We see that n, k and l can be taken as large as required.

Now let $\{Y_s'\}$ be a realisation of a resolution C' of $H^*(X)$, for the same n. According to Lemmas 3.4, 3.5 there is a well-defined map from that part of a spectral sequence defined by $\{Y_s'\}$ to that part of a spectral sequence defined by $\{Y_s\}$. Now, in the latter we have

$$\bigcap_{\substack{t-s=m\\ s\leqslant u}} B^{s,t}(Y) = 0\,.$$

Therefore, if k and l are taken large enough, we have

$$\bigcap_{\substack{t-s=m\\ s\leqslant u}} B^{s,t}(Y') = 0$$

in the former. This concludes the proof in this case.

We next transfer this result to the case in which $H_t(X)$ (though finitely-generated, as always) is not necessarily finite or p-primary. In fact, let

$$x \in \Pi^S_m(X)$$

be an element not in K^m, that is, not of finite order prime to p. Let f be such an integer that the equation $x = p^f y$ has no roots y in $\Pi^S_m(X)$. Let Z be formed by attaching the cone on SX to SX by a map of degree p^f on the suspension coordinate. Then there is an inclusion map $SX \to Z$; this induces a map of spectral sequences. The image of x in $\Pi^S_{m+1}(Z)$ is non-zero; so it does not lie in

$$\bigcap_{\substack{t-s=m+1\\ s\leqslant u}} B^{s,t}(Z)$$

(for a certain u). Therefore x itself does not lie in

$$\bigcap_{\substack{t-s=m\\ s\leqslant u}} B^{s,t}(X) .$$

This argument gives a value for the integer u; for if $\Pi^S_i(X)$ is non-zero for just g values of i with $i \leqslant m$, then it is sufficient to take $u = 2fg$.

At this point we have completed the proof of (vi).

We turn next to the proof of (iv). Here we have to argue from the structure of $\Pi_{n+t-s}(Y_s)$. Now this structure is not invariant, presumably, unless we restrict $\{Y_s\}$. We therefore proceed as follows.

We call a resolution C *minimal* if the numer of A-free generators in $C_{s,t}$ is the least possible, given the structure of $C_{s,t'}$ for $t'<t$ and of $C_{s',t'}$ for $s'<s$. Since each $C_{s,t}$ is finitely-generated, one can prove by induction that if C, C' are minimal resolutions of $H^*(X)$, then any map $f: C \to C'$ (compatible with the identity map of $H^*(X)$) is an isomorphism. It follows from this (using the five-lemma) that if $\{Y_s\}$, $\{Y'_s\}$ are realisations of C, C', and if

$$f: \{Y_s\} \to \{Y'_s\}$$

is a map constructed by Lemma 3.4, then f_* maps $\Pi_{n+t-s}(Y_s)$ isomorphically (for $s \leqslant k$, $t<l-1$). In the same range, $\Pi_{n+t-s}(Y_s)$ is preserved on suspending $\{Y_s\}$.

We suppose, then, that C is a minimal resolution of $H^*(X)$. Let $\{Y_s\}$ be a realisation of C with $k \geqslant 2s+1$, $l>s+1+t$, so that $E^{s,t}_{s+1}$, $E^{s,t}_\infty$ are defined. Let $F_r^{s,t}$ be the p-component of the torsion subgroup of

$$\Pi_{n+t-s-1}(Y_{s+r}) ;$$

let $K_r^{s,t}$ be the image (by d) of $\Pi_{n+t-s}(Y_{s+1}, Y_{s+r})$ in $\Pi_{n+t-s-1}(Y_{s+r})$. Then $K_r^{s,t} \subset F_r^{s,t}$, $F_r^{s,t}$ is finite, and there is a monomorphism (induced by d) from $E_r^{s,t}/E_\infty^{s,t}$ to $F_r^{s,t}/K_r^{s,t}$ (if $r > s$). If we take this monomorphism for two values of r, it commutes with the maps induced by inclusions.

Next take $r = s+1$, and take an element $e \neq 0$ in $E_{s+1}^{s,t}/E_\infty^{s,t}$. For each member x of the corresponding coset e' in $F_{s+1}^{s,t}/K_{s+1}^{s,t}$, the equation $x = p^f y$ is insoluble in $\Pi_{n+t-s-1}(Y_{2s+1})$, provided that f is suitably chosen; for example, let p^f be the order of $F_r^{s,t}$.

It is next necessary to suppose that the realisation $\{Y_s\}$ has n, k and l so large that $E_{s+1+2f(t-s+1)}^{s,t}$ is defined. This is possible, because by our supposition that C is minimal, we may replace $\{Y_s\}$ by another realisation with increased n, k and l, without changing f.

Next note that for suitable n', $\{S^{n'} Y_{s'}\}$ (over $s' \geqslant 2s+1$) is a realisation of a resolution of $H^*(Y_{2s+1})$. Applying to this our results on (vi), we see that for each $x \in e'$, $S^{n'}x$ is not in the image of $\Pi_{n'+n+t-s-1}(S^{n'} Y_{2s+1+2f(t-s+1)})$. Desuspending, we see that for each $x \in e'$, x is not in the image of

$$\Pi_{n+t-s-1}(Y_{2s+1+2f(t-s+1)}) .$$

Therefore e is not in the image of $E_{s+1+2f(t-s+1)}^{s,t}$.

This concludes the proof of (iv), and of Theorem 2.1.

§ 4. Multiplicative Properties of the Spectral Sequence

In this section we prove Theorem 2.2, by establishing the multiplicative properties of the spectral sequence in case $X = S^0$. For clarity, we proceed in slightly greater generality. Let X, X' be spaces, and let $X'' = X \times X'/X \vee X'$. We will show that there is a pairing

$$E_r^{s,t}(X) \otimes E_r^{s',t'}(X') \to E_r^{s+s',t+t'}(X'') .$$

Our method is to take realisations $\{Y_s\}$, $\{Y'_{s'}\}$ corresponding to X, X' and form a realisation $\{Y''_{s''}\}$, using the join operation, so that Y''_0 is the join $Y_0 * Y'_0$. If $Y_0 \sim S^n X$, $Y'_0 \sim S^{n'} X'$, then $Y''_0 \sim S^{n+n'+1}(X'')$. We have a join operation in relative homotopy groups; it will appear that it gives a pairing

$$\Pi_*(Y_s, Y_u) \otimes \Pi_*(Y'_{s'}, Y'_{u'}) \to \Pi_*(Y''_{s''}, Y''_{u''})$$

for $s'' = s + s'$, $u'' = \mathrm{Min}(s + u', u + s')$. Such pairings yield a pairing of spectral sequences, by passing to quotients.

It is convenient to begin the detailed work by describing the cup-product

$$\mathrm{Ext}_A^{s,t}(H^*(X),Z_p)\otimes \mathrm{Ext}_A^{s',t'}(H^*(X'),Z_p) \to \mathrm{Ext}_A^{s+s',t+t'}(H^*(X''),Z_p) \; .$$

This depends on the diagonal map $\Delta: A \to A\otimes A$ of the STEENROD algebra, which is defined as follows. There is one and only one universal formula for expanding the result of an operation a applied to a cup-product uv; let it be

$$a(uv) = \sum_i (-1)^{jq}(a^{L,i}u)(a^{R,i}v)$$

where $a \in A$, $\deg(v) = q$, $a^{L,i} \in A_j$ $(j = j(i))$ and $a^{R,i} \in A$. Then we set

$$\Delta a = \sum_i a^{L,i} \otimes a^{R,i} \; .$$

One verifies that this defines a diagonal map[10]).

We next remark that since our cohomology groups are augmented, we have $H^*(X'') \cong H^*(X)\otimes H^*(X')$; the isomorphism is defined using the external cup-product. The operations of A on $H^*(X'')$ are given by

$$a(u\otimes v) = \sum_i(-1)^{jq}(a^{L,i}u\otimes a^{R,i}v) \; ;$$

here $v \in H^q(X')$, while a, $a^{L,i}$, $a^{R,i}$ and j are as above.

Next suppose that C, C' are resolutions of $H^*(X)$, $H^*(X')$. Then we may make $C\otimes C'$ into an acyclic A-complex, whose homology in dimension $s=0$, $s'=0$ is $H^*(X'')$. In fact, we set

$$d(c\otimes c') = dc\otimes c' + (-1)^s c\otimes dc' \qquad (c \in C_{s,t})$$
$$a(c\otimes c') = \sum_i(-1)^{jt'}(a^{L,i}c\otimes a^{R,i}c') \qquad (c' \in C'_{s',t'}) \; .$$

It follows that if C'' is a resolution of $H^*(X'')$, there is a map $m: C''\to C\otimes C'$. On the other hand, there is a pairing

$$\mu: \; \mathrm{Hom}_A(C,Z_p)\otimes \mathrm{Hom}_A(C',Z_p) \to \mathrm{Hom}_A(C\otimes C',Z_p)$$

defined by $(\mu(h\otimes h'))(c\otimes c') = (hc)(h'c')$. The composite $m^*\mu^*$ yields the required cup-product

$$\mathrm{Ext}_A^{s,t}(H^*(X),Z_p)\otimes \mathrm{Ext}_A^{s',t'}(H^*(X'),Z_p) \to \mathrm{Ext}_A^{s+s',t+t'}(H^*(X''),Z_p) \; .$$

Now that we have considered the acyclic A-complex $C\otimes C'$, we will consider a realisation of it. Let $\{Y_s\}$, $\{Y'_{s'}\}$ be realisations of C, C'. We will define Y''_0 by setting $Y''_0 = Y_0 * Y'_0$, and subcomplexes of Y''_0 by setting

$$Y''_{s''} = \bigcup_{s+s'=s''} Y_s * Y'_{s'} \; .$$

[10]) The products in $A\otimes A$ are defined by $(a\otimes b)(c\otimes d) = (-1)^{il}(ac\otimes bd)$ (for $a \in A_i$, $d \in A_l$).

We may suppose that Y_0, Y'_0 are enumerable CW-complexes; Y''_0 is then an enumerable CW-complex. We repeat that since $Y_0 \sim S^n X$ and $Y'_0 \sim S^{n'} X'$ we have $Y''_0 \sim S^{n+n'+1}(X'')$.

In order to show that $\{Y''_{s''}\}$ is a realisation of $C \otimes C'$, we must display $H^*(Y''_{s''}, Y''_{s''+1})$. Let

$$\begin{aligned} \varphi: \quad & C_{s,t} \to H^{n+t-s}(Y_s, Y_{s+1}) \ , \\ \varphi': \quad & C_{s',t'} \to H^{n'+t'-s'}(Y'_{s'}, Y'_{s'+1}) \end{aligned}$$

be the given isomorphisms[11]). Then we may define a map

$$\varphi'': \quad C_{s,t} \otimes C'_{s',t'} \to H^{n''+t''-s''}(Y''_{s''}, Y''_{s''+1})$$

(where $n'' = n + n' + 1$, $s'' = s + s'$, $t'' = t + t'$) by setting

$$\varphi(u \otimes v) = (-1)^{(t-s)n' + ts'} E(\varphi u \cdot \varphi' v) \ .$$

Here we employ the fact that $Y_0 * Y'_0 \sim S(Y_0 \times Y'_0 / Y_0 \vee Y'_0)$. The element $\varphi u \cdot \varphi' v$ is defined using the exterior cup-product; the map E is defined using excision and suspension. The map

$$\varphi'': \quad \sum_{\substack{s+s'=s'' \\ t+t'=t''}} C_{s,t} \otimes C'_{s',t'} \to H^{n''+s''-t''}(Y''_{s''}, Y''_{s''+1})$$

is an isomorphism. One verifies that it commutes with the operations of A. One also verifies that it satisfies the third condition imposed on a realisation, by making the diagrams for d and ε commutative. (The sign in the definition of φ'' is essential at these points.)

We have now verified that $\{Y''_{s''}\}$ is a realisation of $C \otimes C'$. Maps or homotopies of $\{Y_s\}$ or $\{Y'_{s'}\}$ induce maps or homotopies of $\{Y''_{s''}\}$. It follows that the spectral sequence associated with $\{Y''_{s''}\}$ is well-defined. Since $C \otimes C'$ is acyclic, there is (by Lemmas 3.4, 3.5) a well-defined map from this spectral sequence to that associated with X''. It remains, therefore, to define a pairing

$$E_r^{s,t}(Y) \otimes E_r^{s',t'}(Y') \to E_r^{s+s',t+t'}(Y'') \ .$$

To do this, we now introduce the join operation in relative homotopy groups. This is defined by the join of maps of oriented cells[12]). We obtain a pairing

$$\Pi_m(K, L) \otimes \Pi_{m'}(M, N) \to \Pi_{m+m'+1}(K*M, K*N \cup L*M) \ .$$

This is natural with respect to maps of K, L and M, N. If α, β lie in the groups paired, then

$$jd(\alpha * \beta) = -i(d\alpha * \beta) + (-1)^{m+1} i'(\alpha * d\beta) \ .$$

[11]) In this section we omit to make explicit the finite ranges of dimensions in which these isomorphisms are supposed to hold. The details are similar to those in § 3.

[12]) The join $K * L$ is oriented as $S(K \times L / K \vee L)$, and we suspend over the first coordinate.

Here j, i, i' are the canonical maps with values in $\Pi_{m+m'}(K*N \cup L*M, L*N)$.

The products we require, however, represent composition products, not join products; the two differ by a sign. With this in mind, we define the product

$$\Pi_{n+t-s}(Y_s, Y_u) \otimes \Pi_{n'+t'-s'}(Y'_{s'}, Y'_{u'}) \to \Pi_{n''+t''-s''}(Y''_{s''}, Y''_{u''})$$

by $\alpha \times \beta = (-1)^{(t-s)n'}\alpha * \beta$. (Here we have $n'' = n + n' + 1$, $s'' = s + s'$, $t'' = t + t'$, $u'' = \mathrm{Min}(s + u', u + s')$.) We now have the boundary formula

$$j(-1)^{n''} d(\alpha \times \beta) = i((-1)^n d\alpha \times \beta) + (-1)^{(t-s)} i'(\alpha \times (-1)^{n'} d\beta) .$$

The following statements are now open to verification. Firstly, the pairing of relative homotopy groups passes to quotients, and defines a pairing

$$E_r^{s,t}(Y) \otimes E_r^{s',t'}(Y') \to E_r^{s+s',t+t'}(Y'') .$$

Secondly, the boundary d_r satisfies

$$d_r(uv) = (d_r u)v + (-1)^{(t-s)} u(d_r v) .$$

Thirdly, the products are natural with respect to the isomorphisms

$$E_{r+1}^{s,t} \cong H^{s,t}(E_r; d_r)$$

and with respect to the monomorphisms $E_R^{s,t} \to E_r^{s,t}$ for $s < r < R \leqslant \infty$.

Fourthly, the composite map

$$E_2^{s,t}(X) \otimes E_2^{s',t'}(X') \to E_2^{s+s',t+t'}(X'')$$

coincides, except for the sign $(-1)^{ts'}$, with the pairing

$$\mathrm{Ext}_A^{s,t}(H^*(X), Z_p) \otimes \mathrm{Ext}_A^{s',t'}(H^*(X'), Z_p) \to \mathrm{Ext}_A^{s+s',t+t'}(H^*(X''), Z_p) .$$

(This follows from the description above, on considering the pairing of Π_* and H^*.)

Fifthly, the products are associative, and anticommutative for the grading $(t - s)$. This follows from analogous facts for the join operation, together with naturality arguments. We note that the products in $\mathrm{Ext}_A^*(Z_p, Z_p)$ are associative, and satisfy the anticommutative law

$$uv = (-1)^{ss'+tt'} vu \quad (\text{for } u \in \mathrm{Ext}^{s,t}, \quad v \in \mathrm{Ext}^{s',t'}) .$$

Sixthly, consider the case $X = X' = X'' = S^0$. Then a composition product is defined in $\Pi_*^S(S^0)$. This passes to the quotient and defines products in $E_\infty^{s,t}(S^0)$, which coincide with those considered above. (This follows from the known equivalence between composition products and join products.)

This concludes the proof of Theorem 2.2.

§ 5. The Structure of the STEENROD Algebra

We collect our results on the subalgebras of the (mod p) STEENROD algebra A in Theorem 5.1. This theorem is followed by explanatory definitions, and a great deal of proof. This is followed by two further theorems (5.12, 5.13) giving further information needed in considering $H^*(A)$. One concerns a self-map $A \to A$; the other concerns the commutators in A.

If $p > 2$ we have:

Theorem 5.1$_p$. (*a*) *Any finite set of elements of A generates a finite algebra.*

(*b*) *A contains subalgebras $A^{R,T}$ for each $1 \leqslant R < \infty$, $1 \leqslant T \leqslant \infty$ with the following properties.*

(*c*) *$A^{R,T} = A^{1,T} \cap A^{R,\infty}$. If $Q \leqslant T$, then $A^{1,Q} \subset A^{1,T}$. If $P \leqslant R$, then $A^{P,\infty} \supset A^{R,\infty}$. $A^{R,T}$ is the unit subalgebra if $R > T$. The subalgebra $A^{1,T}$ is that generated by β and P^k for $k < p^{T-1}$. We have $A^{1,\infty} = A$.*

(*d*) *If $R \leqslant T < \infty$, the rank of $A^{R,T}$ is $2^U p^V$, where $U = T - R + 1$, $V = \frac{1}{2}(T-R)(T-R+1)$.*

(*e*) *If $P < Q$, $A^{Q,T}$ is normal in $A^{P,T}$, so that $A^{P,T}//A^{Q,T}$ exists; and $A^{P,T}$ is free, qua (left or right) module over $A^{Q,T}$.*

(*f*) *If $P < Q < R$, then $A^{Q,T}//A^{R,T}$ is embedded monomorphically in $A^{P,T}//A^{R,T}$; the former is normal in the latter, with quotient isomorphic to $A^{P,T}//A^{Q,T}$.*

(*g_p*) *Moreover, if $R \leqslant P + Q$, then $A^{Q,T}//A^{R,T}$ is central in $A^{P,T}//A^{R,T}$, in the sense that if $a_{i,j} \in A^{P,T}$, $b_{k,l} \in A^{Q,T}$, then in $A^{P,T}//A^{R,T}$ we have*

$$\{a_{i,j}\}\{b_{k,l}\} = (-1)^{jl}\{b_{k,l}\}\{a_{i,j}\} .$$

(*h_p*) *If $R \leqslant T$, we have*

$$A^{R,T}//A^{R+1,T} \cong E(\sum_{t=0}^{R-2} p^t, 1) \otimes P(\sum_{t=0}^{R-1} p^t, 0; p^{T-R}) .$$

If $p = 2$ we have:

Theorem 5.1$_2$. *The statements* (*a*) *to* (*f*) *hold word for word on interpreting "p" as 2, "β" as Sq^1 and "P^k" as Sq^{2k}. The statements* (*g_p*), (*h_p*) *become:*

(*g_2*) *Moreover, if $R \leqslant P + Q$, then $A^{Q,T}//A^{R,T}$ is central in $A^{P,T}//A^{R,T}$.*

(*h_2*) *If $R \leqslant T$, we have*

$$A^{R,T}//A^{R+1,T} \cong P(2^R - 1; 2^{T-R+1}) .$$

Explanatory definitions. All our algebras are algebras with unit and with diagonal [4, p. 211] over the field Z_p. They are graded if $p = 2$ and bigraded if $p > 2$. Their components in dimension 0 or $(0,0)$ are their unit subalgebras (here the unit subalgebra has the unit as a Z_p base). All our maps of algebras preserve this structure; in particular, the injections of subalgebras and projections onto quotient algebras do so.

The STEENROD algebra A (over Z_p) is defined as above. If $p>2$ it may be bigraded by setting $\mathrm{Deg}(P_p^k)=(k,0)$, $\mathrm{Deg}(\beta_p)=(0,1)$[13]. The single grading q corresponding to the bigrading (i,j) is given by $q=2(p-1)i+j$.

The other algebras introduced are as follows. The exterior algebra $E(i,j)$ (over Z_p) has a Z_p-base $\{1,f'\}$. The element 1 is the unit; the element f' has bidegree (i,j). The product is given by $(f')^2=0$; the diagonal is given by $\Delta(f')=f'\otimes 1+1\otimes f'$. The integer j must be odd (if $p>2$).

The truncated divided polynomial algebra $P(i,j;k)$ (over Z_p) has a Z_p-base containing one element f_l of bidegree (li,lj) for each l such that $0\leqslant l<k$. The product is given by $f_l\cdot f_m=(l,m)f_{l+m}$; the diagonal is given by

$$\Delta f_l=\sum_{m+n=l} f_m\otimes f_n \,.$$

$\left(\text{Here the binomial coefficients mod}\,p\text{ are defined by }(l,m)=\dfrac{(l+m)!}{l!\,m!}\right)$. The integer j must be even (if $p>2$); and k must be a power of p, or else ∞. If $k=\infty$, the algebra is not truncated. The algebra $P(i;k)$ is similarly defined, but graded instead of bigraded. For the tensor product of algebras, see [4].

If A is a bigraded algebra, as above, we define $I(A)=\sum_{(i,j)\neq(0,0)}A_{i,j}$; similarly for a graded algebra. If A is an algebra containing B as a subalgebra, we call B *normal* in A (cf. [4] p. 349) if $A\cdot I(B)=I(B)\cdot A$; we then define $A//B=A/I(B)\cdot A$.

Since the word "dimension" is already in use for the grading, we speak of the *rank* of a subalgebra, meaning its dimension when considered as a vector space over Z_p.

This concludes the explanatory definitions.

Proof. The proof will proceed in several stages. Following SERRE, THOM and CARTAN, we shall make use of a faithful representation of the STEENROD algebra, obtained by allowing it to operate on a Cartesian product of spaces of type $(Z_p,1)$. We take first the case $p>2$.

Let X be the Cartesian product of $n+n'$ spaces, each of type $(Z_p,1)$; let their fundamental classes be $x_1,\ldots,x_n,x'_1,\ldots,x'_{n'}$. Set

$$y_i=\beta_p(x_i),\quad y'_i=\beta_p(x'_i)\ .$$

Let $u\in H^{2n+n'}(X)$ be the cup-product $y_1\ldots y_n\,x'_1\ldots x'_{n'}$.

$H^*(X)$ is the tensor product of exterior algebras generated by the x_i,x'_i and polynomial algebras generated by the y_i,y'_i. We shall need a notation

[13]) In [2] the second grading is called the "type".

for certain polynomials $D^I_{n,n'}$ lying in $H^*(X)$. Let

$$I = \{\varepsilon_1, \lambda_1, \ldots, \varepsilon_i, \lambda_i, \ldots\}$$

be a sequence of integers with $\lambda_i \geqslant 0$, $\varepsilon_i = 0$ or 1; only a finite number are to be non-zero. Following CARTAN [2], we define the polynomial $D^I_{n,n'}$ as follows. Among the monomials $y_1^{h_1} \ldots y_n^{h_n}$, consider those in which just λ_i exponents are p^i (for each i), the remainder being 1. Let the sum of such monomials be s[14]). Next consider the monomials obtained from $x'_1 \ldots x'_{n'}$ by replacing x'_j by $(y'_j)^{p^i}$ for just ε_{i+1} values of j (this for each i). Each such monomial is to be taken with a sign, namely the signature of a certain permutation ϱ of $1, \ldots, n'$. Here ϱ brings the factors x'_k in the monomial to the left (arranged from left to right in increasing order of k) and the factors $(y'_j)^{p^i}$ to the right (arranged in increasing order of i). Let the sum of such signed monomials be t. Then we define $D^I_{n,n'} = st$.

The elements $D^I_{n,n'}$ in $H^{2n+n'+q}(X)$ generate a submodule $D^{2n+n'+q}$; they are linearly independent if n, n' are sufficiently large (depending on q).

Evidently a Z_p-linear function $\theta\colon A \to H^*(X)$ is defined by $\theta(a) = a(u)$. This is the representation used.

Theorem 5.2. *If n, n' are sufficiently large (depending on q) then $\theta_q : A_q \to H^{2n+n'+q}(X)$ has kernel zero and image $D^{2n+n'+q}$.*

This theorem is due to CARTAN [2]. His proof shows also that θ preserves the bigrading, if the second degree of polynomials $D^I_{n,n'}$ is defined by $j = \sum_i \varepsilon_i$.

We next note that this representation has a convenient relation to the diagonal in A. In fact, if the space Y is defined using $m + m'$ factors of type $(Z_p, 1)$, then $X \times Y$ is homeomorphic to the space Z defined using $(n + m) + (n' + m')$ factors of type $(Z_p, 1)$. If v and w are the analogues of u for Y and Z, then uv corresponds to w in this homeomorphism. Let θ_X, θ_Y, θ_Z be the functions θ for the three spaces. By evaluating $a(uv)$ and $a(w)$ we have the following obvious lemma.

Lemma 5.3. *If $a \in A$ and $\theta_Z a = D^I_{n+m,n'+m'}$ then*

$$(\theta_X \otimes \theta_Y)(\Delta a) = \sum_{J+K=I} (-1)^\varepsilon D^J_{n,n'} \otimes D^K_{m,m'} \,.$$

Here, for sequences $I = \{\varepsilon_i, \lambda_i\}$, $J = \{\eta_i, \mu_i\}$ and $K = \{\zeta_i, \nu_i\}$, the equation $J + K = I$ means that $\eta_i + \zeta_i = \varepsilon_i$ and $\mu_i + \nu_i = \lambda_i$ for each $i \geqslant 1$. The sign is given by

$$\varepsilon = \sum_{j>k} \eta_j \zeta_k \,.$$

---

[14]) Thus, s will be zero if $\sum_i \lambda_i > n$. Similarly, later, for t.

Since Δa has among its components the term $a \otimes 1$, Lemma 5.3 has the following corollary: if $\theta_Z a = D^I_{n+m,n'+m'}$, then $\theta_X a = D^I_{n,n'}$. It follows that we may write $\theta a = D^I$ to mean that $\theta_X a = D^I_{n,n'}$ for all n, n'; similarly for the equation $\theta a = \sum_I \lambda_I D^I$.

Theorem 5.2 allows us to exhibit certain distinguished elements in the STEENROD algebra. In fact, let us define $I(r,k)$ $(r \geqslant 1, k \geqslant 0)$ by $\lambda_r = k$, $\lambda_i = 0$ for $i \neq r$, $\varepsilon_i = 0$ for all i. Define $e_{r,k} \in A$ by $\theta(e_{r,k}) = D^{I(r,k)}$. Define I'_r $(r \geqslant 1)$ by $\lambda_i = 0$ for all i, $\varepsilon_r = 1$, $\varepsilon_i = 0$ for $i \neq r$. Define $e'_r \in A$ by $\theta(e'_r) = D^{I'(r)}$.

These elements have the following properties, which indeed characterise them.

(1) $e_{r,0}$ is 1, the unit.

(2) $$\operatorname{Deg} e_{r,k} = (k \sum_{t=0}^{r-1} p^t, 0) .$$

$$\operatorname{Deg} e'_r = (\sum_{t=0}^{r-2} p^t, 1) .$$

(3) $$\Delta e_{r,k} = \sum_{i+j=k} e_{r,i} \otimes e_{r,j}$$

$$\Delta e'_r = e'_r \otimes 1 + 1 \otimes e'_r .$$

(4) If x, y are the generators of $H^*(Z_p, 1; Z_p)$, then

$$e_{r,k}(x) = 0 \qquad\qquad e'_r(x) = y^{p^r - 1}$$

$$e_{r,k}(y) = \begin{cases} y & (k = 0) \\ y^{p^r} & (k = 1) \\ 0 & \text{otherwise} \end{cases} \qquad\qquad e'_r(y) = 0 .$$

For example, (3) follows immediately from Lemma 5.3 by using Theorem 5.2. From (3) and (4) we deduce the following property by induction:

(5) $$e_{r,k}(y^{p^s}) = \begin{cases} y^{p^s} & (k = 0) \\ y^{p^{r+s}} & (k = p^s) \\ 0 & \text{otherwise} \end{cases} \qquad\qquad e'_r(y^{p^s}) = 0 .$$

Our next theorem will show in what sense the elements $e_{r,k}$, e'_r are generators for A. In order to state it, let us regard e_{r,k_r}, for each $r \geqslant 1$, as an expression in the variable k_r. Let us order together in some fixed order the expressions e_{r,k_r} and e'_r. Let us form monomials by omitting from this ordering all but a finite number of terms, and then inserting integer values $k_r \geqslant 1$ for the remaining variables k_r. These monomials then represent elements of A. The identity element is included, as the empty product. We then have:

Theorem 5.4. *For each fixed ordering, such monomials form a base for A.*

Proof. Let M be a typical monomial, of single grading q. By Theorem 5.2, it is sufficient to show that the elements $M(u)$ form a base for $D^{2n+n'+q}$ (for n, n' sufficiently large). To this end, we order the base D^I of $D^{2n+n'+q}$ by ordering the sequences I. Following CARTAN [2], we order them lexicographically from the right[15]).

Using (3) and (5), we deduce:

$$\begin{aligned} (6)\quad e_{r,k} D^I &= (\lambda_r, k) D^J + \sum_{K>J} \mu_K D^K \\ e'_r D^I &= (-1)^{\varepsilon} D^L \qquad (\text{if } \varepsilon_r = 0) \\ e'_r D^I &= 0 \qquad\qquad (\text{if } \varepsilon_r = 1)\ . \end{aligned}$$

Here $I = \{\varepsilon_i, \lambda_i\}$, $J = I + I(r,k)$, and $L = I + I'(r)$ if $\varepsilon_r = 0$; the sign is given by $\varepsilon = \sum_{i<r} \varepsilon_i$.

Let $I' = \{\varepsilon'_i, \lambda'_i\}$ be another sequence, $J' = I' + I(r,k)$, and $L' = I' + I'(r)$ if $\varepsilon'_r = 0$. Then we have, trivially:

(7) If $I < I'$, then $J < J'$, and $L < L'$ if $\varepsilon_r = \varepsilon'_r = 0$.

From (6) and (7) we deduce, by induction, that:

(8) If M is a typical monomial, then

$$M(u) = \eta D^I + \sum_{J>I} \mu_J D^J \ ,$$

where $\eta = \pm 1$, and the sequence I is determined by M as follows. ε_r is 1 or 0 according as e'_r is in M or omitted; λ_r is 0 if e_{r,k_r} is omitted from M, and otherwise it is the integer substituted for k_r.

We see that there is a $(1-1)$ correspondence between M and I. Therefore the elements $M(u)$ in $D^{2n+n'+q}$ form a base for it. This concludes the proof of Theorem 5.4.

The statement and proof of Theorem 5.4 remain valid if, instead of using expressions e_{r,k_r}, one for each r, we use expressions $(e_{r,p^i})^{d_{r,i}}$, one for each r and each $i \geqslant 0$. We have then to substitute, for the variables $d_{r,i}$, integer values such that $0 < d_{r,i} < p$.

We may now obtain the subalgebras $A^{R,T}$. Given a fixed ordering, as in Theorem 5.4, we may consider the monomials M in which the factors $e_{r,k}$ have $r \geqslant R$, $k < p^{T-r}$ and the factors e'_r have $R \leqslant r \leqslant T$. These form a base for

[15]) However, our argument differs from his in that our inductions (if stated) would proceed in the opposite direction along the ordering. Since the bases ordered are finite (for each q) this is immaterial.

a submodule $A^{R,T}$ of A. This is so whenever $1 \leqslant R < \infty$, $1 \leqslant T \leqslant \infty$; but we note that the base reduces to the unit element if $T < R$. The inclusion and intersection properties in Theorem 5.1 (c) are trivial.

Theorem 5.5. *$A^{R,T}$ is independent of the ordering chosen. It is a subalgebra of A and closed for the diagonal.*

Proof. Consider the polynomials D^I for which the sequence $I = \{\varepsilon_i, \lambda_i\}$ satisfies $\lambda_i < p^{T-i}$ for each i, $\lambda_i = 0$ for $i < R$, and $\varepsilon_i = 0$ unless $R \leqslant i \leqslant T$. These generate a submodule $D^{R,T}$ of $H^*(X)$.

Lemma 5.6. $e_{r,k} D^{R,T} \subset D^{R,T}$ *if* $r \geqslant R$, $k < p^{T-r}$,

$$e'_r D^{R,T} \subset D^{R,T} \quad \textit{if} \quad R \leqslant r \leqslant T.$$

We will defer the proof of this lemma in order to show how the theorem follows from it. In fact, the lemma clearly implies that $\theta(A^{R,T}) \subset D^{R,T}$. But further, if T is finite, $A^{R,T}$ and $D^{R,T}$ have the same rank, namely $2^U p^V$ where $U = T - R + 1$, $V = \frac{1}{2}(T-R)(T-R+1)$. Therefore $\theta(A^{R,T}) = D^{R,T}$ if T is finite; this implies the same equation with T infinite. The equation $\theta(A^{R,T}) = D^{R,T}$ shows that $A^{R,T}$ is independent of the ordering chosen; and with Lemma 5.6, it implies that $A^{R,T}$ is a subring. Lastly, it is clear from Lemma 5.3 and the definition of $D^{R,T}$ that $\theta^{-1}(D^{R,T})$ is closed for the diagonal. This concludes the deduction of the theorem from the lemma. We note that we have proved Theorem 5.1 (d).

Proof of Lemma 5.6. The property (6) above shows trivially that

$$e'_r D^{R,T} \subset D^{R,T} \quad \textit{if} \quad R \leqslant r \leqslant T .$$

Let us take D^I, where $I = \{\varepsilon_i, \lambda_i\}$, and form the expansion

$$e_{r,k} D^I = \sum_K \lambda_K D^K .$$

The sum may be given explicitly as follows, by using (3) and (5). It extends over sequences $J = \{\mu_0, \eta_1, \mu_1, \ldots, \eta_i, \mu_i, \ldots\}$ with $\sum_i (\mu_i + \eta_{i+1}) p^i = k$. The term given by J will correspond to the summands in which, for each i, just μ_i of the factors $y_j^{p^i}$ $(1 \leqslant j \leqslant n)$ and just η_{i+1} of the factors $(y'_j)^{p^i}$ $(1 \leqslant j \leqslant n')$ are operated on by e_{r,p^i}. The sequence $K = \{\zeta_i, \nu_i\}$ is given by

$$\nu_i = \lambda_i - \mu_i + \mu_{i-r} ,$$
$$\zeta_i = \varepsilon_i - \eta_i + \eta_{i-r} .$$

If $\lambda_i - \mu_i < 0$, $\varepsilon_i - \eta_i < 0$ or $\varepsilon_i - \eta_i + \eta_{i-r} = 2$, then the term given by

J is zero. The coefficient λ_K is given as follows.

$$\lambda_K = (-1)^{\varepsilon} \prod_i (\lambda_i - \mu_i, \mu_{i-r})$$

where $\varepsilon = \sum_{i<j<i+r} (\varepsilon_j - \eta_j)\eta_i$.

We now introduce the assumption that $D^I \in D^{R,T}$, $r \geqslant R$, and $k < p^{T-r}$.

In fact, since $D^I \in D^{R,T}$, we have $\lambda_i = 0$ and $\varepsilon_i = 0$ for $i < R$; we also have $r \geqslant R$. Using the formulae for ν_i, ζ_i we see that $\nu_i = 0$ and $\zeta_i = 0$ for $i < R$.

Similarly, since

$$\sum_i (\mu_i + \eta_{i+1}) p^i = k < p^{T-r} ,$$

we have $\eta_{i+1} = 0$ unless $i < T - r$. Thus $\eta_{i-r} = 0$ unless $i \leqslant T$. We also have $\varepsilon_i = 0$ unless $i \leqslant T$; thus $\zeta_i = 0$ unless $i \leqslant T$.

Lastly, we have $\lambda_i < p^{T-i}$, so $\lambda_i - \mu_i < p^{T-i}$. We also have

$$\sum_i (\mu_i + \eta_{i+1}) p^i = k < p^{T-r} ,$$

so $\mu_i < p^{T-r-i}$ and $\mu_{i-r} < p^{T-i}$. It follows that $(\lambda_i - \mu_i, \mu_{i-r}) = 0$ whenever $\lambda_i - \mu_i + \mu_{i-r} \geqslant p^{T-i}$.

These remarks in combination show that if $D^I \in D^{R,T}$, $r \geqslant R$, and $k < p^{T-r}$, then the non-zero terms of the expansion

$$e_{r,k} D^I = \sum_K \lambda_K D^K$$

lie in $D^{R,T}$. The proof of the lemma is complete.

The statement and proof of Theorem 5.5 remain valid if, in defining $A^{R,T}$, we use, instead of the expressions e_{r,k_r}, the expressions

$$(e_{r,p^i})^{d_{r,i}} \quad (0 < d_{r,i} < p) .$$

We have to take those for which $r \geqslant R$ and $i < T - r$.

Corollary 5.7. *$A^{1,T}$ is the subalgebra generated by β and by P^k for $k < p^{T-1}$.*

Proof. Since $\beta = e'_1$, it lies in $A^{1,T}$ for $T \geqslant 1$. Since $P^k = e_{1,k}$, it lies in $A^{1,T}$ for $k < p^{T-1}$. On the other hand, by the remarks immediately above, $A^{1,T}$ admits a system of multiplicative generators in bidegrees $(p^i \sum_0^{r-1} p^t, 0)$ (where $i < T - r$) and $(\sum_0^{r-2} p^t, 1)$ (where $r \leqslant T$). These can be written in terms of the elements β and P^k, and by their dimensions we shall have $k < p^{T-1}$, each k. This concludes the proof of the corollary.

Corollary 5.8. *Any finite set F of elements of A generates a finite subalgebra.*

Proof. The elements of F may be expressed in terms of β and P^k for $k<p^{T-1}$, some finite T. Thus the subalgebra generated by F is contained in the finite subalgebra $A^{1,T}$.

In order to obtain our next result we will consider the representations of A by its operations on a different space. Let Y_Q be the Cartesian product of n lens spaces $Y_1,\ldots,Y_n$ of dimension $2p^Q-1$ and n' lens spaces $Y'_1,\ldots,Y'_{n'}$ of dimension $2p^{Q-1}-1$, so that we have

$$H^*(Y_i)=E(1)\otimes T(2;p^Q) \quad ,$$
$$H^*(Y'_j)=E(1)\otimes T(2;p^{Q-1}) \ ,$$

where T is a truncated polynomial algebra. We introduce notations for the elements of $H^*(Y_Q)$ exactly as before, writing v instead of u, and E^I instead of D^I for the distinguished polynomials. The submodule $E^{R,T}$ of $H^*(Y_Q)$ is defined word for word as $D^{R,T}$ is. Consider the sequences $I=\{\varepsilon_i,\lambda_i\}$ of grading q which satisfy $\varepsilon_i=0$ and $\lambda_i=0$ if $i\geqslant Q$; the corresponding polynomials E^I are linearly independent and form a base of $E^{2n+n'+q}$ (at least if n, n' are sufficiently large, depending on q). All other polynomials E^I of grading q are zero.

Theorem 5.9. *$A^{Q,T}$ is normal in $A^{P,T}$ if $P<Q$. The rule $\theta\{M\}=M(v)$ defines a Z_p-linear function*

$$\theta_q:(A^{P,T}//A^{Q,T})_q\rightarrow H^{2n+n'+q}(Y_Q;Z_p) \ .$$

If n, n' are sufficiently large (depending on q) then θ_q has kernel zero and image $E^{P,T}\cap E^{2n+n'+q}$.

Proof. The proof that $A^{Q,T}$ is normal in $A^{P,T}$ will be as follows. Each element of $A^{P,T}$ gives a Z_p-linear map from $H^*(Y_Q)$ to itself; these maps constitute a quotient ring $A^{P,T}/K$ of $A^{P,T}$. We will show that the kernel K is both $I(A^{Q,T})\cdot A^{P,T}$ and $A^{P,T}\cdot I(A^{Q,T})$.

Let us use the representation of Theorem 5.4, with the ordering e'_1, e_{1,k_1}, e'_2, $e_{2,k_2},\ldots$ Let us divide the monomials of $A^{P,T}$ onto two classes; one, say B, shall consist of monomials formed from factors $e_{r,k}$ with $P\leqslant r<Q$, $k<p^{T-r}$ and e'_r with $P\leqslant r<Q$; the other, say C, shall consist of the remaining monomials. C is a base for a submodule L. Then, by the choice of ordering, $L\subset A^{P,T}\cdot I(A^{Q,T})$. We also note that if $r\geqslant Q$, then

$$e_{r,k}H^*(Y_Q)=0 \ , \quad e'_rH^*(Y_Q)=0 \ .$$

This shows that $A^{P,T}\cdot I(A^{Q,T})\subset K$. On the other hand, let M run through B; then, by property (8) (which remains valid for Y_Q) the elements $M(v)$ are linearly independent elements of $H^*(Y_Q)$. Hence the classes $\{M\}$ are linearly independent elements of $A^{P,T}/K$, and $K = L$. We conclude that

$$A^{P,T} \cdot I(A^{Q,T}) = K \ ;$$

by using the opposite order in the monomials, we see that

$$I(A^{Q,T}) \cdot A^{P,T} = K \ .$$

We have shown that $A^{Q,T}$ is normal in $A^{P,T}$, and incidentally established the representation θ of $A^{P,T}//A^{Q,T}$. The proof of Theorem 5.9 is complete.

Corollary 5.10. *If* $P<Q<R$, *the injection* $A^{Q,T}//A^{R,T} \to A^{P,T}//A^{R,T}$ *is monomorphic.*

Proof. Both quotient algebras are represented by operations on the same module $H^*(Y_R)$.

This corollary, with Theorem 5.9, implies Theorem 5.1 (f).

For Theorem 5.1 (g), it is sufficient to prove the anticommutativity relation when $a_{i,j}$ and $b_{k,l}$ are generators $e_{k,r}$ or e'_r. This is easily done by expanding abv and bav.

It remains to obtain the structure of $A^{R,T}//A^{R+1,T}$ (for $R \leqslant T$). We have $e_{R,k} \in A^{R,T}$ if $k<p^{T-R}$, $e'_R \in A^{R,T}$; let their images in $A^{R,T}//A^{R+1,T}$ be f_k, f'.

Lemma 5.11. *The elements* $f_k(f')^\varepsilon$ ($\varepsilon = 0$ or 1, $0 \leqslant k < p^{T-R}$) *form a base for* $A^{R,T}//A^{R+1,T}$. *The product is given by*

$$f_k(f')^\varepsilon \cdot f_l(f')^\eta = \begin{cases} (k,l) f_{k+l}(f')^{\varepsilon+\eta} & (\varepsilon + \eta = 0 \text{ or } 1) \\ 0 & (\varepsilon + \eta = 2) \end{cases}$$

The diagonal is given by

$$\Delta f_k(f')^\varepsilon = \sum_{\substack{i+j=k \\ \eta+\zeta=\varepsilon}} f_i(f')^\eta \otimes f_j(f')^\zeta \ .$$

Proof. According to Theorem 5.9, the elements $\{M\}$ of $A^{R,T}//A^{R+1,T}$ are faithfully represented by the corresponding elements $M(v)$ of $H^*(Y_{R+1})$. The image module $E^{R,T}$ has as a base the polynomials E^I, where I runs over the sequences $\{\varepsilon_i, \lambda_i\}$ for which $0 \leqslant \lambda_R < p^{T-R}$, $\varepsilon_R = 0$ or 1, and $\lambda_i = 0$, $\varepsilon_i = 0$ if $i \neq R$. By (6), such a polynomial E^I is exactly $e_{R,k}(e'_r)^\varepsilon v$ with $k = \lambda_R$, $\varepsilon = \varepsilon_R$. Therefore the elements $f_k(f')^\varepsilon$ form a base of $A^{R,T}//A^{R+1,T}$.

The product formula for $M_1 M_2$ now follows by expanding $M_1(M_2 v)$, using (6). The diagonal formula follows similarly from Lemma 5.3.

The proof of Theorem 5.1_p is now complete.

The proof of Theorem 5.1_2 is analogous, but somewhat simpler. As already remarked, we interpret "p" as 2, "β_p" as Sq^1 and "P_p^k" as Sq^{2k}. We replace X by the CARTESIAN product of n spaces, each of type $(Z_2, 1)$, and delete all reference to the other n' factors. $H^*(X; Z_2)$ is now a polynomial algebra. In the arguments, we replace each y_i $(1 \leqslant i \leqslant n)$ by x_i, and delete all references to x_i. Thus, we take $u = x_1 x_2 \ldots x_n$, the product of the fundamental classes. Our indices I become sequences $\{\alpha_1, \alpha_2, \ldots, \alpha_i, \ldots\}$; α_i replaces λ_i and ε_i is deleted. D^I is the sum of all monomials (in the x_i) in which exactly α_i exponents are 2^i, the rest being 1. We define $I(r,k)$ by $\alpha_r = k$, $\alpha_i = 0$ for $i \neq r$; we define $e_{r,k} \in A$ by $e_{r,k}(u) = D^{I(r,k)}$. Thus $\operatorname{Deg} e_{r,k} = k(2^r - 1)$. These $e_{r,k}$ replace those defined for $p > 2$; we do not define any e'_r, and delete all references to them. In the proof of Theorem 5.4, we order the sequences $\{\alpha_i\}$ lexicographically from the right. In defining $A^{R,T}$ and $D^{R,T}$, and in all subsequent arguments, we replace T by $T+1$ in each inequality restricting the choice of generators $e_{r,k}$ or of entries α_i in sequences. The rank of $A^{R,T}$ is thus 2^V where $V = \frac{1}{2}(T - R + 1)(T - R + 2)$. We represent $A^{P,T}//A^{Q,T}$ on a CARTESIAN product Y_Q of n real projective spaces of dimension $2^Q - 1$. In Theorem 5.11, we obtain base elements $f_k = \{e_{R,k}\}$ without elements f'.

With these alterations and interpretations, all our intermediate theorems remain valid.

We next pass on to the last theorems of this section. If $p > 2$, we have

Theorem 5.12$_p$. *There is a homomorphism $h: A \to A$ of algebras with diagonal such that*

$$h(e_{r,k}) = \begin{cases} e_{r,k/p} & \text{if } k \equiv 0 \bmod p \\ 0 & \text{otherwise} \end{cases} \qquad h(e'_r) = 0 \ .$$

If $p = 2$, the theorem remains valid on interpreting "p" as 2 and omitting all reference to e'_r.

Proof. Let us take the space X as before, but with $n' = 0$; and in what follows, let us suppose as necessary that n is sufficiently large.

Let I be a sequence. Let the polynomial E^I be formed from D^I by substituting $(y_i)^p$ for y_i $(1 \leqslant i \leqslant n)$; similarly, let $v = y_1^p y_2^p \ldots y_n^p$. Let E be the submodule of $H^*(X)$ generated by the E^I; it is clearly closed under A.

Each element of A induces a linear map of E, and these constitute a quotient ring R of A. We will next show that the linear map $\{a\}$ is determined by $a(v)$. In fact, we have an analogue of Lemma 5.3; if $\Delta a = \sum_i a^{L,i} \otimes a^{R,i}$, then $a(v)$ determines $a^{L,i}(v)$, $a^{R,i}(v)$. This implies the following statement. Suppose that $a(v)$ determines $a(w)$ and $a(z)$ for all $a \in A$ and certain w, z. Then

$a(v)$ determines $a(wz)$ for all $a \in A$. But $a(v)$ determines $a(y_i^p)$ for all a and each i; therefore $a(v)$ determines $a(E^I)$.

This argument shows that in the epimorphism $A \to R$, the diagonal map passes to the quotient.

Next let S be the subalgebra $\sum_i A_{i,0}$ of A. We will identify R with S. In fact, let J run over the sequences in which $\varepsilon_i = 0$ for all i; and define a function $g: R \to S$ by $g\{a\} = b$ where

$$a(v) = \sum_J \lambda_J E^J \quad \text{and} \quad b(u) = \sum_J \lambda_J D^J .$$

It is clear that g commutes with the diagonal. We will show that g is a homomorphism. In fact, by the definition of g, we have the following statement, in which y stands for a fundamental class y_i:

$$\text{If } a \in A, \text{ and } ay^p = \sum_j \lambda_j y^{pj}, \text{ then } (g\{a\})y = \sum_j \lambda_j y^j .$$

From this we deduce, by passing to products, that:

$$\text{If } a \in A, \text{ and } ay^{pk} = \sum_l \mu_l y^{pl}, \text{ then } (g\{a\})y^k = \sum_l \mu_l y^l .$$

It follows that $g\{a\}((g\{b\})y) = (g\{ab\})y$.

Again, since g commutes with the diagonal, the statements

$$g\{a\}g\{b\}w = g\{ab\}w \ , \quad g\{a\}g\{b\}z = g\{ab\}z \quad (\text{for all a}, b)$$

imply $g\{a\}g\{b\}wz = g\{ab\}wz$. Therefore, for the fundamental class u, we have $g\{a\}g\{b\}u = g\{ab\}u$. Thus $g\{a\}g\{b\} = g\{ab\}$, and g is homomorphic.

It is also clear that g is monomorphic.

Lastly, the composite $h: A \to R \xrightarrow{g} S$ satisfies

$$h(e_{r,k}) = \begin{cases} e_{r,k/p} & \text{if } h \equiv 0 \ (\text{mod } p) \\ 0 \text{ otherwise} \end{cases}$$

$$h(e'_r) = 0$$

Thus g is epimorphic. The existence and properties of h are established.

We state explicitly that h does not preserve the grading. We have

$$h(A_{i,j}) = \begin{cases} A_{i/p,0} & \text{if } i \equiv 0 \ (\text{mod } p) \quad \text{and} \quad j = 0 \\ 0 & \text{otherwise} \end{cases}$$

The proof remains valid for $p = 2$ on interpreting "y_i" as x_i and "S" as A.

For the next theorem, we will fix on an ordering of the generators e'_r and $(e_{r,p^i})^{d_{r,i}}$. For definiteness, we take the ordering e'_1, $(e_{1,1})^{d_{1,0}}$, $(e_{1,p})^{d_{1,1}}, \ldots,$ e'_2, $(e_{2,1})^{d_{2,0}}$, $(e_{2,p})^{d_{2,1}}, \ldots, e'_3, \ldots$

Next consider the anticommutator $[e, f] = ef - (-1)^\varepsilon fe$. Here e, f are distinct generators, with indices r, s say: $\varepsilon = 1$ if $e = e'_r$, $f = e'_s$ and $\varepsilon = 0$ otherwise. One of ef, fe is a permitted monomial for the ordering above; the other is not. The anticommutator can be expanded as a sum of permitted monomials. The dimension concerned contains at most one generator g, if any; let its coefficient in this expansion be λ, or if there is no generator in this dimension, set $\lambda = 0$. Since the anticommutator maps to zero in $A^{1,\infty}//A^{r+s,\infty}$ (by Theorem 5.1) its expansion takes the following form:

$$[e,f] = \lambda g + \sum_i M_i N_i \ .$$

Here $M_i \in I(A)$, $g, N_i \in I(A^{r+s,\infty})$, and g is a generator. If $p > 2$ we have the following theorem:

Theorem 5.13$_p$. $\lambda \neq 0$ *(mod p) in the following cases, and in these only.*
(a) *The pair e, f is e_{r,p^i}, $e_{s,p^{r+i}}$ in either order.*
(b) *The pair e, f is e'_r , $e_{s,p^{r-1}}$ in either order.*

If $p = 2$ the theorem remains valid on interpreting "p" as 2 and omitting all references to e'_r, including case (b).

Proof. We may first eliminate the case $e = e'_r$, $f = e'_s$, since there is then no g.

We take next the case $e = e'_r$, $f = e_{s,p^j}$, $g = e'_t$. The dimensions must satisfy

$$\sum_{u=0}^{r-2} p^u + p^j \sum_{u=0}^{s-1} p^u = \sum_{u=0}^{t-2} p^u \ .$$

The only solution is $j = r - 1$, $t = r + s$. We then have $[e,f] = -g$ by direct use of the CARTAN representation.

We take third the case $e = e_{r,p^i}$, $f = e_{s,p^j}$, $g = e_{t,p^k}$. The dimensions must satisfy

$$p^i \sum_{u=0}^{r-1} p^u + p^j \sum_{u=0}^{s-1} p^u = p^k \sum_{u=0}^{t-1} p^u \ .$$

There are only two solutions; one is $k = i$, $j = r + i$, $t = r + s$; the other is obtained by interchanging e and f. It is sufficient to consider the first.

Consider the case $i = 0$. Here $[e,f] = -g$ by direct use of the CARTAN representation.

Consider the case $i > 0$. Form the expansion

$$[e,f] = \lambda g + \sum_l M_l N_l \ .$$

Apply, i times over, the homomorphism h of Theorem 5.12$_p$. We obtain (say)

$$[e',f'] = \lambda g' + \sum_l M'_l N'_l \ .$$

Here the monomials $M'_l N'_l$ are still permitted (by the special choice of ordering); while $e' = e_{r,1}$, $f' = e_{s,p^r}$, $g' = e_{r+s,1}$. We have $\lambda = 1$ by the case $i = 0$.

This concludes the proof in case $p > 2$.

In case $p = 2$, we omit all references to e'_r. We have also to consider two further solutions of the equation for the dimensions.

Case (i) $i = j$, and either $r = 1$ or $s = 1$; say $s = 1$. Then $t = 1$ and $k = r + i$.

Case (ii) $i = j$, but neither $r = 1$ nor $s = 1$. We then have $k = i + 1 = j + 1$, $t = r = s$.

Both cases may be eliminated, since $g \in A^{r+s,\infty}$, so that $t \geqslant r + s$.

This concludes the proof in case $p = 2$. We have now obtained sufficient data on the structure of the STEENROD algebra.

§ 6. The Cohomology of the STEENROD Algebra

In this article we prove Theorem 2.4. The results of this section, therefore, are very far from complete, compared with those of § 5.

We take $p = 2$ throughout this article, and will be free to use the letter p for other purposes. Tensor products are taken over Z_2 unless otherwise stated.

When we wish to display specific elements in $H^*(A)$, we use the notation of the bar construction (see [3] p. 3–09). Thus we define

$$I(A) = \sum_{q>0} A_q\,, \quad (I(A))^0 = Z_2 \quad \text{and} \quad (I(A))^s = I(A) \otimes (I(A))^{s-1}\,.$$

We define $\bar{B}(A) = \sum_{s \geqslant 0} (I(A))^s$, and write a typical element as $[a_1 | a_2 | \ldots | a_s]$. The second grading t of $\bar{B}(A)$ is defined by $t = \sum_1^s q_i$ when $a_i \in A_{q_i}$. A boundary in $\bar{B}(A)$ is defined by

$$d[a_1 | \ldots | a_s] = \sum_{1 \leqslant i \leqslant s} (-1)^i [a_1 | \ldots | a_i a_{i+1} | \ldots | a_s]\,;$$

d is of degree $(-1, 0)$. The cohomology group $H^{s,t}(\bar{B}(A), Z_2)$ is $\mathrm{Ext}_A^{s,t}(Z_2, Z_2)$, that is $H^{s,t}(A)$.

For example, let us calculate $H^1(A)$. (Here, by convention, $H^s(A) = \sum_t H^{s,t}(A)$.) There are no coboundaries (except the zero cochain); while a cocycle is a Z_2-linear function f, with values in Z_2, defined for arguments $[a]\,(a \in I(A))$, and such that $f[a_1 a_2] = 0$ (each $a_1, a_2 \in I(A)$.) By the known structure of the STEENROD algebra, there are unique cocycles f_m of dimension $t = 2^m$

$(m \geqslant 0)$ such that $f_m[Sq^{2^m}] = 1$; and these form a base for the cocycles. Define $h_m = \{f_m\}$; we have a base for $H^1(A)$.

The cup-product of two cocycles is obtained by transposition from the diagonal map

$$\Delta: \ \bar{B}(A) \to \bar{B}(A) \otimes \bar{B}(A) .$$

This is defined by

$$\Delta[a_1|a_2|\dots|a_n] = \sum_{0 \leqslant p \leqslant n} [a_1|\dots|a_p] \otimes [a_{p+1}|\dots|a_n] .$$

As stated in § 2, our methods depend on a spectral sequence ([4], p. 349) relating the cohomology rings $H^*(\Lambda)$, $H^*(\Gamma)$ and $H^*(\Omega)$ of an algebra Γ, a normal subalgebra Λ and the corresponding quotient algebra $\Omega = \Gamma // \Lambda$. It is sufficient to take Λ central in Γ; this has the result that the operations of Ω on $H^*(\Lambda)$ become trivial, and we have $E_2^* \cong H^*(\Lambda) \otimes H^*(\Omega)$.

It is convenient to have a specific construction for the spectral sequence. We may obtain it by filtering $\bar{B}(\Gamma)$. A chain $[a_1|\dots|a_s]$ is of filtration p if $a_i \in I(\Lambda)$ for $(s-p)$ values of i; we thus obtain homology and cohomology spectral sequences, in good duality. The isomorphisms

$$H^q(\Lambda) \leftarrow E_2^{0,q}, \quad E_2^{p,0} \leftarrow H^p(\Omega)$$

are induced by the natural maps $\bar{B}(\Lambda) \to \bar{B}(\Gamma)$, $\bar{B}(\Gamma) \to \bar{B}(\Omega)$. The (cohomology) spectral sequence has good products; these induce the isomorphism

$$H^p(\Omega) \otimes H^q(\Lambda) \to E_2^{p,0} \otimes E_2^{0,q} \to E_2^{p,q} .$$

Since this section is not a final treatment, the reader will perhaps excuse it if we do not give the details more fully.

We next proceed to the details of the calculation. We will write A^r for $A^{r,\infty}$, so that $A^1 = A$. We recall that $A^r // A^{r+1}$ is a divided polynomial algebra, with a Z_2-base $\{f_k\}$. Take a cocycle $f'_{r,i}$ in dimension $(1, (2^r - 1)2^i)$ such that $f'_{r,i}[f_{2^i}] = 1$; let $h_{r,i}$ be its cohomology class. Then $H^*(A^r // A^{r+1})$ is a polynomial algebra with generators $h_{r,i}$ (for $i \geqslant 0$).

Let h_m be the image in $H^*(A^1 // A^r)$ of $h_{1,m}$ in $H^*(A^1 // A^2)$. This is consistent, for the original h_m in $H^*(A^1)$ is the image of $h_{1,m}$. By the same argument as before, $H^1(A^1 // A^r)$ has as a base the elements h_m in it.

Now take $\Lambda = A^{r+1} // A^{r+2}$, $\Gamma = A^1 // A^{r+2}$, $\Omega = A^1 // A^{r+1}$. Let $d^2: E_2^{0,1} \to E_2^{2,0}$ be the differential in the spectral sequence. Since

$$H^1(A^1 // A^{r+2}) \to H^1(A^{r+1} // A^{r+2})$$

is zero (for $r \geqslant 1$), the classes $d^2 h_{r+1,i}$ must be non-zero elements of $H^2(A^1 // A^{r+1})$. Let us write

$$d^2 h_{r+1,i} = g_{r,i} \in H^2(A^1 // A^{r+1}) .$$

In case $r=1$, we have $g_{1,i}=h_i h_{i+1}$. This is proved as follows. In $H_2(A^1//A^2)$ we have one non-zero class in dimension $t = 3.2^i$; it is dual to $h_i h_{i+1}$, and may be represented by the cycle

$$[e_{1,2^i}|e_{1,2^{i+1}}] + [e_{1,2^{i+1}}|e_{1,2^i}] \qquad (= z, \text{ say}).$$

By Theorem 5.13_2, we have in A a relation

$$e_{1,2^i}e_{1,2^{i+1}} + e_{1,2^{i+1}}e_{1,2^i} = \sum_j M_j N_j + e_{2,2^i}$$

where $M_j \in I(A^1)$, $N_j \in I(A^2)$. The chain

$$[e_{1,2^i}|e_{1,2^{i+1}}] + [e_{1,2^{i+1}}|e_{1,2^i}] + \sum_j [M_j|N_j]$$

has as boundary $[e_{2,2^i}]$; it is thus a cycle in $E^2_{2,0}$ and by the map $A^1 \to A^1//A^2$ it passes to z. Thus $d_2\{z\} = \{[e_{2,2^i}]\}$. Transposing into cohomology we have $d^2 h_{2,i} = h_i h_{i+1}$, as asserted.

In case $r > 1$, we must consider the behaviour of the class $g_{r,i}$ in the spectral sequence which arises when $\Lambda = A^r//A^{r+1}$, $\Gamma = A^1//A^{r+1}$, $\Omega = A^1//A^r$. We know that $H^2(A^1//A^{r+1})$ is filtered. The class $g_{r,i}$ will have an image in the first quotient, which is isomorphic to a subgroup of $E_2^{0,2}$; and if this image is zero, then $g_{r,i}$ will have an image in the second quotient, which is isomorphic to a subgroup of $E_2^{1,1}$. Now $E_2^{0,2} \cong H^2(A^r//A^{r+1})$, which we know; the image of $g_{r,i}$ is necessarily zero, by considering the grading t.

We seek, therefore, the image of $g_{r,i}$ in a subgroup of $E_2^{1,1}$. We will show it is exactly

$$h_{r,i+1}h_i + h_{r,i}h_{r+i} ;$$

here the products are formed by considering h_j as an element of $E_2^{1,0}$ and $h_{r,j}$ as an element of $E_2^{0,1}$. In fact, by transposition, it is sufficient to determine the pairing of $g_{r,i}$ with a certain quotient module of $E^2_{1,1}$. We will construct representative cycles for $E^2_{1,1}$. By Theorem 5.13_2 we have relations

$$e_{r,2^{i+1}}e_{1,2^i} + e_{1,2^i}\,e_{r,2^{i+1}} = \sum_j M_j N_j + e_{r+1,2^i}$$
$$e_{r,2^i}\,e_{1,2^{r+1}} + e_{1,2^{r+1}}e_{r,2^i} = \sum_j P_j Q_j + e_{r+1,2^i}$$

where $M_j, P_j \in I(A^1)$, $N_j, Q_j \in I(A^{r+1})$. Thus the chains

$$[e_{r,2^{i+1}}|e_{1,2^i}] + [e_{1,2^i}|e_{r,2^{i+1}}] \qquad (= w, \text{ say})$$
$$[e_{r,2^i}|e_{1,2^{r+i}}] + [e_{1,2^{r+i}}|e_{r,2^i}] \qquad (= z, \text{ say})$$

give classes in $H_2(A^1//A^{r+1})$. These chains are of filtration 1, and yield a base for $E^2_{1,1}$ in dimension $t = (2^{r+1} - 1)2^i$ (as one verifies, knowing the structure of $E_2^{1,1}$). On applying the map d_2: $H_2(A^1//A^{r+1}) \to H_1(A^{r+1}//A^{r+2})$, we have $d_2\{w\} = d_2\{z\} = \{[e_{r+1,2^i}]\}$. (This is proved by exactly the same argument as was used in the case $r = 1$; note that this d_2 lies in the spectral sequence for $\Lambda = A^{r+1}//A^{r+2}$, etc.) We have, then,

$$\{w\}g_{r,i} = \{z\}g_{r,i} = 1 .$$

On transposing, we have in $E_2^{1,1}$ the relation

$$\{g_{r,i}\} = h_{r,i+1}\, h_i + h_{r,i}\, h_{r+i} ,$$

as asserted.

Lemma 6.1. *In $H^2(A^1//A^{r+1})$ the elements $g_{r,i}$, with the elements $h_j h_k$ for which $j \leqslant k - 2$ or $j = k$, are linearly independent. The elements $h_j h_{j+1}$ are zero.*

In fact, if $r > 1$, then $H^2(A^1//A^{r+1})$ is filtered; the elements $g_{r,i}$ map to a linearly independent set in $E_2^{1,1}$, while the elements $h_j h_k$ map to zero. It remains to prove (by induction over r) that the elements $h_j h_k$ ($j \leqslant k - 2$ or $j = k$) are linearly independent. If they are so in $H^2(A^1//A^r)$, then they are so in $H^2(A^1//A^{r+1})$, unless in the spectral sequence concerned some linear combination of elements $h_j h_k$ is equal to $d^2(h_{r,l})$. This is impossible, by considering the grading t. To begin the induction, the elements named are linearly independent in $H^2(A^1//A^2)$.

The elements $h_j h_{j+1}$ are zero in $H^2(A^1//A^3)$, being $d^2(h_{2,j})$.

We next consider again the spectral sequence which arises when

$$\Lambda = A^r//A^{r+1} , \quad \Gamma = A^1//A^{r+1} , \quad \Omega = A^1//A^r .$$

Since $g_{r,i}$ lies in $H^2(A^1//A^{r+1})$, the elements $h_{r,i+1}h_i + h_{r,i}h_{r+i}$ in $E_2^{1,1}$ are cocycles for d^2. Thus the elements $(h_{r,i+1}h_i + h_{r,i}h_{r+i})h_j$ ($= z_{r,i,j}$ say) in $E_2^{2,1}$ are cocycles for d^2.

Lemma 6.2. *If $r \geqslant 2$, the classes $\{z_{r,i,j}\}$ in $E_3^{2,1}$ satisfy the following relations only:*

$$\{z_{r,i,r+i+1} + z_{r,i+1,i}\} = 0 \quad (r \geqslant 2) ,$$
$$\{z_{2,i,i+1}\} = 0 \quad (r = 2) .$$

If $r \geqslant 2$, $E_3^{0,2}$ has as a base the classes $\{(h_{r,i})^2\}$.

Proof. We have to examine $d^2 : E_2^{0,2} \to E_2^{2,1}$. Here $E_2^{0,2}$ has as a base the elements $h_{r,i}h_{r,j}$. In $E_2^{2,1}$ (if $r > 2$) the elements

$$h_{r,i}g_{r-1,j}, \quad h_{r,i}h_j h_k \quad (j \leqslant k - 2 \text{ or } j = k)$$

are linearly independent, while the elements $h_{r,i} h_j h_{j+1}$ are zero; this is by Lemma 6.1. The boundaries are given by

$$d^2(h_{r,i} h_{r,j}) = h_{r,i} g_{r-1,j} + h_{r,j} g_{r-1,i} .$$

From this, the conclusion follows.

In case $r = 2$, $E_2^{2,1}$ has as a base the elements $h_{2,i} h_j h_k$, while the boundaries are given by

$$d^2(h_{2,i} h_{2,j}) = h_{2,i} h_j h_{j+1} + h_{2,j} h_i h_{i+1} .$$

The conclusion is again elementary.

Lemma 6.3. *If* $r \geqslant 2$, *the classes* $\{g_{r,i} h_j\}$ *in* $H^3(A^1//A^{r+1})/p^* H^3(A^1//A^r)$ *satisfy the following relations only:*

(i) $g_{r,i} h_{r+i+1} + g_{r,i+1} h_i = 0 \quad (r \geqslant 2)$,

(ii) $\{g_{2,i} h_{i+1}\} \qquad = 0 \quad (r = 2)$.

Proof. The image of $g_{r,i} h_j$ in $E_3^{2,1}$ is $\{z_{r,i,j}\}$. On applying Lemma 6.2, we obtain the results of this lemma, except for the exact relation (i). This follows immediately from the remark (above) that $h_{r,i+1} h_i + h_{r,i} h_{r+i}$ is a cocycle for d^2.

Lemma 6.4. *In* $H^3(A^1//A^{r+1})$ $(r \geqslant 3)$ *the elements* $h_i h_j h_k$ *are subject to the following relations only.*

(i) $h_i h_{i+1} h_j = 0$.

(ii) $(h_i)^2 h_{i+2} = (h_{i+1})^3$.

(iii) $h_i (h_{i+2})^2 = 0$.

Proof. We know the structure of $H^3(A^1//A^2)$ and will prove the lemma by induction over r. We have to consider the spectral sequence in which $\Lambda = A^{r+1}//A^{r+2}$, $\Gamma = A^1//A^{r+2}$ and $\Omega = A^1//A^{r+1}$. In this we have to consider the differentials $d^2: E_2^{1,1} \to E_2^{3,0}$ and $d^3: E_3^{0,2} \to E_3^{3,0}$.

We have $d^2(h_{r+1,i} h_j) = g_{r,i} h_j$. By Lemma 6.3 this introduces no new relations into $p^* H^3(A^1//A^r)$ if $r > 2$. If $r = 2$, we have only to consider the case $j = i + 1$. We verify that

$$d^2(h_{3,0} h_1) = h_0 (h_2)^2$$

by a direct calculation (given below); the result

$$d^2(h_{3,i} h_{i+1}) = h_i (h_{i+2})^2$$

follows by using the homomorphism of Theorem 5.12_2. If $r = 1$, we have

$$d^2(h_{2,i} h_j) = h_i h_{i+1} h_j .$$

We next consider d_3. By Lemma 6.2, $E_3^{0,2}$ has as a base the classes

$$\{(h_{r+1,i})^2\} .$$

By considering the grading t, we see that d^3 introduces no new relations between the classes $h_i h_j h_k$ unless $r = 1$. We then verify

$$d^3\{(h_{2,0})^2\} = (h_0)^2 h_2 + (h_1)^3$$

by direct calculations (given below); the result

$$d^3\{(h_{2,i})^2\} = (h_i)^2 h_{i+2} + (h_{i+1})^3$$

follows by using the homomorphism of Theorem 5.12_2.

The direct calculations are similar to those above. They are carried out in homology, and are as follows (for brevity we have replaced the symbol $e_{r,2^i}$ by its dimension $(2^r - 1)2^i$.)

$$d\left\{\begin{array}{c}[1|4|4] + [4|1|4] + [4|4|1] \\ + [1|2|6] + [2|1|6] + [2|6|1] \\ + [4|3|2] + [3|4|2] + [3|2|4]\end{array}\right\} = [2|7] + [7|2]$$

$$d\left\{\begin{array}{c}[2|1|3] + [1|2|3] + [1|3|2] \\ + [2|2|2]\end{array}\right\} = [3|3]$$

$$d\left\{\begin{array}{c}[4|1|1] + [1|4|1] + [1|1|4] \\ + [1|2|3] + [2|1|3] + [2|3|1]\end{array}\right\} = [3|3] .$$

This concludes the proof of Lemma 6.4, of Theorem 2.4, and of Theorem 1.1.

REFERENCES

[1] J. ADEM, *The iteration of the* STEENROD *squares in algebraic topology*, Proc. Nat. Acad. Sci. U. S. A. 38, 1952, p. 720–726.

[2] H. CARTAN, *Sur l'itération des opérations de* STEENROD. Comment. Math. Helv. 29, 1955, p. 40–58.

[3] H. CARTAN, *Séminaire* H. CARTAN *de l'Ecole Normale Supérieure*, 1954/55.

[4] H. CARTAN and S. EILENBERG, *Homological Algebra*. Princeton Mathematical Series.

[5] H. TODA, *Le produit de* WHITEHEAD *et l'invariant de* HOPF. C. R. Acad. Sci. (Paris) 241, 1955, p. 849–850.

(Received June 7, 1957)

ANNALS OF MATHEMATICS
Vol. 72, No. 1, July, 1960

ON THE NON-EXISTENCE OF ELEMENTS OF HOPF INVARIANT ONE

BY J. F. ADAMS

(Received April 20, 1958)
(Revised January 25, 1960)

CHAPTER 1. INTRODUCTION

1.1. *Results.* It is the object of this paper to prove a theorem in homotopy-theory, which follows as Theorem 1.1.1. In stating it, we use one definition. A *continuous product with unit* on a space X is a continuous map $\mu : X \times X \to X$ with a point e of X such that $\mu(x, e) = \mu(e, x) = x$. An *H-space* is a space which admits a continuous product with unit. For the remaining notations, which are standard in homotopy-theory, we refer the reader to [15], [16]. In particular, $H^m(Y; G)$ is the mth singular cohomology group of the space Y with coefficients in the group G.

THEOREM 1.1.1. *Unless $n = 1, 2, 4$ or 8, we have the following conclusions*:

(a) *The sphere S^{n-1} is not an H-space.*

(b) *In the homotopy group $\pi_{2n-3}(S^{n-1})$, the Whitehead product $[\iota_{n-1}, \iota_{n-1}]$ is non-zero.*

(c) *There is no element of Hopf invariant one* [17] *in $\pi_{2n-1}(S^n)$.*

(d) *Let $K = S^m \cup E^{m+n}$ be a* CW-*complex formed by attaching an $(m + n)$-cell E^{m+n} to the m-sphere S^m. Then the Steenrod square* [31]

$$\mathrm{Sq}^n : H^m(K; Z_2) \to H^{m+n}(K; Z_2)$$

is zero.

It is a classical result that the four conclusions are equivalent [36], [31]. Various results in homotopy-theory have been shown to depend on the truth or falsity of these conclusions. It is also classical that the conclusions are false for $n = 2, 4$ and 8. In fact, the systems of complex numbers, quaternions and Cayley numbers provide continuous products on the euclidean spaces R^2, R^4 and R^8; from these one obtains products on the unit spheres in these spaces, that is, on S^1, S^3 and S^7. (See [16].) The case $n = 1$ is both trivial and exceptional, and we agree to exclude it from this point on.

The remarks above, and certain other known theorems, may be summarized by the following diagram of implications.

R^n is a normed algebra over the reals $\Longleftrightarrow n = 2, 4$, or 8.

$\Downarrow$

R^n is a division algebra over the reals $\Longrightarrow n = 2^m$

$\Downarrow$

S^{n-1}, with its usual differentiable structure, is parallelisable $\overset{}{\underset{(6)}{\Longrightarrow}}$ $n = 2, 4$, or 8

$\Downarrow$

S^{n-1}, with some (perhaps extraordinary) differentiable structure, is parallelisable

$\Downarrow$ (5)

S^{n-1} is an H-space (also implied directly by "R^n is a division algebra over the reals")

$\Updownarrow$

There is an element of Hopf invariant one in $\pi_{2n-1}(S^n)$

$\Downarrow$ (1) $n = 2$ or $4r$ $\Downarrow$ (2) $n = 2^m$ $\Downarrow$ (3) $n \neq 16$ $\Downarrow$ (4) $n = 2, 4$, or 8.

The implications (1), (2), (3) represent cases of Theorem 1.1.1 which are already known. In fact, (1) is due to G. W. Whitehead [36]; (2) is due to J. Adem [4]; and (3) is due to H. Toda [34], who used an elegant lemma in homotopy-theory and extensive calculations of the homotopy groups of spheres.

The implication (4) is just Theorem 1.1.1. The implication (5) is due to A. Dold, in answer to a question of A. Borel. (We remark that Theorem 1.1.1 implies strong results on the non-parallelizability of manifolds: see Kervaire [22].) The implication (6) was proved independently by M. Kervaire [21] and by R. Bott and J. Milnor [8]. In each case, it was deduced from deep results of R. Bott [7] on the orthogonal groups.

A summary of the present work appeared as [3]. The first draft of this paper was mimeographed by Princeton; I am most grateful to all those who offered criticisms and suggestions, and especially to J. Stasheff.

1.2. *Method*. Theorem 1.1.1 will be proved by establishing conclusion (d). The method may be explained by analogy with Adem's proof [4] in the case $n \neq 2^r$. In case $n = 6$, for example, Adem relies on the relation

$$\mathrm{Sq}^6 = \mathrm{Sq}^2\mathrm{Sq}^4 + \mathrm{Sq}^5\mathrm{Sq}^1 .$$

Now, in a complex $K = S^m \cup E^{m+6}$, the composite operations $\mathrm{Sq}^2\mathrm{Sq}^4$ and $\mathrm{Sq}^5\mathrm{Sq}^1$: $H^m(K; Z_2) \to H^{m+6}(K; Z_2)$ will be zero, since $H^{m+4}(K; Z_2)$ and $H^{m+1}(K; Z_2)$ are zero. Therefore Sq^6: $H^m(K; Z_2) \to H^{m+6}(K; Z_2)$ is zero.

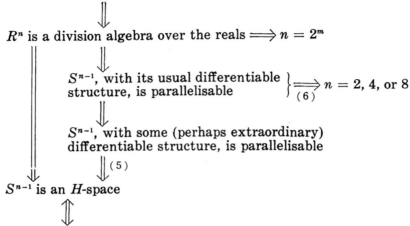

The method fails in the case $n = 2^r$, because in this case Sq^n is not decomposable in terms of operations of the first kind.

We therefore proceed by showing that Sq^n *is decomposable in terms of operations of the second kind* (in case $n = 2^r$, $r \geqq 4$). (cf. [2, § 1]). We next explain what sort of decomposition is meant.

Suppose that $n = 2^{k+1}$, $k \geqq 3$, and that $u \in H^m(X; Z_2)$ is a cohomology class such that $\mathrm{Sq}^{2^i}(u)=0$ for $0 \leqq i \leqq k$. Then certain cohomology operations of the second kind are defined on u; for example,

$$\beta_4(u) \in H^{m+1}(X; Z_2) \,/\, \mathrm{Sq}^1 H^m(X; Z_2)$$

$$\Phi(u) \in H^{m+3}(X; Z_2) \,/\, \mathrm{Sq}^2 H^{m+1}(X; Z_2) + \mathrm{Sq}^3 H^m(X; Z_2) ,$$

and so on. In fact, in § 4.2 of this paper we shall obtain a system of secondary operations $\Phi_{i,j}$, indexed by pairs (i, j) of integers such that $0 \leqq i \leqq j$, $j \neq i + 1$. These operations will be such that (with the data above) $\Phi_{i,j}(u)$ is defined if $j \leqq k$. The value of $\Phi_{i,j}(u)$ will be a coset in $H^q(X; Z_2)$, where $q = m + (2^i + 2^j - 1)$; let us write

$$\Phi_{i,j}(u) \in H^*(X; Z_2) \,/\, Q^*(X; i, j) .$$

We do not need to give the definition of $Q^*(X; i, j)$ here; however, as in the examples above, it will be a certain sum of images of Steenrod operations. (By a Steenrod operation, we mean a sum of composites of Steenrod squares.)

Suppose it granted, then, that we shall define such operations $\Phi_{i,j}$. In § 4.6 we shall also establish a formula, which is the same for all spaces X:-

$$\mathrm{Sq}^n(u) = \textstyle\sum_{i,j;\, j \leq k} a_{i,j,k} \Phi_{i,j}(u) \qquad \text{modulo } \textstyle\sum_{i,j;\, j\leq k} a_{i,j,k} Q^*(X; i, j) .$$

In this formula, each $a_{i,j,k}$ is a certain Steenrod operation. We recall that $n = 2^{k+1}$, $k \geqq 3$.

Suppose it granted, then, that we shall prove such a formula. Then we may apply it to a complex $K = S^m \cup E^{m+n}$. If $u \in H^m(K; Z_2)$, then $\mathrm{Sq}^{2^i}(u) = 0$ for $0 \leqq i \leqq k$. The cosets $\Phi_{i,j}(u)$ will thus be defined for $j \leqq k$; and they will be cosets in zero groups. The formula will be applicable, and will show that $\mathrm{Sq}^n(u)=0$, modulo zero. Theorem 1.1.1 will thus follow immediately.

1.3. *Secondary operations.* It is clear, then, that all the serious work involved in the proof will be concerned with the construction and properties of secondary cohomology operations. Two methods have so far been used to define secondary operations which are stable. The first method is that of Adem [4]. This possesses the advantage that the operations defined are computable, at least in theory. Unfortunately, it does not give us much insight into the properties of such operations. Moreover, not all the operations we need can be defined by this method as it now stands.

The second method, which we shall use, is that of the universal example [6], [30]. This is a theoretical method; it gives us some insight, but it gives us no guarantee that the operations so defined are computable. Similarly, it sometimes shows (for example) that one operation is linearly dependent on certain others, without yielding the coefficients involved.

Both methods show that secondary operations are connected with relations between primary operations. For example, the Bockstein coboundary β_4 is connected with the relation $Sq^1Sq^1 = 0$; the Adem operation Φ is connected with the relation $Sq^2Sq^2 + Sq^3Sq^1 = 0$.

It may appear to the reader that what we say about "relations" in this section is vague and imprecise; however, it will be made precise later by the use of homological algebra [13]; this is the proper tool to use in handling relations, and in handling relations between relations.

In any event, it will be our concern in Chapter 3 to set up a general theory of stable secondary cohomology operations, and to show that *to every "relation" there is associated at least one corresponding secondary operation*. We study these operations, and the relations between them.

This theory, in fact, is not deep. However, it affords a convenient method for handling operations, by dealing with the associated relations instead. For example, we have said that Φ is "associated with" the relation $Sq^2Sq^2 + Sq^3Sq^1 = 0$. We would expect the composite operation $Sq^3\Phi$ to be "associated with" the relation

$$(1) \qquad (Sq^3Sq^2)Sq^2 + (Sq^3Sq^3)Sq^1 = 0 .$$

Similarly, we have said that β_4 is "associated with" the relation $Sq^1Sq^1 = 0$. We would expect the composite operation $Sq^5\beta_4$ to be "associated with" the relation

$$(2) \qquad (Sq^5Sq^1)Sq^1 = 0 .$$

But since $Sq^3Sq^2 = 0$ and $Sq^3Sq^3 = Sq^5Sq^1$, the relations (1), (2) coincide. We would therefore expect to find

$$Sq^3\Phi = Sq^5\beta_4 \qquad \text{(modulo something as yet unknown).}$$

And, in fact, the theory to be presented in Chapter 3 will justify such manipulations, and this is one of its objects.

If we can do enough algebra, then, of a sort which involves relations between the Steenrod squares, we expect to obtain relations of the form

$$\sum_{i,j} a_{i,j}\Phi_{i,j} = 0 \qquad \text{(modulo something as yet unknown).}$$

(Here we use "a" as a generic symbol for a Steenrod operation, and "Φ" as a generic symbol for a secondary operation.) In particular, we shall in fact obtain (in § 4.6) a formula

$$\sum_{i,j;\, j\leq k} a_{i,j,k}\Phi_{i,j}(u) = \lambda \mathrm{Sq}^{2^{k+1}}(u)$$

such as we seek, but containing an undetermined coefficient λ.

To determine the coefficient λ, it is sufficient to apply the formula to a suitable class u in a suitable test-space X. We shall take for X the complex projective space P of infinitely-many dimensions. Our problem, then, reduces to calculating the operations $\Phi_{i,j}$ in this space P. This is performed in § 4.5.

The plan of this paper is then as follows. In Chapter 2 we do the algebraic work; in Chapter 3 we set up a general theory of stable secondary operations; in Chapter 4 we make those applications of the theory which lead to Theorem 1.1.1.

The reader may perhaps like to read § 2.1 first, and then proceed straight to Chapters 3 and 4, referring to Chapter 2 when forced by the applications.

Chapter 2. Homological Algebra

2.1. *Introduction.* In this chapter, we make those applications of homological algebra [13] which are needed for what follows. From the point of view of logic, therefore, this chapter is prior to Chapter 4; but from the point of view of motivation, Chapter 4 is prior to this one.

For an understanding of Chapter 3, only the first article of this chapter is requisite.

The plan of this chapter is as follows. In §§ 2.1, 2.2 we outline what we need from the general theory of homological algebra, proceeding from what is well known to what is less well known. In § 2.4 we state, and begin to use, Milnor's theorem on the structure of the Steenrod algebra A. In § 2.5 we perform the essential step of calculating $\mathrm{Ext}_A^{s,t}(Z_2, Z_2)$ as far as we need it. This work relies on § 2.4, and also relies on a certain spectral sequence in homological algebra. This sequence is set up in § 2.3. In the last section, § 2.6, we calculate $\mathrm{Ext}_A^{s,t}(M, Z_2)$(as far as we need it) for a certain module M that arises in Chapter 4.

We now continue by recalling some elementary algebraic notions.

The letter K will denote a field of coefficients, usually the field Z_p of residue classes modulo a fixed prime p.

A *graded algebra* A over K is an algebra over K; *qua* vector space over K, it is the direct sum of components, $A = \sum_{q \geqq 0} A_q$; and these satisfy $1 \in A_0$, $A_q \cdot A_r \subset A_{q+r}$. The elements lying in one component A_q are called *homogeneous* (of degree q). We shall be particularly concerned with the Steenrod algebra [5], [12]. We shall give a formal, abstract definition of the Steenrod algebra in § 3.5; for present purposes the following description is sufficient. If $p = 2$, the generators are the symbols Sq^k, and the relations are those which hold between the Steenrod squares in the (mod 2) cohomology of every topological space. If $p > 2$, the generators are the symbols β_p and P_p^k, and the relations are those which hold between the Bockstein coboundary and the Steenrod cyclic reduced powers, in the (mod p) cohomology of every topological space. (Here we suppose the Bockstein coboundary defined without signs, so that it anticommutes with suspension.) The Steenrod algebra is graded; the degrees of Sq^k, β_p and P_p^k are k, 1 and $2(p-1)k$.

A *graded left module* M over the graded algebra A is a left module [9] over the algebra A; *qua* vector space over K, it is the direct sum of components, $M = \sum_q M_q$; and these satisfy $A_q \cdot M_r \subset M_{q+r}$. The elements lying in one component M_q are called *homogeneous* (of degree q). We shall write $\deg(m)$ for the degree of a homogeneous element m, and the use of this notation will imply that m is homogeneous. When we speak of *free* graded modules over the graded algebra A, we understand that they have bases consisting of homogeneous elements.

We must also discuss maps between graded modules. A K-linear function $f: M \to M'$ is said to be *of degree* r if we have $f(M_q) \subset M'_{q+r}$. We say that it is a *left* A-map if we propose to write it on the left of its argument, and if it is A-linear in the sense that

$$f(am) = (-1)^{qr} a f(m) \qquad \text{(where } a \in A_q\text{)}.$$

Similarly, we call it a *right* A-map if we propose to write it on the right of its argument, and if it is A-linear in the sense that

$$(am)f = a(mf) .$$

There is, of course, a (1-1) correspondence between left A-maps and right A-maps (of a fixed degree r); it is given by

$$f(m) = (-1)^{qr}(m)f^* \qquad \text{(where } m \in M_q\text{)}.$$

The two notions are thus equivalent.

Sometimes we have to deal with bigraded modules; in that case the degree r which should appear in these signs is the total degree.

It is clear that we can avoid some signs by using right A-maps; and in Chapter 3 we will do this. In the present chapter, however, it is convenient to follow the received notation for the bar construction, so we will use left A-maps. The passage from one convention to the other will cause no trouble, as the applications are in characteristic 2.

Let

$$\begin{array}{ccc} M & \xrightarrow{f} & N \\ {\scriptstyle g}\downarrow & & \downarrow{\scriptstyle g'} \\ P & \xrightarrow{f'} & Q \end{array}$$

be a diagram of A-maps in which f and f' have degree r, while g and g' have degree s. Then we say that the diagram is *anticommutative* if

$$g'f = (-1)^{rs} f'g \ .$$

We now begin to work through the elementary notions of homological algebra, in the case when our modules are graded. We shall suppose given a graded algebra A over K which is locally finite-dimensional; that is, $\sum_{q \leq r} A_q$ is finite-dimensional for each r. We shall also assume that A is connected, that is, $A_0 = K$. Let M be a graded module over A which is locally finitely-generated; this is equivalent to saying that $\sum_{q \leq r} M_q$ is finite-dimensional. A *free resolution* of M consists of the following.

(i) A bigraded module $C = \sum_{s,t} C_{s,t}$ such that $A_q \cdot C_{s,t} \subset C_{s,t+q}$. We set $C_s = \sum_t C_{s,t}$, and require that each C_s is a (locally finitely-generated) free module over A.

(ii) An A-map $d : C \to C$ of bidegree $(-1, 0)$, so that $dC_{s,t} \subset C_{s-1,t}$. (Thus the total degree of d is -1). We write d_s: $C_s \to C_{s-1}$ for the components of d.

(iii) An A-map $\varepsilon : C_0 \to M$ of degree zero. We require that the sequence

$$0 \longleftarrow M \xleftarrow{\varepsilon} C_0 \xleftarrow{d_1} C_1 \longleftarrow \cdots \longleftarrow C_{s-1} \xleftarrow{d_s} C_s \longleftarrow \cdots$$

should be exact, and we regard C as an acyclic chain complex.

Next, let L and N be left and right graded A-modules. We have a group

$$\mathrm{Hom}_A^t(C_s, L)$$

whose elements are the A-maps $\mu : C_s \to L$ of total degree $-(s+t)$, so that $\mu(C_{s,u}) \subset L_{u-t}$. Since C_s is graded, we define

$$\mathrm{Hom}_A(C_s, L) = \sum_t \mathrm{Hom}_A^t(C_s, L)$$

and

$$\mathrm{Hom}_A(C, L) = \textstyle\sum_s \mathrm{Hom}_A(C_s, L) .$$

We also have a group [13]

$$N \otimes_A C = \textstyle\sum_s N \otimes_A C_s;$$

we bigrade it as follows; if $n \in N_u$, $c \in C_{s,t}$, we set

$$n \otimes_A c \in (N \otimes_A C)_{s,t+u} .$$

We may regard $\mathrm{Hom}_A(C, L)$ as a cochain complex and $N \otimes_A C$ as a chain complex, using the boundaries

$$\mathrm{Hom}_A(C_0, L) \longrightarrow \cdots \longrightarrow \mathrm{Hom}_A(C_{s-1}, L) \xrightarrow{(d_s)^*} \mathrm{Hom}_A(C_s, L) \longrightarrow \cdots ,$$

$$N \otimes_A C_0 \longleftarrow \cdots \longleftarrow N \otimes_A C_{s-1} \xleftarrow{(d_s)_*} N \otimes_A C_s \longleftarrow \cdots$$

(Note that we have to define $(d_s)_*$, that is, $1 \otimes d_s$, by the rule

$$(1 \otimes d_s)(n \otimes c) = (-1)^u (n \otimes d_s c)$$

where $n \in N_u$). We may write $\delta = \sum_s (d_s)^*$, $\partial = \sum_s (d_s)_*$.

These complexes are determined up to chain equivalence by L, M, N. In fact, given resolutions C, C' of M, M', and given a map $f : M \rightarrow M'$ we may extend it to a chain map $g : C \rightarrow C'$; moreover, such a map is unique up to chain homotopy. Such chain maps (and homotopies) yield cochain maps (and homotopies) of $\mathrm{Hom}_A(C, L)$. We thus see that the cohomology groups of $\mathrm{Hom}_A(C, L)$ are independent of C (up to a natural isomorphism), and are natural in M. We write $\mathrm{Ext}_A^{s,t}(M, L)$ for $\mathrm{Ker}(d_{s+1})^*/\mathrm{Im}(d_s)^*$ in the sequence

$$\cdots \longrightarrow \mathrm{Hom}_A^t(C_{s-1}, L) \xrightarrow{(d_s)^*} \mathrm{Hom}_A^t(C_s, L) \xrightarrow{(d_{s+1})^*} \mathrm{Hom}_A^t(C_{s+1}, L) \longrightarrow \cdots$$

We also define

$$\mathrm{Ext}_A^s(M, L) = \textstyle\sum_t \mathrm{Ext}_A^{s,t}(M, L) .$$

Similarly for the homology groups of $N \otimes_A C$, which are written $\mathrm{Tor}_{s,t}^A(N, M)$ or $\mathrm{Tor}_s^A(N, M)$.

We shall be particularly concerned with the cases $L = K$, $N = K$. We grade $L = K$ by setting $L_0 = K$, $L_q = 0$ for $q \neq 0$. The structure of $L = K$ as an A-module is thus unique (and trivial) since $a(L_0) = 0$ if $\deg(a) > 0$, while the action of A_0 is determined by that of the unit. Similarly for $N = K$.

In the case $L = K$, $N = K$ we have a formal duality between Tor and Ext. In fact, if V is a finite-dimensional vector space over K, we write V^* for its dual. If V is a locally finite-dimensional graded vector space over K, we set

$$V^* = \sum_q (V_q)^* ,$$

and regard this as the dual in the graded case. Thus $K \otimes_A C_s$ and $\mathrm{Hom}_A(C_s, K)$ are dual (graded) vector spaces over K; the pairing is given by

$$h(k \otimes_A c) = h(kc)$$

for $h \in \mathrm{Hom}_A(C_s, K)$, etc. The maps $(d_s)_*$ and $(d_s)^*$ are dual. Thus

$$\mathrm{Tor}^A_{s,t}(K, M) \quad \text{and} \quad \mathrm{Ext}_A^{s,t}(M, K)$$

are dual vector spaces over K.

We now introduce some further notions which are applicable because $A_0 = K$. We set $I(A) = \sum_{q>0} A_q$; and if N is a (graded) A-module, we set $J(N) = I(A)\cdot N$. (Thus $J(N)$ is the kernel of the usual map $N \to K \otimes_A N$.) We call a map $f: N \to N'$ *minimal* if $\mathrm{Ker}\, f \subset J(N)$. We call a resolution *minimal* if the maps d_s and ε are minimal. The word "minimal" expresses the intuitive notion that in constructing such a resolution by the usual inductive process, we introduce (at each stage) as few A-free generators as possible.

It is easy to show that each (locally finitely-generated) A-module M has a minimal resolution. Any two minimal resolutions of M are isomorphic.

We note that if C is a minimal resolution of M, then

$$\mathrm{Tor}^A_{s,t}(K, M) \cong (K \otimes_A C)_{s,t}$$
$$\mathrm{Ext}_A^{s,t}(M, K) \cong \mathrm{Hom}_A^t(C_s, K) .$$

This is immediate, since the boundary ∂ in $K \otimes_A C$ is zero, and so is the coboundary δ in $\mathrm{Hom}_A(C, K)$.

This concludes our survey of the elementary notions which are needed in Chapter 3.

2.2. *General notions.* In this section we continue to survey the general notions of homological algebra that we shall have occasion to use later.

We begin by setting up a lemma which forms a sort of converse to the last remarks of § 2.1. It arises in the following context. Let

$$M \longleftarrow C_0 \xleftarrow{d_1} C_1 \longleftarrow \cdots \longleftarrow C_{s-1} \xleftarrow{d_s} C_s$$

be a partial resolution of M; and set $Z(s) = \mathrm{Ker}(d_s) \cap J(C_s)$. Then there is a homomorphism

$$\theta : Z(s) \longrightarrow \mathrm{Tor}^A_{s+1}(K, M)$$

defined as follows. Extend the partial resolution by adjoining some C_{s+1},

d_{s+1}. Given $z \in Z(s)$, take $w \in C_{s+1}$ such that $d_{s+1}w = z$, and define

$$\theta(z) = \{1 \otimes_A w\} .$$

We easily verify that θ is well-defined, epimorphic, and natural for maps of M.

Next, let $\{g_i\}$ be a K-base for $\mathrm{Tor}^A_{s+1}(K, M)$; choose $z_i \in Z(s)$ so that $\theta(z_i) = g_i$. Let C_{s+1} be an A-free module on generators c_i in (1-1) correspondence with the z_i, and of the same t-degrees; and define

$$d_{s+1} : C_{s+1} \to C_s$$

by

$$d_{s+1}(c_i) = z_i .$$

Lemma 2.2.1. *The map d_{s+1} is minimal, and if d_s is also minimal, then*

$$C_{s-1} \xleftarrow{d_s} C_s \xleftarrow{d_{s+1}} C_{s+1}$$

is exact. In this case $\{1 \otimes_A c_i\} = g_i$.

We shall use this lemma to construct minimal resolutions in a convenient fashion. In this application, since d_s will be minimal, θ will be defined on $\mathrm{Ker}(d_s)$.

Proof. By the construction, we have $d_s d_{s+1} = 0$. Moreover, if $C_{s-1} \xleftarrow{d_s} C_s \xleftarrow{d_{s+1}} C_{s+1}$ is not exact, we may add further generators to C_{s+1} (obtaining C'_{s+1}, d'_{s+1}, say) so that the sequence becomes exact. By using the definition of θ, the condition $\theta(z_i) = g_i$ now yields

$$\{1 \otimes_A c_i\} = g_i .$$

We will now prove that d_{s+1} is minimal. In fact, take an element

$$z = \sum_i (\lambda_i + a_i) c_i$$

of $\mathrm{Ker}(d_{s+1})$, with $\lambda_i \in K$, $a_i \in I(A)$. Then, on extending our resolution to

$$C_s \xleftarrow{d'_{s+1}} C'_{s+1} \xleftarrow{d'_{s+2}} C'_{s+2} ,$$

we can find w such that $d'_{s+2}w = z$. Hence

$$\partial(1 \otimes_A w) = \sum_i (1 \otimes_A \lambda_i c_i) .$$

That is, $\sum_i \lambda_i g_i = 0$. Thus $\lambda_i = 0$ for each i, and z lies in $J(C_{s+1})$. We have shown that d_{s+1} is minimal.

We now suppose that d_s is minimal, so that $Z(s) = \mathrm{Ker}(d_s)$. We wish to prove the exactness. Suppose, as an inductive hypothesis, that

$$C_{s-1,t} \xleftarrow{d_s} C_{s,t} \xleftarrow{d_{s+1}} C_{s+1,t}$$

is exact for $t < n$; this hypothesis is vacuous for sufficiently small n. One may verify that the kernel of

$$\theta : Z(s) \to \operatorname{Tor}^A_{s+1}(K, M)$$

is $J(\ker(d_s))$. Hence any x in $Z(s) \cap C_{s,n}$ can be written in the form

$$x = \sum_i \lambda_i z_i + \sum_j (-1)^{\varepsilon(j)} a_j x_j$$

with $\lambda_i \in K$, $a_j \in I(A)$, $x_j \in \operatorname{Ker}(d_s)$, $\varepsilon(j) = \deg(a_j)$. By the inductive hypothesis, $x_j = d_{s+1} w_j$, say; hence

$$x = d_{s+1}(\sum_i \lambda_i c_i + \sum_j a_j w_j).$$

This completes the induction, and the proof of Lemma 2.2.1.

We next introduce products into the cohomology groups Ext. One method of doing this is due to Yoneda [37]. Let M, M', M'' be three A-modules, with resolutions C, C', C''. Let

$$f : C_s \to M', \qquad g : C'_{s'} \to M''$$

be A-maps of total degrees $-(s + t)$, $-(s' + t')$, such that $fd_{s+1} = 0$ and similarly for g, so that f, g represent elements of

$$\operatorname{Ext}^{s,t}_A(M, M'), \qquad \operatorname{Ext}^{s',t'}_A(M', M'').$$

Then we may form an anticommutative diagram, as follows.

$$
\begin{array}{c}
M \xleftarrow{\varepsilon} C_0 \leftarrow \cdots \leftarrow C_s \xleftarrow{d_{s+1}} C_{s+1} \leftarrow \cdots \cdots \leftarrow C_{s+s'} \leftarrow \cdots \\
\quad {}^{f}\swarrow \;\; \downarrow f_0 \quad\; \downarrow f_1 \qquad\qquad\qquad\qquad \downarrow f_{s'} \\
M' \xleftarrow{\varepsilon'} C'_0 \xleftarrow{d'_1} C'_1 \leftarrow \cdots \cdots \leftarrow C'_{s'} \leftarrow \cdots \\
\qquad\qquad\qquad\qquad\qquad\qquad\qquad\qquad \downarrow g \\
\qquad\qquad\qquad\qquad\qquad\qquad\qquad\qquad M''
\end{array}
$$

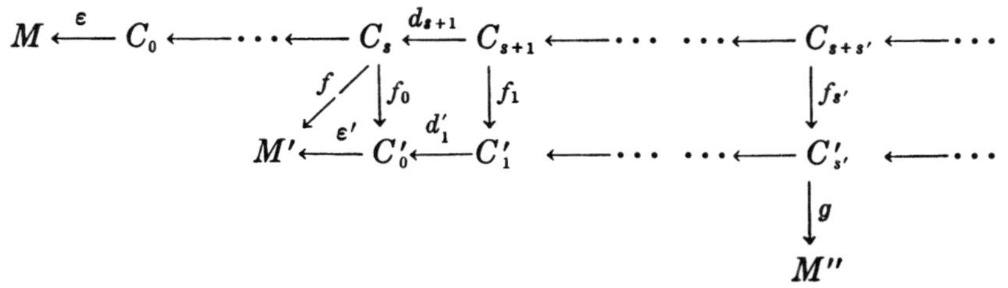

The composite map $(-1)^{(s+t)(s'+t')} g f_{s'}$ represents an element of

$$\operatorname{Ext}^{s+s',t+t'}_A(M, M'').$$

(The sign is introduced for convenience later.) By performing the obvious verifications, we see that this "composition" product gives us an invariantly-defined pairing from $\operatorname{Ext}^{s',t'}_A(M', M'')$ and $\operatorname{Ext}^{s,t}_A(M, M')$ to $\operatorname{Ext}^{s+s',t+t'}_A(M, M'')$. This product is bilinear and associative.

Our next lemma states an elementary relation between this product and the homomorphism

$$\theta : Z(s) \to \operatorname{Tor}^A_{s+1}(K, M)$$

introduced above. We first set up some data.

Suppose given a K-linear function $\alpha : A \to K$, of degree $-t'(t' > 0)$, and such that $\alpha(ab)=0$ if $a \in I(A)$, $b \in I(A)$. That is, α is a primitive element of A^*. It follows that $\alpha \mid I(A)$ is A-linear. Let us take a resolution C' of K, such that $C'_0 = A$ and $\varepsilon' : C'_0 \to K$ is the projection of A on A_0. Then the composite

$$C'_1 \xrightarrow{d'_1} A \xrightarrow{\alpha} K$$

is A-linear, and defines an element

$$h_\alpha \in \operatorname{Ext}^{1,t'}_A(K, K)$$

depending only on α.

Suppose given an element h of $\operatorname{Ext}^{s,t}_A(M, K)$; let C be a partial resolution of M over A, and define $Z(s) = \operatorname{Ker}(d_s) \cap J(C_s)$, as above. Let x be an element of $Z(s) \cap C_{s,t+t'}$, and suppose that x can be written in the form $x = \sum_i a_i c_i$, where $a_i \in I(A)$ and $d_s c_i \in J(C_{s-1})$. For example, if d_{s-1} is minimal, then we can always write x in this form, by taking the elements c_i from an A-base of C_s. In any case, we have

$$\{1 \otimes_A c_i\} \in \operatorname{Tor}^A_s(K, M) \ .$$

LEMMA 2.2.2.

$$(h_\alpha h)(\theta x) = \sum_i (\alpha a_i)(h\{1 \otimes_A c_i\}) \ .$$

Here, of course, the product $(h_\alpha h)(\theta x)$ can be formed because $\operatorname{Ext}^{s+1,t+t'}_A(M, K)$ and $\operatorname{Tor}^A_{s+1,t+t'}(K, M)$ are dual vector spaces; similarly for the product $h\{1 \otimes_A c_i\}$.

PROOF. In order to obtain the product $h_\alpha h$, it is proper to extend the partial resolution C and set up the following diagram, in which f is a representative cocycle for h.

$$\begin{array}{ccccccccc} M & \xleftarrow{\varepsilon} & C_0 & \longleftarrow & \cdots & \cdots & \longleftarrow & C_s & \xleftarrow{d_{s+1}} & C_{s+1} & \longleftarrow & \cdots \\ & & & & & & & {}_f\swarrow \downarrow f_0 & & \downarrow f_1 & & \\ & & & & & & K & \longleftarrow A & \xleftarrow{d'_1} & C'_1 & \longleftarrow & \cdots \\ & & & & & & & {}_\alpha\searrow \downarrow & & & & \\ & & & & & & & K & & & & \end{array}$$

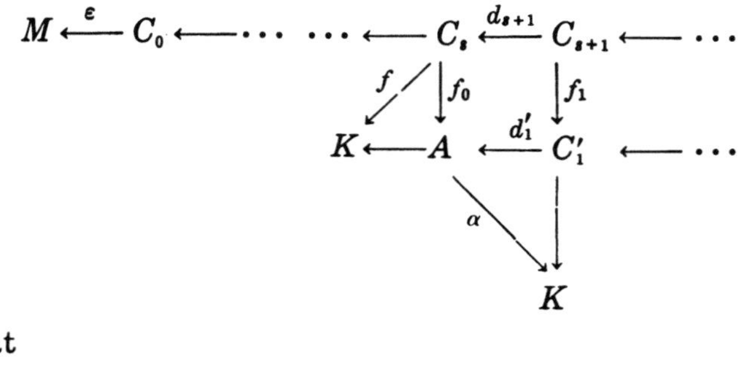

We see that

$$(-1)^{(s+1)(1+t')}(h_a h)(\theta x)$$
$$= \alpha d_1' f_1(d_{s+1}^{-1} x) = (-1)^{s+t}\alpha f_0 x = (-1)^{s+t}\alpha f_0 \sum_i a_i c_i = \alpha \sum_i (-1)^{\varepsilon(i)} a_i (f_0 c_i)$$
$$= \sum_i (-1)^{\varepsilon(i)} (\alpha a_i)(f c_i) \qquad \text{where } \varepsilon(i) = (s+t)(1+\deg(a_i)) .$$

But $h\{1 \otimes_A C_i\} = (fc_i)$, and the only terms which contribute to the sum have $\deg(a_i) = t'$. This proves the lemma.

We next introduce the bar construction, which gives us a standard resolution of K over A. Let us write $M \otimes M'$ for $M \otimes_K M'$. Then we set

$$\bar{A} = A/A_0 , \quad (\bar{A})^0 = K , \quad (\bar{A})^s = \bar{A} \otimes (\bar{A})^{s-1} \qquad (s > 0),$$
$$\bar{B}(A) = \sum_{s \geqq 0} (\bar{A})^s , \qquad B(A) = A \otimes \bar{B}(A) .$$

Thus, $B(A)$ is a free A-module, and $\bar{B}(A) \cong K \otimes_A B(A)$. We write the elements of $(\bar{A})^s$ and $A \otimes (\bar{A})^s$ in the forms

$$[a_1 \mid a_2 \mid \cdots \mid a_s], \qquad a[a_1 \mid a_2 \mid \cdots \mid a_s] .$$

We also write a for $a[\ ]$. We define an augmentation $\varepsilon : B(A) \to K$ by $\varepsilon(1) = 1$, $\varepsilon(I(A)) = 0$, $\varepsilon(A \otimes (\bar{A})^s) = 0$ if $s > 0$. We define a contracting homotopy S in $B(A)$ by $Sa_0[a_1 \mid a_2 \mid \cdots \mid a_s] = 1\,[a_0 \mid a_1 \mid a_2 \mid \cdots \mid a_s]$. We define a boundary d in $B(A)$ by the inductive formulae $d(1) = 0$, $dS + Sd = 1 - \varepsilon$, $d(am) = (-1)^q a(dm)$ $(a \in A_q)$. $B(A)$ thus becomes a free, acyclic resolution of K over A. We take the induced boundary $\bar{d}$ in $\bar{B}(A)$; its homology is therefore $\mathrm{Tor}^A(K, K)$.

Explicit forms for these boundaries are as follows (where $a_i \in I(A)$ for $i \geqq 1$.)

$$da_0[a_1 \mid a_2 \mid \cdots \mid a_s] = (-1)^{\varepsilon(0)} a_0 a_1 [a_2 \mid \cdots \mid a_s]$$
$$+ \sum_{1 \leqq r \leqq s} (-1)^{\varepsilon(r)} a_0 [a_1 \mid \cdots \mid a_r a_{r+1} \mid \cdots \mid a_s] ,$$
$$\bar{d}[a_1 \mid a_2 \mid \cdots \mid a_s] = \sum_{1 \leqq r \leqq s} (-1)^{\eta(r)} [a_1 \mid \cdots \mid a_r a_{r+1} \mid \cdots \mid a_s] ,$$

where

$$\varepsilon(r) = r + \sum_{0 \leqq i \leqq r} \deg(a_i) , \qquad \eta(r) = r + \sum_{1 \leqq i \leqq r} \deg(a_i) .$$

It would therefore be equivalent to set

$$I(A) = \sum_{q>0} A_q , \quad I(A)^0 = K , \quad I(A)^s = I(A) \otimes I(A)^{s-1} , \quad \bar{B}(A) = \sum_{s \geqq 0} I(A)^s$$

and define the boundary $d_s : I(A)^s \to I(A)^{s-1}$ in $\bar{B}(A)$ by the formula given above for $\bar{d}$.

It is now easy to obtain the vector-space dual $(\bar{B}(A))^*$ of $\bar{B}(A)$. Let A^* be the dual of A. We define $\overline{A^*} = A^*/A_0^*$; $\overline{A^*}$ is dual to $I(A)$; $(\overline{A^*})^s$ is dual to $(I(A))^s$. We may define

$$F(A^*) = \sum_{s \geqq 0} (\overline{A^*})^s ;$$

$F(A^*)$ is dual to $\bar{B}(A)$. We write elements of $F(A^*)$ in the form

$$[\alpha_1 \,|\, \alpha_2 \,|\cdots|\, \alpha_s]$$

where $\alpha_i \in A^*$.

From the product map $\varphi : A \otimes A \to A$ we obtain a diagonal map $\psi = \varphi^*$: $A^* \to A^* \otimes A^*$. (Here we define the pairing of $A^* \otimes A^*$ and $A \otimes A$ by $(\alpha \otimes \beta)(a \otimes b) = (\alpha a)(\beta b)$; we thus omit the sign introduced by Milnor [25]. Since the applications are in characteristic 2, this is immaterial). Let us write $\psi(\alpha) = \sum_u \alpha'_u \otimes \alpha''_u$; we may now define the coboundary

$$\bar{d}^s \colon (\overline{A^*})^{s-1} \to (\overline{A^*})^s$$

by

$$\bar{d}^s [\alpha_1 \,|\, \alpha_2 \,|\cdots|\, \alpha_{s-1}] = \sum\nolimits_{1 \leq r \leq s;\, u} (-1)^{\varepsilon(r,u)} [\alpha_1 \,|\cdots|\, \alpha'_{r,u} \,|\, \alpha''_{r,u} \,|\cdots|\, \alpha_{s-1}] \,,$$

where

$$\varepsilon(r, u) = r + \deg(\alpha'_{r,u}) + \sum\nolimits_{1 \leq i < r} \deg(\alpha_i) \,.$$

The coboundary $\bar{d}^s$ is dual to $\bar{d}_s$; the cohomology of $F(A^*)$ is therefore $\mathrm{Ext}_A(K, K)$. Of course, $F(A^*)$ is nothing but the cobar construction [1] on the coalgebra A^*.

There is a second method of introducing products, which uses the bar construction. In fact, we may define a cup-product of cochains in $F(A^*)$ by

$$[\alpha_1 \,|\, \alpha_2 \,|\cdots|\, \alpha_s][\alpha_{s+1} \,|\, \alpha_{s+2} \,|\cdots|\, \alpha_{s+s'}] = [\alpha_1 \,|\, \alpha_2 \,|\cdots|\, \alpha_s \,|\, \alpha_{s+1} \,|\cdots|\, \alpha_{s+s'}] \,.$$

It is clear that it is associative, bilinear and satisfies

$$\delta(xy) = (\delta x)y + (-1)^{s+t} x(\delta y)$$

(where $x = [\alpha_1 \,|\, \alpha_2 \,|\cdots|\, \alpha_s]$, $t = \sum_{1 \leq i \leq s} \deg(\alpha_i)$.) Therefore it induces an associative, bilinear product in the cohomology of $F(A^*)$, that is, in $\mathrm{Ext}_A(K, K)$. Indeed, this cup-product of cochains even allows us to define Massey products [23] [24] [35], etc.

We should show that this product coincides with the previous one (in case $M = M' = M'' = K$.) Let

$$f : A \otimes (\bar{A})^s \to K$$

be an A-linear map of degree $-(s+t)$ such that $f d_{s+1} = 0$. The previous method requires us to construct certain functions

$$f_{s'} : A \otimes (\bar{A})^{s+s'} \to A \otimes (\bar{A})^{s'} \,.$$

We may do this by setting

$$f_{s'}(a_0[a_1 \,|\, a_2 \,|\cdots|\, a_{s+s'}]) = (-1)^{\varepsilon} a_0[a_1 \,|\, a_2 \,|\cdots|\, a_{s'}] f [a_{s'+1} \,|\cdots|\, a_{s+s'}] \,,$$

where $\varepsilon = (s + t)(s' + t')$ and $t' = \sum_{0 \leq i \leq s'} \deg(a_i)$. From this it is immediate that the two products coincide.

A third method of defining products is useful under a different set of conditions. Before proceeding to state them, we give $A \otimes A$ the structure of a graded algebra by setting

$$\deg(a_1 \otimes a_2) = \deg(a_1) + \deg(a_2)$$

and

$$(a_1 \otimes a_2)(a_3 \otimes a_4) = (-1)^{\varepsilon}(a_1 a_3 \otimes a_2 a_4)$$

where

$$\varepsilon = \deg(a_2)\deg(a_3) \; .$$

Similarly, if M_1, M_2 are (graded) A-modules, we give $M_1 \otimes M_2$ the structure of a graded $A \otimes A$-module by setting

$$\deg(m_1 \otimes m_2) = \deg(m_1) + \deg(m_2)$$

and

$$(a_1 \otimes a_2)(m_1 \otimes m_2) = (-1)^{\varepsilon}(a_1 m_1 \otimes a_2 m_2)$$

where

$$\varepsilon = \deg(a_2)\deg(m_1) \; .$$

We now suppose that A is a Hopf algebra [25][26]. That is, there is given a "diagonal" or "co-product" map

$$\psi : A \to A \otimes A$$

which is a homomorphism of algebras. It is required that ψ should be "co-associative", in the sense that the following diagram is commutative.

$$\begin{array}{ccc} A & \xrightarrow{\psi} & A \otimes A \\ {\scriptstyle\psi}\downarrow & & \downarrow{\scriptstyle 1 \otimes \psi} \\ A \otimes A & \xrightarrow{\psi \otimes 1} & A \otimes A \otimes A \end{array}$$

(This diagram is obtained from that which expresses the associativity of a product map, by reversing the arrows.) Lastly, it is required that ψ should have a " co-unit", in a similar sense.

Let C be a resolution of K over A. We may form $C \otimes C$, and give it a first grading by setting

$$(C \otimes C)_{s''} = \sum_{s+s'=s''} C_s \otimes C_{s'} \; .$$

Here, each summand is a (graded) module over $A \otimes A$. We give $C \otimes C$ a boundary by the rule

$$d(x \otimes y) = dx \otimes y + (-1)^{s+t} x \otimes dy \qquad (x \in C_{s,t});$$

the complex $C \otimes C$ is acyclic. We may thus construct a map

$$\Delta : C \to C \otimes C$$

compatible with the map ψ of operations, and with the canonical isomorphism $K \to K \otimes K$. The map Δ induces a product

$$\mu : \mathrm{Hom}_A(C, K) \otimes \mathrm{Hom}_A(C, K) \to \mathrm{Hom}_A(C, K) .$$

By performing the obvious verifications, one sees that this product yields an invariantly-defined pairing from $\mathrm{Ext}_A^{s,t}(K, K)$ and $\mathrm{Ext}_A^{s',t'}(K, K)$ to $\mathrm{Ext}_A^{s+s',t+t'}(K, K)$. This product is bilinear and associative. Moreover, if the diagonal map ψ of A is anticommutative, then one easily sees (as in [13, Chapter XI]) that this product is anticommutative, the sign being $(-1)^{(s+t)(s'+t')}$.

We should next show that this product coincides with the previous one, defined using the bar construction. In fact, if we take $C = B(A)$, we may construct a map Δ by using the contracting homotopy

$$T = S \otimes 1 + \varepsilon \otimes S$$

in $B(A) \otimes B(A)$; we use the inductive formulae

$$\Delta(1) = 1 \otimes 1 , \qquad \Delta S = T\Delta ,$$
$$\Delta(am) = (\psi a)(\Delta m) .$$

Let us write $\psi(a_r) = \sum a_r' \otimes a_r''$, leaving the parameter in this summation to be understood; and suppose $\deg(a_r) > 0$ for $1 \leqq r \leqq q$. Then we find

$$\Delta[a_1 | a_2 | \cdots | a_q] = \sum_{0 \leq r \leq q} (-1)^{\varepsilon}[a_1' | a_2' | \cdots | a_r'] \otimes a_1'' a_2'' \cdots a_r''[a_{r+1} | \cdots | a_q]$$

where $\varepsilon = \sum_{1 \leq i < j \leq r} \deg(a_i'')(1 + \deg(a_j'))$. The resulting product μ in $\mathrm{Hom}_A(B(A), K)$ coincides with the cup-product in $F(A^*)$ given by

$$[\alpha_1 | \alpha_2 | \cdots | \alpha_r][\alpha_{r+1} | \alpha_{r+2} | \cdots | \alpha_q] = [\alpha_1 | \alpha_2 | \cdots | \alpha_r | \alpha_{r+1} | \cdots | \alpha_q] .$$

These two products in $\mathrm{Ext}_A(K, K)$ thus coincide. From this we can make two deductions. First, the product defined using the diagonal map ψ in A is independent of ψ. Secondly, if the algebra A should happen to admit an anticommutative diagonal ψ, then the cup-products defined using the bar construction are anticommutative. This would not be true for a general algebra A.

Let us now assume that the diagonal ψ in A is anticommutative; let us set $C = B(A)$, and let $\rho : C \otimes C \to C \otimes C$ be the map which permutes the two factors and introduces the appropriate sign. We will define an explicit chain homotopy χ between the maps

$$\Delta, \rho\Delta : C \to C \otimes C .$$

We do this by the following inductive formulae:

$$\chi(1) = 0$$
$$\chi S = T(\rho\Delta - \Delta)S - T\chi$$
$$\chi(am) = (-1)^{\deg a}\psi(a)\chi(m) .$$

We will now assume $K = Z_2$, since this is satisfied in the applications; we may thus omit the signs. We have the following explicit formula for χ:

$$\chi[a_1 | a_2 | \cdots | a_q] = \sum_{0 \leq t < r \leq q} [a'_1 | a'_2 | \cdots | a'_t | a'_{t+1}a'_{t+2} \cdots a'_r | a_{r+1} | a_{r+2} | \cdots | a_q] \otimes a''_1 a''_2 \cdots a''_t [a''_{t+1} | a''_{t+2} | \cdots | a''_r] .$$

(Here, of course, we have again used the convention that $\psi(a_r) = \sum a'_r \otimes a''_r$).

Passing to the complex $\bar{B}(A)$, and then to its dual $F(A^*)$, we obtain a product $\smile_1$ in $F(A^*)$, satisfying the usual formula

$$\delta(x \smile_1 y) = \delta x \smile_1 y + x \smile_1 \delta y + x \smile y + y \smile x .$$

To give an explicit form for this product, we recall that if A is a Hopf algebra, then its dual A^* is also a Hopf algebra, with ψ^* for product and φ^* for diagonal. This gives sense to the following explicit formula.

$$[\alpha_1 | \alpha_2 | \cdots | \alpha_p] \smile_1 [\beta_1 | \beta_2 | \cdots | \beta_q] = \sum_{1 \leq r \leq p} [\alpha_1 | \alpha_2 | \cdots | \alpha_{r-1} | \alpha_r^{(1)}\beta_1 | \alpha_r^{(2)}\beta_2 | \cdots | \alpha_r^{(q)}\beta_q | \alpha_{r+1} | \cdots | \alpha_p] .$$

Here we have written the iterated diagonal $\Psi : A^* \to (A^*)^q$ in the form

$$\Psi(\alpha_r) = \sum \alpha_r^{(1)} \otimes \alpha_r^{(2)} \otimes \cdots \otimes \alpha_r^{(q)} ,$$

the parameter in the summation being left to be understood.

In particular, we have

$$[\alpha_1] \smile_1 [\alpha_2] = [\alpha_1\alpha_2]$$
$$[\alpha_1 | \alpha_2] \smile_1 [\alpha_3] = [\alpha_1\alpha_3 | \alpha_2] + [\alpha_1 | \alpha_2\alpha_3]$$
$$[\alpha_1 | \alpha_2] \smile_1 [\alpha_3 | \alpha_4] = \sum [\alpha'_1\alpha_3 | \alpha''_1\alpha_4 | \alpha_2] + \sum [\alpha_1 | \alpha'_2\alpha_3 | \alpha''_2\alpha_4] .$$

This concludes the present survey of general notions.

2.3. *A spectral sequence.* In this section we establish a spectral sequence which is needed in our calculations. It arises in the following situation.

Let Γ be a (connected) Hopf algebra [26] over a field K, and let Λ be a Hopf subalgebra of Γ. We will suppose that Λ is *central* in Γ, in the sense that

$$ab = (-1)^{tu}ba \qquad \text{if } a \in \Lambda_t, b \in \Gamma_u .$$

This already implies that Λ is *normal* in Γ, in the sense that

$$I(\Lambda)\cdot\Gamma = \Gamma\cdot I(\Lambda) ,$$

where $I(\Lambda) = \sum_{t>0}\Lambda_t$ (cf. [13, Chapter XVI, §6]). We define the quotient $\Omega = \Gamma//\Lambda$ by

$$\Gamma//\Lambda = \Gamma/(I(\Lambda)\cdot\Gamma) \qquad \text{(cf. [13], loc. cit.)}$$

To simplify the notation, we define $H^{s,t}(A) = \mathrm{Ext}_A^{s,t}(K, K)$, $H^s(A) = \mathrm{Ext}_A^s(K, K)$, where A is a connected, graded algebra over K.

We now take the (bigraded) cochain complex $F(\Gamma^*)$, with the cup-product defined above. We filter it by setting

$$[\alpha_1 | \alpha_2 | \cdots | \alpha_s] \in F(\Gamma^*)^{(p)}$$

if α_i annihilates $I(\Lambda)$ for p values of i.

THEOREM 2.3.1. *This filtration of $F(\Gamma^*)$ defines a spectral sequence with cup-products, such that:*

(i) *E_∞ gives a composition-series for $H^*(\Gamma)$.*

(ii) $E_2 \cong H^*(\Lambda)\otimes H^*(\Omega)$.

Here the ring-structure of the right-hand side is defined by

$$(x\otimes y)(z\otimes w) = (-1)^{(p+t)(q+u)}(xz\otimes yw) ,$$

where $y \in H^{p,t}(\Omega)$, $z \in H^{q,u}(\Lambda)$.

(iii) *The isomorphism*

$$E_2^{0,q} \cong H^q(\Lambda)\otimes K \cong H^q(\Lambda)$$

is induced by the natural map $F(\Gamma^*) \to F(\Lambda^*)$. *The isomorphism*

$$E_2^{p,0} \cong K\otimes H^p(\Omega) \cong H^p(\Omega)$$

is induced by the natural map $F(\Omega^*) \to F(\Gamma^*)$.

This spectral sequence was used in [2]; the author supposes that it coincides with that given in [13, Chapter XVI], but this is not relevant to the applications.

In [13] it is assumed that Γ is free (or at least projective) as a (left or right) module over Λ. In our case, this follows from the assumption that Γ is a Hopf algebra and Λ a Hopf subalgebra. Thus, if $\{\omega_i\}$ is a K-base for Ω, with $\omega_1 = 1$, and if γ_i is a representative for ω_i in Γ, with $\gamma_1 = 1$, then $\{\gamma_i\}$ is a (left or right) Λ-base for Γ (see [26]). This is the only use we make of the diagonal structure of Γ and Λ.

We begin the proof of the theorem in homology, by considering the (bigraded) chain complex $\bar{B}(\Gamma)$. We filter it by setting

$$[a_1 | a_2 | \cdots | a_s] \in \bar{B}(\Gamma)^{(p)}$$

if $a_i \in I(\Lambda)$ for $(s-p)$ values of i. Each $\bar{B}(\Gamma)^{(p)}$ is closed for $\bar{d}$; we thus obtain a spectral sequence, whose term E^∞ gives a composition series for $H_*(\Gamma) = \mathrm{Tor}^\Gamma(K, K)$. We have to calculate E^2. To this end, we begin by calculating the homology of certain subcomplexes of $B(\Gamma)$.

Let us consider

$$\Lambda \otimes \bar{B}(\Gamma)^{(p)} + \Gamma \otimes \bar{B}(\Gamma)^{(p-1)}, \qquad \Gamma \otimes \bar{B}(\Gamma)^{(p-1)} .$$

Both are closed for d, and the first is also closed for S, hence acyclic. Consider

$$C^{(p)} = \Lambda \otimes \bar{B}(\Gamma)^{(p)} + \Gamma \otimes \bar{B}(\Gamma)^{(p-1)} / \Gamma \otimes \bar{B}(\Gamma)^{(p-1)} .$$

LEMMA 2.3.2.

$$H_s(C^{(p)}) \cong \begin{cases} (\bar{\Omega})^p & (s = p) \\ 0 & (s \neq p) . \end{cases}$$

The isomorphism for $s = p$ is obtained by projecting Λ to K and $(\bar{\Gamma})^p$ to $(\bar{\Omega})^p$.

(Here, of course, the suffix s refers to the first grading.)

PROOF. This is certainly true for $p = 0$, since $\Lambda \otimes \bar{B}(\Gamma)^{(0)} = B(\Lambda)$. As an inductive hypothesis, suppose it true for p. Consider the following chain complexes.

$$C' = (\Gamma \otimes \bar{B}(\Gamma)^{(p)}) / (\Lambda \otimes \bar{B}(\Gamma)^{(p)} + \Gamma \otimes \bar{B}(\Gamma)^{(p-1)})$$

$$C'' = (\Lambda \otimes \bar{B}(\Gamma)^{(p+1)} + \Gamma \otimes \bar{B}(\Gamma)^{(p)}) / (\Lambda \otimes \bar{B}(\Gamma)^{(p)} + \Gamma \otimes \bar{B}(\Gamma)^{(p-1)})$$

$$C^{(p+1)} = (\Lambda \otimes \bar{B}(\Gamma)^{(p+1)} + \Gamma \otimes \bar{B}(\Gamma)^{(p)}) / (\Gamma \otimes \bar{B}(\Gamma)^{(p)})$$

We have an exact sequence

$$0 \to C' \to C'' \to C^{(p+1)} \to 0$$

and hence an exact homology sequence. We know that $H_*(C'') = 0$; we may find $H_*(C')$ by expressing C' as a direct sum. Let $\{\gamma_i\}$ be a right Λ-base for Γ, with $\gamma_1 = 1$, as above; we may define an antichain map $f_i : C^{(p)} \to C'$ by

$$f_i(x) = \gamma_i x;$$

f_i is monomorphic (if $i > 1$) and

$$C' = \sum_{i>1} f_i(C^{(p)}) .$$

We deduce that

$$H_s(C') \cong \begin{cases} (\bar{\Omega})^{p+1} & (s = p) \\ 0 & (s \neq p) . \end{cases}$$

The isomorphism for $s = p$ is obtained by projecting Γ to $\bar{\Omega}$ and $(\bar{\Gamma})^p$ to $(\bar{\Omega})^p$. Now, in the exact homology sequence, we have

$$d\colon H^{s+1}(C^{(p+1)}) \cong H^s(C') \,;$$

from this we see that the lemma is true for $(p+1)$. This proves the lemma, by induction over p.

We now note that $C^{(p)}$ is a free (left) Λ-module, and

$$\bar{B}(\Gamma)^{(p)}/\bar{B}(\Gamma)^{(p-1)} \cong K \otimes_\Lambda C^{(p)};$$

this isomorphism is a chain map. Therefore the term E^1 of the spectral sequence is given by

$$\begin{aligned} E^1_{p,q} &= H_{p+q}(\bar{B}(\Gamma)^{(p)}/\bar{B}(\Gamma)^{(p-1)}) \\ &\cong H_{p+q}(K \otimes_\Lambda C^{(p)}) \\ &\cong \operatorname{Tor}^\Lambda_q(K, (\bar{\Omega})^p) \\ &\cong \operatorname{Tor}^\Lambda_q(K, K) \otimes (\bar{\Omega})^p \,. \end{aligned}$$

The last step uses the fact that $(\bar{\Omega})^p$, qua $H_p(C^{(p)})$, has trivial operations from Λ (see Lemma 2.3.2.)

In order to calculate $E^2_{p,q}$, we need explicit chain equivalences between

$$\bar{B}(\Lambda) \otimes \bar{\Omega}^p \quad \text{and} \quad \bar{B}(\Gamma)^{(p)}/\bar{B}(\Gamma)^{(p-1)} \,.$$

In one direction the map is easy. Let $\pi : \Gamma \to \Omega$ be the projection. Define a map

$$\nu\colon C^{(p)} \to B(\Lambda) \otimes (\bar{\Omega})^p$$

by

$$\nu a[a_1 \,|\, a_2 \,|\cdots|\, a_{p+q}] = a[a_1 \,|\cdots|\, a_q] \otimes [\pi a_{q+1} \,|\cdots|\, \pi a_{p+q}] \,.$$

Then ν is a Λ-map, a chain map (if the boundary in $B(\Lambda) \otimes (\bar{\Omega})^p$ is $d \otimes 1$) and induces the isomorphism of homology established above. By the uniqueness theorem in homological algebra, the induced map

$$\nu_*\colon \bar{B}(\Gamma)^{(p)}/\bar{B}(\Gamma)^{(p-1)} \to \bar{B}(\Lambda) \otimes (\bar{\Omega})^p$$

is a chain equivalence.

We have not yet made any essential use of the fact that Λ is central in Γ. However, this fact is required; it ensures that $H_*(\Lambda)$ (which is analogous to the homology of the fibre) has simple operations from Ω. We use it in constructing the equivalence in the other direction.

We define a product

$$\mu\colon B(\Lambda) \otimes B(\Gamma) \to B(\Gamma)$$

as follows, using shuffles [14]. Take elements

$$a[a_1 | a_2 | \cdots | a_q] \in B(\Lambda)$$
$$b[b_1 | b_2 | \cdots | b_p] \in B(\Gamma) .$$

Let $c_1, c_2, \cdots, c_{p+q}$ consist of the a_i and b_j in some order; we require that the a_i occur in their correct relative order and that the b_j occur in their correct relative order. We call such a set $c = \{c_k\}$ a *shuffle*. Let us write $\sum n \mid P$ instead of $\sum_P n$ if the proposition P has a complicated form. Then we define the *signature* $(-1)^{\varepsilon(c)}$ of a shuffle by

$$\varepsilon(c) = \sum (1 + \deg(a_i))(1 + \deg(b_j)) \mid a_i = c_k,\ b_j = c_l,\ k > l .$$

We also set

$$\eta = \sum_i (1 + \deg(a_i))(\deg(b)) .$$

We now define

$$\mu(a[a_1 | a_2 | \cdots | a_q] \otimes b[b_1 | b_2 | \cdots | b_p]) = \sum_c (-1)^{\varepsilon(c)+\eta} ab[c_1 | c_2 | \cdots | c_{p+q}] .$$

LEMMA 2.3.3. $d\mu(x \otimes y) = \mu(dx \otimes y) + (-1)^{q+t}\mu(x \otimes dy)$ (*where* $x \in B(\Lambda)_{q,t}$).

This lemma, of course, is the usual one for shuffle-products (see [14]); it depends on the fact that Λ is central in Γ.

If we restrict y to lie in $(\bar{\Gamma})^p$, and pass to the tensor-product with K, we obtain an induced map

$$\mu_* : \bar{B}(\Lambda) \otimes (\bar{\Gamma})^p \rightarrow \bar{B}(\Gamma)^{(p)} .$$

Its explicit form is

$$\mu_*([a_1 | a_2 | \cdots | a_q] \otimes [b_1 | b_2 | \cdots | b_p]) = \sum_c (-1)^{\varepsilon(c)} [c_1 | c_2 | \cdots | c_{p+q}] .$$

It satisfies

$$\bar{d}\mu_*(x \otimes y) = \mu_*(\bar{d}x \otimes y) + (-1)^{q+t}\mu_*(x \otimes \bar{d}y) .$$

We define a K-map $l : \bar{\Omega} \rightarrow \bar{\Gamma}$ by $l(\omega_i) = \gamma_i$, where $\{\omega_i\}$, $\{\gamma_i\}$ are bases for Ω, Γ over K, Λ, as above. We may now define

$$\mu' : B(\Lambda) \otimes (\bar{\Omega})^p \rightarrow C^{(p)}$$

by

$$\mu'(x \otimes y) = \{\mu(x \otimes l^p y)\} .$$

Then μ' is a Λ-map, a chain map, and induces the correct isomorphism of homology. On passing to tensor products with K, we obtain an induced map

$$\mu'_* : \bar{B}(\Lambda)\otimes(\bar{\Omega})^p \to \bar{B}(\Gamma)^{(p)}/\bar{B}(\Gamma)^{(p-1)}$$

which, as before, is a chain equivalence. We even have $\nu_*\mu'_* = 1$.

The equivalence μ'_*, then, is just the composite

$$\bar{B}(\Lambda)\otimes(\bar{\Omega})^p \xrightarrow{1\otimes l^p} \bar{B}(\Lambda)\otimes(\bar{\Gamma})^p \xrightarrow{\mu_*} \bar{B}(\Gamma)^{(p)} \longrightarrow \bar{B}(\Gamma)^{(p)}/\bar{B}(\Gamma)^{(p-1)} .$$

LEMMA 2.3.4. *The following diagram is commutative.*

$$\begin{array}{ccc} E^1_{p,q} & \xrightarrow{\quad d^1 \quad} & E^1_{p-1,q} \\ \uparrow{\scriptstyle \mu'_{**}} & & \downarrow{\scriptstyle \nu_{**}} \\ H_q(\bar{B}(\Lambda)\otimes(\bar{\Omega})^p) & \xrightarrow{(-1)^{q+t}(1\otimes\bar{d})} & H_q(\bar{B}(\Lambda)\otimes(\bar{\Omega})^{p-1}) \end{array}$$

(It is, of course, implied that this ν_{**} is the one defined for dimension $(p-1)$.)

PROOF. Take $x \in \bar{B}(\Lambda)_{q,t}$ such that $\bar{d}x = 0$, and $y \in (\bar{\Omega})^p$, so that $x\otimes y$ is a representative for an element of $H_q(B(\Lambda)\otimes(\bar{\Omega})^p)$. Then the following elements represent $\nu_{**}d^1\mu'_{**}\{x\otimes y\}$:

$$\begin{aligned} \nu_*\bar{d}\mu_*(1\otimes l^p)(x\otimes y) &= \nu_*\mu_*((-1)^{q+t}x\otimes\bar{d}l^p y) \\ &= (-1)^{q+t}(1\otimes\pi^{p-1})(x\otimes\bar{d}l^p y) \\ &= (-1)^{q+t}(x\otimes\bar{d}\pi^p l^p y) \\ &= (-1)^{q+t}(x\otimes\bar{d}y) . \end{aligned}$$

This proves the lemma.

We conclude that

$$E^2_{p,q} \cong H_q(\bar{B}(\Lambda))\otimes H_p(\bar{B}(\Omega)) .$$

Inverse isomorphisms are induced by μ'_{**} and ν_{**}. We note that the isomorphisms

$$\begin{aligned} H_q(\bar{B}(\Lambda)) &\cong H_q(\bar{B}(\Lambda)\otimes K) \to E^2_{0,q} \\ E^2_{p,0} \to K\otimes H_p(B(\bar{\Omega})) &\cong H_p(B(\bar{\Omega})) \end{aligned}$$

are induced by the natural maps $\bar{B}(\Lambda)\to\bar{B}(\Gamma)$, $\bar{B}(\Gamma)\to\bar{B}(\Omega)$.

Let us now pass to the vector-space dual of this spectral sequence. It is obtained by giving $F(\Gamma^*)$ a filtration in which $F(\Gamma^*)^{(p)}$ is the annihilator of $\bar{B}(\Gamma)^{(p-1)}$. This is the filtration originally described. The cup-products satisfy

$$F(\Gamma^*)^{(p)}\cdot F(\Gamma^*)^{(p')} \subset F(\Gamma^*)^{(p+p')} .$$

We thus have a spectral sequence with products. We have obtained the whole of Theorem 2.3.1, except that part which relates to the ring-structure of E_2.

To obtain this ring-structure, we consider the isomorphism

$$E_1^{p,q} \cong H^q(F(\Lambda^*)) \otimes (\bar{\Omega}^*)^p$$

dual to μ'_{**} and ν_{**}. It may be described as follows. Let x be a cocycle of dimension q in $F(\Lambda^*)$, and let y be a cochain in $(\bar{\Omega}^*)^p$. Let x' be a cochain in $F(\Gamma^*)$ such that $i^*x' = x$, where $i : \Lambda \to \Gamma$ is the inclusion. Then $\{x' \cdot \pi y\}$ is an element of $E_1^{p,q}$, independent of the choice of x'; and ν^{**}, the dual of ν_{**}, maps $\{x\} \otimes y$ to $\{x' \cdot \pi^* y\}$.

Next, let x, z be cocycles in $F(\Lambda^*)$, representing elements of $H^{q',u'}(\Lambda)$, $H^{q,u}(\Lambda)$. Let y, w be cochains in $F(\Omega^*)$, of bidegrees (p, t), (p', t'). Let x', z' be cochains in $F(\Gamma^*)$ such that $i^*x' = x$, $i^*z' = z$. Then

$$\nu^{**}(\{x\} \otimes y) \cdot \nu^{**}(\{z\} \otimes w) = \{x' \cdot \pi^* y \cdot z' \cdot \pi^* w\} .$$

Now let X be a cycle in $\bar{B}(\Lambda)$, of dimension $q+q'$, and let Y be a chain in $\bar{B}(\Omega)$, of dimension $p+p'$. Inspecting the definition of the shuffle-product, we see

$$\{x' \cdot \pi^* y \cdot z' \cdot \pi^* w\} \cdot \mu'_{**}(\{X\} \otimes Y) = (-1)^{(p+t)(q+u)}(\{xz\}\ \{X\})((yw)Y) .$$

That is,

$$\mu'_{**}(\nu^{**}(\{x\} \otimes y) \cdot \nu^{**}(\{z\} \otimes w)) = (-1)^{(p+t)(q+u)}\{xz\} \otimes yw .$$

We have shown that the isomorphism

$$E_1^{p,q} \cong H^q(F(\Lambda^*)) \otimes (\bar{\Omega}^*)^p$$

preserves the ring-structure. Therefore the induced isomorphism of $E_2^{p,q}$ does so. This completes the proof of Theorem 2.3.1.

2.4. *Milnor's description of A.* In this section we recall J. Milnor's elegant description of the Steenrod algebra A [25], and begin to deduce from it the results we shall need later.

We recall that the mod p Steenrod algebra A is a Hopf algebra; that is, besides having a product map $\varphi : A \otimes A \to A$, it has a diagonal map $\psi : A \to A \otimes A$, and these satisfy certain axioms. The diagonal ψ may be described as follows. We have an isomorphism

$$\nu : H^*(X \times Y; Z_p) \to H^*(X; Z_p) \otimes H^*(Y; Z_p)$$

given by the external cup-product. The left-hand side admits operations from A; the right-hand side admits operations from $A \otimes A$, defined as in § 2.3. There is one and only one function

$$\psi : A \to A \otimes A$$

such that

$$\nu(ah) = \psi(a)\nu(h)$$

for all X, Y, a and h. (This may be shown, for example, by the method of § 3.9.)

We make the identification $(A \otimes A)^* = A^* \otimes A^*$; A^* thus becomes a Hopf algebra, whose product and diagonal maps are the duals of ψ and φ. We now quote Milnor's theorem [25] on the structure of A^*, in the case $p = 2$.

THEOREM 2.4.1.

(i) *A^* is a polynomial algebra on generators ξ_i, $i=1, 2, \cdots$, of grading $2^i - 1$.*

(ii) $\varphi^* \xi_k = \sum_{i+j=k} \xi_j^{2^i} \otimes \xi_i$ *(where $\xi_0 = 1$).*

(iii) $\xi_1^k(\mathrm{Sq}^k) = 1$ *and* $m(\mathrm{Sq}^k) = 0$ *for any other monomial m in the ξ_i.*

It is possible to describe the elements ξ_i very simply. In fact, consider $H^*(\pi, 1; Z_2)$ for $\pi = Z_2$; this is a polynomial algebra on one generator x of dimension 1. If $a \in A$, then ax is primitive, so that

$$ax = \sum_{i \geqq 0} \lambda_i x^{2^i} \qquad \text{with } \lambda_i \in Z_2 .$$

We define ξ_i by $\xi_i(a) = \lambda_i$. The elements ξ_i are thus closely connected with the Thom-Serre-Cartan representation [30], [12] of A. In fact, consider $H^*(\pi, 1; Z_2)$, where π is a finite vector space over Z_2; this is a polynomial algebra on generators $x_1, x_2, \cdots, x_n$ of dimension 1. If $a \in A$, we have

$$a(x_1 x_2 \cdots x_n) = \sum_{i,j,\ldots,l} ((\xi_i \xi_j \cdots \xi_l)a) x_1^{2^i} x_2^{2^j} \cdots x_n^{2^l} .$$

We now pass on to the study of certain quotient algebras of A. It is immediate that the generators ξ_i of A^* which satisfy $1 \leqq i \leqq n$ generate a Hopf subalgebra of A^*; call it A_n^*. The dual of A_n^* is a Hopf algebra Q_n, which is a quotient of A. We have

$$H^{s,t}(Q_n) \cong H^{s,t}(A) \qquad \text{if } t \leqq 2^{n+1} - 2 .$$

For each n, Q_n is a quotient of Q_{n+1}.

Now, suppose we are given an epimorphism $\pi : \Gamma \to \Omega$ of connected Hopf algebras, and suppose that the dual monomorphism $\pi^* : \Omega^* \to \Gamma^*$ embeds Ω^* as a normal subalgebra of Γ^* (see § 2.3). Then we may form the quotient $\Gamma^*//\Omega^*$, which is again a Hopf algebra; let us call its dual Λ, so that $\Lambda^* = \Gamma^*//\Omega^*$; then Λ is embedded monomorphically in Γ.

LEMMA 2.4.2. *If Λ, Γ, Ω are as above, then Λ is normal in Γ, and $\Gamma//\Lambda \cong \Omega$.*

This lemma is an exercise in handling Hopf algebras [26]; we only sketch the proof. It is trivial that

$$I(\Lambda)\cdot\Gamma \subset \operatorname{Ker}\pi\,, \qquad \Gamma\cdot I(\Lambda) \subset \operatorname{Ker}\pi;$$

we have to prove the opposite inclusions. We see that Λ is the kernel of the composite map

$$\Gamma \xrightarrow{\ \psi\ } \Gamma\otimes\Gamma \longrightarrow \Gamma\otimes\bar{\Omega};$$

call this composite χ. We see that if

$$\chi(x) \in \textstyle\sum_{r\geqq m}\Gamma_r\otimes\bar{\Omega}\,,$$

then

$$\chi(x) \in \Lambda_m\otimes\bar{\Omega} + \textstyle\sum_{r>m}\Gamma_r\otimes\bar{\Omega}\,.$$

Now suppose $\pi x = 0$. By an inductive process, using $\chi(x)$, we may subtract from x products yz with $y \in \Lambda_m$, $m>0$; we finally obtain a new x' with $\chi(x') = 0$. This shows that $\operatorname{Ker}\pi \subset I(\Lambda)\cdot\Gamma$; similarly for $\Gamma\cdot I(\Lambda)$.

We may apply this lemma to the epimorphism $Q_n \to Q_{n-1}$ introduced above. We see that in this application, Γ^* and Ω^* become A_n^* and A_{n-1}^*. Thus $\Lambda^* = \Gamma^*//\Omega^*$ becomes $A_n^*//A_{n-1}^*$, that is, a polynomial algebra on one generator ξ_n, whose diagonal is given by

$$\psi\xi_n = \xi_n\otimes 1 + 1\otimes\xi_n\,.$$

We write K_n for the corresponding algebra A; it is a divided polynomial algebra.

Since we propose to apply the spectral sequence of § 2.3 to the case $\Lambda = K_n$, $\Gamma = Q_n$, we should show that K_n is central in Q_n. We may proceed in the duals, by showing that the following diagram is commutative.

$$\begin{array}{ccc}
A_n^* & \xrightarrow{\ \psi\ } & A_n^*\otimes A_n^*//A_{n-1}^* \\
 & \searrow{\scriptstyle\psi} & \downarrow{\scriptstyle\rho} \\
 & & A_n^*//A_{n-1}^*\otimes A_n^*
\end{array}$$

(Here, of course, ρ is the map which permutes the two factors.) Since each map is multiplicative, we need only check the commutativity on the generators, for which it is immediate.

The reader may care to compare the work of this section with that of [2, § 5].

2.5. *Calculations*. In this section we prove what we need about the cohomology of the (mod 2) Steenrod algebra A. The result is:-

THEOREM 2.5.1.

(0) *$H^0(A)$ has as a base the unit element* 1.

(1) *$H^1(A)$ has as a base the elements $h_i = \{[\xi_1^{2^i}]\}$ for $i = 0, 1, 2, \cdots$.*

(2) *In* $H^2(A)$ *we have* $h_{i+1}h_i = 0$. $H^2(A)$ *admits as a base the products* $h_j h_i$ *for which* $j \geqq i \geqq 0$ *and* $j \neq i+1$.

(3) *In* $H^3(A)$ *we have the relations*

$$h_{i+2}h_i^2 = h_{i+1}^3, \qquad h_{i+2}^2 h_i = 0 .$$

If we take the products $h_k h_j h_i$ *for which* $k \geqq j \geqq i \geqq 0$ *and remove the products*

$$h_{j+1}h_j h_i , \qquad h_k h_{i+1} h_i , \qquad h_{i+2}h_i^2 , h_{i+2}^2 h_i ,$$

then the remaining products are linearly independent in $H^3(A)$.

We propose to prove this theorem by considering a family of spectral sequences. The nth spectral sequence, say ${}_nE_r^{p,q}$, will be obtained by applying § 2.3 to the algebras

$$\Lambda = K_n , \qquad \Gamma = Q_n , \qquad \Omega = Q_{n-1} .$$

In these spectral sequences, we know $H^*(K_n)$; we have $Q_1 = K_1$, so that $H^*(Q_1)$ is known; we propose to obtain information about $H^*(Q_n)$ by induction over n.

We will now give names to the cohomology classes which will appear in our calculations. Let $h_{n,i}$ be the generator $\{[\xi_n^{2^i}]\}$ in $H^1(K_n)$; $H^*(K_n)$ is thus a polynomial algebra on the generators $h_{n,i}$, where $i = 0, 1, 2, \cdots$. We shall write h_i (instead of $h_{1,i}$) for the generator $\{[\xi_1^{2^i}]\}$ in $H^1(Q_1)$, or for its image in $H^1(Q_n)$.

We define the class $g_{n-1,i}$ in $H^2(Q_{n-1})$ by

$$g_{n-1,i} = {}_n\tau h_{n,i} ,$$

where ${}_n\tau : H^1(K_n) \to H^2(Q_{n-1})$ is the transgression. The class $g_{n-1,i}$ can be represented by the explicit cocycle

$$\delta[\xi_n^{2^i}] = \sum [\xi_j^{2^{k+i}} \mid \xi_k^{2^i}]$$

where the sum extends over $j + k = n$, $j > 0$, $k > 0$.

Similarly, we can define a class $f_{n-1,i}$ in $H^3(Q_{n-1})$ by

$$f_{n-1,i} = {}_n\tau h_{n,i}^2 ,$$

since $h_{n,i}^2$ is clearly transgressive. The class $f_{n-1,i}$ can be represented by the explicit cocycle

$$\delta(x \smile x + \delta x \smile_1 x) = \delta x \smile_1 \delta x ,$$

where $x = [\xi_n^{2^i}]$, so that δx is the cocycle obtained above. The $\smile_1$ product $\delta x \smile_1 \delta x$ can be expanded by the formula given at the end of § 2.2.

LEMMA 2.5.2.

(i) *We have* $g_{1,i} = h_{i+1}h_i$, $f_{1,i} = h_{i+2}h_i^2 + h_{i+1}^3$.

(ii) *If* $n > 1$, *then* $g_{n,i}$ *is of filtration* 1 *in* $H^*(Q_n)$, $h_{n,i+1}h_i + h_{n,i}h_{n+i}$ *is a cycle in* ${}_nE_2^{1,1}$, *and in* ${}_nE_\infty^{1,1}$ *we have*

$$\{g_{n,i}\} = \{h_{n,i+1}h_i + h_{n,i}h_{n+i}\} .$$

(iii) *If* $n>1$, *then* $f_{n,i}$ *is of filtration* 1 *in* $H^*(Q_n)$, $h^2_{n,i+1}h_{i+1}+h^2_{n,i}h_{n+i+1}$ *is a cycle in* ${}_nE_2^{1,2}$ *and in* ${}_nE_3^{1,2}$, *and in* ${}_nE_\infty^{1,2}$ *we have*

$$\{f_{n,i}\} = \{h^2_{n,i+1}h_{i+1} + h^2_{n,i}h_{n+i+1}\} .$$

These conclusions follow from the explicit cocycles given above. For example, the explicit cocycle for $g_{1,i}$ is

$$[\xi_1^{2^{i+1}} \mid \xi_1^{2^i}] ;$$

the explicit cocycle for $f_{1,i}$ is

$$[\xi_1^{2^{i+2}} \mid \xi_1^{2^i} \mid \xi_1^{2^i}] + [\xi_1^{2^{i+1}} \mid \xi_1^{2^{i+1}} \mid \xi_1^{2^{i+1}}] ;$$

the explicit cocycle for $g_{n,i}$ differs from

$$[\xi_n^{2^{i+1}} \mid \xi_1^{2^i}] + [\xi_1^{2^{n+i}} \mid \xi_n^{2^i}]$$

by a cochain of filtration 2; and so on.

We will now begin the calculations.

LEMMA 2.5.3.

(i) *The elements* h_i *in* $H^1(Q_n)$ *are linearly independent.*

(ii) *If* $n > 1$, *the elements* $\{g_{n,i}\}$ *in* ${}_nE_\infty^{1,1}$ *are linearly independent.*

(iii) *In* $H^2(Q_n)$, *the elements* $g_{n,k}$ *and* h_jh_i (*where* $j \geqq i \geqq 0, j \neq i + 1$) *are linearly independent.*

(iv) $H^1(Q_n)$ *is spanned by the elements* h_i.

PROOF. Part (i) is immediate, since no differential maps into ${}_nE_r^{1,0}$.

Part (ii) follows from Lemma 2.5.2 (ii), since no differential maps into ${}_nE_r^{1,1}$.

Part (iii) is true for $n = 1$, since $g_{1,k} = h_{k+1}h_k$. We proceed by induction over n; let us assume that part (iii) holds for Q_{n-1}. We must examine the differential

$${}_nd_2 \colon {}_nE_2^{0,1} \longrightarrow {}_nE_2^{2,0} .$$

It is described by

$${}_nd_2(h_{n,i}) = g_{n-1,i} .$$

We conclude that:-

(a) ${}_nE_\infty^{0,1} = 0$.

(b) The classes $\{h_jh_i\}$ (for $j \geqq i \geqq 0$, $j \neq i + 1$) are linearly independent in ${}_nE_\infty^{2,0}$.

Using part (ii), we see that part (iii) holds for Q_n.

Part (iv) follows immediately from the fact that ${}_nE_\infty^{0,1} = 0$. This completes the proof.

LEMMA 2.5.4. *In $H^2(Q_n)$ for $n \geqq 2$ we have*

(i) $h_{i+1}h_i = 0$

(ii) $\langle h_i, h_{i+1}, h_i\rangle = h_{i+1}^2$

(iii) $\langle h_{i+1}, h_i, h_{i+1}\rangle = h_{i+2}h_i$.

In $H^2(Q_2)$ we have

(iv) $\langle h_{i+2}, h_{i+1}, h_i\rangle = g_{2,i}$

(v) $h_{i+2}^2h_i = g_{2,i}h_{i+1}$.

PROOF. The following formulae show that $h_{i+1}h_i = 0$, $h_ih_{i+1} = 0$ in $H^2(Q_n)$ for $n \geqq 2$:

$$\delta[\xi_2^{2^i}] = [\xi_1^{2^{i+1}} \mid \xi_1^{2^i}]$$

$$\delta[\xi_2^{2^i} + \xi_1^{3\cdot 2^i}] = [\xi_1^{2^i} \mid \xi_1^{2^{i+1}}] .$$

These formulae will also help us to write down explicit cocycles representing the Massey products mentioned in the lemma. For example, $\langle h_{i+2}, h_{i+1}, h_i\rangle$ is represented by the explicit cocycle

$$[\xi_2^{2^{i+1}} \mid \xi_1^{2^i}] + [\xi_1^{2^{i+2}} \mid \xi_2^{2^i}] ,$$

which coincides with that given above for $g_{2,i}$. This proves (iv).

To prove (ii) and (iii), we may quote the formula

$$\langle x, y, x\rangle = (x \smile_1 x)y ;$$

or by substituting appropriate values in the proof of this formula, we obtain the following:-

$$\delta[\xi_1^{2^i}\xi_2^{2^i}] = [\xi_2^{2^i} + \xi_1^{3\cdot 2^i} \mid \xi_1^{2^i}] + [\xi_1^{2^i} \mid \xi_2^{2^i}] + [\xi_1^{2^{i+1}} \mid \xi_1^{2^{i+1}}]$$

$$\delta[\xi_1^{2^{i+1}}\xi_2^{2^i}] = [\xi_2^{2^i} \mid \xi_1^{2^{i+1}}] + [\xi_1^{2^{i+1}} \mid \xi_2^{2^i} + \xi_1^{3\cdot 2^i}] + [\xi_1^{2^{i+2}} \mid \xi_1^{2^i}] .$$

These formulae prove (ii) and (iii).

Since we now know $H^1(Q_n)$, it it easy to check that the Massey products considered above are defined modulo zero.

To prove (v), it is sufficient to make the following manipulation:

$$\begin{aligned} h_{i+2}^2h_i &= h_{i+2}\langle h_{i+1}, h_i, h_{i+1}\rangle \\ &= \langle h_{i+2}, h_{i+1}, h_1\rangle h_{i+1} = g_{2,i}h_{i+1} . \end{aligned}$$

Alternatively, by substituting appropriate values in the proof of the relation

$$a\langle b, c, d\rangle = \langle a, b, c\rangle d ,$$

we obtain the following:-

$$\begin{aligned}\delta[\xi_2^{2^{i+1}} \mid \xi_2^{2^i} + \xi_1^{3\cdot 2^i}] &+ \delta[\xi_1^{2^{i+2}} \mid \xi_1^{2^{i+1}}\xi_2^{2^i}] \\ &= [\xi_1^{2^{i+2}} \mid \xi_1^{2^{i+2}} \mid \xi_1^{2^i}] + [\xi_2^{2^{i+1}} \mid \xi_1^{2^i} \mid \xi_1^{2^{i+1}}] + [\xi_1^{2^{i+2}} \mid \xi_2^{2^i} \mid \xi_1^{2^{i+1}}] .\end{aligned}$$

This formula proves (v). This completes the proof.

LEMMA 2.5.5.

(i) *If $n > 1$, the elements $h_{n,i}^2$ form a base for ${}_nE_3^{0,2}$.*

(ii) *If $n > 1$, the elements $\{f_{n,i}\}$ in ${}_nE_\infty^{1,2}$ are linearly independent.*

(iii) *If $n > 2$, the elements $\{g_{n,j}h_i\}$ for which $j \neq i+1$ are linearly independent in ${}_nE_\infty^{2,1}$. The same is true for $n = 2$, provided we exclude also the elements $\{g_{2,i}h_{i+1}\}$.*

PROOF. To prove part (i), it is sufficient to note that the differential

$${}_nd_2 : {}_nE_2^{0,2} \longrightarrow {}_nE_2^{2,1}$$

is described by

$${}_nd_2(h_{n,j}h_{n,i}) = h_{n,j}g_{n-1,i} + h_{n,i}g_{n-1,j} .$$

Part (ii) follows from Lemma 2.5.2 (iii), since no differential maps into ${}_nE_n^{1,2}$.

To prove part (iii), we note that the following formula holds in ${}_nE_\infty^{2,1}$:

$$\{g_{n,j}h_i\} = \{h_{n,j+1}h_jh_i + h_{n,j}h_{n+j}h_i\} .$$

Moreover, the only differential mapping into ${}_nE_r^{2,1}$ is

$${}_nd_2 : {}_nE_2^{0,2} \longrightarrow {}_nE_2^{2,1} ,$$

which has just been described. It is now easy to obtain part (iii). This completes the proof.

LEMMA 2.5.1.

(i) *If $n > 1$, the following elements are linearly independent in $H^3(Q_n)$: the elements $f_{n,i}$; the elements $g_{n,j}h_i$ for which $j \neq i+1$; the products $h_kh_jh_i$ for which $k \geqq j \geqq i \geqq 0$, with the following exceptions:*

$$h_{j+1}h_jh_i , \quad h_kh_{i+1}h_i , \quad h_{i+2}h_i^2 , \quad h_{i+2}^2h_i .$$

The same conclusion remains true for $n = 1$ if we include the products $h_{i+2}^2h_i$.

(ii) *$H^2(Q_n)$ is spanned by the elements $g_{n,k}$ and h_jh_i (where $j \geqq i \geqq 0$, $j \neq i+1$).*

PROOF. It is elementary to check part (i) for $n = 1$. We proceed by induction over n; let us assume that part (i) holds for Q_{n-1}. We must examine the differentials

$${}_nd_2 : {}_nE_2^{1,1} \longrightarrow {}_nE_2^{3,0} , \qquad {}_nd_3 : {}_nE_3^{0,2} \longrightarrow {}_nE_3^{3,0} .$$

These are described by

$$ {}_nd_2(h_{n,j}h_i) = g_{n-1,j}h_i \,, \qquad {}_nd_3(h_{n,i}^2) = f_{n-1,i} \,. $$

We obtain the following conclusions.

(a) ${}_nE_3^{1,1}$ has as a base the elements

$$ \{g_{n,i}\} = \{h_{n,i+1}h_i + h_{n,i}h_{n+i}\} \,. $$

(b) ${}_nE_4^{0,2} = 0$.

(c) In ${}_nE_4^{3,0}$ the products $h_k h_j h_i$ named above are linearly independent. In the case $n = 2$, this conclusion remains true when we include also the products $h_{i+2}^2 h_i$.

Using Lemma 2.5.5 parts (ii) and (iii), we see that part (i) holds for Q_n. (If $n = 2$, we need to know that $g_{2,i}h_{i+1} = h_{i+2}^2 h_i$; this was proved in Lemma 2.5.4.) This completes the proof of part (i).

Part (ii) follows immediately from the facts (a) and (b) established during the proof of part (i).

Since $H^{s,t}(Q_n) \to H^{s,t}(A)$ as $n \to \infty$ (for fixed s and t), the work which we have done completes the proof of Theorem 2.5.1.

2.6. *More calculations.* In this section we shall calculate $\mathrm{Ext}_A^{s,t}(M, Z_2)$ for a certain module M which arises in the applications (and for a limited range of s and t). The results are stated in Theorem 2.6.2.

We first obtain a lemma which is true for a general algebra A over Z_2. Suppose given a primitive element α in A_n^* $(n > 0)$, that is, an element α such that

$$ \varphi^*\alpha = \alpha \otimes 1 + 1 \otimes \alpha \,. $$

According to § 2.2, it defines an element h_α in $\mathrm{Ext}_A^{1,n}(Z_2, Z_2)$, which for example may be written $\{[\alpha]\}$, using the cobar construction.

In terms of α, we define a module $M = M(\alpha)$ as follows. *Qua* vector space over Z_2, it has a base containing two elements m_1, m_2 of degrees 0, n. The operations A are defined by

$$ am_1 = (\alpha a)m_2 \qquad (a \in A_n) \,. $$

These operations do give M the structure of an A-module, since α annihilates all decomposable elements of A.

The element m_2 generates a submodule M_2 of M isomorphic with Z_2; we define $M_1 = M/M_2$, so that $M_1 \cong Z_2$. We agree to write $x^{(i)}$ $(i = 1, 2)$ for the element of $\mathrm{Ext}_A^s(M_i, Z_2)$ corresponding to the element x of $\mathrm{Ext}_A^s(Z_2, Z_2)$.

From the exact sequence

$$0 \longrightarrow M_2 \longrightarrow M \longrightarrow M_1 \longrightarrow 0$$

we obtain an exact sequence

$$\cdots \leftarrow \mathrm{Ext}_A^{s+1,t}(M_1, Z_2) \xleftarrow{\ \delta\ } \mathrm{Ext}_A^{s,t}(M_2, Z_2) \leftarrow \mathrm{Ext}_A^{s,t}(M, Z_2) \leftarrow \cdots$$

which one might use to calculate $\mathrm{Ext}_A^{s,t}(M, Z_2)$.

LEMMA 2.6.1. *The coboundary* δ *is given by*

$$\delta(x^{(2)}) = (xh_\alpha)^{(1)}$$

(*where* $x \in \mathrm{Ext}_A^s(Z_2, Z_2)$).

PROOF. Let us take two resolutions C, C' of Z_2 over A, as follows:

$$\begin{array}{l} Z_2 \xleftarrow{\ \varepsilon\ } A \xleftarrow{\ d\ } C_1 \longleftarrow \cdots \quad \cdots \longleftarrow C_q \longleftarrow \cdots \\ Z_2 \xleftarrow{\ \varepsilon'\ } A \xleftarrow{\ d'\ } C'_1 \longleftarrow \cdots \quad \cdots \longleftarrow C'_q \longleftarrow \cdots . \end{array}$$

To calculate the cup-product with h_α, we must construct a diagram, as considered in § 2.2.

$$\begin{array}{ccccccccc} Z_2 & \xleftarrow{\ \varepsilon\ } & A & \xleftarrow{\ d\ } & C_1 & \xleftarrow{\ d\ } & C_2 \longleftarrow \cdots & \cdots \longleftarrow & C_{q+1} \longleftarrow \cdots \\ \downarrow{\scriptstyle \alpha} & \swarrow{\scriptstyle f} & & & \downarrow{\scriptstyle f_0} & & \downarrow{\scriptstyle f_1} & & \downarrow{\scriptstyle f_q} \\ Z_2 & \xleftarrow[\ \varepsilon'\ ]{} & A & & \xleftarrow[\ d'\ ]{} & & C'_1 \longleftarrow \cdots & \cdots \longleftarrow & C'_q \longleftarrow \cdots \end{array}$$

Let us now define a boundary d on $C + C'$ and an augmentation ε: $C + C' \rightarrow M$ by setting

$$d(x, y) = (dx, d'y + f_{q-1}(x)) \qquad (x \in C_q, q > 0)$$

$$\varepsilon(1, 0) = m_1 , \qquad \varepsilon(0, 1) = m_2 .$$

$C + C'$ is a chain complex in which C' is embedded, with quotient C; the exact cohomology sequence shows that $C + C'$ is acyclic. $C + C'$ is thus a resolution of M; it contains C', which is a resolution of M_2, and the quotient is C, which is a resolution of M_1. We obtain an exact sequence of cochain complexes

$$0 \leftarrow \mathrm{Hom}_A(C', Z_2) \leftarrow \mathrm{Hom}_A(C + C', Z_2) \leftarrow \mathrm{Hom}_A(C, Z_2) \leftarrow 0 .$$

The corresponding exact cohomology sequence is the one required. The coboundary in

$$\mathrm{Hom}_A(C + C', Z_2) \cong \mathrm{Hom}_A(C, Z_2) + \mathrm{Hom}_A(C', Z_2)$$

is given by

$$\delta(x, y) = (\delta x + yf_{q-1}, \delta y)$$

(if $y \in \mathrm{Hom}_A(C'_{q-1}, Z_2)$). It follows immediately that in the exact cohomology sequence we have

$$\delta(\{y\}^{(2)}) = \{yf_{q-1}\}^{(1)} = (\{y\}h_\alpha)^{(1)} .$$

This proves Lemma 2.6.1.

We now suppose that A is the (mod 2) Steenrod algebra. Take an integer $k \geqq 2$; we define a module $M = M(k)$ as follows. *Qua* vector space over Z_2, it has a base containing three elements m_1, m_2, m_3 of degrees 0, 2^k, 2^{k+1}. The operations from A are defined by

$$\left.\begin{aligned} am_1 &= (\xi_1^{2^k} a)m_2 \\ am_2 &= 0 \end{aligned}\right\} \quad (\text{if } \deg(a) = 2^k)$$

$$am_1 = (\xi_1^{2^{k+1}} a)m_3 \quad (\text{if } \deg(a) = 2^{k+1}) .$$

The elements m_2, m_3 generate submodules M_2, M_3 isomorphic with Z_2; we write i_2, i_3 for their injections. We define $M_1 = M/(M_2 + M_3)$, so that $M_1 \cong Z_2$; we have an exact sequence

$$0 \longrightarrow M_1 + M_2 \xrightarrow{i_2 + i_3} M \xrightarrow{j} M_1 \longrightarrow 0 .$$

We continue the previous convention about $x^{(i)}$.

THEOREM 2.6.2. *In dimensions $t < 3\cdot 2^k$, $\mathrm{Ext}^1_A(M, Z_2)$ has as a base the elements*

$$\begin{array}{ll} j^*h_i^{(1)} & (0 \leqq i \leqq k-1) \\ (i_2^*)^{-1}h_{k-1}^{(2)} & \end{array}$$

and $\mathrm{Ext}^2_A(M, Z_2)$ has as a base the elements

$$\begin{array}{ll} j^*(h_ih_l)^{(1)} & (0 \leqq i \leqq l \leqq k-1,\ l \neq i+1) \\ (i_2^*)^{-1}(h_ih_{k-1})^{(2)} & (0 \leqq i \leqq k-1,\ i \neq k-2) \\ (i_2^*)^{-1}(h_{k-2}h_k)^{(2)} . & \end{array}$$

PROOF. We have a diagram

$$\begin{array}{ccccccccc} 0 & \longrightarrow & M_2 + M_3 & \longrightarrow & M & \longrightarrow & M_1 & \longrightarrow & 0 \\ & & \downarrow & & \downarrow & & \downarrow & & \\ 0 & \longrightarrow & M_2 & \longrightarrow & M/M_3 & \longrightarrow & M_1 & \longrightarrow & 0 . \end{array}$$

Now, Lemma 2.6.1 applies to M/M_3, with $\alpha = \xi_1^{2^k}$; thus $\delta(x^{(2)}) = (xh_k)^{(1)}$. By naturality, the same formula holds in the exact sequence

$$\cdots \xleftarrow{\delta} \mathrm{Ext}_A^{s,t}(M_2 + M_3, Z_2) \longleftarrow \mathrm{Ext}_A^{s,t}(M, Z_2) \xleftarrow{j^*} \mathrm{Ext}_A^{s,t}(M_1, Z_2) \xleftarrow{\delta} \cdots .$$

Similarly,

$$\delta(x^{(3)}) = (xh_{k+1})^{(1)} .$$

It follows (using Theorem 2.5.1) that in this exact sequence, Coker δ and Ker δ have the following bases, at least in degrees $t < 3\cdot 2^k$.

$s = 1$, Coker δ: $h_i^{(1)}$ $(0 \leqq i \leqq k-1)$.

$s = 1$, Ker δ: $h_{k-1}^{(2)}$.

$s = 2$, Coker δ: $(h_i h_l)^{(1)}$ $(0 \leqq i \leqq l \leqq k-1,\ l \neq i+1)$.

$s = 2$, Ker δ: $(h_i h_{k-1})^{(2)}$ $(0 \leqq i \leqq k-1,\ i \neq k-2)$,

$(h_{k-2} h_k)^{(2)}$.

This proves Theorem 2.6.2.

We have now completed all the homological algebra which is necessary for our applications.

CHAPTER 3. SECONDARY COHOMOLOGY OPERATIONS

3.1. *Introduction.* In this chapter we shall develop the general theory of stable secondary operations. The results at issue are not deep; the author hopes that this fact will not be obscured by the language necessary to express them in the required generality.

The plan of this chapter is as follows. In § 3.6 we give axioms for the sort of operations we shall consider; in Theorems 3.6.1, 3.6.2 we prove the existence and uniqueness of operations satisfying these axioms. In § 3.7 (and in Theorems 3.7.1, 3.7.2. in particular) we consider relations between composite operations. In § 3.9 (and in Theorem 3.9.4 in particular) we consider Cartan formulae for such operations. All the theorems mentioned above are essential for the applications.

In an attempt to arrange the proofs of these theorems lucidly, we begin by giving a formal status to some of the ideas involved. These ideas concern cohomology operations (of kinds higher than the first), universal examples for such operations, the suspension of such operations, and so on. Of course, these notions are common property; but by giving a connected account, we can build up the lemmas which we need later. This preliminary work occupies §§ 3.2 to 3.5.

In § 3.8 we give a short account of the application of homological algebra to the study of stable secondary operations.

In this chapter, the following conventions will be understood. We shall assume that all cohomology groups have coefficients in a fixed finitely-generated group G; in § 3.8 we assume that G is a field, and in § 3.9 that G is the field Z_p. We shall omit the symbol for the fixed coefficient group G, except where special emphasis is needed.

We shall assume that all the spaces considered are arcwise-connected. There is nothing essential in this assumption, but it serves to simplify some statements. Symbols such as x_0, y_0 will denote base-points in the corresponding spaces X, Y. It is understood that the base-points in CW-complexes are chosen to be vertices.

The study of stable operations forces us to work with suspension. We therefore agree that in §§ 3.2 to 3.8 the symbols $H^*(X)$, $H^n(X)$ denote augmented cohomology (with coefficients in G).

In § 3.9, however, we have to work with products. In this section, therefore, the symbol $H^n(X)$ denotes ordinary cohomology. We write $H^+(X)$ for $\sum_{n>0} H^n(X)$; since our spaces are arcwise-connected, we may use H^+ instead of augmented cohomology.

We have to take a little care with signs. It is usual to write cohomology operations on the left of their arguments; we shall follow Milnor [25] in taking the signs which arise naturally from this convention. We shall also try to keep our theoretical work as free from signs as possible. For this purpose it seems best to write our homomorphisms on the right of their arguments, accepting the signs which arise naturally from this convention.

In particular, we introduce a "right" coboundary

$$\delta^*: H^n(Y) \to H^{n+1}(X, Y)$$

whose definition in terms of the usual coboundary δ is

$$(h)\delta^* = (-1)^n \delta(h) \qquad \text{(where } h \in H^n(Y)\text{)}.$$

We shall use this signed coboundary in discussing suspension. For example, let Y be a space with base-point y_0; let $\Omega Y \longrightarrow LY \xrightarrow{\pi} Y$ be the path-space fibering introduced by Serre [29]. Then we define the "suspension" homomorphism

$$\sigma : H^{n+1}(Y) \to H^n(\Omega Y)$$

to be the composite map

$$H^{n+1}(Y) \xleftarrow{\cong} H^{n+1}(Y, y_0) \xrightarrow{\pi^*} H^{n+1}(LY, \Omega Y) \xleftarrow{\delta^*} H^n(\Omega Y) .$$

Similarly for the transgression τ.

3.2. *Theory of universal examples.* It is the object of this section to set up a general theory of universal examples for cohomology operations of higher kinds. This is done by making the obvious changes in the corresponding theory for primary operations, which is due to Serre [30]. The considerations which guide our definitions are the following. We expect our operations to be defined on a subset of all cohomology classes; we expect their values to lie in a quotient set of cohomology classes; we expect them to be natural.

It would be convenient, in some ways, if we set up our theory for the category of CW-complexes. However, it would be inconvenient in other ways, since we need to use fiberings (in the sense of Serre). The difficulty could be avoided by working in the category of CSS-complexes; but it seems preferable to work, at first, with concepts as geometrical as possible. We therefore work in the category of all spaces. This forces us to use the device of replacing a space by a weakly equivalent CW-complex. We recall that a map $f : X \to Y$ (between arcwise-connected spaces) is said to be a weak homotopy equivalence if it induces isomorphisms of all homotopy groups. Two spaces X, Y are said to be weakly equivalent if they can be connected by a finite chain of weak homotopy equivalences. These notions have the following properties. For each space Y, there is a CW-complex X and a map $f : X \to Y$ which is a weak homotopy equivalence; it is sufficient, for example, to take X to be a geometrical realisation of the singular complex of Y, or of a minimal complex. Let $\mathrm{Map}(X, Y)$ denote the set of homotopy classes of maps from X to Y; if $f : X \to Y$ is a weak homotopy equivalence and W is a CW-complex, then the induced function $f_* : \mathrm{Map}(W, X) \to \mathrm{Map}(W, Y)$ is a (1-1) correspondence.

Our first definition is phrased so as to cover the case of cohomology operations in several variables; the reader may prefer to consider first the case of one variable. Let J be a set of indices j. We call S a *natural subset of cohomology* (in J variables, of degrees n_j) if S associates with each space X a set $S(X)$ of J-tuples $\{x_j\}$ (where $x_j \in H^{n_j}(x)$) and satisfies the following axioms.

Axiom 1. If $f: X \to Y$ is a map and $\{y_j\} \in S(Y)$, then $\{y_j f^*\} \in S(X)$.

Axiom 2. If $f : X \to Y$ is a weak homotopy equivalence and $\{y_j f^*\} \in S(x)$, then $\{y_j\} \in S(Y)$.

Although we shall not assume that the indexing set J is finite, we shall assume that for each integer N we have $n_j < N$ for only a finite number of j.

For the next definition, we suppose given such a natural subset S. We call Φ a *cohomology operation* (defined on S, and with values of degree m) if Φ associates with each space X and each J-tuple $\{x_j\}$ in $S(X)$ a non-empty subset $\Phi\{x_j\} \subset H^m(X)$ which satisfies the following axioms.

Axiom 3. If $f : X \to Y$ is a map and $\{y_j\} \in S(Y)$, then

$$(\Phi\{y_j\})f^* \subset \Phi\{y_j f^*\} .$$

Axiom 4. If $f : X \to Y$ is a weak homotopy equivalence and $\{y_j\} \in S(Y)$, then

$$(\Phi\{y_j\})f^* = \Phi\{y_j f^*\} .$$

There is nothing in our definition to limit the size of the subsets $\Phi\{x_j\}$. We therefore make the definitions which follow. They refer to cohomology operations defined on a fixed natural subset S. We write $\Phi \subset \Psi$ if we have $\Phi\{x_j\} \subset \Psi\{x_j\}$ for each X and each $\{x_j\}$ in $S(X)$. We call Ψ *minimal* if there is no operation Φ such that $\Phi \subset \Psi$ and $\Phi \neq \Psi$. We are mainly interested in operations which are minimal.

We now introduce the notion of a universal example. Let S be a natural subset of cohomology (in J variables, of degrees n_j). Let U be a space and $\{u_j\}$ a J-tuple in which $u_j \in H^{n_j}(U)$. We say that $(U, \{u_j\})$ is a *universal example* for S if $\{u_j\} \in S(U)$ and the following axiom is satisfied.

Axiom 5. For each CW-complex X and each J-tuple $\{x_j\}$ in $S(X)$ there is a map $g : X \to U$ such that $\{u_j g^*\} = \{x_j\}$.

LEMMA 3.2.1. *For each space U and each J-tuple $\{u_j\}$ there is one and only one natural subset S admitting $(U, \{u_j\})$ as a universal example.*

PROOF. The uniqueness of S is immediate; for if X is a CW-complex, $S(x)$ is the set of J-tuples $\{u_j g^*\}$, where g runs over all maps from X to U; while if X is a general space, we may take a weak homotopy equivalence $f: W \to X$ such that W is a CW-complex; then $S(X)$ is the set of J-tuples $\{x_j\}$ such that $\{x_j f^*\} \in S(W)$.

This procedure also shows how to construct S from $(U, \{u_j\})$; it is not hard to verify that the S so constructed is well-defined, is such that $\{u_j\} \in S(U)$ and satisfies Axioms 1 and 2.

For our next definition, let S be a natural subset admitting a universal example $(U, \{u_j\})$. Let Φ be a cohomology operation defined on S, and with values of degree m; let v be a class in $H^m(U)$. We say that $(U, \{u_i\}, v)$ is a *universal example* for Φ if $v \in \Phi\{u_j\}$ and the following axiom is satisfied.

Axiom 6. For each CW-complex X, each J-tuple $\{x_j\}$ in $S(X)$ and each class y in $\Phi\{x_j\}$ there is a map $g : X \to U$ such that $\{u_j g^*\} = \{x_j\}$, $vg^* = y$.

LEMMA 3.2.2. *For each space U, each J-tuple $\{u_j\}$ and each class v there is one and only one cohomology operation Φ admitting $(U, \{u_j\}, v)$ as a universal example.*

The proof is closely similar to that of Lemma 3.2.1.

If the space U is understood, we may write "$\{u_j\}$ is a universal example for S"; similarly, if U and $\{u_j\}$ are understood, we may write "v is a universal example for Φ".

For our last lemma, suppose that S is a natural subset admitting a universal example $\{u_j\}$, and that Ψ is an operation defined on S.

LEMMA 3.2.3. *If $v \in \Psi\{u_j\}$ and Φ is the operation given by the universal example v, then $\Phi \subset \Psi$.*

The proof is obvious.

This lemma shows that, if we wish to study the operations defined on such a natural subset S, it is sufficient to consider the ones given by universal examples.

3.3. *Construction of universal examples.* It is the object of this section to construct the universal examples on which our theory of secondary operations will be based. In fact, our secondary operations will be defined on natural subsets which can be described using primary operations. We shall show that these natural subsets admit universal examples, in the sense of § 3.2.

Our universal examples will be fiberings in which both base and fibre are weakly equivalent to Cartesian products of Eilenberg-MacLane spaces. It will be clear that the method of this section is only the beginning of an obvious induction; we might equally well construct an example-space of the $(n+1)^{\text{th}}$ kind as a fibering with the same sort of fibre, but with an example-space of the nth kind as a base. However, we shall not do this.

We say that Φ is a *primary* operation (acting on J variables of degrees n_j, and with values of degree m) if it has the following properties.

(1) $\Phi\{x_j\}$ is defined for every J-tuple $\{x_j\}$ such that $x_j \in H^{n_j}(X)$.

(2) The values $\Phi\{x_j\}$ of Φ are single elements of $H^m(X)$.

Let K be a set of indices k. When we speak of a *K-tuple $\{a_k\}$ of primary operations*, we shall understand that each a_k acts on J variables whose degrees n_j do not depend on k. We shall also suppose that for each integer M we have $m_k < M$ for only a finite number of k in K.

We define *the natural subset T determined by $\{a_k\}$* as follows: $T(X)$ is the set of J-tuples $\{x_j\}$ such that $x_j \in H^{n_j}(X)$ and $a_k\{x_j\} = 0$ for each k in K. It is clear that T is indeed a natural subset.

We shall call a cohomology operation Φ *secondary* if it is defined on a natural subset T of this kind. We shall prove that every natural subset T of this kind admits a universal example; this is formally stated as Theorem 3.3.7.

We must begin by considering universal examples for primary operations. Suppose given integers $n_j > 0$, for $j \in J$. Then we can form a Cartesian product

$$X = \times_{j \in J} X_j$$

in which X_j is an Eilenberg-MacLane space of type (G, n_j). Let x_j be the fundamental class in $H^{n_j}(X_j; G)$; let $\pi_j \colon X \to X_j$ be the projection map onto the jth factor; then we have classes $x_j \pi_j^*$ in $H^*(X)$. Suppose given a space Y and classes y_j in $H^{n_j}(Y)$. We will say that Y is a *generalised Eilenberg-MacLane space* (with fundamental classes y_j) if Y is weakly equivalent to a product such as X in such a way that the classes y_j correspond to the classes $x_j \pi_j^*$.

LEMMA 3.3.1. *If Y is a generalised Eilenberg-MacLane space, W is a* CW-*complex and $w_j \in H^{n_j}(W)$ for each j, then there is one and only one homotopy class of maps $f \colon W \to Y$ such that $y_j f^* = w_j$ for each j.*

This follows immediatly from the corresponding fact for Eilenberg-MacLane spaces.

Next, let Y and W be generalised Eilenberg-MacLane spaces, for the same integers n_j, and with fundamental classes y_j, w_j. We call a map $f \colon Y \to W$ a *canonical equivalence* if $w_j f^* = y_j$ for each j. Such a map is necessarily a weak homotopy equivalence.

Let Y be a generalised Eilenberg-MacLane space, for integers $n_j > 1$; Y is thus weakly equivalent to a product $X = \times_{j \in J} X_j$ of Eilenberg-MacLane spaces. Then, if the base points are chosen consistently, the loop-space ΩY is weakly equivalent to ΩX, which is homeomorphic to $\times_{j \in J} \Omega X_j$; thus ΩY is a generalised Eilenberg-MacLane space. In each loop-space ΩX_j we take the fundamental class x_j^Ω given by

$$x_j^\Omega = (x_j)\sigma$$

where σ denotes "suspension". In ΩX we have fundamental classes $x_j^\Omega \pi_j^*$, and in ΩY we have the corresponding fundamental classses y_i^Ω.

We have said that we propose to construct our universal examples as fiberings. All our fiberings will be fiberings in the sense of Serre [29]. We must recall that the notion of an induced fibering is valid in this context.

Let X, B be spaces with base-points x_0, b_0. Let $f \colon X, x_0 \to B, b_0$ be a map, and let $\pi \colon E \to B$ be a fibering (in the sense of Serre). Let $f^- E$ be the subspace of $X \times E$ consisting of pairs (x, e) such that $fx = \pi e$. We define maps $\varpi \colon f^- E \to X$ and $f_- \colon f^- E \to E$ by $\varpi(x, e) = x$, $f_-(x, e) = e$;

we thus obtain the following commutative diagram.

$$\begin{array}{ccc} f^{-}E & \xrightarrow{f_-} & E \\ \varpi\big\downarrow & & \big\downarrow\pi \\ X & \xrightarrow{f} & B \end{array}$$

LEMMA 3.3.2. *The map* ϖ *is a fibering (in the sense of Serre). The map* f_- *maps the fibre* $\varpi^{-1}x_0$ *of* ϖ *homeomorphically onto the fibre* $\pi^{-1}b_0$ *of* π.

The verification is trivial. The original reference for this lemma, so far as the author knows, is [10].

The next lemma states that this construction is natural, in an obvious sense. Suppose given the following diagram.

$$\begin{array}{ccc} E & \xrightarrow{\varepsilon} & E' \\ \pi\big\downarrow & & \big\downarrow\pi' \\ B, b_0 & \xrightarrow{\beta} & B', b_0' \\ f\big\uparrow & & \big\uparrow f' \\ X, x_0 & \xrightarrow{\xi} & X', x_0' \end{array}$$

LEMMA 3.3.3. *We can define a map*

$$\xi \otimes \varepsilon : f^{-}E \rightarrow (f')^{-}E'$$

by the rule $(\xi \otimes \varepsilon)(x, e) = (\xi x, \varepsilon e)$. *This map makes the following diagram commutative.*

$$\begin{array}{ccc} E & \xrightarrow{\varepsilon} & E' \\ f_-\big\uparrow & & \big\uparrow f'_- \\ f^{-}E & \xrightarrow{\xi\otimes\varepsilon} & (f')^{-}E' \\ \varpi\big\downarrow & & \big\downarrow\varpi' \\ X & \xrightarrow{\xi} & X' \end{array}$$

The verification is trivial. The diagram shows in particular, that the effect of $\xi \otimes \varepsilon$ on the fibres is the same as that of ε, up to the homeomorphisms of Lemma 3.3.2.

We can now describe the fibre-spaces which we shall use as universal examples. Suppose given, as above, a fixed K-tuple $\{a_k\}$ of primary operations, such that each a_k acts on J variables of degrees n_j and has values of degree m_k. We suppose $n_j > 0$, $m_k > 1$. We call $\varpi: E \rightarrow B$ a *canonical fibering associated with* $\{a_k\}$ if it satisfies the following conditions.

(1) The fibering $\varpi : E \to B$ is induced by a map μ: $B, b_0 \to Y, y_0$ from the fibering $\pi : LY \to Y$.

(2) The spaces B and Y are generalised Eilenberg-MacLane spaces with fundamental classes b_j and y_k of degrees n_j and m_k.

(3) We have $y_k\mu^* = a_k\{b_j\}$.

We remark that if these conditions are fulfilled, then the fibre $F = \varpi^{-1}b_0$ of ϖ is a generalised Eilenberg-MacLane space. In fact, the map μ_- maps F homeomorphically onto ΩY, by Lemma 3.3.2; and by our remarks above, ΩY is a generalised Eilenberg-MacLane space, with fundamental classes y_k^{Ω} of degree $m_k - 1$. Therefore F is also a generalised Eilenberg-MacLane space, with fundamental classes f_k corresponding to the y_k^{Ω} under the map μ_-.

LEMMA 3.3.4. *The class* $f_k\delta^*$ *in* $H^*(E, F)$ *is the image of* $a_k\{b_j\}$ *under the composite homomorphism*

$$H^*(B) \xleftarrow{\cong} H^*(B, b_0) \xrightarrow{\varpi^*} H^*(E, F) .$$

This is immediate, by naturality.

LEMMA 3.3.5. *For each* $\{a_k\}$ *and each corresponding space* Y *there exists a canonical fibering associated with* $\{a_k\}$ *in which the space* B *is a* CW-*complex.*

This is clear; for we can construct B to be a CW-complex, and there is then a map $\mu : B \to Y$ of the sort required.

Our next theorem will assert that all the canonical fiberings associated with $\{a_k\}$ are "equivalent", in a suitable sense. In fact, let $\varpi: E \to B$ and $\varpi': E' \to B'$ be two such fiberings; then we define an equivalence between them to be a diagram

$$\begin{array}{ccc} F & \xrightarrow{\varphi} & F' \\ \downarrow & & \downarrow \\ E & \xrightarrow{\varepsilon} & E' \\ \varpi \downarrow & & \downarrow \varpi' \\ B & \xrightarrow{\beta} & B' \end{array}$$

in which $\varphi : F \to F'$ and $\beta : B, b_0 \to B', b_0'$ are canonical equivalences, in the sense explained above. Such a diagram, of course, induces an isomorphism of the exact homotopy sequences, so that ε will be a weak homotopy equivalence.

THEOREM 3.3.6. *Any two canonical fiberings associated with the same* $\{a_k\}$ *may be connected by a finite chain of equivalences.*

PROOF. Let E', E'' be two such canonical fiberings. Let $\varpi : E \to B$ be another canonical fibering, constructed using spaces B and Y which are CW-complexes. It is sufficient to show how to connect E to E' by a chain of equivalences, since we may connect E to E'' similarly. We may construct canonical equivalences

$$\beta : B, b_0 \to B', b_0'. , \qquad \eta : Y, y_0 \to Y', y_0' ,$$

so that $b_j'\beta^* = b_j$, $y_k'\eta^* = y_k$. It follows that $\mu'\beta \sim \eta\mu$. Let $h : I \times B/I \times b_0 \to Y'$ be a homotopy between them; we may write $hi' = \mu'\beta$, $hi = \eta\mu$, where i and i' are the embeddings of B in $I \times B/I \times b_0$. Using Lemma 3.3.3 we obtain the following chain of equivalences.

$$E = \mu^- LY \xrightarrow{i \otimes L(\eta)} h^- LY' \xleftarrow{i' \otimes 1} (\mu'\beta)^- LY' \xrightarrow{\beta \otimes 1} (\mu')^- LY' = E'$$

Thus E, E' can be connected as required. This completes the proof of Theorem 3.3.6.

For our next theorem, let $\varpi : E \to B$ be a canonical fibering associated with $\{a_k\}$; we define a J-tuple $\{e_j\}$ of classes in $H^*(E)$ by $e_j = b_j\varpi^*$. Let T be the natural subset determined by $\{a_k\}$, as defined at the beginning of this section.

THEOREM 3.3.7. *The natural subset* T *admits the universal example* $\{e_j\}$.

PROOF. We must begin by proving that $\{e_j\} \in T(E)$. In fact, we have

$$a_k\{e_j\} = a_k\{b_j\varpi^*\} = (a_k\{b_j\})\varpi^* = y_k\mu^*\varpi^* = y_k\pi^*(\mu_-)^* .$$

But since LY is acyclic, we have $y_k\pi^* = 0$ and hence $a_k\{e_j\} = 0$. Thus $\{e_j\} \in T(E)$.

It remains to show that if X is a CW-complex and $\{x_j\} \in T(x)$, then there is a map $g : X \to E$ such that $\{e_j g^*\} = \{x_j\}$. For this purpose we introduce the following lemma.

LEMMA 3.3.8. *If* X *is a* CW-*complex and* $f : X \to B$ *is a map such that* $a_k\{b_j f^*\} = 0$, *then there is a map* $g : X \to E$ *such that* $\varpi g = f$.

From this lemma, the theorem follows immediately. In fact, suppose we have a CW-complex X and a J-tuple $\{x_j\}$ such that $a_k\{x_j\} = 0$; then we can construct $f : X \to B$ such that $b_j f^* = x_j$ and $g : X \to E$ such that $\varpi g = f$; it follows that

$$e_j g^* = b_j \varpi^* g^* = b_j f^* = x_j \ .$$

PROOF OF LEMMA 3.3.8. By Theorem 3.3.6, it is sufficient to consider the case in which Y is a Cartesian product of Eilenberg-MacLane spaces; say $Y = \times_{k \in K} Y_k$, where Y_k is of type (G, m_k). We can now consider the canonical fibering E in a different way. We have a map $\mu: B \rightarrow Y$; its kth component is $\pi_k \mu: B \rightarrow Y_k$. We may define $E_k = (\pi_k \mu)^- L Y_k$; the product $E' = \times_{k \in K} E_k$ is fibred over the base $B' = \times_{k \in K} B$. Let $\Delta : B \rightarrow B'$ be the diagonal map; then $\Delta^- E'$ coincides with E, up to a homeomorphism. It is therefore sufficient to lift the map f in each factor E_k separately. But to this there is only one obstruction; and up to a sign, the obstruction is

$$y_k \mu^* f^* = (a_k \{b_j\}) f^* = a_k \{b_j f^*\} = 0 \ .$$

The lifting is therefore possible. This completes the proof of Lemma 3.3.8. and of Theorem 3.3.7.

3.4. *Suspension.* In this section we discuss the suspension of cohomology operations, and show that if Φ admits a universal example, so does its suspension. This is formally stated as Theorem 3.4.6. We also show that the application of this principle does not enlarge the class of universal examples considered in § 3.3.

In this section the symbol sX will denote the suspension of X, so that $sX = I \times X/(0 \times X, 1 \times X)$. We shall also use s to denote the canonical isomorphism

$$s : H^{n+1}(sX) \rightarrow H^n(X)$$

which we define to be the following composite map.

$$H^{n+1}(sX) \xleftarrow{\cong} H^{n+1}(sX, s_0) \longrightarrow H^{n+1}(tX, X) \xleftarrow{\delta^*} H^n(X) \ .$$

Here, tX is the cone on X, so that $tX = I \times X/0 \times X$; the embedding of X in tX, the base-point s_0 in sX and the map from tX to sX are the obvious ones.

Let S be a natural subset of cohomology (in J variables, of degrees n_j). Then we can define a natural subset S^s (in J variables, of degrees $n_j - 1$) by taking $S^s(X)$ to be the set of J-tuples $\{x_j s\}$, where $\{x_j\} \in S(sX)$. We call S^s the *suspension* of S.

Similarly, let Φ be a cohomology operation defined on S, and with values of degree m. Then we can construct an operation Φ^s, defined on S^s and with values of degree $(m - 1)$, by setting

$$\Phi^s \{x_j s\} = (\Phi\{x_j\}) s \ .$$

We call Φ^s the *suspension* of Φ.

For our first lemma, let $\{a_k\}$ be a K-tuple of primary operations, and let T be the natural subset determined by $\{a_k\}$.

LEMMA 3.4.1. *The natural subset determined by* $\{a_k^s\}$ *is* T^s.

The verification is trivial.

LEMMA 3.4.2. *If S is a natural subset and* $\{x_j\} \in S^s(X)$, *then* $\{-x_j\} \in S^s(X)$.

If Φ is a cohomology operation defined on S, while $\{x_j\} \in S^s(X)$ *and* $y \in \Phi^s\{x_j\}$, *then* $-y \in \Phi^s\{-x_j\}$.

To prove this lemma, one merely considers the map $\nu : sX \to sX$ defined by

$$\nu(t, x) = (1 - t, x) .$$

For the sake of similar arguments later on, we set down the following lemma, which is well known.

LEMMA 3.4.3. *The set of homotopy classes*

$$\text{Map}\,[sX, sx_0; Y, y_0]$$

is a group. Let φ be the function which assigns to each homotopy class of maps $g : sX, sx_0 \to Y, y_0$ *the induced map* $g^* : H^*(Y) \to H^*(sX)$; *then φ is a homomorphism.*

Our next results relate the notions of "suspension" and "universal example".

LEMMA 3.4.4. *There is a* (1-1) *correspondence between maps* $f : X, x_0 \to \Omega Y, \omega_0$ *and maps* $g : sX, sx_0 \to Y, y_0$. *For corresponding maps we have* $g^*s = \sigma f^* : H^*(Y) \to H^*(X)$.

This lemma is well-known. The (1-1) correspondence is set up by the equation

$$(f(x))(u) = g(u, x) \qquad \text{(where } u \in I.)$$

The equation $g^*s = \sigma f^*$ is proved by passing to cohomology from the following diagram.

$$\begin{array}{ccccc} X & \longrightarrow & tX & \longrightarrow & sX \\ {\scriptstyle f}\big\downarrow & & {\scriptstyle h}\big\downarrow & & {\scriptstyle g}\big\downarrow \\ \Omega Y & \longrightarrow & LY & \longrightarrow & Y \end{array}$$

Here, the map h is defined by

$$(h(u, x))(v) = g(uv, x) \qquad \text{(where } u, v \in I.)$$

For our next lemma, we suppose that the space X is 1-connected, so that ΩX is 0-connected. Similarly, in Theorem 3.4.6 we shall assume that U is 1-connected.

LEMMA 3.4.5. *If* $\{x_j\} \in S(x)$, then $\{x_j\sigma\} \in S^s(\Omega X)$. *If (further)* Φ *is defined on* S, *then*

$$(\Phi\{x_j\})\sigma \subset \Phi^s\{x_j\sigma\} \ .$$

This is immediate, by considering the map $g : s\Omega X \to X$ corresponding to $f = 1 : \Omega X \to \Omega X$.

THEOREM 3.4.6. *If the natural subset* S *admits the universal example* $(U, \{u_j\})$, *then* S^s *admits the universal example* $(\Omega U, \{u_j\sigma\})$. *In this case* $S^s(X)$ *is a subgroup of* $\prod_{j \in J} H^{n_j - 1}(X)$.

If the cohomology operation Φ *admits the universal example* $(U, \{u_j\}, v)$, *then* Φ^s *admits the universal example* $(\Omega U, \{u_j\sigma\}, v\sigma)$. *In this case* $\Phi^s\{0\}$ *is a subgroup of* $H^{m-1}(X)$, $\Phi^s\{x_j\}$ *is a coset of* $\Phi^s\{0\}$ *and* Φ^s *is a homomorphism.*

This theorem follows easily from Lemmas 3.4.3 to 3.4.5.

Our next theorem will show that if U lies in the class of universal examples considered in § 3.3, then the universal example ΩU lies in the same class. For this purpose we need two lemmas.

LEMMA 3.4.7. *If* $\pi : E, e_0 \to B, b_0$ *is a fibering (in the sense of Serre) then* $\Omega\pi : \Omega E \to \Omega B$ *is a fibering (in the sense of Serre) with fibre* ΩF.

The verification is trivial. The next lemma concerns induced fiberings, so we adopt the notation of Lemma 3.3.2.

LEMMA 3.4.8. *There is a canonical homeomorphism* $h : \Omega(f^-E) \to (\Omega f)^-(\Omega E)$ *which makes the following diagram commutative.*

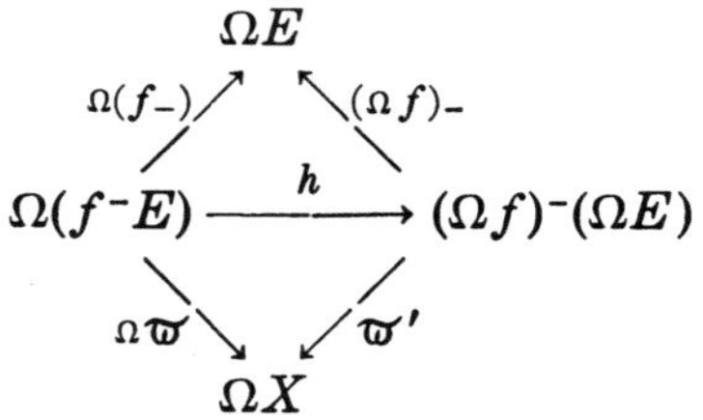

The verification is trivial.

THEOREM 3.4.9. *If* $\varpi: E, e_0 \to B, b_0$ *is a canonical fibering associated with* $\{a_k\}$, *and if we take in* ΩB, ΩF *the fundamental classes* $b_j\sigma$, $-f_k\sigma$, *then* $\Omega\varpi : \Omega E \to \Omega B$ *is (up to a canonical homeomorphism) a canonical fibering associated with* $\{a_k^s\}$.

PROOF. Suppose that E is induced by a map $\mu : B, b_0 \to Y, y_0$. There is an obvious homeomorphism between the fiberings $\Omega(LY) \xrightarrow{\Omega\pi} \Omega Y$ and $L(\Omega Y) \xrightarrow{\pi'} \Omega Y$; both fiberings have fibre $\Omega^2 Y$, but the homeomorphism induces a non-trivial automorphism α of $\Omega^2 Y$, defined by

$$[(\alpha\omega)(u)](v) = [\omega(v)](u) \qquad \text{(where } u, v \in I\text{)}.$$

Under this automorphism, the class $-y_k\sigma^2$ in the fibre of $\Omega\pi$ corresponds to the class $y_k\sigma^2$ in the fibre of π'.

We now remark that the fiberings induced by $\Omega\mu$ from $\Omega\pi$ and from π' must still be homeomorphic. The first is homeomorphic to $\Omega\varpi : \Omega E \to \Omega B$, by Lemma 3.4.8. The second is a canonical fibering, induced by $\Omega\mu$, and satisfying

$$(y_k\sigma)(\Omega\mu)^* = a_k^s\{b_j\sigma\}$$

(as we see using Lemma 3.4.5.) It is thus a canonical fibering associated with $\{a_k^s\}$. It is now easy to check that the kth fundamental class in its fibre corresponds to $-f_k\sigma$ in ΩF. This completes the proof.

3.5. *Stable operations*. In this section we shall study stable operations, and prove two lemmas needed in § 3.6.

In § 3.4 we defined the suspension of natural subsets and of cohomology operations. We use this notion to make the following definitions.

A *stable natural subset* S associates to each (positive or negative) integer l a natural subset S^l in such a way that $S^l = (S^{l+1})^s$. We may take our notations for degrees so that the variables in S^l are of degrees $n_j + l$. We admit, of course, that the natural subsets S^l may be trivial if l is large and negative. A similar remark applies to the next definition.

A *stable cohomology operation* (defined on such an S) associates to each integer l a cohomology operation Φ^l defined on S^l in such a way that $\Phi^l = (\Phi^{l+1})^s$. We may take our notations for degrees so that Φ^l has values of degree $m + l$.

We also allow ourselves to write $S(X) = \bigcup_l S^l(X)$, and to regard Φ as a function defined on $S(X)$ by the rule $\Phi | S^l = \Phi^l$. This is done in order to preserve the analogy between Φ and symbols such as Sq^i, which denote operations applicable in each dimension. We may call S^l, Φ^l the *components* of S, Φ.

As a particular case of the above, we have the notion of a stable primary operation a in one variable. Its lth component is a primary operation

$$a^l : H^{n+l}(X) \to H^{m+l}(X);$$

we assign to a the degree $(m - n)$. By Theorem 3.4.6, a^l is a homomorphism. Since natural homomorphisms can be composed and added, we easily see that the set of stable primary operations in one variable is a graded ring. We write A for this ring, or A_G if we wish to emphasise its dependence on the coefficient group; we call A the *Steenrod ring*. If X is any space, then $H^*(X)$ is a graded module over the graded ring A. If $f: X \to Y$ is a map, then $f^*: H^*(Y) \to H^*(X)$ is an A-map (of degree zero).

Let X be an Eilenberg-MacLane space of type (G, n), with fundamental class x; and let C be a free A-module, on one generator c of degree n. We can define an A-map $\theta : C \to H^*(X)$ by $c\theta = x$. It is both clear and well-known that $\theta | C_m : C_m \to H^m(X)$ is an isomorphism if $m < 2n$ (and a monomorphism if $m = 2n$).

Similarly, let Y be a generalised Eilenberg-MacLane space, with fundamental classes y_j of degrees n_j. Set $\nu = \text{Min}_{j \in J} n_j$. Let C be a free A-module on generators c_j of dimension n_j. We can define an A-map $\theta : C \to H^*(Y)$ by $c_j\theta = y_j$. Then, as before, $\theta | C_m : C_m \to H^m(Y)$ is an isomorphism if $m < 2\nu$ (and a monomorphism if $m = 2\nu$).

It follows, incidentally, that every stable primary cohomology operation in J variables is of the form $a\{x_j\} = \sum_{j \in J} a_j x_j$, where the sum is finite and the coefficients a_j lie in the Steenrod ring.

We next take a K-tuple $\{a_k\}$ of stable primary operations. Each stable operation a_k has components a_k^l; we shall suppose that a_k^l acts on J variables of degrees $n_j + l$ and has values of degree $m_k + l$. (We suppose, as always, that for each integer N we have $n_j < N$ for only a finite number of j and $m_k < N$ for only a finite number of k.)

Such a K-tuple evidently determines a stable natural subset T; we define T^l to be the natural subset determined by $\{a_k^l\}$. We shall call a stable cohomology operation Φ *secondary* if it is defined on a stable subset T of this kind.

In considering such stable secondary operations, it is natural to introduce a sequence of canonical fiberings E_l in which E_l is associated (in the sense of § 3.3) with the K-tuple $\{a_k^l\}$. For such fiberings to exist we require $l + n_j > 0$, $l + m_k > 1$; so we should assume that $l > -\nu$, where $\nu = \text{Min}_{j \in J, k \in K}(n_j, m_k - 1)$.

The next lemma will show that if a relation (between stable secondary operations) holds in the canonical fibering E_l (where l is sufficiently large), then it holds universally.

We will assume that T^l is the natural subset determined by $\{a_k^l\}$, as above, and that χ is a stable operation such that χ^l is defined on T^l and has values of degree $q + l$. We will also assume that χ satisfies the following axiom.

Axiom 1. If $g : X \to Y$ is a map such that $g^*: H^r(Y) \to H^r(X)$ is an isomorphism for $r \leqq q + l$, and if $\{y_j\} \in T^l(Y)$, then

$$(\chi^l \{y_j\})g^* = \chi^l \{y_j g^*\} .$$

Of course, this axiom is slightly stronger than Axiom 4, § 3.2; however, it is satisfied in the applications.

Let E_l be a canonical fibering associated with $\{a_k^l\}$, and let e_j^l be the fundamental classes in E_l.

LEMMA 3.5.1. *If* $0 \in \chi^\lambda \{e_j^\lambda\}$ *for one value* λ *of* l *such that* $\lambda >$ Max $(-\nu, q - 2\nu)$, *then* $0 \in \chi^l \{x_j\}$ *for all* l, *all* X *and all* $\{x_j\}$ *in* $T^l(X)$.

PROOF. We first note that if $0 \in \chi^l \{e_j^l\}$, then $0 \in \chi^l \{x_j\}$ for all X and all $\{x_j\}$ in $T^l(X)$; this is immediate, by naturality. We will show that if this holds for some l (where $l \geqq \lambda$) then it holds also for $l + 1$. We may find a space X and a map $g : sX \to E_{l+1}$ such that

$$g^*: H^r(E_{l+1}) \to H^r(sX)$$

is an isomorphism for $r \leqq q + l + 1$; for it is sufficient to take X to be ΩE_{l+1}, and g to be the map which corresponds to $f = 1$ in Lemma 3.4.3. We now have

$$\begin{aligned} 0 \in \chi^l \{e_j^{l+1} g^* s\} &= (\chi^{l+1} \{e_j^{l+1} g^*\})s \\ &= (\chi^{l+1} \{e_j^{l+1}\})g^* s \qquad \text{(by Axiom 1.)} \end{aligned}$$

Hence $0 \in \chi^{l+1} \{e_j^{l+1}\}$.

This proves that $0 \in \chi^l \{x_j\}$ if $l \geqq \lambda$. The corresponding result for $l < \lambda$ follows immediately, by suspension.

Our next lemma gives information about the group $H^{l+q}(E_l)$, at least if l is sufficiently large. We suppose given a K-tuple $\{a_k\}$ of stable primary operations, as described above; we may express each a_k in the form

$$a_k \{x_j\} = \sum_{j \in J} \beta_{k,j} x_j$$

where the sum is finite and the coefficients $\beta_{k,j}$ lie in the Steenrod ring A. Let C_0, C_1 be free A-modules on generators $c_{0,j}$, $c_{1,k}$ of degrees n_j, m_k. We can define an A-map $d : C_1 \to C_0$ by setting

$$c_{1,k} d = \sum_{j \in J} \beta_{k,j} c_{0,j} .$$

Let us take a value of l such that $l > -\nu$, and let $\varpi: E \to B$ be a canonical fibering associated with the K-tuple $\{a_k^l\}$. We can define A-maps

$$\theta_0: C_0 \to H^*(B) , \qquad \theta_1: C_1 \to H^*(F)$$

by setting

$$c_{0,j}\theta_0 = b_j\,, \quad c_{1,k}\theta_1 = (-1)^l f_k\,.$$

We take the "total degree" of $c_{s,t}$ to be $t - s (s = 0, 1)$, so that both θ_0 and θ_1 have degree l. The sign in the definition of θ_1 is essential in order that θ_1 should be compatible with σ; see Theorem 3.4.9.

Finally, we recall the following convention. Let

$$\begin{array}{ccc} M & \xrightarrow{f} & N \\ {\scriptstyle g}\downarrow & & \downarrow{\scriptstyle g'} \\ P & \xrightarrow{f'} & Q \end{array}$$

be a diagram of A-maps in which f and f' have total degree r, while g and g' have total degree s. Then we say that the diagram is *anticommutative* if

$$fg' = (-1)^{rs} gf'\,.$$

Lemma 3.5.2. *With the above data, we have the following anticommutative diagram.*

$$\begin{array}{ccccccc} C_{1,q} & & & \xrightarrow{\qquad\qquad\theta_1\qquad\qquad} & & & H^{l+q-1}(F) \\ {\scriptstyle d}\downarrow & & & & & & \downarrow{\scriptstyle \delta^*} \\ C_{0,q} & \xrightarrow{\theta_0} & H^{l+q}(B) & \xleftarrow{\cong} & H^{l+q}(B, b_0) & \xrightarrow{\varpi^*} & H^{l+q}(E, F) \end{array}$$

If $q < l + 2\nu$, we have also the following anticommutative diagram.

$$\begin{array}{ccccccc} \cdots\longrightarrow & H^{l+q}(B) & \xrightarrow{\varpi^*} & H^{l+q}(E) & \xrightarrow{i^*} & H^{l+q}(F) & \xrightarrow{\tau} & H^{l+q+1}(B) \\ & {\scriptstyle\theta_0}\uparrow{\scriptstyle\cong} & & & & {\scriptstyle\theta_1}\uparrow{\scriptstyle\cong} & & \uparrow{\scriptstyle\theta_0} \\ \cdots\longrightarrow & C_{0,q} & & & & C_{1,q+1} & \xrightarrow{d} & C_{0,q+1} \end{array}$$

The horizontal sequence in this diagram is exact; the maps θ_0 and θ_1 marked as isomorphisms are such; the remaining map θ_0 is a monomorphism.

Proof. We take first the anticommutativity of the first diagram. By using Lemma 3.3.4 and the definitions of the various homomorphisms, it is easy to check that the two ways of chasing c_{1k} round the diagram agree, up to the sign $(-1)^l$. The corresponding result for a general element of C_1 now follows by linearity over A.

In the second diagram, the squares are provided by the first diagram. We have already noted the behaviour of maps such as θ_0 and θ_1. The exact sequence is due to Serre [29]; it is valid up to $H^{2l+2\nu-1}(F)$ because

B and F are $(l + \nu - 1)$-connected. We can add the last τ, since it is defined on the whole of $H^{2l+2\nu-1}(F)$. This completes the proof.

3.6. *Axiom for stable secondary operations.* In §§ 3.2, 3.4 we were concerned with operations in general. It is the object of this section to give a system of axioms for stable secondary operations. This work is essential for the applications. After giving the axioms, we state Theorems 3.6.1 and 3.6.2, which assert the existence and (essential) uniqueness of operations satisfying the axioms. We then give some explanation of the axioms. Finally, we prove the two theorems.

It is generally understood that a secondary operation corresponds to a relation between primary operations. For example, the Massey product [23] [24] [35] corresponds to the relation $(uv)w = u(vw)$; the Adem operation [4] corresponds to the relation $\mathrm{Sq}^2\mathrm{Sq}^2 + \mathrm{Sq}^3\mathrm{Sq}^1 = 0$; and so on. We aim to get a hold on stable secondary operations by dealing with their associated relations. The essential feature of our axioms is that they axiomatise the connection between the secondary operation and its associated relation.

The notion of a "relation" between primary operations will be formalised in a suitable way. In fact, we shall replace the notion of a "relation" by the notion of a pair (d, z), of the following algebraic nature. The first entry d is to be a map $d: C_1 \to C_0$. Here, the objects C_0 and C_1 are to be graded modules over the Steenrod ring A (see §§ 2.1, 3.5); they are to be locally finitely-generated and free, and d is to be a right A-map such that $(C_{0,t})d \subset C_{1,t}$. Following § 3.5, we ascribe to $c_{s,t}$ the "total degree" $t - s$. The second entry z is to be a homogeneous element of Ker d.

We must next explain the connection between pairs (d, z) and relations in the intuitive sense. The equations

$$\mathrm{Sq}^1\mathrm{Sq}^1 = 0\ , \quad \mathrm{Sq}^2\mathrm{Sq}^2 + \mathrm{Sq}^3\mathrm{Sq}^1 = 0\ ,$$
$$\mathrm{Sq}^1\mathrm{Sq}^4 + \mathrm{Sq}^2\mathrm{Sq}^3 + \mathrm{Sq}^4\mathrm{Sq}^1 = 0$$

are relations in the intuitive sense. More generally, suppose given an integer q and a finite number of elements α_k, β_k in A such that

$$\textstyle\sum_{k \in K} \alpha_k \beta_k = 0\ , \quad \deg(\alpha_k) + \deg(\beta_k) = q + 1\ .$$

Then the equation $\sum_{k \in K} \alpha_k \beta_k = 0$ is a (homogeneous) relation in the intuitive sense; we shall associate it with the pair (d, z) constructed in the following way. We take C_0 to be free on one generator c_0 in $C_{0,0}$; we take C_1 to be free on generators $c_{1,k}$ in $C_{1,t(k)}$, where $t(k) = \deg(\beta_k)$. We define $d: C_1 \to C_0$ by $c_{1,k}d = \beta_k c_0$; we define z by $z = \sum_{k \in K} \alpha_k c_{1,k}$. We thus have $zd = 0$, $z \in C_{1,q+1}$.

Our axioms will ensure that if an operation Φ is associated with such a pair (d, z), then it is defined on classes x in $H^l(X)$ such that $\beta_k(x) = 0$ for each k, and has values in

$$H^{l+q}(X)/\sum_{k \in K} \alpha_k H^{l+q-r_k}(X)$$

where $r_k = \deg(\alpha_k)$. (Note that a relation of degree $(q+1)$ corresponds to an operation of degree q.)

The reader may like to keep in mind some explicit examples of pairs (d, z), to illustrate the considerations of this section.

We next remark that, according to our axioms, stable secondary operations are defined, not on J-tuples $\{x_j\}$ of cohomology classes, but on right A-maps $\varepsilon : C_0 \rightarrow H^*(X)$. There is no essential difference hence; if we take a base of elements $c_{0,j}$ in the free module C_0, then an A-map $\varepsilon : C_0 \rightarrow H^*(X)$ is determined uniquely by giving the images $(c_{0,j})\varepsilon$ of the base elements; and these classes $(c_{0,j})\varepsilon$ in $H^*(X)$ may be chosen at will, provided that they have the correct degrees. We set up a (1-1) correspondence between J-tuples $\{x_j\}$ and maps ε by writing

$$x_j = (c_{0,j})\varepsilon \ .$$

It is always to be understood that operations $\Phi\{x_j\}$ are to be identified with operations $\Phi(\varepsilon)$ in this way.

We now give the axioms. We will say that Φ is a *stable secondary operation associated with the pair* (d, z) if it satisfies the following axioms.

Axiom 1. $\Phi(\varepsilon)$ is defined if and only if $\varepsilon : C_0 \rightarrow H^*(X)$ is a right A-map such that $d\varepsilon = 0$.

For the next axiom, suppose that the total degrees of ε, z are l, q. Let $f : C_1 \rightarrow H^*(X)$ run over the right A-maps of total degree l, and let $Q^{l+q}(z, X)$ be the set of elements zf in $H^{l+q}(X)$.

Axiom 2. $\Phi(\varepsilon) \in H^{l+q}(X)/Q^{l+q}(z, X)$.

For the next axiom, let $g : X \rightarrow Y$ be a map, and let $\varepsilon : C_0 \rightarrow H^*(Y)$ be a right A-map such that $d\varepsilon = 0$.

Axiom 3. $(\Phi(\varepsilon))g^* = \Phi(\varepsilon g^*)$.

It is understood that the g^* on the left-hand side of this equation denotes a homomorphism of quotient groups, induced by the homomorphism g^* of cohomology groups.

For the next axiom, let sX be the suspension of X, and let s: $H^*(sX) \rightarrow$

$H^*(X)$ be the suspension isomorphism, as in § 3.4. Let $\varepsilon\colon C_0 \to H^*(sX)$ be a right A-map such that $d\varepsilon = 0$.

Axiom 4. $(\Phi(\varepsilon))s = \Phi(\varepsilon s)$.

The s on the left-hand side of this equation is to be interpreted like the g^* in Axiom 3.

For the next axiom, let (X, Y) be a pair of spaces, and let $\varepsilon : C_0 \to H^*(X)$ be a right A-map of degree l such that $d\varepsilon = 0$ and $\varepsilon i^* = 0$. We can now find right A-maps $\eta : C_0 \to H^*(X, Y)$ and $\zeta : C_1 \to H^*(Y)$(of total degree l) to complete the following anticommutative diagram.

$$\begin{array}{ccccccccc}
H^*(Y) & \xleftarrow{i^*} & H^*(X) & \xleftarrow{j^*} & H^*(X, Y) & \xleftarrow{\delta^*} & H^*(Y) & \xleftarrow{i^*} & H^*(X) \\
 & & \nwarrow{\scriptstyle\varepsilon} & & \uparrow{\scriptstyle\eta} & & \uparrow{\scriptstyle\zeta} & & \\
 & & & & C_0 & \xleftarrow{\quad d \quad} & C_1 & &
\end{array}$$

Axiom 5. $(\Phi(\varepsilon))i^* = [z\zeta] \mod (Q^{l+q}(z, X))i^*$.
It is understood that $[z\zeta]$ denotes the coset containing $z\zeta$. We easily check that this coset is independent of the choice of η and ζ.

The following theorems may help to justify this set of axioms.

THEOREM 3.6.1. *For each pair* (d, z), *there is at least one associated operation* Φ.

THEOREM 3.6.2. *If* Φ, Ψ *are two operations associated with the same pair* (d, z), *then they differ by a primary operation, in the sense that there is an element* c *in* (Coker $d)_q$ *such that* $\Phi(\varepsilon) - \Psi(\varepsilon) = [c\varepsilon]$.

We note that these theorems do not depend on any choice of bases in C_0 and C_1; however, we may of course use bases in the proofs.

We will now comment on the effect of these axioms. Let us take bases $c_{0,j}$, $c_{1,k}$ in C_0, C_1; suppose that the total degrees of $c_{0,j}, c_{1,k}$ are n_j, $m_k - 1$. We may write

$$(c_{1,k})d = \sum_{j \in J} \beta_{k,j} c_{0,j}\,, \quad z = \sum_{k \in K} \alpha_k c_{1,k}$$

where α_k, $\beta_{k,j}$ lie in A. We may define a stable primary operation a_k (in J variables) by

$$a_k\{x_j\} = \sum_{j \in J} \beta_{k,j} x_j\,.$$

Then (as we easily check) Axiom 1 is equivalent to saying that Φ is defined on J-tuples $\{x_j\}$ such that $x \in H^{l+n_j}(X)$ for each $j \in J$ and $a_k\{x_j\} = 0$ for each $k \in K$.

Axiom 2 states that the "indeterminacy" of Φ is $Q^{l+q}(z, X)$; with the above notation, we have

$$Q^{l+q}(z, X) = \sum_{k \in K} \alpha_k H^{l+m_k-1}(X) .$$

It is easy to see that any operation whose indeterminacy is given in this way satisfies Axiom 1, § 3.5.

Axiom 3 states that Φ is natural. Axiom 4 states that Φ is stable, in the sense of § 3.5. It is now clear that every Φ satisfying our axioms is a stable secondary operation in the sense of § 3.5.

Axiom 5 may be regarded in two ways. On the one hand, it is a version of one of the Peterson-Stein relations [28], and is of some use in applications. On the other hand, it serves to prescribe the universal example for Φ, without making explicit mention of any such thing. This is made precise by Lemma 3.6.3; we shall need the following notation. Let Φ be an operation satisfying Axioms 1–4, and let a_k be as above. Let Φ^l, a_k^l be the lth components of the stable operations Φ, a_k (as in § 3.5); and let E be a canonical fibering associated with the K-tuple $\{a_k^l\}$, as in §§ 3.3, 3.5. (For this purpose we assume that $l > -\nu$, where ν is given by $\nu = \mathrm{Min}_{j \in J,\, k \in K}(n_j, m_k - 1)$.) We may refer to E as a "canonical fibering associated with d". Let the maps θ_0, θ_1 be as in Lemma 3.5.2; and let $\varepsilon_E \colon C_0 \to H^*(E)$ be the A-map corresponding to the J-tuple $\{e_j\}$, so that $\varepsilon_E = \theta_0 \varpi^*$. We may regard ε_E as analogous to the "fundamental class" in an Eilenberg-MacLane space. We have $a_k\{e_j\} = 0$, or equivalently $d\varepsilon_E = 0$, so that $\Phi^l(\varepsilon_E)$ is defined.

LEMMA 3.6.3. *Φ^l satisfies Axiom 5 if and only if*

$$z\theta_1 \in (\Phi^l(\varepsilon_E))i^* .$$

PROOF. Suppse Φ^l satisfies Axiom 5; then we may apply Axiom 5 to the following diagram (in which the square is provided by Lemma 3.5.2):

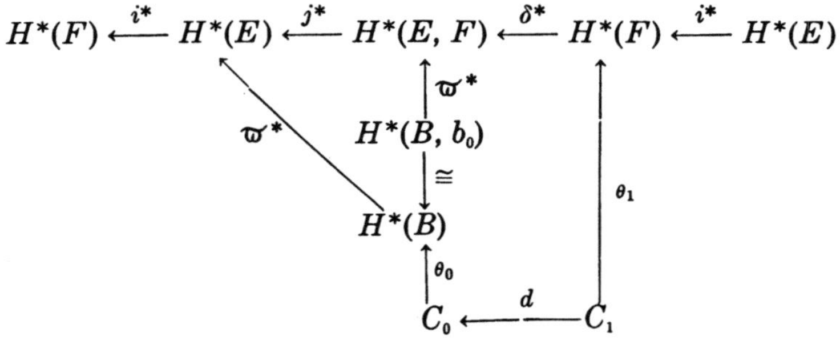

The conclusion which we obtain is

$$z\theta_1 \in (\Phi^l(\varepsilon_E))i^* .$$

Conversely, suppose that Φ^l satisfies this condition; we have to verify Axiom 5. It is sufficient to do so when the pair (X, Y) is a pair of CW-complexes. In this case, suppose given ε and $\eta : C_0 \to H^*(X, Y)$, as in Axiom 5. Since B is a generalised Eilenberg-MacLane space, we can construct a map $f : X, Y \to B, b_0$ so that the composite map

$$C_0 \xrightarrow{\theta_0} H^*(B) \xleftarrow{\cong} H^*(B, b_0) \xrightarrow{f^*} H^*(X, Y)$$

coincides with η. By Lemma 3.3.8, we can lift f to $g : X, Y \to E, F$. We can now take ζ to be the composite map

$$C_1 \xrightarrow{\theta_1} H^*(F) \xrightarrow{g^*} H^*(Y) .$$

But with this choice of ζ, we have

$$z\zeta \in (\Phi^l(\varepsilon))i^* ,$$

by applying g^* to the original condition. This completes the proof.

PROOF OF THEOREM 3.6.1. Suppose given a pair (d, z). Let us take bases $c_{0,j}$, $c_{1,k}$ in C_0, C_1; and let us keep the other notations introduced in the comments on the axioms, so that the K-tuple $\{a_k^l\}$ is as above. Let E_l be a canonical fibering associated with the K-tuple $\{a_k^l\}$. Let us fix on a value λ of l such that $\lambda > \mathrm{Max}(-\nu, q - 2\nu)$. By Lemma 3.5.2, we may choose a class v in $H^{\lambda+q}(E_\lambda)$ so that

$$vi^* = z\theta_1 .$$

Since we shall later wish to quote the part of the argument which starts at this point, we give it the status of a lemma.

LEMMA 3.6.4. *With the data above, there is at least one operation Φ associated with (d, z) (in the sense of Axioms 1-5) and such that $(E_\lambda, \{e_j^\lambda\}, v)$ is a universal example for Φ^λ.*

It is clear that this lemma implies the theorem.

PROOF OF LEMMA 3.6.4. We have canonical fiberings $\varpi_l: E_l \to B_l$ associated with K-tuples $\{a_k^l\}$ (at least for $l > -\nu$). By Theorem 3.4.9, $\Omega\varpi_{l+1} : \Omega E_{l+1} \to \Omega B_{l+1}$ is a canonical fibering associated with $\{a_k^l\}$; by Theorem 3.3.6, it is equivalent to $\varpi_l : E_l \to B_l$. We choose a finite chain of equivalences connecting them. Our next step is to choose a sequence of classes $v^l \in H^{l+q}(E_l)$ so that $v^\lambda = v$ and so that the class $v^{l+1}\sigma$ in $H^{l+q}(\Omega E_{l+1})$ corresponds to v^l in $H^{l+q}(E_l)$ under the finite chain of equivalences. This choice is clearly possible and unique, because the suspension

$$\sigma : H^{r+1}(E_{l+1}) \to H^r(\Omega E_{l+1})$$

is an isomorphism for $r < 2(l + \nu)$. Let Φ^l(for $l > -\nu$) be the operation given by the universal example $(E_l, \{e_j^l\}, v^l)$. Then it is clear that the operation given by v^l(namely Φ^l) coincides with that given by $v^{l+1}\sigma$ (namely $(\Phi^{l+1})^s$). To obtain a stable operation Φ (in the sense of § 3.5) we need only define its components Φ^l for $l \leqq -\nu$ by the inductive formula

$$\Phi^l = (\Phi^{l+1})^s \qquad \text{(for each } l).$$

We have now defined the operation Φ required; it remains to verify that it has the desired properties. We have ensured that v is a universal example for Φ^λ. Further, we have satisfied Axioms 3 and 4 (since Φ is natural and stable) and also Axiom 1 (since Φ^l is defined on the natural subset determined by $\{a_k^l\}$).

We next consider the equation

$$v^l i^* = z\theta_1 .$$

It holds for $l = \lambda$, by hypothesis; we may deduce that it holds for all l (such that $l > -\nu$), since the suspension

$$\sigma: H^{r+1}(F_{l+1}) \to H^r(\Omega F_{l+1})$$

is an isomorphism for $r < 2(l + \nu)$.

We can now verify Axiom 2. By suspension, it is sufficient to do this for $l > -\nu$. In this case, Theorem 3.4.6 shows that the values of Φ^l are cosets of $\Phi^l(0)$; we will show that $\Phi^l(0) = Q^{l+q}(z, X)$. It is sufficient to do this in the case when X is a CW-complex. In this case, any element of $\Phi^l(0)$ can be written in the form $v^l g^*$, where $g : X \to E_l$ is a map such that $\{e_j^l g^*\} = 0$. Since $e_j^l = b_j^l \varpi^*$, we have $\{b_j^l \varpi^* g^*\} = 0$, and therefore $\varpi g : X \to B_l$ is homotopic to the constant map at b_0. Covering this homotopy, we find a map $h : X \to F_l$ such that $g \sim ih: X \to E_l$. We now have

$$v^l g^* = v^l i^* h^* = z\theta_1 h^* .$$

Since $\theta_1 h^*: C_1 \to H^*(X)$ is an A-map of degree l, we have shown that $\Phi^l(0) \subset Q^{l+q}(z, X)$. Since any A-map $f : C_1 \to H^*(X)$ of degree l may be written in the form $\theta_1 h^*$ by a suitable choice of h, we easily see that $\Phi^l(0) \supset Q^{l+q}(z, X)$. This completes the proof of Axiom 2.

It remains only to verify Axiom 5. But this follows immediately from Lemma 3.6.3, at least if $l > -\nu$; and the case $l \leqq -\nu$ may be deduced by suspension. This completes the proof of Lemma 3.6.4 and of Theorem 3.6.1.

PROOF OF THEOREM 3.6.2. Suppose that Φ, Ψ are two operations associated with the same pair (d, z). Let us keep the general notation used above; let λ be a value of l such that $\lambda > \mathrm{Max}(-\nu, q - 2\nu)$, and let E be a canonical fibering associated with $\{a_k^\lambda\}$. By Axiom 5 and Lemma 3.6.3, we have

$$z\theta_1 \in (\Phi(\varepsilon_E))i^* .$$

Let v be a class in $\Phi(\varepsilon_E)$ such that $vi^* = z\theta_1$. Similarly, let w be a class in $\Psi(\varepsilon_E)$ such that $wi^* = z\theta_1$; then $(v-w)i^* = 0$. By Lemma 3.5.2, there is an element c in $(C_0)_q$ such that $v - w = c\theta_0\varpi^*$. This shows that

$$\Phi(\varepsilon_E) - \Psi(\varepsilon_E) - [c\varepsilon_E] = 0 .$$

Let us define a stable operation χ by

$$\chi(\varepsilon) = \Phi(\varepsilon) - \Psi(\varepsilon) - [c\varepsilon] ;$$

then χ satisfies the conditions of Lemma 3.5.1; therefore $\chi(\varepsilon) = 0$. This completes the proof of Theorem 3.6.2.

It is clear that operations satisfying Axioms 1–5 are linear, in the sense that

$$\Phi(\varepsilon + \varepsilon') = \Phi(\varepsilon) + \Phi(\varepsilon') .$$

In fact, Theorem 3.4.6 shows that this is true for operations constructed by the method of Lemma 3.6.4; and Theorem 3.6.2 enables us to deduce the corresponding result for any operation Φ.

3.7. *Properties of the operations.* In this section we shall prove certain properties of the stable secondary operations described in the last section. In content these properties are relations which hold between certain sums of composite operations, such as $\sum_{r \in R} a_r \Phi_r$ and $\sum_{r \in R} \Phi_r a_r$; here, the a_r are primary operations, and the Φ_r are secondary operations. We shall give these properties a form in keeping with our algebraic machinery. They are stated as Theorems 3.7.1 and 3.7.2; these results are essential for the applications.

We shall not discuss operations of the form $\Phi\Psi$, since such operations are tertiary, not secondary; $\Phi\Psi\{x_j\}$ is only defined if $a_k \Psi\{x_j\} = 0$ for various a_k.

We first suppose given an A-map $d : C_1 \to C_0$ (of the usual kind) and finitely many elements z_r in Ker d and of degrees q_r. Suppose that

$$z = \sum_{r \in R} a_r z_r$$

where $a_r \in A$ and $\deg(a_r) + q_r = q$. Suppose that operations Φ_r correspond to the pairs (d, z_r); then we have the following result.

THEOREM 3.7.1. *There is an operation Φ associated with (d, z) and such that*

$$\sum_{r\in R}[a_r\Phi_r(\varepsilon)] = [\Phi(\varepsilon)] \quad \mathrm{mod} \sum_r a_r Q^{l+q_r}(z_r, X)$$

for each X and each A-map $\varepsilon : C_0 \to H^*(X)$ (of degree l) such that $d\varepsilon=0$.

We remark that $Q^{l+q}(z, X) \subset \sum_r a_r Q^{l+q_r}(z_r, X)$; this is easily verified. We may call the group $\sum_r a_r Q^{l+q_r}(z_r, X)$ the *total indeterminacy* of the operations $a_r\Phi_r$ and Φ; the expressions in square brackets denote cosets of this group. It is worth noting, for the applications, that if the set R contains only one member r, then

$$Q^{l+q}(z, X) = a_r Q^{l+q_r}(z_r, X).$$

PROOF OF THEOREM 3.7.1. Let $E = E_\lambda$ be a canonical fibering associated with d, for some $\lambda > \mathrm{Max}(-\nu, q-2\nu)$. Let v_r be an element in $\Phi_r(\varepsilon_E)$ such that $v_r i^* = z_r\theta_1$. Define v by $v = \sum_{r\in R} a_r v_r$; then $v \in H^{\lambda+q}(E)$ and $vi^* = z\theta_1$. By Lemma 3.6.4 there is an operation Φ associated with (d, z) such that $v \in \Phi(\varepsilon_E)$. We thus have

$$[\Phi(\varepsilon_E)] - \sum_{r\in R}[a_r\Phi_r(\varepsilon_E)] = 0 \ .$$

By Lemma 3.5.1 we have

$$[\Phi(\varepsilon)] - \sum_{r\in R}[a_r\Phi_r(\varepsilon)] = 0 \ .$$

This completes the proof.

For the next theorem, we suppose given the following anticommutative diagram, in which d and d' are A-maps of the usual kind, while ρ_0 and ρ_1 are A-maps of degree r.

$$\begin{array}{ccc} C_1 & \xrightarrow{\rho_1} & C_1' \\ {\scriptstyle d}\downarrow & & \downarrow{\scriptstyle d'} \\ C_0 & \xrightarrow{\rho_0} & C_0' \end{array}$$

Let z be an element of Ker d, of total degree q, and let Φ be an operation associated with (d, z). Then we have the following result.

THEOREM 3.7.2. *There is an operation Φ' associated with $(d', z\rho_1)$ such that*

$$\Phi(\rho_0\varepsilon') = [\Phi'(\varepsilon')] \quad \mathrm{mod}\ Q^{l+q+r}(z, X)$$

for each X and each A-map ε': $C_0' \to H^*(X)$ (of degree l) such that $d'\varepsilon'=0$.

We remark that $\Phi(\rho_0\varepsilon')$ is defined and that $Q^{l+q+r}(z', X) \subset Q^{l+q+r}(z, X)$; this is easily verified.

EXAMPLE. Consider the case in which C_0, C'_0 are free on single generators c_0, c'_0 of degree zero, and ρ_0 is defined by $c_0\rho_0 = ac'_0$, where $a \in A_r$. We may replace the map ε' by a class x' in $H^*(X)$, and our conclusion becomes

$$\Phi(ax') = [\Phi'(x')] .$$

For later use, we give the first step in the proof the status of a lemma. Let λ be a value of l such that $\lambda > \mathrm{Max}(-\nu', q + r - 2\nu')$; let $E' = E'_\lambda$ be a canonical fibering associated with d'.

LEMMA 3.7.3. *There is a class v' in $H^{\lambda+q+r}(E')$ such that*

$$v' \in \Phi(\rho_0\varepsilon_{E'}) , \quad v'i^* = z\rho_1\theta'_1 .$$

This is immediate, by applying Axiom 5 (for Φ) to the following anticommutative diagram.

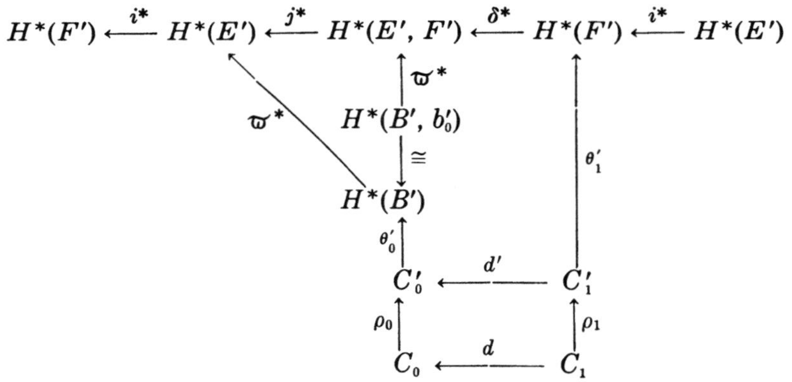

PROOF OF THEOREM 3.7.2. Let v' be as in the lemma. By Lemma 3.6.4, there is an operation Φ' associated with $(d', z\rho_1)$ and such that $v' \in \Phi'(\varepsilon_{E'})$. It is now easy to check that the operation

$$\chi(\varepsilon') = \Phi(\rho_0\varepsilon') - [\Phi'(\varepsilon')]$$

satisfies the conditions of Lemma 3.5.1. Therefore

$$\Phi(\rho_0\varepsilon') - [\Phi'(\varepsilon')] = 0 .$$

This completes the proof.

Theorem 3.7.2 has a converse, which we state as Lemma 3.7.4. Suppose given the following diagram, in which d and d' are A-maps of the usual kind, while ρ_0 is an A-map of degree r.

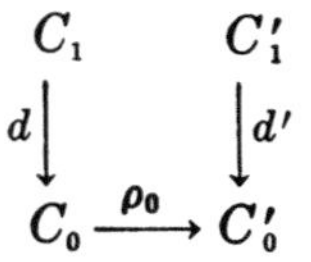

Let z, z' be elements of Ker d, Ker d' of total degrees $q, q+r$, and let Φ, Φ' be operations associated with the pairs $(d, z), (d', z')$. Assume the following conditions:-

(i) Wherever $\Phi'(\varepsilon')$ is defined, $\Phi(\rho_0\varepsilon')$ is defined.

(ii) For one value λ of l such that $\lambda > \mathrm{Max}(-\nu', q+r-2\nu')$ we have

$$\Phi(\rho_0\varepsilon') \supset \Phi'(\varepsilon')$$

for each A-map $\varepsilon': C_0 \to H^*(X)$ of degree λ such that $\Phi'(\varepsilon')$ is defined.

Then we have the following conclusions.

LEMMA 3.7.4. *With the above data, there is an A-map $\rho_1: C_1 \to C_1'$ (of degree r) such that $\rho_1 d' = (-1)^r d\rho_0$ and $z' = z\rho_1$. Moreover, we have*

$$\Phi(\rho_0\varepsilon') \supset \Phi'(\varepsilon')$$

for A-maps ε' of any degree.

PROOF. Our first step is to deduce from (i) that $\mathrm{Im}(d\rho_0) \subset \mathrm{Im}\, d'$. Let $E' = E_l'$ be a canonical fibering associated with d'; then $d'\varepsilon_{E'} = 0$, and hence $d\rho_0\varepsilon_{E'} = 0$; using Lemma 3.5.2, we see that

$$(\mathrm{Im}\, d\rho_0)_t \subset (\mathrm{Im}\, d')_t$$

if $t < l + 2\nu'$. Since l is arbitrary, we have $\mathrm{Im}(d\rho_0) \subset \mathrm{Im}\, d'$.

We can now construct some map $\rho_1': C_1 \to C_1'$ (of degree r) such that $\rho_1 d' = (-1)^r d\rho_0$. Let $E' = E_\lambda'$ be a canonical fibering associated with d'. By Lemma 3.7.3, there is a class v' in $H^{\lambda+q+r}(E')$ such that

$$v' \in \Phi(\rho_0\varepsilon_{E'}) , \quad v'i^* = z\rho_1'\theta_1' .$$

On the other hand, there is a class w' in $H^{\lambda+q+r}(E')$ such that

$$w' \in \Phi'(\varepsilon_{E'}) \subset \Phi(\rho_0\varepsilon_{E'}) , \quad w'i^* = z'\theta_1' .$$

By Axiom 2 for Φ, we have an A-map $f: C_1 \to H^*(E')$ such that $w' - v' = zf$; thus $(z' - z\rho_1')\theta_1' = zfi^*$. Using Lemma 3.5.2, we can define an A-map $g: C_1 \to \mathrm{Ker}\, d'$ such that

$$g\theta_1' | C_{1,t} = fi^* | C_{1,t} \qquad \text{for} \quad t \leqq q+1 .$$

We now have $z' - z\rho_1' = zg$. We can take $\rho_1 = \rho_1' + g$.

It is now clear that $Q^{l+q+r}(z', X) \subset Q^{l+q+r}(z, X)$ for each l, so we may apply Lemma 3.5.1 to the operation

$$\chi(\varepsilon') = \Phi(\rho_0\varepsilon') - [\Phi'(\varepsilon')]$$

and show that it is zero. This completes the proof.

We now give a subsidiary result, which may however serve to justify some of our concepts. We shall suppose that the coefficient group G is a field, that Φ is an operation associated with a pair (d, z), and that $l > \text{Max}(-\nu, q - 2\nu)$ (with the notations used above); in other words, we shall only prove this result in a stable range of dimensions.

LEMMA 3.7.5. *If the map $d : C_1 \to C_0$ is minimal (in the sense of § 2.1) then the lth component Φ^l of Φ is minimal (in the sense of § 3.2).*

PROOF. Let $E = E_l$ be a canonical fibering associated with d. If there is an operation Ψ such that $\Psi \subset \Phi^l$, choose a class v in $\Psi(\varepsilon_E)$; let χ^l be the operation determined by the universal example v; then $\chi^l \subset \Psi \subset \Phi^l$, by Lemma 3.2.3. By Lemma 3.6.4, χ^l is one component of a stable operation χ associated with some pair (d, z'). By Lemma 3.7.4, there is an A-map $\rho_1: C_1 \to C_1$ such that $\rho_1 d = d$ and $z' = z\rho_1$. Since d is minimal, ρ_1 is an isomorphism. Therefore $Q^{l+q}(z, X) = Q^{l+q}(z', X)$, and $\chi^l = \Psi = \Phi^l$. This completes the proof.

3.8. *Outline of applications.* Throughout this section we shall assume that the coefficient group G is a field. Under this condition, we shall give a general scheme for applying the results of §§ 3.6, 3.7. We wish to show, in particular, how homological algebra helps us to find secondary operations to serve given purposes, and to find relations between such operations. For example, if a class x in $H^n(X)$ generates a sub-A-module M of $H^*(X)$, we shall be led to consider operations Φ_r defined on x and in (1-1) correspondence with a base of $\text{Tor}_2^A(G, M)$. The particular formulae used later were found by applying the principles outlined in this section.

Suppose given a sub-A-module M of $H^*(X)$ which is locally finite-dimensional. Suppose that we wish to study the stable secondary operations Φ which are defined on J-tuples $\{x_j\}$ of classes in M. It is equivalent to say that such operations Φ are defined on A-maps $\varepsilon : C_0 \to M$. Each such Φ will be associated with a pair (d, z) such that $d\varepsilon = 0$.

It is sufficient to consider one particular pair of A-maps $d: C_1 \to C_0$ and $\varepsilon : C_0 \to M$ such that

$$C_1 \xrightarrow{d} C_0 \xrightarrow{\varepsilon} M \longrightarrow 0$$

is exact. For suppose that $d': C_1' \to C_0'$ and $\varepsilon': C_0' \to M$ are some other A-maps such that $d'\varepsilon' = 0$, and let Φ' be an operation corresponding to a pair (d', z'). Then we can form the following diagram.

$$\begin{array}{ccccc} C_1' & \xrightarrow{d'} & C_0' & \xrightarrow{\varepsilon'} & M \\ \downarrow{\scriptstyle \rho_1} & & \downarrow{\scriptstyle \rho_0} & & \downarrow{\scriptstyle 1} \\ C_1 & \xrightarrow{d} & C_0 & \xrightarrow{\varepsilon} & M \end{array}$$

By Theorem 3.7.2, we have

$$\Phi'(\varepsilon') = [\Phi(\varepsilon)]$$

for some Φ associated with $(d, z'\rho_1)$.

We may therefore suppose that the A-maps $d : C_1 \to C_0$ and $\varepsilon : C_0 \to M$ considered are the beginning of a minimal resolution, in the sense of § 2.1; let its first few terms be

$$C_3 \xrightarrow{d_3} C_2 \xrightarrow{d_2} C_1 \xrightarrow{d_1} C_0 \xrightarrow{\varepsilon} M .$$

We may now consider a subset of the operations Φ. Take an A-base of elements $c_{2,r}$ in C_2; set $z_r = c_{2,r}d_2$; let Φ_r be an operation corresponding to (d_1, z_r). It is a property of the operations Φ_r that the other operations Φ are linearly dependent on them, in a suitable sense. To be precise, let Φ be an operation associated with a pair (d_1, z). Since $zd_1 = 0$ and Ker $d_1 = \text{Im}\, d_2$, we have $z = \sum_r a_r z_r$ for some a_r in A.
By Theorems 3.7.1, 3.6.2 we have

$$[\Phi(\varepsilon)] = [c\varepsilon] + \sum_r [a_r \Phi_r(\varepsilon)]$$

(modulo the total indeterminacy involved). It is therefore sufficient to consider the operations $\Phi_r(\varepsilon)$, provided that their indeterminacies are small enough for our purposes.

By § 2.1, the basic operations Φ_r are in (1-1) correspondence with a G-base of $\text{Tor}_2^A(G, M)$. It may happen that we can calculate $\text{Tor}_2^A(G, M)$ without using resolutions; if so, we can count how many basic operations Φ_r are needed.

We have now shown how we may consider a set of basic operations; we proceed to show how we may consider relations between them.

First take an element c_3 of C_3. We may write

$$c_3 d_3 = \sum_r a_r c_{2,r} \qquad \text{(where } a_r \in A\text{)} .$$

Applying d_2, we have $\sum_r a_r z_r = 0$. By Theorems 3.7.1, 3.6.2 we have $\sum_r [a_r \Phi_r(\varepsilon)] = [c\varepsilon]$ (modulo the total indeterminacy involved). Now let c_3 run over an A-base of C_3; we obtain basic relations between the Φ_r, in (1-1) correspondence with a base of $\text{Tor}_3^A(G, M)$. As before, it may happen that we can calculate $\text{Tor}_3^A(G, M)$ without using resolutions. In this case we can count how many basic relations are available.

We will now consider a slightly different application, which concerns composite operations of the form Φa. Suppose that M, M' are (locally finite-dimensional) submodules of $H^*(X)$ such that $M' \subset M$. We can take minimal resolutions of M, M' and form the following diagram.

$$\begin{array}{ccccccc} C_2' & \xrightarrow{d_2'} & C_1' & \xrightarrow{d_1'} & C_0' & \xrightarrow{\varepsilon'} & M' \\ \downarrow{\scriptstyle \rho_2} & & \downarrow{\scriptstyle \rho_1} & & \downarrow{\scriptstyle \rho_0} & & \downarrow{\scriptstyle i} \\ C_2 & \xrightarrow{d_2} & C_1 & \xrightarrow{d_1} & C_0 & \xrightarrow{\varepsilon} & M \end{array}$$

Let Ψ' be a basic operation corresponding to a pair $(d_1', c_2'd_2')$; we seek to evaluate $\Psi'(\varepsilon')$ in terms of basic operations $\Phi_r(\varepsilon)$. For example, if M is generated (*qua* A-module) by one generator x, and if M' is generated by ax (where $a \in A$), then the problem is equivalent to evaluating $\Psi'(ax)$ in terms of operations $\Phi_r(x)$.

Let us write

$$c_2'\rho_2 = \sum_r b_r c_{2,r} \qquad \text{(where } b_r \in A) .$$

Then we have

$$c_2'd_2'\rho_1 = \sum_r b_r z_r .$$

By Theorems 3.6.2, 3.7.1, 3.7.2 we have

$$[\Psi'(\varepsilon')] = [c\varepsilon] + \sum_r [b_r \Phi_r(\varepsilon)]$$

(modulo the total indeterminacy involved).

It may happen that this formula is useful to us only if the coefficients b_r are of positive degree. To locate the Ψ' which admit a formula of this sort, one should find the kernel of

$$i_*\colon \operatorname{Tor}_2^A(G, M') \to \operatorname{Tor}_2^A(G, M);$$

for the coefficients b_r of degree zero in $c_2'\rho_2 = \sum_r b_r c_{2,r}$ are determined by i_*.

This concludes our outline of the use of homological algebra in searching for operations and relations to serve given purposes.

3.9. *The Cartan formula.* Throughout this section, we shall assume that the coefficient domain G is the field Z_p of integers modulo p, where p is a prime. Under this condition, we shall prove the existence of a Cartan formula [11] for expanding $\Phi(xy)$, where xy is a cup-product and Φ is an operation of the sort considered in § 3.6. The expansion which we obtain is of the form

$$\sum_r (-1)^{\eta(r)} \Phi_r'(x) \Phi_r''(y) .$$

where the signs are given by

$$\eta(r) = \deg(x) \deg(\Phi_r'') .$$

We can give one elementary example of the sort of Cartan formula at issue; for the Bockstein coboundary β_{p^f} is a stable operation, and is a secondary operation if $f = 2$; it satisfies the formula

$$\beta_{p^f}(xy) \supset (\beta_{p^f}x)y + (-1)^{\eta}x(\beta_{p^f}y)$$

where $\eta = \deg(x)$.

The precise result we require is stated as Theorem 3.9.4; the remainder of this section is devoted to proving it. The proof uses the method of the universal example. The obvious universal example for this purpose is a Cartesian product, as considered by Serre [30]. However, we are particularly concerned with stable operations; we must therefore show how our Cartan formulae behave under suspension. For this purpose we use a "product" more conveniently related to the suspension. Let Y', Y'' be enumerable CW-complexes with base-points y'_0, y''_0; we may form the "reduced product" [18]

$$Y' \times Y'' = Y' \times Y''/(Y' \times y''_0 \cup y'_0 \times Y'') .$$

This is again an enumerable CW-complex. The quotient map $q\colon Y' \times Y'' \to Y' \times Y''$ induces a homomorphism

$$q^+\colon H^+(Y' \times Y'') \to H^+(Y' \times Y'') ,$$

and q^+ embeds $H^+(Y' \times Y'')$ as a direct summand of $H^+(Y' \times Y'')$, complementary to $H^+(Y')$ and $H^+(Y'')$. If $y' \in H^+(Y')$ and $y'' \in H^+(Y'')$, we have the "external" cup-product $y' \times y''$ in $H^+(Y' \times Y'')$ and a "reduced" cup-product $y' \times y''$ in $H^+(Y' \times Y'')$ defined by

$$(y' \times y'')q^+ = y' \times y'' .$$

We now set up some more notation. If K, L are subsets of $H^q(X)$, $H^r(X)$ then we define KL to be the set of cup-products kl, where $k \in K$, $l \in L$; thus $KL \subset H^{q+r}(X)$. If K, L are subsets of $H^q(X)$ and λ, μ lie in Z_p, then we define $\lambda K + \mu L$ to be the set of linear combinations $\lambda k + \mu l$, where $k \in K$, $l \in L$. These definitions give a precise sense to formulae such as

$$\textstyle\sum_r (-1)^{\eta(r)} \Phi'_r(x)\Phi''_r(y);$$

an expression of this sort denotes some set of cohomology classes.

If K, L are subsets of $H^q(Y')$, $H^q(Y'')$ then we define the subset $K \times L$ of $H^{q+r}(Y' \times Y'')$ in a similar way.

We will now use the reduced product to study Cartan formulae valid in a fixed pair of dimensions. We will suppose that each of S, S', S'' is a natural subset of cohomology, in one variable, whose dimension is l, l', l'' in the three cases. Let R be a finite set of indices r; let Φ, Φ'_r, Φ''_r be operations defined on S, S', S'', and of degrees (say) q, q'_r, q''_r. We will suppose that

$$l' + l'' = l , \quad q'_r + q''_r = q \qquad \text{(for each } r)$$

and that our operations have arguments and values of positive degree.

We will say that Φ *can be expanded on* S', S'' *in terms of* Φ'_r, Φ''_r if the following two conditions hold. First, for each space X and each $x' \in S'(X)$, $x'' \in S''(X)$ we have

(i) $x'x'' \in S(X)$, and $\Phi(x'x'')$ has a non-empty intersection with

$$\sum_r (-1)^{\eta(r)} \Phi'_r(x') \Phi''_r(x'')$$

where $\eta(r) = l'q''_r$.

Secondly, whenever Y', Y'' are enumerable CW-complexes and $y' \in S'(Y')$, $y'' \in S''(Y'')$ we have

(ii) $y' \times y'' \in S(Y' \times Y'')$, and $\Phi(y' \times y'')$ has a non-empty intersection with

$$\sum_r (-1)^{\eta(r)} \Phi'_r(y') \times \Phi''_r(y'') .$$

For our first lemma, we suppose that we can choose enumerable CW-complexes Y', Y'' and classes $y' \in H^+(Y')$, $y'' \in H^+(Y'')$ so that (Y', y'), (Y'', y'') are universal examples for S', S''. (We can certainly do this if the subsets S', S'' are determined by K-tuples of primary operations, since we can replace the canonical fiberings of § 3.3 by weakly equivalent enumerable CW-complexes.)

LEMMA 3.9.1. *If condition* (ii) *above holds for one such pair of universal examples* (Y', y') *and* (Y'', y''), *then conditions* (i) *and* (ii) *hold in general, so that* Φ *can be expanded on* S', S'' *in terms of* Φ'_r, Φ''_r.

This lemma is immediate, by naturality.

For our next lemma, we suppose that not only the subsets S', S'' but also the subsets $(S')^s$, $(S'')^s$ admit universal examples which are enumerable CW-complexes.

LEMMA 3.9.2. *If* Φ *can be expanded on* S', S'' *in terms of* Φ'_r, Φ''_r *then* Φ^s *can be expanded on* $(S')^s$, S'' *in terms of* $(\Phi'_r)^s$, Φ''_r *and on* S', $(S'')^s$ *in terms of* Φ'_r, $(\Phi''_r)^s$.

PROOF. It is sufficient to prove that half of the lemma which relates to $(S')^s$, S'', since the other half may be proved similarly,

If Y is a CW-complex, we may interpret the suspension sY to be the reduced product $S^1 \times Y$, since this is homotopy-equivalent to the ordinary suspension in this case. Suppose that $y \in H^n(Y) (n > 0)$; let s^1 be the generator of $H^1(S^1)$; and let s be the suspension isomorphism. Then we have the equation

$$ys^{-1} = (-1)^n s^1 \times y .$$

If Y', Y'' are enumerable CW-complexes, we have the "associativity" formula

$$(S^1 \times Y') \times Y'' = S^1 \times (Y' \times Y'') .$$

In $H^*(S^1 \times Y' \times Y'')$ we have

$$(y' \times y'')s^{-1} = (-1)^{n''}(y's^{-1}) \times y''$$

(where $n'' = \deg(y'')$); this follows easily from the equation above and the associativity of the cup-product.

Now let $(Y', y'), (Y'', y'')$ be universal examples for $(S')^s, S''$. Since $(-1)^{l''}y' \in (S')^s(Y')$, we have $(-1)^{l''}y's^{-1} \in S'(S^1 \times Y')$; also $y'' \in S''(Y'')$. If Φ can be expanded on S', S'', we have

$$(-1)^{l''}(y's^{-1}) \times y'' \in S(S^1 \times Y' \times Y'') ,$$

whence $(y' \times y'')s^{-1} \in S(S^1 \times Y' \times Y'')$ and $y' \times y'' \in S^s(Y' \times Y'')$.

Again, if Φ can be expanded on S', S'', we have

$$\Phi\big((-1)^{l''}(y's^{-1}) \times y''\big) \cap \textstyle\sum_r (-1)^{\eta(r)}\Phi'_r\big((-1)^{l''}y's^{-1}\big) \times \Phi''_r(y'') \neq 0 ,$$

where $\eta(r) = l'q''_r$. In this expression we have

$$\Phi'_r\big((-1)^{l''}y's^{-1}\big) = \big((\Phi'_r)^s((-1)^{l''}y')\big)s^{-1} = (-1)^{l''}\big((\Phi'_r)^s(y')\big)s^{-1} ,$$

by Lemma 3.4.2. A little manipulation now shows that

$$\Phi^s(y' \times y'') \cap \textstyle\sum_r (-1)^{\zeta(r)}(\Phi'_r)^s(y') \times \Phi''_r(y'') \neq 0 ,$$

where $\zeta(r) = (l' - 1)q''_r$. By Lemma 3.9.1, Φ^s can be expanded on $(S')^s$, S'' in terms of $(\Phi'_r)^s$, Φ''_r. This completes the proof.

It will be convenient if we now set up the data for Theorem 3.9.4. Let C_0 be a free A-module on one generator c_0 of degree zero. We suppose given three A-maps $d: C_1 \to C_0$, $d': C'_1 \to C'_0$ and $d'': C''_1 \to C''_0$, as in § 3.6. We shall suppose that $C'_{1,q+1} = 0$ for $q < 0$ and that $d' \mid C'_{1,1}: C'_{1,1} \to C'_{0,1}$ is monomorphic; this assumption is automatically satisfied if d' is minimal. The result of this assumption about d' is that if E'_l is a canonical fibering associated with d', as in § 3.6, then E'_l is $(l-1)$-connected and $H^l(E'_l) = Z_p$, generated by the "fundamental class" e^l. We make the corresponding assumption about d''.

Corresponding to d, d', d'' we have stable natural subsets T, T', T'' in one variable. We now make an essential assumption restricting d, d', d''; we suppose that for each space X and for each $x' \in T'(X)$, $x'' \in T''(X)$ the cup-product $x'x''$ lies in $T(X)$.

LEMMA 3.9.3.

(a) *If Y', Y'' are enumerable* CW-*complexes and $y' \in T'(Y')$, $y'' \in T''(Y'')$ then $y' \times y'' \in T(Y' \times Y'')$.*

(b) *If X is any space, then*

$$T'(X) \subset T(X) , \quad T''(X) \subset T(X) .$$

PROOF OF PART (a). By our assumptions, the external cup-product $y' \times y''$ lies in $T(Y' \times Y'')$; and if a_k is a primary operation such that

$a_k(y' \times y'') = 0$, then $a_k(y' \mathbf{\times} y'') = 0$, since q^+ is a monomorphism.

PROOF OF PART (b). It is sufficient to prove one inclusion, say the second. If $l > 0$, let (Y'', y'') be a universal example for $(T'')^l$ such that Y'' is an enumerable CW-complex; then $y'' \in (T'')^l(Y'')$. The element $(-1)^l s^1$ in $H^1(S^1)$ certainly lies in $(T')^1(S^1)$, since $a(-1)^l s^1 = 0$ for each primary operation a of positive degree. Therefore $(-1)^l s^1 \mathbf{\times} y'' \in T^{l+1}(S^1 \mathbf{\times} Y'')$; since $y'' s^{-1} = (-1)^l s^1 \mathbf{\times} y''$ and T is stable, we have $y'' \in T^l(Y'')$. The inclusion $(T'')^l(X) \subset T^l(X)$ for a general space X now follows by naturality.

We will now state Theorem 3.9.4. This theorem will allow us to expand $\Phi(x'x'')$ in the form

$$\Phi'_\alpha(x')x'' + \textstyle\sum_{r \neq \alpha, \omega} (-1)^{\eta(r)} \Phi'_r(x')\Phi''_r(x'') + (-1)^{\eta(\omega)} x'\Phi''_\omega(x'')$$

whenever $x' \in T'(X)$ and $x'' \in T''(X)$. It will also give us some information about the operations Φ'_α, Φ''_ω which occur in the end terms of this expansion; they are essentially the same as Φ.

THEOREM 3.9.4. *If d, d', d'', T, T' and T'' are as above, and if Φ is an operation of degree $q > 0$ associated with a pair (d, z), then there are operations Φ'_r, Φ''_r which are associated with pairs (d', z'_r), (d'', z''_r) and satisfy the following conditions.*

(i) *For each pair of dimensions (l, m), the component Φ^{l+m} can be expanded on $(T')^l$, $(T'')^m$ in terms of $(\Phi'_r)^l$, $(\Phi''_r)^m$.*

(ii) *There are two values α, ω of r such that Φ''_α, Φ'_ω are identity operations. For $r \neq \alpha$, ω we have*

$$0 < \deg(\Phi'_r) < q\,, \quad 0 < \deg(\Phi''_r) < q\,.$$

(iii) *There is an A-map $\rho'_1\colon C_1 \to C'_1$ such that $\rho'_1 d' = d$ and $z\rho'_1 = z'_\alpha$. For each space X and each $x' \in T'(X)$ we have*

$$\Phi'_\alpha(x') \subset \Phi(x')\,.$$

(iv) *There is an A-map $\rho''_1\colon C_1 \to C''_1$ such that $\rho''_1 d'' = d$ and $z\rho''_1 = z''_\omega$. For each space X and each $x'' \in T''(X)$ we have*

$$\Phi''_\omega(x'') \subset \Phi(x'')\,.$$

We require a further lemma, which is a converse of Lemma 3.9.2. We shall suppose that Φ, Φ'_r, Φ''_r are stable operations associated with pairs (d, z), (d', z'_r), (d'', z''_r); we set $q = \deg(\Phi)$.

LEMMA 3.9.5. *If Φ^{l+m} can be expanded on $(T')^l$, $(T'')^m$ in terms of $(\Phi'_r)^l$, $(\Phi''_r)^m$ and if $l > q$, $m > q$ then Φ^{l+m+1} can be expanded on $(T')^{l+1}$, $(T'')^m$ in terms of $(\Phi'_r)^{l+1}$, $(\Phi''_r)^m$ and on $(T')^l$, $(T'')^{m+1}$ in terms of $(\Phi'_r)^l$, $(\Phi''_r)^{m+1}$.*

Roughly speaking, the effect of Lemmas 3.9.2 and 3.9.5 is that in order to prove Theorem 3.9.4, it is sufficient to consider a single pair of dimensions (l, m).

PROOF OF LEMMA 3.9.5. As for Lemma 3.9.2, it is sufficient to prove that half of the lemma which passes from dimensions (l, m) to dimensions $(l + 1, m)$. Let (Y', y'), (Y'', y'') be universal examples for $(T')^{l+1}$, $(T'')^m$ such that Y', Y'' are enumerable CW-complexes. As in Lemma 3.5.1, we can take a space X and a map $g : sX \to Y'$ such that

$$g^*: H^r(Y') \to H^r(sX)$$

is an isomorphism for $r \leqq l + q + 1$; we may suppose that X is an enumerable CW-complex, and that sX is the reduced product $S^1 \times X$. We can now form the map

$$g \times 1: S^1 \times X \times Y'' \to Y' \times Y'' ;$$

this induces isomorphisms of cohomology up to dimension $l + m + q + 1$ at least. It is now easy to check that

$$\Phi^{l+m+1}(y' \times y'') \cap \sum_r (-1)^{\eta(r)} (\Phi'_r)^{l+1}(y') \times (\Phi''_r)^m(y'') \neq 0$$

by applying $(g \times 1)^+$ and using the data. The conclusion now follows by Lemma 3.9.1. This completes the proof.

PROOF OF THEOREM 3.9.4. We begin by fixing attention on a pair of dimensions (l, m) such that $l > q$, $m > q$. Let E'_l, E''_m be canonical fiberings associated with d', d'', as in § 3.6. We may take weakly equivalent enumerable CW-complexes Y', Y''; we write y', y'' for the classes corresponding to the fundamental classes e', e''. Since $H^*(Y')$, $H^*(Y'')$ are locally finite-dimensional, the reduced cup-product

$$\mu : H^+(Y') \otimes H^+(Y'') \to H^+(Y' \times Y'')$$

is an isomorphism. The coset $\Phi(y' \times y'')$ is defined; we may choose a class v in $\Phi(y' \times y'')$, and expand v in the form

$$v = \sum_r (-1)^{\eta(r)} v'_r \times v''_r$$

where $v'_r \in H^+(Y')$, $v''_r \in H^+(Y'')$ and $\eta(r) = l(\deg(v''_r) - m)$. By our original data, Y' is $(l - 1)$-connected and $H^l(Y') = Z_p$, generated by y'; similarly for Y''. We may therefore take our expansion so that $\deg(v'_r) > l$ and $\deg(v''_r) > m$, except that $v'_\omega = y'$ and $v''_\alpha = y''$. By Lemma 3.6.4, there are operations Φ'_r (for $r \neq \omega$), Φ''_r (for $r \neq \alpha$) which are associated with pairs (d', z'_r), (d'', z''_r) and are such that v'_r, v''_r are universal examples for $(\Phi'_r)^l$, $(\Phi''_r)^m$. We define Φ'_ω and Φ''_α directly, defining them to be identity operations. By Lemma 3.9.1, Φ^{l+m} can be expanded on $(T')^l$, $(T'')^m$

in terms of $(\Phi'_r)^l$, $(\Phi''_r)^m$. By Lemmas 3.9.2 and 3.9.5, a similar conclusion follows in every pair of dimensions. This establishes parts (i) and (ii) of the theorem.

We will now examine the class v''_ω which occurs in the above expansion. Let us take $W = S^l$, so that $H^l(W) = Z_p$, generated by w. We may take a map $f: W \to Y$ such that $y'f^+ = (-1)^{lm}w$; thus $f^+: H^r(Y') \to H^r(W)$ is zero for $r > l$. We have

$$\sum_r (-1)^{\eta(r)} v'_r \times v''_r \in \Phi^{l+m}(y' \times y'');$$

applying $(f \times 1)^+$, we find

$$(-1)^{l(q+m)} w \times v''_\omega \in \Phi^{l+m}((-1)^{lm} w \times y'')$$

in $W \times Y''$. But $W \times Y''$ is the l-fold suspension of Y'', and we have the equation

$$ys^{-l} = (-1)^{lt} w \times y \qquad \text{(for } y \in H^t(Y'')\text{)}.$$

Since Φ is stable, we deduce $v''_\omega \in \Phi^m(y'')$. Lemma 3.2.3 now shows that

$$(\Phi''_\omega)^m (x'') \subset \Phi^m(x'') \qquad \text{(for } x'' \in (T'')^m(X)\text{)}.$$

Lemma 3.7.4 now shows the existence of an A-map $\rho''_1: C_1 \to C''_1$ such that $\rho''_1 d'' = d$ and $z\rho''_1 = z''_\omega$; it also guarantees that

$$\Phi''_\omega(x'') \subset \Phi(x'')$$

for classes x'' of any degree. This establishes part (iv) of the theorem; we may establish part (iii) similarly. The proof of Theorem 3.9.4 is complete.

CHAPTER 4. PARTICULAR OPERATIONS

4.1. *Introduction.* In this chapter we shall use the theory of Chapter 3 to define and study a particular set of secondary cohomology operations $\Phi_{i,j}$. These operations act on cohomology with mod 2 coefficients; they will be defined in § 4.2. The object of our work is to prove the formula

$$\sum_{i,j} a_{i,j,k} \Phi_{i,j}(u) = [\mathrm{Sq}^{2^{k+1}}(u)]$$

mentioned in Chapter 1; this formula is proved in § 4.6. The line of proof is as follows. We first apply the theory of Chapter 3 to prove a formula

$$\sum_{i,j} a_{i,j,k} \Phi_{i,j}(u) = [\lambda \mathrm{Sq}^{2^{k+1}}(u)]$$

containing an undetermined coefficient λ. We then determine the coefficient λ by applying the formula to a suitable class u in the cohomology of a suitable space. For this purpose we use complex projective space of infinitely-many dimensions, which we shall call P. We therefore need to know the values of the operations $\Phi_{i,j}$ in P; these are found in § 4.5. It turns out that an inductive calculation is possible; there are many relations between the different operations in P, and these enable us to deduce, from the value of one selected operation, the values of all the others. In § 4.4 we find the value of this one operation. In § 4.3 we apply the theory of Chapter 3 to prove those relations between the operations which are needed for the calculation in § 4.5. This work, therefore, will complete the proof of Theorem 1.1.1.

In this chapter we have to use the Steenrod squares Sq^k for values of k which may have a complicated form. We therefore make a convention, by which we write $\mathrm{Sq}(k)$ instead of Sq^k in such cases. Similarly, we may write $\xi_i(k)$ instead of ξ_i^k in dealing with A^* (see § 2.4). Again, we write $H^m(X)$ instead of $H^m(X; Z_2)$, and $H^*(X)$ instead of $H^*(X; Z_2)$, since we shall not have to deal with any coefficients except Z_2.

4.2. *The operations* $\Psi(u)$ *and* $\Phi_{i,j}(u)$. In this section we apply the theory of Chapter 3 to define certain particular secondary operations, acting on cohomology with mod 2 coefficients. These will be operations on one variable.

To define our first operation, we have to give a pair (d, z) (see § 3.1). We take C_0 to be A-free on one generator c of degree zero; we take C_1 to be A-free on three generators c_1, c_3, c_4 of degrees 1, 3, 4. We define d by

$$c_i d = \mathrm{Sq}^i c \qquad (i = 1, 3, 4).$$

We define z by

$$z = \mathrm{Sq}^4 c_1 + \mathrm{Sq}^2 c_3 + \mathrm{Sq}^1 c_4 .$$

This pair (d, z) corresponds, of course, to the relation

$$\mathrm{Sq}^4\mathrm{Sq}^1 + \mathrm{Sq}^2\mathrm{Sq}^3 + \mathrm{Sq}^1\mathrm{Sq}^4 = 0 .$$

We note that $(C_0/dC_1)_4 = 0$; so by Theorems 3.6.1, 3.6.2, there is a unique operation Ψ (of degree 4) associated with this pair (d, z). This is the first operation we require.

To define further operations, we introduce further pairs (d, z). We begin by constructing the first terms

$$C_1 \xrightarrow{d} C_0 \xrightarrow{\varepsilon} Z_2$$

of a minimal resolution (see § 2.1) of Z_2 over A. We may do this as follows. We take C_0 to be A-free on one generator c of degree zero, and define $c\varepsilon = 1$. We take C_1 to be A-free on generators c_i of degrees 2^i, for $i = 0, 1, 2, \cdots$. We define d by setting

$$c_i d = \mathrm{Sq}(2^i)c \ .$$

It is clear that $\mathrm{Im}(d) = \mathrm{Ker}(\varepsilon)$, since the elements $\mathrm{Sq}(2^i)$ are multiplicative generators for A (see [4]). It is also easy to show that d is minimal. In fact, to do this, we should take an element $x = \lambda c_i + \sum_{j<i} a_j c_j$ of degree 2^i in C_1, assume $xd=0$, and deduce that $\lambda=0$. This is immediate from the equation

$$\xi_1(2^i)((\lambda c_i + \textstyle\sum_{j<i} a_j c_j)d) = \lambda \ .$$

Next, we make use of the epimorphism $\theta\colon \mathrm{Ker}(d) \to \mathrm{Tor}_2^A(Z_2, Z_2)$ introduced in § 2.2. It was shown in § 2.5 that the elements $h_i h_j$ (with $0 \leqq i \leqq j$, $j \neq i+1$) in $\mathrm{Ext}_A^2(Z_2, Z_2)$ form a base for it. For $0 \leqq i \leqq j$, $j \neq i+1$, then, we may take cycles $z_{i,j}$ (of degree $2^i + 2^j$) in $\mathrm{Ker}(d)$ such that

$$(h_i h_j)(\theta z_{i,j}) = 1 \ .$$

Let $C_1(j)$ be the submodule of C_1 generated by $c_0, c_1, \cdots, c_j$. Then the cycle $z_{i,j}$ lies in $C_1(j)$. This is clear from the degrees if $i < j$; if $i = j$, it follows by using also the fact that d is minimal. We set $d(j) = d \mid C_1(j)$.

By Theorem 3.6.1, once $z_{i,j}$ is chosen, there is an operation $\Phi_{i,j}$ (of degree $2^i + 2^j - 1$) associated with the pair $(d(j), z_{i,j})$ (where $0 \leqq i \leqq j$, $j \neq i+1$). Such an operation is unique, by Theorem 3.6.2, since $(C_0/dC_1(j))_n = 0$ if $n < 2^{j+1}$. These operations $\Phi_{i,j}$ are the ones which we require. They are defined on classes u such that

$$\mathrm{Sq}(2^r)(u) = 0 \qquad \text{for } 0 \leqq r \leqq j \ .$$

The indeterminacy of $\Phi_{i,j}$ may depend on the choice of the cycle $z_{i,j}$, and *a fortiori* the operation $\Phi_{i,j}$ may do so. However, all the propositions that we shall state about the operations $\Phi_{i,j}$ remain equally true, whatever choice of the $z_{i,j}$ is made. We shall therefore only need to suppose that the $z_{i,j}$ are chosen in some fixed fashion.

For completeness, we should perhaps consider the operation $\Phi'_{i,j}$ associated with the pair $(d(k), z_{i,j})$ for some $k > j$. It has the same indeterminacy as $\Phi_{i,j}$, but is defined on fewer classes u. Moreover, by Theorem 3.7.2, we have

$$\Phi'_{i,j}(u) = \Phi_{i,j}(u)$$

whenever $\Phi'_{i,j}(u)$ is defined. Thus, in what follows, we shall not need to distinguish $\Phi'_{i,j}$ from $\Phi_{i,j}$ by a separate symbol.

It may be of interest to display a particular relation, holding in A, which corresponds to a cycle which one might choose for $z_{i,j}$. We consider first the case $i < j$, so that $i \leqq j-2$. Then the Adem relations [4] for $\mathrm{Sq}(2^i)\,\mathrm{Sq}(2^j)$ and $\mathrm{Sq}(2^{i+1})\mathrm{Sq}(2^j - 2^i)$ both contain the term $\mathrm{Sq}(2^i + 2^j)$. Their sum is therefore an equation of the form

$$\mathrm{Sq}(2^i)\mathrm{Sq}(2^j) = \sum_{0<k<2^j} \lambda_k \mathrm{Sq}(2^i + 2^j - k)\mathrm{Sq}(k)$$

with certain coefficients λ_k. We may use Adem's method [4] to express $\mathrm{Sq}(k)$ in the form

$$\mathrm{Sq}(k) = \sum_{0 \leq l < j} a_{k,l} \mathrm{Sq}(2^l)$$

with certain coefficients $a_{k,l}$ in A. Substituting, we obtain an equation of the form

$$\mathrm{Sq}(2^i)\mathrm{Sq}(2^j) = \sum_{0 \leq l < j} b_l \mathrm{Sq}(2^l)$$

with certain coefficients b_l in A. Hence the expression

$$z_{i,j} = \mathrm{Sq}(2^i)c_j + \sum_{0 \leq l < j} b_l c_l$$

is a cycle in $C_1(j)$. The fact that it satisfies the equation

$$(h_i h_j)(\theta z_{i,j}) = 1$$

follows from Lemma 2.2.2.

The case $i = j$ may be treated similarly, but even more simply, using the Adem relation for $\mathrm{Sq}(2^j)\mathrm{Sq}(2^j)$ (cf. [4]).

It may be remarked that the above process allows us to choose the cycles $z_{i,j}$ in a way which is quite definite, if this should be required. In fact, we have only to remark that Adem's method for reducing $\mathrm{Sq}(k)$ to a sum of products of the generators $\mathrm{Sq}(2^r)$ leads to a well-determined answer. And this is clear, since it proceeds by a well-determined reductive process, using at each step a well-determined substitution.

We conclude by remarking that the operations $\Phi_{0,0}$, $\Phi_{1,1}$ and $\Phi_{0,2}$ do not depend on the choice of the cycles $z_{0,0}$, $z_{1,1}$ and $z_{0,2}$. In fact, it is easy to see that there is only one choice for the cycle $z_{0,0}$, namely $\mathrm{Sq}^1 c_0$. There are only two choices for the cycle $z_{1,1}$, namely

$$(\mathrm{Sq}^2 c_1 + \mathrm{Sq}^3 c_0) + \lambda \mathrm{Sq}^2(\mathrm{Sq}^1 c_0) \qquad (\lambda = 0, 1) .$$

These two cycles can be mapped into one another by an automorphism $\rho_1 \colon C_1(1) \to C_1(1)$ defined as follows:

$$c_0 \rho_1 = c_0$$
$$c_1 \rho_1 = c_1 + \mathrm{Sq}^1 c_0 .$$

By Theorem 3.7.2, the operations corresponding to the two cycles coincide.

Similarly, there are only four choices for the cycle $z_{0,2}$, namely

$$(\mathrm{Sq}^1 c_2 + \mathrm{Sq}^2\mathrm{Sq}^1 c_1 + \mathrm{Sq}^4 c_0) + \lambda \mathrm{Sq}^1(\mathrm{Sq}^2 c_1 + \mathrm{Sq}^3 c_0) + \mu \mathrm{Sq}^3(\mathrm{Sq}^1 c_0) \qquad (\lambda, \mu \in Z_2).$$

As above, the choice does not affect $\Phi_{0,2}$.

4.3. *Relations between the operations Ψ and $\Phi_{i,j}$.* In this section we shall obtain those relations between the operations Ψ and $\Phi_{i,j}$ which we need in § 4.5.

For our first lemma, let $u \in H^m(X)(m>0)$ be a class such that $\mathrm{Sq}^1(u)=0$, $\mathrm{Sq}^3(u) = 0$, $\mathrm{Sq}^4(u) = 0$.

LEMMA 4.3.1. *There is a formula*

$$\Phi_{0,2}\mathrm{Sq}^4\mathrm{Sq}^2(u) = [\mathrm{Sq}^6\Psi(u) + \lambda\mathrm{Sq}^{10}(u)]$$

valid in

$$H^{m+10}(X)/ (\mathrm{Sq}^1H^{m+9}(X) + \mathrm{Sq}^2\mathrm{Sq}^1H^{m+7}(X) + \mathrm{Sq}^4H^{m+6}(X))$$

for a fixed $\lambda \in Z_2$.

We require this formula in order to apply it to the fundamental class in complex projective space. The actual value of λ is not relevant, although at a later stage in our calculations it would be possible to show that $\lambda = 0$.

PROOF. Informally, the proof consists in showing that the relations.

$$(\mathrm{Sq}^1\mathrm{Sq}^4 + \mathrm{Sq}^2\mathrm{Sq}^1\mathrm{Sq}^2 + \mathrm{Sq}^4\mathrm{Sq}^1)\mathrm{Sq}^4\mathrm{Sq}^2 = 0 \ ,$$

$$\mathrm{Sq}^6(\mathrm{Sq}^1\mathrm{Sq}^4 + \mathrm{Sq}^2\mathrm{Sq}^3 + \mathrm{Sq}^4\mathrm{Sq}^1) = 0$$

are the "same". Formally, we shall obtain this lemma as an application of Theorem 3.7.2, and we use the notation of that Theorem. In particular, we take $d: C_1 \to C_0$ to be the map $d(2)$, as used to define $\Phi_{0,2}$ in § 4.2; thus, we have

$$c_i d = \mathrm{Sq}(2^i)c \qquad (i = 0, 1, 2) \ .$$

Again, we take $d': C'_1 \to C'_0$ to be the d used in defining Ψ. That is, we take

$$c'_i d' = \mathrm{Sq}^i c' \qquad (i = 1, 3, 4) \ .$$

We define an A-linear map $\rho_0: C_0 \to C'_0$(of degree 6) by $c\rho_0 = \mathrm{Sq}^4\mathrm{Sq}^2c'$. To apply Theorem 3.7.2, we have to construct a map $\rho_1: C_1 \to C'_1$ such that $\rho_1 d' = d\rho_0$. It is sufficient to take

$$\begin{aligned} c_0\rho_1 &= \mathrm{Sq}^4c'_3 \\ c_1\rho_1 &= \mathrm{Sq}^7c'_1 + \mathrm{Sq}^4c'_4 \\ c_2\rho_1 &= \mathrm{Sq}^6\mathrm{Sq}^3c'_1 + (\mathrm{Sq}^7 + \mathrm{Sq}^6\mathrm{Sq}^1)c'_3 \ . \end{aligned}$$

Taking

$$z = \mathrm{Sq}^4c_0 + \mathrm{Sq}^2\mathrm{Sq}^1c_1 + \mathrm{Sq}^1c_2$$

we find

$$z\rho_1 = \mathrm{Sq}^6(\mathrm{Sq}^4 c_1' + \mathrm{Sq}^2 c_3' + \mathrm{Sq}^1 c_4') .$$

Theorems 3.7.2, 3.6.2 now yield the conclusion, since $(\mathrm{Coker}\ d')_{10} = Z_2$, generated by the image of $\mathrm{Sq}^{10}c'$.

For our next lemma, let

$$C_1 \xrightarrow{d} C_0 \xrightarrow{\varepsilon} Z_2$$

be the first part of a minimal resolution of z_2 over A, as constructed in § 4.2. Let $z \in C_1$ be such that $zd = 0$ and $\deg(z) \leqq 2^k$; thus $z \in C_1(k-1)$. Let χ be an operation associated with the pair $(d(k-1), z)$; thus $\deg(\chi) < 2^k$. Let u be a class such that $\mathrm{Sq}(2^r)u = 0$ for $0 \leqq r < k$. Then we have the following conclusion.

LEMMA 4.3.2. *If $\Phi_{i,j}(u)$ has zero indeterminacy, and is zero, for each pair (i, j) with $0 \leqq i \leqq j < k$, $j \neq i+1$, then $\chi(u)$ has zero indeterminacy and is zero.*

PROOF. We first define C_2 to be A-free on generators $c_{i,j}$ of degree $2^i + 2^j$ with $c_{i,j}d = z_{i,j}$; then by Lemma 2.2.1, the terms

$$C_2 \xrightarrow{d} C_1 \xrightarrow{d} C_0 \xrightarrow{\varepsilon} Z_2$$

form part of a minimal resolution of Z_2 over A. Since $z \in C_1$ and $zd = 0$, we have

$$z = (\textstyle\sum_{i,j} a_{i,j} c_{i,j})d .$$

By considering degrees, all the terms in this sum have $j < k$. Thus we have

$$z = \textstyle\sum_{i,j;\, j<k} a_{i,j} z_{i,j} .$$

We may apply Theorems 3.7.1, 3.6.2; since $(\mathrm{Coker}\ d(k-1))_n = 0$ if $n < 2^k$, we find

$$\begin{aligned}[\chi(u)] &= \textstyle\sum_{i,j;\, j<k} a_{i,j} \Phi_{i,j}(u) \\ &= 0 \quad (\mathrm{mod\ zero}) .\end{aligned}$$

This proves the lemma.

For the next lemma, we take an integer $k \geqq 2$ and suppose that $u \in H^m(X)(m > 0)$ is a class such that $\mathrm{Sq}(2^r)u = 0$ for $0 \leqq r \leqq k$ and $\mathrm{Sq}(2^k)\mathrm{Sq}(2^{k+1})(u) = 0$.

LEMMA 4.3.3. *There is a formula*

$$\begin{aligned}\Phi_{0,k+1}\mathrm{Sq}(2^{k+1})(u) &= \sum_{\substack{0\leq i\leq j\leq k \\ j\neq i+1}} a_{i,j,k}\Phi_{i,j}(u) \\ &\quad + \sum_{\substack{0\leq i\leq k \\ i\neq k-1}} a_{i,k}\Phi_{i,k}\mathrm{Sq}(2^{k+1})(u) + \lambda_k \mathrm{Sq}(2^{k+2})(u)\end{aligned}$$

in which $a_{i,j,k} \in A$, $a_{i,k} \in A$ and $\lambda_k \in Z_2$. It holds modulo the total indeterminacy of both sides. The coefficient $a_{0,k,k}$ of $\Phi_{0,k}$ satisfies

$$(\xi_1(3\cdot 2^k) + \xi_2(2^k))a_{0,k,k} = 1 .$$

We require this formula in order to apply it to a power y^{2^k} of the fundamental class y in complex projective space. The actual value of λ_k is not relevant, although at a later stage in our calculations it would be possible to show that $\lambda_k = 0$.

This lemma should be considered as strictly analogous to Lemma 4.3.1; it is another application of the theorems of § 3.7. The only difference is that we do not propose to carry out the calculations explicitly.

We begin by constructing a partial minimal resolution

$$C_2'' \xrightarrow{d_2''} C_1'' \xrightarrow{d_1''} C_0'' \xrightarrow{\varepsilon''} M .$$

Here M is the module of Theorem 2.6.2, except that the integer k of that theorem is replaced by $(k+1)$. Thus, M is a module whose Z_2-base consists of three elements m, $\mathrm{Sq}(2^{k+1})m$, $\mathrm{Sq}(2^{k+2})m$.

We take C_0'' to be free on one generator c'', and define ε'' by $c''\varepsilon'' = m$. We know, by Theorem 2.6.2, that C_1'' will require generators $c_i''(0 \leqq i \leqq k)$, c_*'' of degrees 2^i, $3\cdot 2^k$, plus other generators of degrees $t \geqq 3\cdot 2^{k+1}$. By Lemma 2.2.1, it is sufficient to specify $c_i''d_1''$, $c_*''d_1''$, etc., in a suitable fashion. We may take

$$c_i''d_1'' = \mathrm{Sq}(2^i)c'',\ c_*''d_1'' = \mathrm{Sq}(2^k)\mathrm{Sq}(2^{k+1})c'' .$$

The choice of d_1'' on the other generators does not concern us.

We postpone the construction of d_2'', in order to indicate how we propose to apply Theorem 3.7.2. Using the notations of that theorem, we shall take $d\colon C_1 \to C_0$ to be $d(k+1)$, as used in § 4.2 to define $\Phi_{0,k+1}$. We shall take C_1' to be the submodule of C_1'' generated by the $c_i''(0 \leqq i \leqq k)$ and c_*''; we take $C_0' = C_0''$ and $d' = d_1'' \mid C_1'$.

The map $\rho_0\colon C_0 \to C_0'$ will be defined by

$$c\rho_0 = \mathrm{Sq}(2^{k+1})c'' .$$

Since this induces a map from C_0/dC_1 to M_1 it is possible to construct a map $\rho_1\colon C_1 \to C_1''$ so that $\rho_1 d_1'' = d\rho_0$. By considering dimensions we see that ρ_1 will map into C_1'.

The map ρ_1 may be taken in any way; we only require the following property.

LEMMA 4.3.4. *If*

$$c_0\rho_1 = \sum\nolimits_{0\leq i\leq k} a_i c_i'' \qquad (a_i \in A)$$

then

$$\xi_1(2^{k+1})a_0 = 1 .$$

PROOF. Set $h = 2^{k+1}$. If $a, b \in I(A)$, we have

$$\xi_1^{h+1}(ab) = (\xi_1^h a)(\xi_1 b) + (\xi_1 a)(\xi_1^h b) .$$

We define a function $\xi\colon C_0' \to Z_2$ by setting

$$\xi(ac'') = \xi_1^{h+1}a .$$

Thus

$$\begin{aligned}\xi(c_0\rho_1 d') &= \xi\left(\textstyle\sum_{0\leq i\leq k} a_i \mathrm{Sq}(2^i)c''\right) \\ &= \xi_1^h a_0 .\end{aligned}$$

But

$$\begin{aligned}\xi(c_0 d\rho_0) &= \xi(\mathrm{Sq}^1\mathrm{Sq}^h c'') \\ &= 1 .\end{aligned}$$

This proves the lemma.

We now revert to the construction of C_2''. We know by Theorem 2.6.2, that C_2'' will require generators

$$\begin{array}{lll} c''_{i,j} & (0 \leqq i \leqq j \leqq k, j \neq i+1) & \text{of degrees } 2^i + 2^j \\ c''_{i,*} & (0 \leqq i \leqq k, i \neq k-1) & \text{of degrees } 2^i + 3\cdot 2^k \\ c''_{*,*} & & \text{of degree } \ 2^{k-1} + 2^{k+2} \end{array}$$

plus other generators of degrees $t \geqq 3\cdot 2^{k+1}$. Using Lemma 2.2.1, we may construct d_2'' as follows. Define an embedding $e\colon C_1(k) \to C_1''$ by $c_i e = c_i''$. Then we may take

$$c''_{i,j}d''_2 = z_{i,j}e , \quad c''_{i,*}d''_2 = z_{i,k}\rho_1 ,$$
$$c''_{*,*}d''_2 = z_{k-1,k+1}\rho_1 .$$

The choice of d_2'' on the other generators does not concern us.

We now observe that $z_{0,k+1}\rho_1$ is a cycle in C_1''. Thus it lies in Im d_2'', and must have the following form:

$$z_{0,k+1}\rho_1 = \sum_{\substack{0\leq i\leq j\leq k \\ j\neq i+1}} a_{i,j,k}(z_{i,j}e) + \sum_{\substack{0\leq i\leq k \\ i\neq k-1}} a_{i,k}(z_{i,k}\rho_1) . \tag{4.3.5}$$

Since $k \geqq 2$, there is no term in $z_{k-1,k+1}\rho_1$, by considering degrees. Applying Theorems 3.7.2, 3.7.1, 3.6.2, we obtain the formula which was to be proved.

It remains only to obtain the required information about the coefficient $a_{0,k,k}$. To this end, we define a function $\xi\colon C_1' \to Z_2$ by setting

$$\xi(a'_* c''_* + \sum_{0 \leq i \leq k} a'_i c''_i) = (\xi_1(2^k)\xi_2(2^k))a'_0 .$$

We shall apply ξ to both sides of the equation (4.3.5). We first note that

$$\begin{aligned}\psi(\xi_1(2^k)\xi_2(2^k)) = \xi_1(2^k)\xi_2(2^k) \otimes 1 + (\xi_1(3\cdot 2^k) + \xi_2(2^k)) \otimes \xi_1(2^k) \\ + \xi_1(2^{k+1}) \otimes \xi_1(2^{k+1}) + \xi_1(2^k) \otimes \xi_2(2^k) + 1 \otimes \xi_1(2^k)\xi_2(2^k) .\end{aligned}$$

In particular, $\xi_1(2^k)\xi_2(2^k)(ab) = 0$ unless $\deg(a)$ and $\deg(b)$ are both divisible by 2^k.

Let us expand $z_{0,k+1}$ in the form

$$z_{0,k+1} = \sum_{0 \leq i \leq k+1} b_i c_i \qquad (b_i \in A) .$$

Then we have

$$\xi(z_{0,k+1}\rho_1) = \xi\left(\sum_i b_i(c_i\rho_1)\right) .$$

But $\deg(b_i) = 2^0 + 2^{k+1} - 2^i$, which is odd unless $i = 0$. Let us write $c_0\rho_1 = \sum_i a_i c''_i$. Then

$$\begin{aligned}\xi(z_{0,k+1}\rho_1) &= \xi(b_0(c_0\rho_1)) \\ &= (\xi_1(2^{k+1})b_0)(\xi_1(2^{k+1})a_0) \\ &= 1 ,\end{aligned}$$

using Lemma 2.2.2. (for b_0) and Lemma 4.3.4 (for a_0).

Let us apply ξ to the right-hand side of the equation (4.3.5). We have

$$\begin{aligned}\deg(a_{i,j,k}) &= 2^0 + 2^{k+2} - 2^i - 2^j \\ \deg(a_{i,k}) &= 2^0 + 2^k - 2^i\end{aligned}$$

and these are odd, except in the cases with $i = 0$. Moreover,

$$\deg(a_{0,j,k}) = 2^{k+2} - 2^j ,$$

and this is not divisible by 2^k, except in the case $j = k$. Thus it remains only to evaluate

$$\xi(a_{0,k,k}(z_{0,k}e)) , \quad \xi(a_{0,k}(z_{0,k}\rho_1)) .$$

We deal with the latter first. In C_1, let us write $z_{0,k} = \sum_{0 \leq i \leq k} b_i c_i$, with a new set of coefficients b_i in A. Let us write

$$c_i\rho_1 = b_{i,*}c''_* + \sum_{0 \leq j \leq k} b_{i,j}c''_j .$$

Then

$$\xi(a_{0,k}(z_{0,k}\rho_1)) = (\xi_1(2^k)\xi_2(2^k))(a_{0,k}\sum_{0 \leq i \leq k} b_i b_{i,0}) .$$

Here we have

$$\begin{aligned}\deg(a_{0,k}) = 2^k, \quad \deg(b_i) &= 2^0 + 2^k - 2^i , \\ \deg(b_{i,0}) &= 2^i + 2^{k+1} - 2^0 .\end{aligned}$$

Thus

$$\xi(a_{0,k}(z_{0,k}\rho_1)) = (\xi_1(2^k)a_{0,k})\sum_{0\le i\le k}(\xi_2(2^k)b_i b_{i,0}) = 0 .$$

Lastly, we consider $\xi(a_{0,k,k}(z_{0,k}e))$. Let us write $z_{0,k}e = \sum_{0\le i\le k} b_i c_i''$. Then

$$\begin{aligned}\xi(a_{0,k,k}(z_{0,k}e)) &= ([\xi_1(3\cdot 2^k) + \xi_2(2^k)]a_{0,k,k})(\xi_1(2^k)b_0)\\ &= (\xi_1(3\cdot 2^k) + \xi_2(2^k))a_{0,k,k} .\end{aligned}$$

We conclude that

$$(\xi_1(3\cdot 2^k) + \xi_2(2^k))a_{0,k,k} = 1 .$$

This completes the proof of Lemma 4.3.3.

4.4. *The operation Ψ in P^6.* In this section we find the value of that operation which is needed to start the induction in § 4.5.

Let P be complex projective space of infinitely-many dimensions; let y be a generator of $H^2(P)$, so that $H^*(P)$ is a polynomial algebra (over Z_2) generated by y. Let Ψ be the operation defined in § 4.2.

THEOREM 4.4.1. $\Psi(y) = y^3$.

The operation Ψ is defined on y because the elements $\mathrm{Sq}^1 y$, $\mathrm{Sq}^3 y$ and $\mathrm{Sq}^4 y$ are zero. It is defined modulo zero because the elements $\mathrm{Sq}^2 y^2$ and $\mathrm{Sq}^4 y$ are zero.

Before proving Theorem 4.4.1, we insert some remarks on its proof. It is easy to show (by considering the universal example) that if $u \in H^2(X)$ is any class such that $\mathrm{Sq}^1(u) = 0$, we have

$$\Psi(u) = [\lambda u^3]$$

for some fixed coefficient λ. Theorem 4.4.1 is therefore equivalent to the proposition that if u is any class such that $\mathrm{Sq}^1 u = 0$, we have

$$\Psi(u) = [u^3] .$$

It would be desirable, in some ways, to prove this latter proposition by arguments lying wholly inside homology-theory. This is indeed possible, by using the methods of Steenrod and Adem (see [32], [5]) to give a construction for Ψ and to discuss its properties. However, to employ such methods here would lengthen the present paper by a chapter; for brevity, therefore, we make an *ad hoc* application of the methods of homotopy theory.

In fact, the space P may be decomposed as a CW-complex $S^2 \cup E^4 \cup E^6 \cup \cdots \cup E^{2n} \cup \cdots$, where the subcomplex $S^2 \cup E^4 \cup E^6 \cup \cdots \cup E^{2n}$ is just P^{2n}, the complex projective space of n complex dimensions. The stable cohomology operations in $H^*(P)$ depend on the attaching maps of these cells, or rather, on their stable or S-homotopy classes.

The attaching map for E^4 is just the Hopf map $\eta : S^3 \to S^2$. Similarly, the attaching map $\varpi : S^5 \to P^4$ for E^6 is just the usual fibering, with fibre S^1. Let $S\varpi : S^6 \to SP^4$ be the suspension of ϖ, and let $Si: S^3 \to SP^4$ be the suspension of the embedding $i: S^2 \to P^4$.

LEMMA 4.4.2. *In $\pi_6(SP^4)$ we have*

$$\{S\varpi\} = (Si)_* \omega ,$$

where ω is a generator of $\pi_6(S^3)$.

This lemma is essentially due to H. Toda [33, Chapter 7]; but unfortunately, he does not state it explicitly. A variety of proofs are available; the neatest I have seen is the following, for which I am indebted to Dr. I. M. James. It depends on the following lemma.

LEMMA 4.4.3. *Let $B = S^q \cup E^n \cup E^{n+q}$ be a q-sphere bundle over S^n, decomposed into cells in the obvious way. Let $\alpha \in \pi_{n-1}(R_{q+1})$ be the characteristic element for B, and let $\beta \in \pi_{n+q-1}(S^q \cup E^n)$ be the attaching element for E^{n+q}. Then*

$$S\beta = \pm (Si)_*(\omega) ,$$

where

$$(Si)_* : \pi_{n+q}(S^{q+1}) \to \pi_{n+q}(S(S^q \cup E^n))$$

is the injection and ω is obtained from α by the Hopf construction.

This lemma is cognate with the work done in [20] (see § 7 in particular).

For the application, we take $q = 2$, $n = 4$, and take B to be the standard fibering

$$S^2 \to P^6 \to S^4 .$$

The element α is a generator of $\pi_3(R_3)$, and hence ω is a generator of $\pi_6(S^3)$.

We now proceed to deduce Theorem 4.4.1 from Lemma 4.4.2. We need one more lemma. Let $K = S^n \cup E^{n+4}$ be a complex (with $n \geqq 5$) in which the class of the attaching map is $2r\nu$, where ν is a generator of $\pi_{n+3}(S^n)$ and r is an integer.

LEMMA 4.4.4. *$\Psi: H^n(K) \to H^{n+4}(K)$ is zero if r is even, non-zero if r is odd.*

PROOF. We first observe that Ψ is defined, and is defined modulo zero, because

$$\mathrm{Sq}^4 : H^n(K) \to H^{n+4}(K)$$

is zero. We now construct a space X, equivalent to K, as follows; take the mapping-cylinder of a map $f: S^{n+3} \to S^n$ representing ν; then attach a cell E^{n+4} to S^{n+3} by a map of degree $2r$. Let Y be the subspace $E^{n+4} \cup S^{n+3}$. We shall apply Axiom 5, § 3.6, to the pair X, Y. Inspecting the exact cohomology sequence of this pair, we see that we may take generators as follows.

$$\begin{aligned} u &\in H^n(X, Y), & j^*u &\in H^n(X) \\ v &\in H^{n+3}(Y), & \delta v &\in H^{n+4}(X, Y) \\ w &\in H^{n+4}(X), & i^*w &\in H^{n+4}(Y). \end{aligned}$$

We have $\mathrm{Sq}^1 u = 0$, $\mathrm{Sq}^3 u = 0$, $\mathrm{Sq}^4 u = \delta v$, and $\mathrm{Sq}^1 v = r(i^*w)$. By Axiom 5, $i^*\Psi(j^*u) = r(i^*w)$ and hence $\Psi(j^*u) = rw$. This proves Lemma 4.4.4.

We now prove Theorem 4.4.1. Consider S^3P^6, the threefold suspension of P^6. By Lemma 4.4.2, the attaching map of S^3E^6 lies in the class $(S^3 i)_*(S^2\omega)$, where ω is some generator of $\pi_6(S^3)$. But $S^3\omega = 2r\nu$ with r odd. Let K be a complex, as considered in Lemma 4.4.4, for $n = 5$ and this value of r; then there is a map $f: K^5 \to S^3P^6$ inducing isomorphisms of H^5, H^9. Since Ψ is non-zero in K by Lemma 4.4.4, it is non-zero in S^3P^6. Since Ψ commutes with suspension, it is non-zero in P^6, and hence in P. This completes the proof of Theorem 4.4.1.

We remark that the operation Ψ is by no means the only secondary operation in P which we can evaluate directly. In particular, one can evaluate $\Phi_{0,2}(y^{4t})$ using James's results on the attaching maps in quaternionic projective spaces—see (2.10a) of [19].

4.5. *The operations $\Phi_{i,j}$ in P.* In this section we shall obtain the values of the operations $\Phi_{i,j}$ when they act in complex projective space of infinitely-many dimensions. We write P for this projective space, and write y for the generator of $H^2(P)$, so that $H^*(P)$ is a polynomial algebra (over Z_2) generated by y. The Steenrod squares in $H^*(P)$ are easily calculated; we have

$$\begin{aligned} \mathrm{Sq}^{2k+1}(y^t) &= 0 \\ \mathrm{Sq}^{2k}(y^t) &= (t-k, k) y^{t+k}. \end{aligned}$$

(Here (h, k) stands for the (mod 2) binomial coefficient $(h+k)!/h!\,k!$).

Let χ be an operation associated with a pair $(d(j), z)$ (using the notation of § 4.2). Then χ is defined on y^t if and only if $\mathrm{Sq}(2^r)(y^t)=0$ for $0 \leqq r \leqq j$. For this, it is necessary and sufficient that $t \equiv 0 \bmod 2^j$. Now set $\deg(\chi) = n$; and suppose that $n < 2^{j+1}$. If n is odd, then $\chi(y^t)$ is a coset in a zero group. If n is even, we should examine the Steenrod operations a_r which enter into the indeterminacy of χ (and are defined by $z = \sum_r a_r c_r$). We see

$$\deg(a_r) = n + 1 - 2^r.$$

If $r>0$, this degree is odd, so that a_r contributes nothing to the indeterminacy. On the other hand, if $r = 0$, then this degree is n; and if $t \equiv 0 \bmod 2^j$, then

$$a_0 H^{2t}(P) = 0$$

for any a_0 in A such that $\deg(a_0) = n < 2^{j+1}$. We conclude that $\chi(y^t)$, if defined at all, has zero indeterminacy.

In particular, we conclude that $\Phi_{i,j}(y^t)$ is defined if and only if $t \equiv 0 \bmod 2^j$; that its indeterminacy is then zero; and that it is zero (since of odd degree) unless $i = 0$, $j \geqq 2$.

For our next result, which gives the values of the $\Phi_{0,j}$, we set $h = 2^j$.

THEOREM 4.5.1. $\Phi_{0,j}(y^{ht}) = t y^{h(t+1/2)}$.

PROOF. We first obtain the case $j = 2$, $t = 1$. In fact, by applying Lemma 4.3.1 to the case $u = y$ and using Theorem 4.4.1, we see

$$\Phi_{0,2}(y^4) = \Phi_{0,2}(\mathrm{Sq}^4\mathrm{Sq}^2 y) = \mathrm{Sq}^6 \Psi y + \lambda \mathrm{Sq}^{10} y = \mathrm{Sq}^6 y^3 = y^6 .$$

This case serves to start an induction. Suppose, as an inductive hypothesis, that we have established the result for all $j<k$ (where $k \geqq 2$) and for the case $j = k$, $t = 1$. We now note that if χ is any secondary operation associated with $d(k)$ such that $\deg(\chi) < h = 2^k$, then $\chi(y^{ht}) = 0$ (modulo zero). This is immediate by Lemma 4.3.2, using the inductive hypothesis. We also note that it is possible to apply the Cartan formula (Theorem 3.9.4) in case $d = d' = d'' = d(k)$ (with the notations of §§ 3.9, 4.2). We will verify the main condition on d, d' and d'' imposed in § 3.9. In fact, if u, v are such that $\mathrm{Sq}(2^r)u = 0$, $\mathrm{Sq}(2^r)v = 0$ for $0 \leqq r \leqq k$, then $\mathrm{Sq}^i(u) = 0$, $\mathrm{Sq}^i(v) = 0$ for $1 \leqq i \leqq 2^k$; by the ordinary Cartan formula [11], we deduce $\mathrm{Sq}^i(uv) = 0$ for $1 \leqq i \leqq 2^k$; *a fortiori* $\mathrm{Sq}(2^r)(uv)=0$ for $0 \leqq r \leqq k$.

We can thus obtain the result for $j = k$ and any t, by induction over t, using Theorem 3.9.4. In fact, suppose

$$\Phi_{0,k}(y^{ht}) = t y^{h(t+1/2)} \qquad \text{(where } h = 2^k\text{)}.$$

Then

$$\begin{aligned}\Phi_{0,k}(y^{h(t+1)}) &= \Phi_{0,k}(y^{ht} \cdot y^h) \\ &= \Phi_{0,k}(y^{ht}) \cdot y^t + y^{ht} \cdot \Phi_{0,k}(y^h)\end{aligned}$$

(by Theorem 3.9.4, since the intermediate terms yield zero). That is

$$\Phi_{0,k}(y^{h(t+1)}) = (t+1) y^{h(t+3/2)} .$$

This completes the induction over t; we have obtained the result for $j \leqq k$ and all t.

We now apply Lemma 4.3.3 to the class $u = y^h$, where $h = 2^k$. The left-hand side of the formula yields $\Phi_{0,k+1}(y^{2h})$, modulo zero. On the right, the term $\lambda_k \mathrm{Sq}(2^{k+2})u$ yields zero. The term $a_{i,k}\Phi_{i,k}\mathrm{Sq}(2^{k+1})u$ yields zero, modulo zero, by what we have already proved. The term $a_{i,j,k}\Phi_{i,j}u$ yields zero, modulo zero, except in the case $i = 0$, $j = k$. In this case, $\Phi_{0,k}(u)$ becomes $y^{3h/2}$, modulo zero. Now, we easily see that if $a \in A_{3h}$, then

$$a \cdot y^{3h/2} = ([\xi_1(3\cdot 2^k) + \xi_2(2^k)]a)y^{3h} .$$

Using the last part of Lemma 4.3.3, we conclude that the term $a_{0,k,k}\Phi_{0,k}(u)$ yields y^{3h}, modulo zero, and

$$\Phi_{0,k+1}(y^{2h}) = y^{3h} .$$

We have proved the result for the case $j = k + 1$, $t = 1$. This completes the induction over k; Theorem 4.5.1 is proved.

4.6. *The final relation.* In this section we obtain the relation required to carry out the argument indicated in Chapter 1.

Take an integer $k \geqq 3$. Let $u \in H^m(X)(m > 0)$ be a class such that $\mathrm{Sq}(2^r)u = 0$ for $0 \leqq r \leqq k$.

THEOREM 4.6.1. *There is a relation*

$$\sum_{\substack{0 \leq i \leq j \leq k \\ j \neq i+1}} a_{i,j,k}\Phi_{i,j}(u) = [\mathrm{Sq}(2^{k+1})u]$$

(independent of X) which holds modulo the total indeterminacy of the left-hand side.

PROOF. Let us consider the first few terms

$$C_2 \xrightarrow{d_2} C_1 \xrightarrow{d_1} C_0 \xrightarrow{\varepsilon} Z_2$$

of a minimal resolution of Z_2, as constructed in §§ 4.2, 4.3. Let us choose a cycle z in C_2 such that $(h_0 h_k^2)(\theta z) = 1$; this is possible by Theorem 2.5.1, since $k \geqq 3$. Let us write $z = \sum_{i,j} a_{i,j,k} c_{i,j}$. By considering degrees and using the fact that d_1 is minimal, we see that this sum consists of terms with $j \leqq k$. By Lemma 2.2.2, we have

$$\xi_1(2^k)a_{0,k,k} = 1 .$$

Since z is a cycle, we have

$$0 = zd = (\textstyle\sum_{i,j} a_{i,j,k} c_{i,j} d) = \sum_{i,j} a_{i,j,k} z_{i,j} .$$

Now, this relation holds in the submodule $C_1(k)$ of C_1. Appealing to the theorems of §§ 3.6, 3.7, we find the required relation

$$\sum_{\substack{0 \leq i \leq j \leq k \\ j \neq i+1}} a_{i,j,k} \Phi_{i,j}(u) = [\lambda \mathrm{Sq}(2^{k+1})u] .$$

It remains only to determine the coefficient λ.

To do this, we apply both sides to a power y^h (where $h = 2^k$) of the fundamental class y in $H^2(P)$. The total indeterminacy of the left-hand side is then zero. The right-hand side yields λy^{2h}. By § 4.5 (and Theorem 4.5.1 in particular), each term on the left-hand side yields zero, modulo zero, except the term

$$a_{0,k,k}\Phi_{0,k}(y^h) = a_{0,k,k} y^{3h/2} \qquad \text{(mod zero)} .$$

But we see that for any $a \in A_h$, we have

$$a y^{3h/2} = (\xi_1^h a) y^{2h} .$$

Since $\xi_1^h a_{0,k,k} = 1$, we have $\lambda = 1$, and the proof is complete.

This establishes the decomposability of Sq^i for $i = 2^r$, $r \geqq 4$, which implies that the corresponding groups $\pi_{2i-1}(S^i)$ contain no elements of Hopf invariant one.

ADDENDUM

1. The paper to which this is an addendum makes use of the following as a key lemma:

LEMMA 1. *Let P denote complex projective space of infinitely-many dimensions; let u be the generator of $H^2(P; Z_2)$; let Ψ be the secondary operation associated with the relation*

$$\mathrm{Sq}^4\mathrm{Sq}^1 + \mathrm{Sq}^2\mathrm{Sq}^3 + \mathrm{Sq}^1\mathrm{Sq}^4 = 0 .$$

Then we have

$$\Psi(u) = u^3 .$$

(See Theorem 4.4.1).

It is the object of this addendum to give a simple proof of this lemma. We will actually prove:

LEMMA 2. *Let $u \in H^2(X; Z_2)$ be a cohomology class such that* $\mathrm{Sq}^1 u = 0$. *Then the coset Ψu contains the element u^3.*

The proof to be given employs a method which I owe to A. Liulevicius; he uses it in his treatment of the problem of "elements of Hopf invariant one mod p." I am most grateful to him for interesting letters on this subject. Liulevicius, in turn, ascribes the basic idea of his method to W. Browder.

2. The basic idea of the method is as follows. One considers a universal example consisting of a space E and a class $v \in H^*(E; Z_p)$, as in §3.3.

Then the loop-space ΩE and the suspension $\sigma v \in H^*(\Omega E; Z_p)$ constitute another universal example, in which the dimensions have been decreased by one. It is possible that the space ΩE may be equivalent to a Cartesian product $X \times Y$, although the space E does not split in the same way. If this happens, then the Pontrjagin product in $H_*(\Omega E; Z_p)$ gives us a ring-structure on

$$H_*(X \times Y; Z_p) = H_*(X; Z_p) \otimes H_*(Y; Z_p) ;$$

in general, this ring-structure does not split as the tensor-product of ring-structures on $H_*(X; Z_p)$ and $H_*(Y; Z_p)$. Since the element $\sigma v \in H^*(\Omega E; Z_p)$ is primitive, it is possible to make deductions about its value.

3. We will now apply this method to our case. In what follows, all cohomology groups have coefficients in the group Z_2. In order to prove Lemma 2, it is sufficient to prove it when u is the fundamental class in a suitable universal example. The universal examples we must consider are those used to define Ψ; they can be constructed as follows.

For any positive integer n, let $K(Z_2, n)$, $K(Z_2, n+1)$, $K(Z_2, n+3)$ and $K(Z_2, n+4)$ be Eilenberg-MacLane spaces of the types indicated; we suppose that the first is a CW-complex, and write b^n for its fundamental class. Then there is a map

$$m : K(Z_2, n) \longrightarrow K(Z_2, n+1) \times K(Z_2, n+3) \times K(Z_2, n+4)$$

which maps the fundamental classes on the right-hand side into $Sq^1 b^n$, $Sq^3 b^n$, $Sq^4 b^n$. The map m induces fibre-space over $K(Z_2, n)$ with fibre $K(Z_2, n) \times K(Z_2, n+2) \times K(Z_2, n+3)$; this fibre-space we call E_n. If we write $f^{n,0}, f^{n,2}, f^{n,3}$ for the fundamental classes in the fibre of E_n, then we have

$$\begin{aligned} \tau f^{n,0} &= Sq^1 b^n \\ \tau f^{n,2} &= Sq^3 b^n \\ \tau f^{n,3} &= Sq^4 b^n . \end{aligned}$$

The class $v^{n+4} \in H^{n+4}(E_n)$ which serves as a universal example for Ψ satisfies

$$i^* v^{n+4} = Sq^4 f^{n,0} + Sq^2 f^{n,2} + Sq^1 f^{n,3} .$$

This condition defines v^{n+4} uniquely, so long as n is sufficiently large.

It would be equivalent, however, to induce our fibering in two stages; first induce a fibering E'_n with fibre $K(Z_2, n)$ over $K(Z_2, n)$; then induce a fibering E''_n with fibre $K(Z_2, n+2) \times K(Z_2, n+3)$ over E'_n. If we adopt this procedure, the first stage evidently gives $E'_n = K(Z_4, n)$; therefore we may regard E_n as a fibering with base $K(Z_4, n)$ and fibre

$$K(Z_2, n+2) \times K(Z_2, n+3) .$$

Let us re-appropriate the symbols b^n, $f^{n,2}$, $f^{n,3}$ for the fundamental classes in these spaces; we have

$$\tau f^{n,2} = \mathrm{Sq}^3 b^n$$
$$\tau f^{n,3} = \mathrm{Sq}^4 b^n$$
$$i^* v^{n+4} = \mathrm{Sq}^2 f^{n,2} + \mathrm{Sq}^1 f^{n,3} .$$

We will now examine what happens to E_n when n is small. Let us take $n=3$; we find that E_3 is equivalent to a product $X \times K(Z_2, 6)$, owing to the fact that Sq^4 vanishes on classes of dimension 3. We can therefore choose a class $g^6 \in H^6(E_3)$ whose image in the fibre $K(Z_2, 5) \times K(Z_2, 6)$ is the fundamental class in the second factor. Similarly, let us take $n=2$; we find that E_2 is equivalent to a product $K(Z_4, 2) \times K(Z_2, 4) \times K(Z_2, 5)$, owing to the fact that Sq^3 and Sq^4 vanish on classes of dimension 2. We can therefore choose classes $g^4 \in H^4(E_2)$, $g^5 \in H^5(E_2)$ whose images in the fibre $K(Z_2, 4) \times K(Z_2, 5)$ are the fundamental classes. Let $g^2 \in H^2(E_2)$ be the fundamental class. We can now write down the following base for $H^6(E_2)$:

$$(g^2)^3 , \quad (\beta_4 g^2) , \quad g^2 g^4 , \quad \mathrm{Sq}^2 g^4 , \quad \mathrm{Sq}^1 g^5 .$$

We wish to know how the element v^6 can be expressed in terms of this base.

Since E_2 is equivalent to ΩE_3, we are precisely in the situation envisaged in § 2. In fact, v^6 is primitive, since by construction it corresponds to σv^7 in the equivalence $E_2 \sim \Omega E_3$. Let μ denote the product: then the homomorphism μ^* of $H^6(E_2)$ is determined by the following equations.

(1) $\mu^* g^2 = g^2 \otimes 1 + 1 \otimes g^2$

(2) $\mu^* g^4 = g^4 \otimes 1 + g^2 \otimes g^2 + 1 \otimes g^4$

(3) $\mu^* g^5 = g^5 \otimes 1 + 1 \otimes g^5$.

Here the equation (1) holds for dimensional reasons; while (3) holds provided we choose g^5 to correspond to σg^6, as we evidently may. As for (2), the only alternative is to suppose that g^4 is primitive; and if it were primitive, then (for dimensional reasons) it would be the suspension of some element $\gamma^5 \in H^5(E_3)$; this would satisfy $i^*\gamma^5 = f^{3,2}$, contradicting the fact that $\tau f^{3,2} = \mathrm{Sq}^3 b^3 \neq 0$.

It is now easy to calculate such values as

$$\mu^*(g^2)^3 = (g^2)^3 \otimes 1 + (g^2)^2 \otimes g^2 + g^2 \otimes (g^2)^2 + 1 \otimes (g^2)^3$$
$$\mu^*(\mathrm{Sq}^2 g^4) = \mathrm{Sq}^2 g^4 \otimes 1 + (g^2)^2 \otimes g^2 + g^2 \otimes (g^2)^2 + 1 \otimes \mathrm{Sq}^2 g^4 .$$

We find the following base of primitive elements in $H^6(E_2)$:

$$(\beta_4 g^2)^2 , \qquad (g^2)^3 + \mathrm{Sq}^2 g^4 , \qquad \mathrm{Sq}^1 g^5 .$$

Since

$$i^* v^6 = \mathrm{Sq}^2 f^{2,2} + \mathrm{Sq}^1 f^{2,3} ,$$

we have

$$v^6 = \lambda(\beta_4 g^2)^2 + (g^2)^3 + \mathrm{Sq}^2 g^4 + \mathrm{Sq}^1 g^5$$

for some $\lambda \in Z_2$.

Now, the indeterminacy of $\Psi(g^2)$ is a subgroup Q of $H^6(E_2)$ which has the following base:

$$\mathrm{Sq}^2 g^4 , \qquad (\beta_4 g^2)^2 , \qquad \mathrm{Sq}^1 g^5 .$$

Moreover, $\Psi(g^2)$ is by definition that coset of Q which contains v^6. Therefore $\Psi(g^2)$ contains $(g^2)^3$. This completes the proof.

TRINITY HALL, CAMBRIDGE

REFERENCES

1. J. F. ADAMS, *On the cobar construction*, Proc. Nat. Acad. Sci. USA 42 (1956), 409–412.
2. ———, *On the structure and applications of the Steenrod algebra*, Comment. Math. Helv. 32 (1958), 180–214.
3. ———, *On the non-existence of elements of Hopf invariant one*, Bull. Amer. Math. Soc. 64 (1958), 279–282.
4. J. ADEM, *The iteration of the Steenrod squares in algebraic topology*, Proc. Nat. Acad. Sci. USA 38 (1952), 720–726.
5. ———, The relations on Steenrod powers of cohomology classes, in Algebraic Geometry and Topology, Princeton, 1957, pp. 191–238.
6. A. L. BLAKERS AND W. S. MASSEY, *Products in homotopy theory*, Ann. of Math. 58 (1953), 295–324.
7. R. BOTT, *The stable homotopy of the classical groups*, Proc. Nat. Acad. Sci. USA 43 (1957), 933–935.
8. ——— AND J. MILNOR, *On the parallelizability of the spheres*, Bull. Amer. Math. Soc. 64 (1958), 87–89.
9. N. BOURBAKI, Algèbre Linéaire, Hermann et Cie, Paris 1947.
10. ———, Petits bouts de topologie (mimeographed notes).
11. H. CARTAN, *Une théorie axiomatique des carrés de Steenrod*, C. R. Acad. Sci. (Paris) 230 (1950), 425–429.
12. ———, *Sur l'iteration des opérations de Steenrod*, Comment. Math. Helv. 29 (1955), 40–58.
13. ——— AND S. EILENBERG, Homological Algebra, Princeton, 1956.
14. S. EILENBERG AND S. MACLANE, *On the groups $H(\pi, n)$*, I, Ann. of Math. 58 (1953), 55–106.
15. S. EILENBERG AND N. E. STEENROD, Foundations of Algebraic Topology, Princeton, 1952.
16. P. J. HILTON, An Introduction to Homotopy Theory, Cambridge, 1953.
17. H. HOPF, *Über die Abbildungen von Sphären auf Sphären niedriger Dimension*, Fund. Math. 25 (1935), 427–440.

18. I. M. JAMES, *Reduced product spaces*, Ann. of Math. 62 (1955), 170–197.

19. ———, *Spaces associated with Stiefel manifolds*, Proc. London Math. Soc. 9 (1959), 115–140.

20. ——— AND J. H. C. WHITEHEAD, *The homotopy theory of sphere bundles over spheres*, I, Proc. London Math. Soc. 4 (1954), 196–218.

21. M. A. KERVAIRE, *Non-parallelizability of the n-sphere for $n > 7$*, Proc. Nat. Acad. Sci USA 44 (1958), 280–283.

22. ———, *An interpretation of G. W. Whitehead's generalisation of H. Hopf's invariant*, Ann. of Math. 69 (1959), 345–365.

23. W. S. MASSEY, Some higher order cohomology operations, in Proc. Int. Symp. on Algebraic Topology, Mexico, 1956.

24. ———, *On the cohomology ring of a sphere bundle*, J. Math. Mech. 7 (1958), 265–290.

25. J. MILNOR, *The Steenrod algebra and its dual*, Ann. of Math. 67 (1958), 150–171.

26. ——— AND J. C. MOORE, *On the structure of Hopf algebras*, to appear.

27. F. P. PETERSON, *Functional cohomology operations*, Trans. Amer. Math. Soc. 86 (1957), 197–211.

28. ——— AND N. STEIN, *Secondary cohomology operations: two formulas*, Amer. Jour. Math. 81 (1959), 281–305.

29. J.-P. SERRE, *Homologie singulière des espaces fibrés*, Ann. of Math. 54 (1951), 425–505.

30. ———, *Cohomologie modulo 2 des complexes d'Eilenberg-MacLane*, Comment. Math. Helv. 27 (1953), 198–232.

31. N. E. STEENROD, *Products of cocycles and extensions of mappings*, Ann. of Math. 48 (1947), 290-320.

32. ———, *Cohomology operations derived from the symmetric group*, Comment. Math. Helv. 31 (1957), 195–218.

33. H. TODA, *Generalised Whitehead products and homotopy groups of spheres*, Jour. Inst. Polytech. Osaka City Univ. 3 (1952), 43–82.

34. ———, *Le produit de Whitehead et l'invariant de Hopf*, C. R. Acad. Sci. (Paris) 241 (1955), 849–850.

35. H. UEHARA AND W. S. MASSEY, The Jacobi identity for Whitehead products, in Algebraic Geometry and Topology, Princeton, 1957, pp. 361-377.

36. G. W. WHITEHEAD, *A generalization of the Hopf invariant*, Ann. of Math. 51 (1950), 192-237.

37. N. YONEDA, *On the homology theory of modules*, Jour. Fac. Sci. Univ. Tokyo Sect. I, 7 (1954), 193–227.

APPLICATIONS OF THE GROTHENDIECK–ATIYAH–HIRZEBRUCH FUNCTOR $K(X)$

By J. F. ADAMS

I shall begin by stating three results; then I shall comment on their authorship and history, and finally I shall try to show how they can be fitted into a general theory.

Let $O(n)$ be the orthogonal group; then we can consider the coset space $O(n)/O(n-k)$ and the fibering

$$O(n)/O(n-k) \to O(n)/O(n-1) = S^{n-1}. \quad (1)$$

The classical problem about vector-fields on spheres used to ask: for what values of n and k does this fibering admit a cross-section? The answer is as follows.

THEOREM 1. *The fibering* (1) *admits a cross-section if and only if n is divisible by N_k, where N_k is the integer defined below.*

We define $N_k = 2^{a(k)}$, where $a(k)$ is the number of integers r such that $1 \leqslant r \leqslant k-1$ and $r \equiv 0$, 1, 2 or 4 mod 8.

Similarly, let $U(n)$ be the unitary group; then we can consider the coset space $U(n)/U(n-k)$ and the fibering

$$U(n)/U(n-k) \to U(n)/U(n-1) = S^{2n-1}. \quad (2)$$

Again we ask: when is there a cross-section?

THEOREM 2. *The fibering* (2) *admits a cross-section if and only if n is divisible by M_k, where M_k is the integer defined below.*

We define $\nu_p(n)$ to be the exponent of the prime p in the decomposition of n into prime powers, so that

$$n = 2^{\nu_2(n)} 3^{\nu_3(n)} 5^{\nu_5(n)} \ldots$$

We define M_k as follows:

$$\nu_p(M_k) = 0 \quad \text{if} \quad p > k,$$

$$\nu_p(M_k) = \text{Sup}\ (r + \nu_p(r)) \quad \text{if} \quad p \leqslant k,$$

where the integer r runs over the range

$$1 \leqslant r \leqslant \frac{k-1}{p-1}.$$

At the last congress we heard a lecture by J. Milnor [10], which was in part about the J-homorphism of H. Hopf and G. W. Whitehead. I recall that this is a map

$$J : \pi_l(SO(n)) \to \pi_{n+l}(S^n).$$

I shall suppose that $l = 4k-1$ and $n > 4k$, so that we are dealing with the "stable J-homomorphism", and J is defined on a cyclic infinite group. The problem is to describe the image of J. Following Milnor and Kervaire [10], we define $m(k)$ to be the denominator of $B_k/4k$, when this fraction is expressed in its lowest terms. Here B_k is the kth Bernoulli number, so that

$$\frac{x}{e^x-1} = 1 - \frac{x}{2} + \sum_{k=1}^{\infty} (-1)^{k-1} B_k \frac{x^{2k}}{(2k)!}.$$

THEOREM 3. *If $l = 4k-1$ and $n > 4k$, then*

$$\operatorname{Im} J = \begin{cases} Z_{m(k)} & \text{if } 4k \equiv 4 \bmod 8, \\ \text{either } Z_{m(k)} \text{ or } Z_{2m(k)} & \text{if } 4k \equiv 0 \bmod 8. \end{cases}$$

In Theorem 1, the existence of a cross-section when n is divisible by N_k is classical; the non-existence of a cross-section when n is not divisible by N_k is proved in [1].

The two halves of Theorem 2 were obtained in the reverse order. The non-existence of a cross-section when n is not divisible by M_k is due to Atiyah and Todd [7]; the existence of a cross-section when n is divisible by M_k is proved in [3].

In Theorem 3, the fact that the order of $\operatorname{Im} J$ is a multiple of $m(k)$ is the result of Milnor and Kervaire [10] as improved by Atiyah and Hirzebruch [5]. The fact that the order of $\operatorname{Im} J$ divides $m(k)$or $2m(k)$, as the case may be, is proved in [2].

I remark that by re-proving the result of Milnor and Kervaire, one can extract more information. One can show [2] that there is a homomorphism

$$e : \pi_{n+4k-1}(S^n) \to Z_{m(k)},$$

such that the composite eJ is an epimorphism (if $n > 4k$). It follows that if $4k \equiv 4 \bmod 8$, then $\operatorname{Im} J$ is a direct summand in $\pi_{n+4k-1}(S^n)$. If $4k \equiv 0 \bmod 8$, then a similar conclusion follows except for the 2-component. The homomorphism e has other interesting properties, on which I shall not dwell.

In order to prove these theorems one makes use of the "extraordinary cohomology theory" $K(X)$ of Grothendieck–Atiyah–Hirzebruch [5, 6]. I will now recall how this is constructed. Let Λ denote either the real field R or the complex field C. Let X be a "good" space, e.g. a finite connected CW-complex. If $\Lambda = R$ we take all orthogonal bundles over X; if $\Lambda = C$ we take all unitary bundles over X. In either case we divide them into isomorphism classes $\{\xi\}$. We take these classes as the generators for a free abelian group $F_\Lambda(X)$. We shall define

$$K_\Lambda(X) = F_\Lambda(X)/T_\Lambda(X),$$

so that $K_\Lambda(X)$ is given by the generators $\{\xi\}$ and certain relations; we define $T_\Lambda(X)$ to be the subgroup of $F_\Lambda(X)$ generated by all elements of the form

$$\{\xi \oplus \eta\} - \{\xi\} - \{\eta\},$$

where $\xi \oplus \eta$ denotes the Whitney sum of ξ and η. The group $K_\Lambda(X)$, then, is obtained by taking the vector bundles over X and forcing them to generate an abelian group under the Whitney sum operation. The elements of $K_\Lambda(X)$ may be called "virtual bundles".

It is possible to use the groups $K_\Lambda(X)$ to prove non-existence results in just the same way that one is accustomed to use ordinary cohomology groups. Thus, if X, A is a pair and A is a retract of X, it follows that $K_\Lambda(A)$ is a direct summand in $K_\Lambda(X)$; and if we find that $K_\Lambda(A)$ is not a direct summand in $K_\Lambda(X)$, then we can conclude that A is not a retract of X. The non-existence proof in [1] is presented in this way.

However, just as in ordinary cohomology we often need to use cohomology operations, so here we need to use cohomology operations in $K_\Lambda(X)$.

The first such operation is a cup-product. We can define the tensor product of two vector-spaces over Λ; therefore we can define the tensor product $\xi \otimes \eta$ of two vector bundles over X; one shows that this defines a product in $K_\Lambda(X)$.

Similarly, we can define the dual of a vector-space over Λ; therefore we can define the dual ξ^* of a vector bundle over X; one shows that this defines an operation in $K_\Lambda(X)$.

Again, we can define the ith exterior power of a vector space over Λ; therefore we can define the ith exterior power $\lambda^i(\xi)$ of a bundle over X. It is possible to extend the definition of λ^i from bundles to virtual bundles in a unique way so as to preserve the following familiar property:

$$\lambda^i(\xi + \eta) = \sum_{j+k=i} \lambda^j(\xi) \otimes \lambda^k(\eta).$$

All this is due to Grothendieck.

Unfortunately, the formal properties of the λ^i are not very convenient. It is possible to obtain operations with better formal properties by an algebraic device. Consider

$$(x_1)^k + (x_2)^k + \ldots + (x_k)^k;$$

this is a symmetric polynomial in $x_1, x_2, \ldots, x_k$; therefore it can be written as a polynomial in the elementary symmetric functions σ_i of $x_1, x_2, \ldots, x_k$; say

$$(x_1)^k + (x_2)^k + \ldots + (x_k)^k = Q^k(\sigma_1, \sigma_2, \ldots, \sigma_k).$$

Now define

$$\Psi^k(\xi) = \begin{cases} Q^k(\lambda^1\xi, \lambda^2\xi, \ldots, \lambda^k\xi), & (k > 0), \\ \Psi^{-k}(\xi^*), & (k < 0). \end{cases}$$

The functions $\Psi^{\cdot}$ are ring homomorphisms from $K_\Lambda(X)$ to $K_\Lambda(X)$.

In order to unify the three theorems with which I started, one makes use of the groups $J(X)$ of Atiyah [4]. First I define the notion of fibre homotopy equivalence. Let ξ, η be sphere bundles over X, with total spaces E_ξ, E_η; then we say that a map $f: E_\xi \to E_\eta$ is "fibrewise" if it covers the identity map of X; we say that ξ, η are fibre homotopy equivalent if there exist fibrewise maps $f: E_\xi \to E_\eta$, $g: E_\eta \to E_\xi$ such that $gf \sim 1$ through fibrewise maps of E_ξ, and similarly for fg. We shall define

$$J_\Lambda(X) = K_\Lambda(X)/U_\Lambda(X),$$

so that $J_\Lambda(X)$ is given by the generators $\{\xi\}$ and certain relations; we define $U_\Lambda(X)$ to be the subgroup of $K_\Lambda(X)$ generated by all elements of the form

$$\{\xi\} - \{\eta\},$$

where ξ, η are fibre homotopy equivalent.

Since all our groups are functorial, we can write

$$K_\Lambda(X) = K_\Lambda(P) + \tilde{K}_\Lambda(X),$$

$$J_\Lambda(X) = J_\Lambda(P) + \tilde{J}_\Lambda(X);$$

here P denotes a point, and these equations are supposed to define the summands $\tilde{K}_\Lambda(X)$, $\tilde{J}_\Lambda(X)$ complementary to $K_\Lambda(P)$, $J_\Lambda(P)$. Atiyah's group $J(X)$ is the one I have called $\tilde{J}_R(X)$.

According to Atiyah [4], if you can compute $J_R(RP^{k-1})$, you can prove Theorem 1; if you can compute $J_C(CP^{k-1})$, you can prove Theorem 2; and we have

$$J_R(S^l) = J(\pi_{l-1}(SO(n))) \quad (n > l).$$

We therefore face the general problem: "compute $J_\Lambda(X)$".

Half of the problem consists in giving a lower bound for $J_\Lambda(X)$, and half of it consists in giving an upper bound for $J_\Lambda(X)$. I start with the lower bound.

It is sometimes easy to prove that two bundles ξ, η are not fibre homotopy equivalent by using the Stiefel-Whitney classes, which are fibre homotopy invariants. The reason why they are fibre homotopy invariants is that they can be defined in a particular way. Suppose given a sphere bundle ξ over B with total space E; we can embed the space E in the corresponding bundle of unit solid balls, say $\bar{E}$. Then in cohomology we have the Thom isomorphism

$$\varphi_H : H^*(B; Z_2) \to H^*(\bar{E}, E; Z_2).$$

We can consider the following diagram.

$$\begin{array}{ccc} H^*(\bar{E}, E; Z_2) & \xrightarrow{Sq = \sum_{i=0}^{\infty} Sq^i} & H^*(\bar{E}, E; Z_2) \\ \varphi_H \uparrow & & \uparrow \varphi_H \\ H^*(B; Z_2) & & H^*(B; Z_2) \end{array}$$

The total Stiefel-Witney class is given by

$$w(\xi) = \varphi_H^{-1} Sq \varphi_H 1$$

We can copy this procedure using the K-cohomology theory. For example, suppose that ξ is a unitary bundle, and let the other notation remain as before. The one can define a Thom isomorphism

$$\varphi_K : K_C(B) \to K_C(\bar{E}, E),$$

where the relative K groups are defined by

$$K_\Lambda(X, Y) = \tilde{K}_\Lambda(X/Y).$$

One can consider the following diagram.

$$\begin{array}{ccc} K_C(\bar{E}, E) & \xrightarrow{\Psi^k} & K_C(\bar{E}, E) \\ \varphi_K \uparrow & & \uparrow \varphi_K \\ K_C(\mathrm{B}) & & K_C(B) \end{array}$$

By analogy, we define $\varrho^k(\xi)=\varphi_K^{-1}\Psi^k\varphi_k\, 1$.

We next seek to extend the definition of ϱ^k to virtual bundles, so as to preserve the following property.

$$\varrho^k(\xi+\eta)=\varrho^k(\xi)\varrho^k(\eta).$$

The extension is possible and unique, provided we intepret $\varrho^k(\xi)$ as an element of the group

$$K_C(B)\otimes Q_k,$$

where Q_k denotes the additive group of fractions a/k^b. It is easy to see that we are forced to introduce these denominators. In fact, we have

$$\varrho_k(1)=k,$$

so

$$\varrho_k(-1)=1/k.$$

For completeness I add that one can also adopt an intermediate approach, and consider (for example) the following diagram.

$$\begin{array}{ccc} K_C(\bar{E}, E) & \xrightarrow{ch} & H^*(\bar{E}, E; Q) \\ \varphi_K \uparrow & & \uparrow \varphi_H \\ K_C(B) & & H^*(B; Q) \end{array}$$

This method yields criteria which can be stated in terms of characteristic classes. However, it is not likely to be adequate if B has torsion; and it also fails to give best possible results for such torsion-free spaces as S^{8m+2}, CP^{4m+1}.

I therefore adopt the following definition of a quotient group $J'_\Lambda(X)$ of $J_\Lambda(X)$, which will serve as a lower bound for $J_\Lambda(X)$. I define

$$J'_\Lambda(X)=K_\Lambda(X)/V,$$

where $x\in V$ if and only if there exists y in $\tilde{K}_\Lambda(X)$ such that

$$\varrho^k(x)=\frac{\Psi^k(1+y)}{1+y} \quad \text{for all } k.$$

(The experts will understand that in the case $\Lambda=R$ we impose also the conditions $w_1(x)=0$ and $w_2(x)=0$, in order that $\varrho^k(x)$ should be defined. Compare [8].)

The reason for adopting a definition of this form is that when one tries to prove that $\varrho^k(x)$ is a fibre homotopy invariant, one finds only that it is an invariant up to multiplication by a factor of the form

$$\frac{\Psi^k(1+y)}{1+y}.$$

I will now pass on to discuss upper bounds for $J_\Lambda(x)$. For this purpose we need a result which will prove that two sphere bundles ξ and η are fibre homotopy equivalent, although they are not isomorphic. I offer the following.

THEOREM 4. *Suppose that ξ, η are sphere bundles over a finite CW-complex and that there is a fibrewise map*

$$f : E_\xi \to E_\eta$$

of degree k on each fibre. Then there exists an integer e such that the Whitney multiples $|k^e|\xi$, $|k^e|\eta$ are fibre homotopy equivalent.

If we put $k = 1$ this is a theorem of Dold [9]. Therefore one may regard this theorem as a mod k analogue of Dold's theorem. The proof of Dold may be summarised by saying that we consider the space of homotopy equivalences from S^{n-1} to S^{n-1}, and treat it seriously as a "structural group". My proof may be summarised by saying that we take the space of all maps from S^{n-1} to S^{n-1}, and treat it similarly.

By applying Theorem 4, I prove the following.

THEOREM 5. *Suppose that k is given, that $y \in K_C(X)$ (where X is a finite CW-complex) and either* (i) *y is a linear combination of complex line bundles over X, or* (ii) $X = S^{2n}$.
Then there exists an integer $e = e(k, y)$ such that

$$k^e(\Psi^k - 1)y$$

maps to zero in $J_C(X)$.

Following the hint contained in Theorem 5, I define a group $J''_\Lambda(X)$ which will act as a sort of conditional upper bound for X; that is to say, if the conclusion of Theorem 5 holds for every pair (k, y), then $J_\Lambda(X)$ will be a homomorphic image of $J''_\Lambda(X)$, no matter how large the integers $e(k, y)$ turn out to be. I define

$$J''_\Lambda(X) = K_\Lambda(X)/W,$$

where $x \in W$ if and only if for every function $e(k, y)$ there exists a function $a(k, y)$, such that

$$x = \sum_{k,y} a(k, y)\, k^{e(k,y)}(\Psi^k - 1)y.$$

It is understood that the functions $e(k, y)$ and $a(k, y)$ are defined for all pairs consisting of an integer k and an element $y \in K_C(X)$; the values of $e(k, y)$ are non-negative integers; $a(k, y)$ takes integer values, and is zero except for a finite number of pairs (k, y).

If we hope to estimate $J_\Lambda(X)$ by means of an upper bound and a lower bound, it is desirable to have these two bounds close together. If $X = RP^{k-1}$, CP^{k-1} or S^l, then

$$J''_R(X) = J'_R(X);$$

and this completes what I want to say about Theorems 1, 2, 3.

It would appear that we have

$$J''_R(X) = J'_R(X)$$

for any finite CW-complex X.

Problem 1. Does the conclusion of Theorem 5 hold for each finite CW-complex X and each element y in $K_R(X)$?

Problem 2. Can Theorem 4 be used to answer Problem 1?

Problem 3. Suppose given two inequivalent representations of $O(n)$, so that it acts on Euclidean spaces V, V' of the same dimension. When can one find an equivariant map $f: V \to V'$ which maps the unit sphere of V onto that of V' with degree k?

REFERENCES

[1]. Adams, J. F., Vector fields on spheres. *Ann. Math.*, 75 (1962), 603–632.

[2]. —— On the J-homomorphism. (To appear.)

[3]. Adams, J. F. & Walker, G., On complex Stiefel manifolds. (To appear.)

[4]. Atiyah, M. F., Thom complexes. *Proc. London Math. Soc.* (*3*), 11 (1961), 291–310.

[5]. Atiyah, M. F. & Hirzebruch, F., Riemann-Roch theorems for differentiable manifolds. *Bull. Amer. Math. Soc.*, 65 (1959), 276–281.

[6]. Atiyah, M. F., & Hirzebruch, F., Vector bundles and homogeneous spaces. *Proc. Symp. Pure Maths. 3. Differential Geometry*, 7–38. Amer. Math. Soc., 1961.

[7]. Atiyah, M. F. & Todd, J. A., On complex Stiefel manifolds. *Proc. Cambridge Philos. Soc.*, 56 (1960), 342–353.

[8]. Bott, R., A note on the KO-theory of sphere bundles. *Bull. Amer. Math. Soc.*, 68 (1962), 395–400.

[9]. Dold, A., Über fasernweise Homotopieäquivalenz von Faserräume. *Math. Z.*, 62 (1955), 111–136.

[10]. Milnor, J. & Kervaire, M. A., Bernoulli numbers, homotopy groups, and a theorem of Rohlin. *Proc. Internat. Congress Math. Edinburgh*, 1958, 454–458. Cambridge Univ. Press, 1960.

ANNALS OF MATHEMATICS
Vol. 75, No. 3, May, 1962

VECTOR FIELDS ON SPHERES

BY J. F. ADAMS

(Received November 1, 1961)

1. Results

The question of vector fields on spheres arises in homotopy theory and in the theory of fibre bundles, and it presents a classical problem, which may be explained as follows. For each n, let S^{n-1} be the unit sphere in euclidean n-space R^n. A vector field on S^{n-1} is a continuous function v assigning to each point x of S^{n-1} a vector $v(x)$ tangent to S^{n-1} at x. Given r such fields $v_1, v_2, \cdots, v_r$, we say that they are linearly independent if the vectors $v_1(x), v_2(x), \cdots, v_r(x)$ are linearly independent for all x. The problem, then, is the following: for each n, what is the maximum number r of linearly independent vector fields on S^{n-1}? For previous work and background material on this problem, we refer the reader to [1, 10, 11, 12, 13, 14, 15, 16]. In particular, we recall that if we are given r linearly independent vector fields $v_i(x)$, then by orthogonalisation it is easy to construct r fields $w_i(x)$ such that $w_1(x), w_2(x), \cdots, w_r(x)$ are orthonormal for each x. These r fields constitute a cross-section of the appropriate Stiefel fibering.

The strongest known positive result about the problem derives from the Hurwitz-Radon-Eckmann theorem in linear algebra [8]. It may be stated as follows (cf. James [13]). Let us write $n = (2a + 1)2^b$ and $b = c + 4d$, where a, b, c and d are integers and $0 \leqq c \leqq 3$; let us define $\rho(n) = 2^c + 8d$. Then there exist $\rho(n) - 1$ linearly independent vector fields on S^{n-1}.

It is the object of the present paper to prove that the positive result stated above is best possible.

THEOREM 1.1. *If $\rho(n)$ is as defined above, then there do not exist $\rho(n)$ linearly independent vector fields on S^{n-1}.*

Heuristically, it is plausible that the "depth" of this result increases with b (where $n = (2a + 1)2^b$, as above). For $b \leqq 3$, the result is due to Steenrod and Whitehead [15]. For $b \leqq 10$, the result is due to Toda [16].

The theorem, as stated, belongs properly to the theory of fibre bundles. However, we shall utilise a known reduction of the problem to one in homotopy theory, concerning real projective spaces. We write RP^q for real projective q-space, although this notation is not consistent with that employed by James and Atiyah [12,1]. If $p < q$, then RP^p is imbedded in RP^q, and we write RP^q/RP^p for the quotient space obtained from RP^q by identifying RP^p to a single point. Our main task is to prove the following theorem.

THEOREM 1.2. *$RP^{m+\rho(m)}/RP^{m-1}$ is not co-reducible; that is, there is no map*

$$f\colon RP^{m+\rho(m)}/RP^{m-1} \longrightarrow S^m$$

such that the composite

$$S^m = RP^m/RP^{m-1} \xrightarrow{i} RP^{m+\rho(m)}/RP^{m-1} \xrightarrow{f} S^m$$

has degree 1.

Theorem 1.1 has the following corollary in homotopy theory.

COROLLARY 1.3. *The Whitehead product $[\iota_{n-1}, \iota_{n-1}]$ in $\pi_{2n-3}(S^{n-1})$ is a $(\rho(n)-1)$-fold suspension but not a $\rho(n)$-fold suspension.*

It is also more or less well known that Theorem 1.1 is relevant to the study of the stable J-homomorphism (cf. [16]). More precisely, one should consider the map

$$J \otimes Z_2\colon \pi_r(\mathrm{SO}) \otimes Z_2 \longrightarrow \pi_{n+r}(S^n) \otimes Z_2 \qquad (n-1 > r > 0)\ .$$

One should deduce from Theorem 1.1 or Theorem 1.2 that $J \otimes Z_2$ is monomorphic. However, it appears to the author that one can obtain much better results on the J-homomorphism by using the methods, rather than the results, of the present paper. On these grounds, it seems best to postpone discussion of the J-homomorphism to a subsequent paper.

A summary of the present paper will be found at the end of § 2.

2. Methods

The proof of Theorem 1.2 will be formulated in terms of the "extraordinary cohomology theory" $K(X)$ of Grothendieck, Atiyah and Hirzebruch [2, 3]. We propose to introduce "cohomology operations" into the "cohomology theory" K; these operations will be functions from $K(X)$ to $K(X)$ which are natural for maps of X. If the space X is reducible or co-reducible, then the corresponding group $K(X)$ will split as a direct sum, and the homomorphisms of the splitting will commute with our operations. We shall find that $K(RP^{m+\rho(m)}/RP^{m-1})$ does not admit any splitting of the sort required.

The author hopes that this line of proof is self-justifying; however, a few historical remarks may serve to put it in perspective. The author's original approach to the present problem was directly inspired by the work of Steenrod and Whitehead [15], and consisted of an attempt to replace the Steenrod squares used in [15] by cohomology operations of higher kinds. This attempt is reasonable, but it involves several difficulties; the first of these is the selection of cohomology operations well-adapted to the solution of this particular problem. The author's work on this topic may be left in decent obscurity, like the bottom nine-tenths of an iceberg. However, it led to the following conclusions.

(1) The required operations should be constructed from universal examples.

(2) The universal examples should be fiberings induced by certain hypothetical maps $f\colon \mathrm{BO} \to \mathrm{BO}$. (Here BO denotes the classifying space of the infinite orthogonal group.)

(3) The hypothetical maps f should satisfy certain stringent algebraic specifications.

At this point the advisability of reformulating matters in terms in the K-theory became evident. The hypothetical maps f led immediately to the notion of cohomology operations in the K-theory. The algebraic conditions mentioned in (3) led easily to the correct operations.

The remainder of this paper is organised as follows. In § 3 we define the ring $K_\Lambda(X)$. Since our cohomology operations are defined with the aid of group representations, we also define the ring $K'_\Lambda(G)$ of virtual representations of G. The remainder of the section is devoted to necessary preliminaries. In § 4 we define and study the virtual representations which we need; in § 5 they are applied to construct our cohomology operations. In § 6 we present further material on $K_\Lambda(X)$, needed for § 7. In § 7 we compute the values of our operations in projective spaces. In § 8 we complete the proof of Theorem 1.2, by the method indicated above. In § 9 we deduce Theorem 1.1 and Corollary 1.3, by citing appropriate references.

3. The ring $K_\Lambda(X)$

In this section we shall define the cohomology ring $K_\Lambda(X)$ and the representation ring $K'_\Lambda(G)$. We proceed to discuss composition, and this leads to the basic lemma which will enable us to define operations in K_Λ. This lemma is stated as Lemma 3.8, near the end of the section.

We begin by defining $K_\Lambda(X)$. (Throughout this paper, the symbols Λ, Λ' will denote either the real field R or the complex field C.) Suppose given a finite CW-complex X; we consider the Λ-vector-bundles over X. (That is, we consider real vector bundles or complex vector bundles according to the choice of Λ. It is immaterial whether the group of our bundles is the full linear group $\mathrm{GL}(n, \Lambda)$ or a compact subgroup $\mathrm{O}(n)$ or $\mathrm{U}(n)$; but for definiteness we suppose it is $\mathrm{GL}(n, \Lambda)$. If X were not connected, we would allow our bundles to have fibres of different dimensions over the different components of X; however, for our purposes it will suffice to consider only connected complexes X.) We divide the Λ-vector bundles ξ over X into equivalence classes $\{\xi\}$, and take these classes as generators for a free abelian group $F_\Lambda(X)$. For each pair of bundles ξ, η over X we form the element $t = \{\xi \oplus \eta\} - \{\xi\} - \{\eta\}$, where $\oplus$ denotes the Whitney sum. We write $T_\Lambda(X)$ for the subgroup of $F_\Lambda(X)$ generated by such elements t; we define $K_\Lambda(X)$ to be the quotient group $F_\Lambda(X)/T_\Lambda(X)$.

We proceed to define $K'_\Lambda(G)$ in a closely analogous way. Suppose given a topological group G. A representation α of G (of degree n, over Λ) is a continuous function $\alpha: G \rightarrow \mathrm{GL}(n, \Lambda)$ which preserves products. Two such representations are equivalent if they coincide up to an inner automorphism of $\mathrm{GL}(n, \Lambda)$. We divide the representations α of G over Λ into equivalence classes $\{\alpha\}$, and take these classes as generators for a free abelian group $F'_\Lambda(G)$. For each pair of representations α, β we form the element $t = \{\alpha \oplus \beta\} - \{\alpha\} - \{\beta\}$, where $\oplus$ denotes the direct sum of representations. We write $T'_\Lambda(G)$ for the subgroup of $F'_\Lambda(G)$ generated by such elements t. We define $K'_\Lambda(G)$ to be the quotient group $F'_\Lambda(G)/T'_\Lambda(G)$. An element of $K'_\Lambda(G)$ is called a virtual representation (of G, over Λ).

It is clear that we can define a homomorphism from $F'_\Lambda(G)$ to the integers which assigns to each representation its degree. This homomorphism passes to the quotient, and defines the virtual degree of a virtual representation.

We next discuss composition. It will shorten explanations if we adopt a convention. The letters f, g, h will denote maps of complexes such as X. The letters ξ, η, ζ will denote bundles, and the letters κ, λ, μ will denote elements of $K_\Lambda(X)$. The letters α, β, γ will denote representations, and the letters θ, φ, ψ will denote virtual representations.

The basic sorts of composition are easily enumerated if we interpret a bundle ξ as a classifying map $\xi: X \rightarrow \mathrm{BGL}(n, \Lambda)$ and a representation α as a map of classifying spaces. We have to define compositions

$$\beta \cdot \alpha, \ \alpha \cdot \xi, \ \xi \cdot f, \ f \cdot g \ .$$

(Composition will be written with a dot, to distinguish it from any other product.) The formal definitions are as follows.

If $\alpha: G \rightarrow \mathrm{GL}(n, \Lambda)$ and $\beta: \mathrm{GL}(n, \Lambda) \rightarrow \mathrm{GL}(n', \Lambda')$ are representations, then $\beta \cdot \alpha$ is their composite in the usual sense. If ξ is a bundle over X with group $\mathrm{GL}(n, \Lambda)$ and $\alpha: \mathrm{GL}(n, \Lambda) \rightarrow \mathrm{GL}(n', \Lambda')$ is a representation, then $\alpha \cdot \xi$ is the induced bundle, defined by using the same coordinate neighbourhoods and applying α to the coordinate transformation functions. If ξ is a bundle over Y and $f: X \rightarrow Y$ is a map, then $\xi \cdot f$ is the induced bundle over X, defined by applying f^{-1} to the coordinate neighbourhoods and composing the coordinate transformation functions with f. If $f: X \to Y$ and $g: Y \to Z$ are maps, then $g \cdot f$ is their composite in the usual sense.

We next wish to linearize over the first factor.

LEMMA 3.1. *It is possible to define composites of the form $\varphi \cdot \alpha$, $\theta \cdot \xi$ and $\kappa \cdot f$, so that $\varphi \cdot \alpha$ lies in the appropriate group $K'_\Lambda(G)$, $\theta \cdot \xi$ and $\kappa \cdot f$ lie in appropriate groups $K_\Lambda(X)$, and they have the following properties.*

(i) *Each composite is linear in its first factor.*

(ii) *If we replace φ, θ or κ by β, α or ξ (respectively), then these composites reduce to those considered above.*

(iii) *The following associativity formulae hold.*

$$(\psi \cdot \beta) \cdot \alpha = \psi \cdot (\beta \cdot \alpha) , \qquad (\varphi \cdot \alpha) \cdot \xi = \varphi \cdot (\alpha \cdot \xi) ,$$
$$(\theta \cdot \xi) \cdot f = \theta \cdot (\xi \cdot f) , \qquad (\kappa \cdot g) \cdot f = \kappa \cdot (g \cdot f) .$$

(iv) *If $\alpha = 1$, then $\varphi \cdot \alpha = \varphi$. If $f = 1$, then $\kappa \cdot f = \kappa$.*

PROOF. The required composites are defined by (i) and (ii), and it is easy to check that they are well defined. The formulae (iii), (iv) follow by linearity from those that hold before linearizing.

We next wish to linearize over the second factor. For this purpose we require a first factor which can act on $\mathrm{GL}(n, \Lambda)$ for any n. We therefore introduce the notion of a sequence $\Theta = (\theta_n)$, where, for each n, θ_n is a virtual representation of $\mathrm{GL}(n, \Lambda)$ over Λ'. We reserve the letters Θ, Φ, Ψ for such sequences. In order to linearize over the second factor, we require a linearity condition on the first factor. In order to state this condition, we write

$$\pi: \mathrm{GL}(n, \Lambda) \times \mathrm{GL}(m, \Lambda) \longrightarrow \mathrm{GL}(n, \Lambda) ,$$
$$\varpi: \mathrm{GL}(n, \Lambda) \times \mathrm{GL}(m, \Lambda) \longrightarrow \mathrm{GL}(m, \Lambda)$$

for the projections of $\mathrm{GL}(n, \Lambda) \times \mathrm{GL}(m, \Lambda)$ onto its two factors. These projections are representations.

DEFINITION 3.2. The sequence $\Theta = (\theta_n)$ is additive if we have

$$\theta_{n+m} \cdot (\pi \oplus \varpi) = (\theta_n \cdot \pi) + (\theta_m \cdot \varpi)$$

for all n, m.

LEMMA 3.3. *Suppose Θ is additive. Then for any two representations $\alpha: G \to \mathrm{GL}(n, \Lambda)$, $\beta: G \to \mathrm{GL}(m, \Lambda)$ we have*

$$\theta_{n+m} \cdot (\alpha \oplus \beta) = (\theta_n \cdot \alpha) + (\theta_m \cdot \beta) .$$

Moreover, for any two bundles ξ, η over X with groups $\mathrm{GL}(n, \Lambda)$, $\mathrm{GL}(m, \Lambda)$ we have

$$\theta_{n+m} \cdot (\xi \oplus \eta) = (\theta_n \cdot \xi) + (\theta_m \cdot \eta) .$$

We postpone the proof for a few lines.

LEMMA 3.4. *If Θ, Φ, Ψ run over additive sequences, then it is possible to define composites of the form $\Phi\cdot\Theta$, $\Phi\cdot\theta$ and $\Theta\cdot\kappa$ so that $\Phi\cdot\Theta$ is an additive sequence, $\Phi\cdot\theta$ lies in the appropriate group $K'_\Lambda(G)$, $\Theta\cdot\kappa$ lies in the appropriate group $K_\Lambda(X)$, and they have the following properties.*

(i) *Each composite is bilinear in its factors.*

(ii) $(\Phi\cdot\Theta)_n = \Phi\cdot\theta_n$. *If $\alpha\colon G\to \mathrm{GL}(n,\Lambda)$ is a representation then $\Phi\cdot\alpha = \varphi_n\cdot\alpha$ (in the sense of Lemma 3.1). If ξ is a $\mathrm{GL}(n,\Lambda)$-bundle over X then $\Theta\cdot\xi = \theta_n\cdot\xi$ (in the sense of Lemma 3.1).*

(iii) *The following associativity formulae hold.*

$$(\Psi\cdot\Phi)\cdot\Theta = \Psi\cdot(\Phi\cdot\Theta)\,, \qquad (\Phi\cdot\Theta)\cdot\kappa = \Phi\cdot(\Theta\cdot\kappa)\,,$$
$$(\Theta\cdot\kappa)\cdot f = \Theta\cdot(\kappa\cdot f)\,.$$

(iv) *If 1 denotes the additive sequence of identity maps $1_n\colon \mathrm{GL}(n,\Lambda)\to \mathrm{GL}(n,\Lambda)$, then*

$$1\cdot\Theta = \Theta\,, \qquad \Phi\cdot 1 = \Phi\,, \qquad 1\cdot\kappa = \kappa\,.$$

PROOF OF LEMMA 3.3. Suppose given two bundles ξ, η over X, with groups $\mathrm{GL}(n,\Lambda)$, $\mathrm{GL}(m,\Lambda)$. Then we can define a bundle $\xi\times\eta$ over X, with group $\mathrm{GL}(n,\Lambda)\times\mathrm{GL}(m,\Lambda)$. (The coordinate neighbourhoods are the intersections of those in ξ and those in η; the coordinate transformation functions are obtained by lumping together those in ξ and those in η.) We have $(\pi\oplus\varpi)\cdot(\xi\times\eta) = \xi\oplus\eta$, $\pi\cdot(\xi\times\eta)=\xi$, $\varpi\cdot(\xi\times\eta)=\eta$. If Θ is additive we have

$$\theta_{n+m}\cdot(\pi\oplus\varpi)\cdot(\xi\times\eta) = (\theta_n\cdot\pi + \theta_m\cdot\varpi)\cdot(\xi\times\eta)\,,$$

that is,

$$\theta_{n+m}\cdot(\xi\oplus\eta) = \theta_n\cdot\xi + \theta_m\cdot\eta\,.$$

The proof for representations is analogous, but slightly more elementary.

PROOF OF LEMMA 3.4. The required composites are defined by (i) and (ii), and it is trivial to check that they are well defined, given the conclusion of Lemma 3.3. The associative laws are preserved at each step of the construction; finally, conclusion (iv) is trivial.

We next recall that the tensor product of bundles defines a product in $K_\Lambda(X)$ (cf. [2, 3]). Similarly, the tensor product of representations defines a product in $K'_\Lambda(G)$; thus $K_\Lambda(X)$ and $K'_\Lambda(G)$ become commutative rings with unit. Composition behaves well for tensor products of the first factor, as is shown by the following formulae.

$$(\varphi \otimes \theta) \cdot \alpha = (\varphi \cdot \alpha) \otimes (\theta \cdot \alpha) ,$$
$$(\varphi \otimes \theta) \cdot \xi = (\varphi \cdot \xi) \otimes (\theta \cdot \xi) ,$$
$$(\kappa \otimes \lambda) \cdot f = (\kappa \cdot f) \otimes (\lambda \cdot f) .$$

Here, for example, the third formula states that the products in $K_\Lambda(X)$ are natural for maps of X. These formulae are deduced by linearity from the corresponding ones for representations and bundles.

In order to ensure that composition behaves well for tensor products of the second factor, we require a condition on the first factor. In order to state this condition, we re-adopt the notation of Definition 3.2.

DEFINITION 3.5. The sequence $\Theta = (\theta_n)$ is multiplicative if we have

$$\theta_{nm} \cdot (\pi \otimes \varpi) = (\theta_n \cdot \pi) \otimes (\theta_m \cdot \varpi)$$

for all n, m.

LEMMA 3.6. *Suppose Θ is multiplicative. Then for any two representations $\alpha: G \to \mathrm{GL}(n, \Lambda)$, $\beta: G \to \mathrm{GL}(m, \Lambda)$ we have*

$$\theta_{nm} \cdot (\alpha \otimes \beta) = (\theta_n \cdot \alpha) \otimes (\theta_m \cdot \beta) .$$

Moreover, for any two bundles ξ, η over X with groups $\mathrm{GL}(n, \Lambda)$, $\mathrm{GL}(m, \Lambda)$ we have

$$\theta_{nm} \cdot (\xi \otimes \eta) = (\theta_n \cdot \xi) \otimes (\theta_m \cdot \eta) .$$

The proof is closely similar to that of Lemma 3.3.

LEMMA 3.7. *If Ψ is both additive and multiplicative, then we have*

$$\Psi \cdot (\theta \otimes \varphi) = (\Psi \cdot \theta) \otimes (\Psi \cdot \varphi) ,$$
$$\Psi \cdot (\kappa \otimes \lambda) = (\Psi \cdot \kappa) \otimes (\Psi \cdot \lambda) .$$

This follows from Lemma 3.6 by linearity.

We now restate our main results in one omnibus lemma.

LEMMA 3.8. *Suppose given an additive sequence $\Theta = (\theta_n)$, where $\theta_n \in K_{\Lambda'}^{\Lambda}(\mathrm{GL}(n, \Lambda))$. Then the function $\Theta \cdot \kappa$ of κ gives (for each X) a group homomorphism*

$$\Theta: K_\Lambda(X) \to K_{\Lambda'}(X)$$

with the following properties.

(i) *Θ is natural for maps of X; that is, if $f: X \to Y$ is a map, then the following diagram is commutative.*

$$\begin{array}{ccc} K_\Lambda(Y) & \xrightarrow{\Theta} & K_{\Lambda'}(Y) \\ {\scriptstyle f^*}\downarrow & & \downarrow{\scriptstyle f^*} \\ K_\Lambda(X) & \xrightarrow{\Theta} & K_{\Lambda'}(X) . \end{array}$$

(ii) *If the sequence* Θ *is multiplicative as well as additive, then*

$$\Theta\colon K_\Lambda(X) \longrightarrow K_{\Lambda'}(X)$$

preserves products.

(iii) *If* θ_1 *has virtual degree* 1, *then*

$$\Theta\colon K_\Lambda(X) \longrightarrow K_{\Lambda'}(X)$$

maps the unit in $K_\Lambda(X)$ *into the unit in* $K_{\Lambda'}(X)$.

PROOF. Except for (iii), this is merely a restatement of what has been said above; thus, (i) is the associativity law $(\Theta \cdot \kappa) \cdot f = \Theta \cdot (\kappa \cdot f)$ of Lemma 3.4, and (ii) is contained in Lemma 3.7. As for (iii), the unit in $K_\Lambda(X)$ is the trivial bundle with fibres of dimension 1. Any representation of $\mathrm{GL}(1, \Lambda)$ will map this into a trivial bundle of the appropriate dimension; hence $\theta_1 \cdot 1 = d$, where d is the virtual degree of θ_1. This completes the proof.

As a first application of Lemma 3.8 (which, however, is hardly necessary in so trivial a case) we consider the following sequences.

(i) The sequence $c = (c_n)$, where

$$c_n\colon \mathrm{GL}(n, R) \longrightarrow \mathrm{GL}(n, C)$$

is the standard injection. (The letter "c" for "complexification" is chosen to avoid confusion with other injections.)

(ii) The sequence $r = (r_n)$, where

$$r_n\colon \mathrm{GL}(n, C) \longrightarrow \mathrm{GL}(2n, R)$$

is the standard injection.

(iii) The sequence $t = (t_n)$, where

$$t_n\colon \mathrm{GL}(n, C) \longrightarrow \mathrm{GL}(n, C)$$

is defined by $t_n(M) = \bar{M}$, and $\bar{M}$ is the complex conjugate of the matrix M.

All these sequences are additive, while c and t are multiplicative. We therefore obtain the following natural group-homomorphisms:

$$\begin{aligned} c&\colon K_R(X) \longrightarrow K_C(X) , \\ r&\colon K_C(X) \longrightarrow K_R(X) , \\ t&\colon K_C(X) \longrightarrow K_C(X) . \end{aligned}$$

The functions c and t are homomorphisms of rings.

LEMMA 3.9. *We have*

$$rc = 2 \quad : K_R(X) \longrightarrow K_R(X) ,$$
$$cr = 1 + t : K_C(X) \longrightarrow K_C(X) .$$

This follows immediately from the corresponding fact for representations. (Cf. [7], Proposition 3.1.)

4. Certain virtual representations

In this section we shall define and study the virtual representations which we need. It is a pleasure to acknowledge at this point helpful conversations with A. Borel and Harish-Chandra; the former kindly read a draft of this section.

The result which we require is stated as Theorem 4.1; the rest of the section is devoted to proving it.

THEOREM 4.1. *For each integer k (positive, negative or zero) and for $\Lambda = R$ or C, there is a sequence Ψ^k_Λ such that this system of sequences has the following properties.*

(i) *$\psi^k_{\Lambda,n}$ is a virtual representation of* GL(n, Λ) *over Λ, with virtual degree n.*

(ii) *The sequence Ψ^k_Λ is both additive and multiplicative (in the sense of* § 3).

(iii) *$\psi^k_{\Lambda,1}$ is the kth power of the identity representation of* GL$(1, \Lambda)$. *(For $k \geqq 0$, the kth power is taken in the sense of the tensor product. The kth power also makes sense for $k < 0$, since 1-dimensional representations are invertible.)*

(iv) *If c is the sequence of injections c_n:* GL$(n, R) \to$ GL(n, C) *(as in* § 3) *then*

$$\Psi^k_C \cdot c = c \cdot \Psi^k_R .$$

(v) $\Psi^k_\Lambda \cdot \Psi^l_\Lambda = \Psi^{kl}_\Lambda$.

(vi) *Let G be a topological group (with typical element g) and let θ be a virtual representation of G over Λ; then the following formula holds for the characters χ.*

$$\chi(\Psi^k_\Lambda \cdot \theta)g = \chi(\theta)g^k .$$

(vii) *$\psi^1_{\Lambda,n}$ is the identity representation. $\psi^0_{\Lambda,n}$ is the trivial representation of degree n. $\psi^{-1}_{\Lambda,n}$ is the representation defined by*

$$\psi^{-1}_{\Lambda,n}(M) = ({}^TM)^{-1} ,$$

where ${}^TM$ is the transpose of the matrix M.

PROOF. We begin by recalling the definition of the rth exterior power.

$$\begin{array}{ccc} K_R(X) & \xrightarrow{\Psi_R^k} & K_R(X) \\ \downarrow c & & \downarrow c \\ K_O(X) & \xrightarrow{\Psi_O^k} & K_O(X) . \end{array}$$

(v) $\Psi_\Lambda^k(\Psi_\Lambda^l(K)) = \Psi_\Lambda^{kl}(K)$.

(vi) *If* $\kappa \in K_O(X)$ *and* $\mathrm{ch}^q \kappa$ *denotes the* $2q$*-dimensional component of the Chern character* $\mathrm{ch}(\kappa)$ [2, 3, 4], *then*

$$\mathrm{ch}^q(\Psi_O^k \kappa) = k^q \mathrm{ch}^q(\kappa) .$$

(vii) Ψ_Λ^1 *and* Ψ_R^{-1} *are identity functions.* Ψ_Λ^0 *is the function which assigns to each bundle over* X *the trivial bundle with fibres of the same dimension.* Ψ_O^{-1} *coincides with the operation* t *considered in* § 3.

PROOF. Parts (i) and (ii) of the theorem follow directly from Lemma 3.8 and Theorem 4.1 (parts (i), (ii)). By using the results of § 3 where necessary, parts (iii), (iv), (v) and (vii) of the theorem follow from the correspondingly-numbered parts of Theorem 4.1, except that it remains to identify Ψ_Λ^{-1}. If $\Lambda = R$, then any n-plane bundle is equivalent to one with structural group $\mathrm{O}(n)$, and for $M \in \mathrm{O}(n)$ we have $({}^T M)^{-1} = M$. If $\Lambda = C$, then any n-plane bundle is equivalent to one with structural group $\mathrm{U}(n)$, and for $M \in \mathrm{U}(n)$ we have $({}^T M)^{-1} = \bar{M}$. This completes the identification of Ψ_Λ^{-1}.

It remains to prove (vi). We first recall the basic facts about the Chern character. If $\Lambda = C$ and ξ is a bundle over X, then $\mathrm{ch}^q(\xi)$ is a characteristic class of ξ lying in $H^{2q}(X; Q)$ (where Q denotes the rationals). The main properties of $\mathrm{ch} = \sum_{q=0}^{\infty} \mathrm{ch}^q$ are as follows.

(i) ch defines a ring homomorphism from $K_O(X)$ to $H^*(X; Q)$.

(ii) ch is natural for maps of X.

(iii) If ξ is the canonical line bundle over CP^n, then $\mathrm{ch}\,\xi = e^{-x}$, where x is the generator of $H^2(CP^n; Z)$ and e^{-x} is interpreted as a power series.

We now turn to the proof. Let $\mathrm{T} \subset \mathrm{U}(n)$ be a (maximal) torus consisting of the diagonal matrices with diagonal elements of unit modulus. The classifying space BT is a product of complex projective spaces CP^∞. Let $Y \subset \mathrm{BT}$ be the corresponding product of complex projective spaces CP^N, where $N \geqq q$; let $x_1, x_2, \cdots, x_n$ be the cohomology generators. We may evidently imbed Y in a finite CW-complex X, and extend the inclusion $i: Y \to \mathrm{BU}(n)$ to a map $f: X \to \mathrm{BU}(n)$, so that f is an equivalence up to any required dimension. We see (by naturality and linearity) that it is sufficient to prove (vi) when the space concerned is X and the element κ is the bundle ξ over X induced by f from the canonical $\mathrm{U}(n)$-bundle over $\mathrm{BU}(n)$.

$$\chi(\psi^k_{\Lambda,n})M = \operatorname{Tr}(M^k) ,$$

where $\operatorname{Tr}(M^k)$ *denotes the trace of* M^k.

PROOF. We begin by recalling the basic facts about characters. If α is a representation of G, then its character $\chi(\alpha)$ is defined by

$$\chi(\alpha)g = \operatorname{Tr}(\alpha g) \qquad (g \in G) .$$

We have

$$\chi(\alpha \oplus \beta) = \chi(\alpha) + \chi(\beta) ,$$
$$\chi(\alpha \otimes \beta) = \chi(\alpha)\chi(\beta) .$$

By linearity, one defines the character $\chi(\theta)$ of a virtual representation θ.

We next remark that, in proving Proposition 4.3, it is sufficient to consider the case $\Lambda = C$. In fact, if α is a real representation it is clear that $\chi(c \cdot \alpha) = \chi(\alpha)$; hence if θ is a virtual representation we have $\chi(c \cdot \theta) = \chi(\theta)$. This remark, together with Proposition 4.2, enables one to deduce the case $\Lambda = R$ from the case $\Lambda = C$.

Let us suppose (to begin with) that M is a diagonal matrix with non-zero complex entries $x_1, x_2, \cdots, x_n$. Then $E^r_C(M)$ is a diagonal matrix with entries $x_{i_1}x_{i_2}\cdots x_{i_r} (i_1 < i_2 < \cdots < i_r)$. Hence $\chi(E^r_C)M$ is the rth elementary symmetric function σ_r of the x_i. It follows that if Q is any polynomial in n variables, we have

$$\chi(Q(E^1_C, E^2_C, \cdots, E^n_C))M = Q(\sigma_1, \sigma_2, \cdots, \sigma_n) .$$

Substituting $Q = Q^k_n$, we find

$$\chi(\psi^k_{C,n})M = \textstyle\sum_{1 \leq i \leq n} (x_i)^k = \operatorname{Tr}(M^k) \qquad (k \geqq 0) .$$

This result depends only on the conjugacy class of M in $\mathrm{GL}(n, C)$, and is therefore true for any M conjugate to a diagonal matrix. But such M are everywhere dense in $\mathrm{GL}(n, C)$, and both sides of the equation are continuous in M; therefore the result holds for all M in $\mathrm{GL}(n, C)$, at least if $k \geqq 0$. It remains only to note that

$$\begin{aligned}\chi(\psi^{-k}_{C,n})M &= \chi(\psi^k_{C,n})({}^TM)^{-1} \\ &= \operatorname{Tr}({}^TM)^{-k} \\ &= \operatorname{Tr}(M^{-k}) .\end{aligned}$$

The proof is complete.

PROPOSITION 4.4. *For each representation* $\alpha\colon G \to \mathrm{GL}(n, \Lambda)$ *and each* $g \in G$ *we have*

$$\chi(\psi^k_{\Lambda,n} \cdot \alpha)g = \chi(\alpha)g^k .$$

PROOF. Substitute $M = \alpha g$ in Proposition 4.3.

If V is a vector space over Λ, then the rth exterior power $E^r(V)$ is a vector space over Λ given by generators and relations. The generators are symbols $v_1 \wedge v_2 \wedge \cdots \wedge v_r (v_i \in V)$; the relations state that these symbols are multilinear and anti-symmetric in their arguments. Since $E^r(V)$ is a covariant functor, any automorphism of V induces one of $E^r(V)$. Let us choose a base $v_1, v_2, \cdots, v_n$ in V and take as our base in $E^r(V)$ the elements $v_{i_1} \wedge v_{i_2} \wedge \cdots \wedge v_{i_r} (i_1 < i_2 < \cdots < i_r)$; we obtain a definite representation

$$E^r_\Lambda \colon \mathrm{GL}(n, \Lambda) \longrightarrow \mathrm{GL}(m, \Lambda) ,$$

where $m = n!/(r!(n-r)!)$. These representations are evidently compatible with "complexification", in the sense that

$$E^r_C \cdot c_n = c_m \cdot E^r_R .$$

We next consider the polynomial $\sum_{1 \leq i \leq n} (x_i)^k$ in the variables $x_1, x_2, \cdots, x_n$. Since this polynomial is symmetric, it can be written as a polynomial in the elementary symmetric functions $\sigma_1, \sigma_2, \cdots, \sigma_n$ of $x_1, x_2, \cdots, x_n$; say

$$\sum_{1 \leq i \leq n} (x_i)^k = Q^k_n(\sigma_1, \sigma_2, \cdots, \sigma_n) .$$

For $k \geqq 0$, we now define

$$\psi^k_{\Lambda, n} = Q^k_n(E^1_\Lambda, E^2_\Lambda, \cdots, E^n_\Lambda) .$$

(The polynomial is evaluated in the ring $K'_\Lambda(\mathrm{GL}(n, \Lambda))$.) To obtain the virtual degree of our representations, we substitute $x_1 = 1, x_2 = 1, \cdots, x_n = 1$; we find that the virtual degree of $\psi^k_{\Lambda, n}$ is n.

As trivial cases, we see that Ψ^1_Λ and Ψ^0_Λ are as described in conclusion (vii), while $\psi^k_{\Lambda, 1}$ is as described in conclusion (iii) for $k \geqq 0$.

We next define Ψ^{-1}_Λ to be as described in conclusion (vii); that is, $\psi^{-1}_{\Lambda, n}$ is the representation defined by

$$\psi^{-1}_{\Lambda, n}(M) = ({}^T M)^{-1} .$$

We define ψ^{-k}_Λ for $k > 1$ by setting

$$\psi^{-k}_{\Lambda, n} = \psi^k_{\Lambda, n} \cdot \psi^{-1}_{\Lambda, n} .$$

It is clear that conclusion (iii) holds for $k < 0$.

PROPOSITION 4.2. $\psi^k_{C, n} \cdot c_n = c \cdot \psi^k_{R, n}$.

PROOF. Since "complexification" commutes with exterior powers as well as with sums and products, this is obvious for $k \geqq 0$. It is clear for $k = -1$, and the case $k < -1$ follows.

PROPOSITION 4.3. *For each matrix M in $\mathrm{GL}(n, \Lambda)$, we have*

value of $P^{-1}\bar{P}$.) This completes the proof of Proposition 4.5.

Since the sequences Ψ^k_Λ have now been shown to be additive, the various compositions written in Theorem 4.1 are well-defined. Conclusion (iv) is a restatement of Proposition 4.2, and conclusion (vi) follows from Proposition 4.4 by linearity. It remains only to prove the following.

PROPOSITION 4.6. $\Psi^k_\Lambda \cdot \psi^l_{\Lambda,n} = \psi^{kl}_{\Lambda,n}$.

PROOF. We begin by checking that the characters of the two sides agree. An obvious calculation, based on Proposition 4.3 and conclusion (vi), shows that

$$\begin{aligned}\chi[\Psi^k_\Lambda \cdot \psi^l_{\Lambda,n}]M &= \operatorname{Tr}(M^{kl}) \\ &= \chi(\psi^{kl}_{\Lambda,n})M\ .\end{aligned}$$

The proof is completed as for Proposition 4.5.

This completes the proof of Theorem 4.1.

REMARK. Grothendieck has considered abstract rings which admit "exterior power" operations λ^i. It is evidently possible to define operations Ψ^k (for $k \geqq 0$) in such rings.

5. Cohomology operations in $K_\Lambda(X)$

In this section we shall use the results of §§ 3, 4 to construct and study certain natural cohomology operations defined in $K_\Lambda(X)$. It would perhaps be interesting to determine the set of all such operations (as defined by some suitable set of axioms); but for our present purposes this is not necessary.

By applying Lemma 3.8 to the sequences of Theorem 4.1, we obtain operations

$$\Psi^k_\Lambda\colon K_\Lambda(X) \longrightarrow K_\Lambda(X)\ ,$$

where k is any integer (positive, negative or zero) and $\Lambda = R$ or C.

THEOREM 5.1. *These operations enjoy the following properties.*

(i) Ψ^k_Λ *is natural for maps of* X.

(ii) Ψ^k_Λ *is a homomorphism of rings with unit.*

(iii) *If* ξ *is a line bundle over* X, *then* $\Psi^k_\Lambda \xi = \xi^k$.

(A line bundle is a bundle with fibres of dimension 1. For $k \geqq 0$ the kth power ξ^k is taken in the sense of the tensor product. The kth power also makes sense for $k < 0$, since line bundles are invertible.)

(iv) *The following diagram is commutative.*

PROPOSITION 4.5. *The sequence Ψ^k_Λ is additive and multiplicative.*

PROOF. Let

$$\pi\colon \mathrm{GL}(n, \Lambda) \times \mathrm{GL}(m, \Lambda) \longrightarrow \mathrm{GL}(n, \Lambda)\ ,$$
$$\varpi\colon \mathrm{GL}(n, \Lambda) \times \mathrm{GL}(m, \Lambda) \longrightarrow \mathrm{GL}(m, \Lambda)$$

be the projections of $\mathrm{GL}(n, \Lambda) \times \mathrm{GL}(m, \Lambda)$ onto its two factors, as in § 3. We have to prove

(i) $\psi^k_{\Lambda,n+m}\cdot(\pi \oplus \varpi) = (\psi^k_{\Lambda,n}\cdot\pi) + (\psi^k_{\Lambda,m}\cdot\varpi)$,

(ii) $\psi^k_{\Lambda,nm}\cdot(\pi \otimes \varpi) = (\psi^k_{\Lambda,n}\cdot\pi) \otimes (\psi^k_{\Lambda,m}\cdot\varpi)$.

We begin by checking that in each equation, the characters of the two sides agree. An obvious calculation, based on Proposition 4.4, shows that

$$\begin{aligned}\chi[\psi^k_{\Lambda,n+m}\cdot(\pi \oplus \varpi)]g &= (\chi(\pi) + \chi(\varpi))g^k \\ &= \chi[(\psi^k_{\Lambda,n}\cdot\pi) + (\psi^k_{\Lambda,m}\cdot\varpi)]g\ .\end{aligned}$$

Similarly for equation (ii), the common answer being $(\chi(\pi)\chi(\varpi))g^k$.

Consider the case $\Lambda = C$, and examine the subgroup $\mathrm{U}(n) \times \mathrm{U}(m) \subset \mathrm{GL}(n, C) \times \mathrm{GL}(m, C)$. This subgroup is compact, and therefore two virtual representations coincide on it if and only if they have the same characters. By transporting negative terms to the opposite side of our equations, we now face the following situation: two representations are defined on $\mathrm{GL}(n, C) \times \mathrm{GL}(m, C)$ and agree on $\mathrm{U}(n) \times \mathrm{U}(m)$; we wish to show that they agree on $\mathrm{GL}(n, C) \times \mathrm{GL}(m, C)$. Now, it is a theorem that two analytic representations which are defined on $\mathrm{GL}(n, C)$ and agree on $\mathrm{U}(n)$ agree also on $\mathrm{GL}(n, C)$. (Such analytic representations define C-linear maps of Lie algebras. The Lie algebra of $\mathrm{GL}(n, C)$ is the space of all $n \times n$ complex matrices; the Lie algebra of $\mathrm{U}(n)$ is the space of skew-hermitian matrices; two C-linear maps which agree on the latter agree on the former. The map of the Lie algebra determines the map of the Lie group.) The same argument clearly applies to the subgroup $\mathrm{U}(n)\times\mathrm{U}(m)$ in $\mathrm{GL}(n, C)\times\mathrm{GL}(m, C)$. Moreover, all our representations are clearly analytic. This completes the proof in the case $\Lambda = C$.

We now consider the case $\Lambda = R$. We face the following situation. Two real representations are given over $\mathrm{GL}(n, R) \times \mathrm{GL}(m, R)$; it has been proved that after composing with c ("complexifying") they become equivalent. We wish to show that the real representations are equivalent. Now, it is a theorem that if two real representations α, β of G are equivalent over C, then they are equivalent over R. (Suppose given a complex non-singular matrix P such that $P\alpha(g) = \beta(g)P$ for all $g \in G$. Then for any complex number λ, the matrix $Q = \lambda P + \bar{\lambda}\bar{P}$ is real and such that $Q\alpha(g) = \beta(g)Q$ for all $g \in G$. In order to ensure the non-singularity of $Q = P(\lambda\bar{\lambda}^{-1}I + P^{-1}\bar{P})\bar{\lambda}$, it is sufficient to ensure that $-\lambda\bar{\lambda}^{-1}$ is not an eigen-

Let $i: Y \to X$ be the inclusion. According to Borel [5], the map

$$i^*: H^{2q}(X; Q) \longrightarrow H^{2q}(Y; Q)$$

is a monomorphism. Moreover,

$$i^*\xi = \xi_1 \oplus \xi_2 \oplus \cdots \oplus \xi_n ,$$

where $\xi_1, \xi_2, \cdots, \xi_n$ are line-bundles induced from the canonical line-bundles over the factors of Y. We now have

$$\begin{aligned} i^* \operatorname{ch}(\Psi_C^k \xi) &= \operatorname{ch} \Psi_C^k(i^*\xi) \\ &= \operatorname{ch} \Psi_C^k(\xi_1 \oplus \xi_2 \oplus \cdots \oplus \xi_n) \\ &= \operatorname{ch}(\Psi_C^k \xi_1 + \Psi_C^k \xi_2 + \cdots + \Psi_C^k \xi_n) \\ &= \operatorname{ch}((\xi_1)^k + (\xi_2)^k + \cdots + (\xi_n)^k) \\ &= (\operatorname{ch} \xi_1)^k + (\operatorname{ch} \xi_2)^k + \cdots + (\operatorname{ch} \xi_n)^k \\ &= e^{-kx_1} + e^{-kx_2} + \cdots + e^{-kx_n} . \end{aligned}$$

Similarly,

$$\begin{aligned} i^* \operatorname{ch} \xi &= \operatorname{ch} i^*\xi \\ &= \operatorname{ch}(\xi_1 \oplus \xi_2 \oplus \cdots \oplus \xi_n) \\ &= e^{-x_1} + e^{-x_2} + \cdots + e^{-x_n} . \end{aligned}$$

Comparing the components in dimension $2q$, we find the required result.

This completes the proof of Theorem 5.1.

In order to state the next corollary, we recall some notation [3]. Let P denote a point. Then, for any X, $K_\Lambda(P)$ is a direct summand of $K_\Lambda(X)$. We write $\tilde{K}_\Lambda(X)$ for the complementary direct summand. Evidently our operations Ψ_Λ^k act on $\tilde{K}_\Lambda(X)$. (Actually $K_\Lambda(P) = Z$, and $\Psi_\Lambda^k: K_\Lambda(P) \longrightarrow K_\Lambda(P)$ is the identity.)

COROLLARY 5.2. *The operations*

$$\Psi_C^k: \tilde{K}_C(S^{2q}) \longrightarrow \tilde{K}_C(S^{2q}) ,$$
$$\Psi_R^k: \tilde{K}_R(S^{2q}) \longrightarrow \tilde{K}_R(S^{2q}) \qquad (q \text{ even})$$

are given by

$$\Psi_\Lambda^k(\kappa) = k^q \kappa .$$

PROOF. It is a well-known corollary of Bott's work that

$$\operatorname{ch}^q: \tilde{K}_C(S^{2q}) \longrightarrow H^{2q}(S^{2q}; Q)$$

maps $\tilde{K}_C(S^{2q})$ isomorphically onto the image of $H^{2q}(S^{2q}; Z)$. (See, for example, [4] Proposition 2.2.) The result for $\Lambda = C$ now follows from Theorem 5.1 (vi).

It also follows from Bott's work that if q is even, then

$$c\colon Z = \tilde{K}_R(S^{2q}) \longrightarrow \tilde{K}_C(S^{2q}) = Z$$

is monomorphic. (In fact, $\operatorname{Im} c = Z$ if $q \equiv 0 \bmod 4$, while $\operatorname{Im} c = 2Z$ if $q \equiv 2 \bmod 4$. This can be derived from the corresponding result for the homomorphism

$$c_*\colon \pi_{2q-1}(\mathrm{O}) \longrightarrow \pi_{2q-1}(\mathrm{U})\ ;$$

and this in turn can be obtained from the results on U/O given in [6, p. 315].) The case $\Lambda = R$ therefore follows from the case $\Lambda = C$.

In order to state our next corollary, we recall that according to [7, Theorem 1] we have the following isomorphisms.

$$I\colon \tilde{K}_C(X) \xrightarrow{\cong} \tilde{K}_C(S^2X)\ ,$$
$$J\colon \tilde{K}_R(X) \xrightarrow{\cong} \tilde{K}_R(S^8X)\ .$$

Here S^rX denotes the rth suspension of X, and I, J are defined as follows. The groups $\tilde{K}_C(S^2X)$, $\tilde{K}_R(X^8X)$ are represented as direct summands in $\tilde{K}_C(S^2 \times X)$, $\tilde{K}_R(S^8 \times X)$. We now define

$$I(\kappa) = \pi^*\lambda \otimes \varpi^*\kappa\ ,$$
$$J(\kappa) = \pi^*\mu \otimes \varpi^*\kappa\ .$$

Here π, ϖ denote the projections of $S^2 \times X$, resp. $S^8 \times X$ on its factors, and the elements $\lambda \in \tilde{K}_C(S^2)$, $\mu \in \tilde{K}_R(S^8)$ are generators.

COROLLARY 5.3. *The following diagrams are* NOT *commutative.*

$$\begin{array}{ccc} \tilde{K}_C(X) & \xrightarrow{I} & \tilde{K}_C(S^2X) \\ \Psi^k_C \downarrow & & \downarrow \Psi^k_C \\ \tilde{K}_C(X) & \xrightarrow{I} & \tilde{K}_C(S^2X)\ , \end{array} \qquad \begin{array}{ccc} \tilde{K}_R(X) & \xrightarrow{J} & \tilde{K}_R(S^8X) \\ \Psi^k_R \downarrow & & \downarrow \Psi^k_R \\ \tilde{K}_R(X) & \xrightarrow{J} & \tilde{K}_R(S^8X)\ . \end{array}$$

In fact, we have

$$\Psi^k_C \cdot I = kI \cdot \Psi^k_C\ , \qquad \Psi^k_R \cdot J = k^4 J \cdot \Psi^k_R\ .$$

PROOF. Consider the case $\Lambda = C$. Using the previous corollary, we have

$$\begin{aligned} \Psi^k_C \cdot I(\kappa) &= \Psi^k_C(\pi^*\lambda \otimes \varpi^*\kappa) \\ &= (\pi^*\Psi^k_C\lambda) \otimes (\varpi^*\Psi^k_C\kappa) \\ &= k(\pi^*\lambda) \otimes (\varpi^*\Psi^k_C\kappa) \\ &= kI \cdot \Psi^k_C(\kappa)\ . \end{aligned}$$

Similarly for the case $\Lambda = R$.

6. A spectral sequence

In this section we recall certain extra material on the groups $K(X)$.

To begin with, recall from [3] that one can define the groups of a "cohomology theory" as follows. Let Y be a subcomplex of X, and let X/Y be the space obtained by identifying Y with a newly-introduced base-point. Define

$$K_\Lambda^{-n}(X, Y) = \tilde{K}_\Lambda(S^n(X/Y)) .$$

If Y is empty we have $K_\Lambda^0(X, \varphi) \cong K_\Lambda(X)$.

Using the Bott periodicity (as at the end of § 5) one shows that

$$K_G^{-n-2}(X, Y) \cong K_G^{-n}(X, Y) ,$$
$$K_R^{-n-8}(X, Y) \cong K_R^{-n}(X, Y) .$$

One may use these equations to define the abelian groups $K_\Lambda^n(X, Y)$ for positive values of n.

Nota bene. Owing to the state of affairs revealed in Corollary 5.3, we shall be most careful not to identify $K_G^{-n-2}(X, Y)$ with $K_G^{-n}(X, Y)$ or $K_R^{-n-8}(X, Y)$ with $K_R^{-n}(X, Y)$. We therefore regard $K_\Lambda^n(X, Y)$ as graded over Z, not over Z_2 or Z_8. Given this precaution one can define operations Ψ_Λ^k in $K_\Lambda^n(X, Y)$ for $n \leqq 0$; however, we shall not need such operations. We shall use operations Ψ_Λ^k only in $K_\Lambda(X)$ and $\tilde{K}_\Lambda(X)$ (that is, in dimension $n = 0$); the groups $K_\Lambda^n(X, Y)$ with $n \neq 0$ will be used only to help in calculating the additive structure of $K_\Lambda^0(X, Y)$. This will avoid any confusion.

We next recall from [3] that one can define induced maps and coboundary maps between the groups $K^n(X, Y)$, so that these groups verify all the Eilenberg-Steenrod axioms [9] except for the dimension axiom. (If one chose to introduce operations Ψ_Λ^k into the groups $K_\Lambda^n(X, Y)$ for $n \leqq 0$, then these operations would commute with induced maps and coboundary maps, because both are defined in terms of induced maps of $\tilde{K}_\Lambda$.)

We next recall from [3, § 2] the existence of a certain spectral sequence. Let X be a finite CW-complex, and let X^p denote its p-skeleton. Then each pair X^p, X^q yields an exact sequence of groups K_Λ^n. These exact sequences yield a spectral sequence. The E_∞ term of the spectral sequence is obtained by filtering $K_\Lambda^*(X) = \sum_{n=-\infty}^{+\infty} K_\Lambda^n(X)$. The E_1 and E_2 terms of the spectral sequence are given by

$$E_1^{p,q} \cong C^p(X; K_\Lambda^q(P)) , \qquad E_2^{p,q} \cong H^p(X; K_\Lambda^q(P)) ,$$

where P is a point. The values of $K_\Lambda^q(P)$ are given by the homotopy groups of BO or BU; they are as follows, by [6, p. 315].

$q \equiv 0,$	-1	-2	-3	-4	-5	-6	-7	-8	mod 8
$K_C^q(P) = Z$	0	Z	0	Z	0	Z	0	Z	
$K_R^q(P) = Z$	Z_2	Z_2	0	Z	0	0	0	Z.	

In what follows it will sometimes be useful to know that the spectral sequence is defined in this particular way. For example, $\tilde{K}_\Lambda(X)$ is filtered by the images of the groups $\tilde{K}_\Lambda(X/X^{p-1})$; and if we are given an explicit element κ in $\tilde{K}_\Lambda(X/X^{p-1})$, then we can take the image of κ in $\tilde{K}_\Lambda(X^{p-1+r}/X^{p-1})$, and so, by passing to quotients, obtain an explicit element κ_r in $E_r^{p,-p}$ (for $1 \leqq r \leqq \infty$), so that $d_r\kappa_r = 0$ and the homology class of κ_r is κ_{r+1}, while κ_∞ is the class containing κ. Again, an element κ in $\tilde{K}_C(X^p/X^{p-1})$ gives an element in $E_1^{p,-p}$; the space X^p/X^{p-1} is a wedge-sum of spheres S^p, and we can tell whether κ is a generator or not by examining ch κ (cf. the proof of Lemma 5.2).

On the other hand, the work that follows has been so arranged that we do not need any theorem concerning the identification of the differentials in the spectral sequence in terms of cohomology operations.

7. Computations for projective spaces

It is a pleasure to acknowledge at this point my indebtedness to J. Milnor, who read a draft of the following section and suggested several improvements.

In this section we shall calculate the various rings $K_\Lambda(X)$ which we require, together with their operations Ψ_Λ^k. Our plan is to obtain our results in the following order.

(i) Results on complex projective spaces for $\Lambda = C$.
(ii) Results on real projective spaces for $\Lambda = C$.
(iii) Results on real projective spaces for $\Lambda = R$.

These results are stated as Theorems 7.2, 7.3 and 7.4. For completeness, every stunted projective space is considered, whether or not it arises in the applications. The theorems are preceded by one lemma, which we need in order to specify generators in our rings.

LEMMA 7.1. *Let ξ be the canonical real line-bundle over RP^{2n-1}; let η be the canonical complex line-bundle over CP^{n-1}; let π: $RP^{2n-1} \to CP^{n-1}$ be the standard projection. Then we have*

$$c\xi = \pi^*\eta .$$

PROOF. Complex line bundles are classified by their first Chern class c_1. In our case this lies in $H^2(RP^{2n-1}; Z)$, which is Z_2, at least if $n > 1$ (the case $n = 1$ being trivial). We have $c_1\pi^*\eta = \pi^*c_1\eta \neq 0$. It is therefore sufficient to show that the bundle $c\xi$ is non-trivial. Let w denote the total Stiefel-Whitney class, let x be the generator of $H^1(RP^{2n-1}; Z_2)$, and let r be as in Lemma 3.9. Then we have $rc\xi = \xi \oplus \xi$ and $w(rc\xi) = 1 + x^2$. This shows that $c\xi$ is non-trivial, and completes the proof.

In terms of the canonical line-bundles we introduce the following elements λ, μ, ν.

$$\lambda = \xi - 1 \in \tilde{K}_R(RP^n) ,$$
$$\mu = \eta - 1 \in \tilde{K}_C(CP^n) ,$$
$$\nu = c\lambda = \pi^*\mu \in \tilde{K}_C(RP^n) .$$

In terms of these elements we may write polynomials $Q(\lambda)$, $Q(\mu)$, $Q(\nu)$.

THEOREM 7.2. *$K_C(CP^n)$ is a truncated polynomial ring (over the integers) with one generator μ and one relation $\mu^{n+1} = 0$. The operations are given by*

$$\Psi_C^k \cdot \mu^s = ((1 + \mu)^k - 1)^s .$$

The projection $CP^n \to CP^n/CP^m$ maps $\tilde{K}_C(CP^n/CP^m)$ isomorphically onto the subgroup of $K_C(CP^n)$ generated by $\mu^{m+1}, \mu^{m+2}, \cdots, \mu^n$.

Note. If k is negative, $(1 + \mu)^k$ may be interpreted by means of the binomial expansion

$$1 + \kappa\mu + \frac{k(k-1)}{2!}\mu^2 + \cdots .$$

This expansion terminates because $\mu^{n+1} = 0$.

PROOF. So far as the additive and multiplicative structures go, this result is due to Atiyah and Todd; see [4, Propositions 2.3, 3.1 and 3.3]. In any case, it is almost evident. The spectral sequence of § 6 shows that $K_C^s(CP^n, CP^m)$ is zero for s odd, and free abelian on $n - m$ generators for s even. Let us examine matters more closely, and suppose as an inductive hypothesis that $K_C(CP^{n-1})$ is as stated (this is trivial for $n = 1$). Then the elements $1, \mu, \mu^2, \cdots, \mu^{n-1}$ in $K_C(CP^n)$ project into a Z-base for $K_C(CP^{n-1})$. Moreover, the element μ^n in $K_C(CP^n)$ projects into zero in $K_C(CP^{n-1})$, so it must come from $\tilde{K}_C(CP^n/CP^{n-1})$. Let $y \in H^2(CP^n; Z)$ be the cohomology generator; then $\text{ch}\,\mu = y + y^2/2 + \cdots$ and $\text{ch}\,\mu^n = y^n$; hence μ^n comes from a generator of $\tilde{K}_C(CP^n/CP^{n-1}) = Z$. Using the exact sequence of the pair CP^n, CP^{n-1}, we see that $1, \mu, \mu^2, \cdots, \mu^n$ form a Z-base for $K_C(CP^n)$. It is now clear that

$$\text{ch}\colon K_C(CP^n) \longrightarrow H^*(CP^n; Q)$$

is monomorphic. (This also follows from a general theorem; see [3, § 2.5], [4, Prop. 2.3].) Since $\text{ch}(\mu^{n+1}) = 0$, we have $\mu^{n+1} = 0$. This completes the induction, and establishes the result about $K_C(CP^n)$. The result about

$\tilde{K}_G(CP^n/CP^m)$ follows immediately from the exact sequence of the pair CP^n, CP^m.

It remains to calculate the operations. According to Theorem 5.1 (iii), we have $\Psi^k_G(\eta) = \eta^k$; that is

$$\Psi^k_G(1+\mu) = (1+\mu)^k .$$

Hence

$$\Psi^k_G(\mu) = (1+\mu)^k - 1$$

and

$$\Psi^k_G(\mu^s) = \big((1+\mu)^k - 1\big)^s .$$

This completes the proof.

In order to state the next theorem, we define certain generators. We write $\mu^{(s+1)}$ for the element in $\tilde{K}_G(CP^N/CP^s)$ which maps into μ^{s+1} in $K_G(CP^N)$ (see Theorem 7.2). (It is clear that as we alter N the resulting elements $\mu^{(s+1)}$ map into one another; this justifies us in not displaying N in the notation.)

The standard projection π: $RP^{2N+1} \rightarrow CP^N$ factors to give

$$\bar{\varpi}: RP^{2N+1}/RP^{2s+1} \longrightarrow CP^N/CP^s ,$$
$$\varpi: RP^{2N+1}/RP^{2s} \longrightarrow CP^N/CP^s .$$

We write $\bar{\nu}^{(s+1)} = \bar{\varpi}^*\mu^{(s+1)}$, $\nu^{(s+1)} = \varpi^*\mu^{(s+1)}$. It is clear that $\bar{\nu}^{(s+1)}$ maps into $\nu^{(s+1)}$, and $\nu^{(s+1)}$ in turn maps into the element ν^{s+1} in $K_G(RP^{2N+1})$; this explains the notation. (As above, the dependence of these elements on N is negligible.)

THEOREM 7.3. *Assume $m = 2t$. Then we have $\tilde{K}_G(RP^n/RP^m) = Z_{2^f}$, where f is the integer part of $\frac{1}{2}(n-m)$. If $m = 0$ then $K_G(RP^n)$ may be described by the generator ν and the two relations*

$$\nu^2 = -2\nu , \qquad \nu^{f+1} = 0$$

(so that $2^f\nu = 0$). Otherwise $\tilde{K}_G(RP^n/RP^m)$ is generated by $\nu^{(t+1)}$ (where $t = \frac{1}{2}m$, as above); and the projection $RP^n \rightarrow RP^n/RP^m$ maps $\tilde{K}_G(RP^n/RP^m)$ isomorphically onto the subgroup of $K_G(RP^n)$ generated by ν^{t+1}.

In the case when m is odd, we have

$$\tilde{K}_G(RP^n/RP^{2t-1}) = Z + \tilde{K}_G(RP^n/RP^{2t}) ,$$

where the first summand is generated by $\bar{\nu}^{(t)}$, and the second is imbedded by the projection $RP^n/RP^{2t-1} \rightarrow RP^n/RP^{2t}$.

The operations are given by the following formulae.

$$\text{(i)}\quad \Psi_O^k \nu^{(t+1)} = \begin{cases} 0 & (k \text{ even}) \\ \nu^{(t+1)} & (k \text{ odd}) \end{cases},$$

$$\text{(ii)}\quad \Psi_O^k \bar{\nu}^{(t)} = k^t \bar{\nu}^{(t)} + \begin{cases} \frac{1}{2} k^t \nu^{(t+1)} & (k \text{ even}) \\ \frac{1}{2}(k^t - 1)\nu^{(t+1)} & (k \text{ odd}) \end{cases}.$$

Note 1. As usual, the symbol Z denotes a cyclic infinite group, and the symbol Z_{2^f} denotes a cyclic group of order 2^f.

Note 2. So far as the additive structure of $K_O(RP^n)$ goes, the result is due to J. Milnor (unpublished).

Note 3. The factor $\frac{1}{2}$ in the final formula will be vitally important in what follows; the reader is advised to satisfy himself as to its correctness.

PROOF. We begin by establishing the relation $\nu^2 = -2\nu$ in $K_O(RP^n)$; for this purpose we begin work in $K_R(RP^n)$. A real line-bundle is equivalent to one with structural group $O(1) = \{+1, -1\}$; it is therefore directly obvious that, for any real line-bundle ξ, we have $\xi \otimes \xi = 1$. (Alternatively, this may be deduced from the fact that real line-bundles are characterized by their first Stiefel-Whitney class w_1.) Taking ξ to be the canonical real line-bundle over RP^n, we have $\xi^2 = 1$, that is, $(1 + \lambda)^2 = 1$ or $\lambda^2 = -2\lambda$. Applying c, we find $(\pi^*\eta)^2 = (c\xi)^2 = 1$ and $\nu^2 = -2\nu$. (Alternatively, the former equation may be deduced from the fact that complex line-bundles are characterized by their first Chern class.)

The relation $\nu^{f+1} = 0$ follows from the fact that ν^{f+1} is the image of $\bar{\nu}^{f+1} \in \tilde{K}_O(RP^N/RP^{2f+1})$ and $2f + 1 \geqq m$.

We now apply the spectral sequence of § 6 to the space $X = RP^n/RP^m$. If n and m are even the group $H^p(X; Z)$ is Z_2 for even p such that $m < p \leqq n$; otherwise it is zero. If $m = 2t - 1$ we obtain an extra group $H^{2t}(X; Z) = Z$. If n is odd we obtain an extra group $H^n(X; Z) = Z$. Let f be the integral part of $\frac{1}{2}(n - 2t)$, where $m = 2t$ or $2t - 1$. Then the elements $\bar{\nu}^{(t+i)} \in \tilde{K}_O(RP^n/RP^{2t+2i-1})$ $(i = 1, 2, \cdots, f)$ yield generators for the f groups Z_2 in our E_2 term, and survive to E_∞ (as explained in § 6). Again, if $m = 2t - 1$, the element $\bar{\nu}^{(t)}$ yields a generator for the corresponding group Z in our E_2 term; this also survives to E_∞. If n is odd the group $H^n(X; Z) = Z$ has odd total degree, and all differentials vanish on it for dimensional reasons. Our spectral sequence is therefore trivial. This leads to the following conclusions.

(i) If $m = 2t$, $\tilde{K}_O(X)$ can be filtered so that the successive quotients are f copies of Z_2, whose generators are the images of $\bar{\nu}^{(t+1)}, \bar{\nu}^{(t+2)}, \cdots, \bar{\nu}^{(t+f)}$.

(ii) If $m = 2t - 1$, we have an exact sequence

$$0 \longleftarrow Z = \widetilde{K}_O(RP^{2t}/RP^{2t-1}) \longleftarrow \widetilde{K}_O(RP^n/RP^{2t-1}) \longleftarrow \widetilde{K}_O(RP^n/RP^{2t}) \longleftarrow 0 ,$$

in which $\bar{\nu}^{(t)}$ maps to a generator of Z.

It is now evident that $\widetilde{K}_O(RP^n/RP^{2t})$ is monomorphically imbedded in $\widetilde{K}_O(RP^n)$.

We have next to determine the group extensions involved in (i) above, in the case $t = 0$. If $t = 0$ then the generators of the successive quotients become $\nu, \nu^2, \cdots, \nu^f$, and the relation $\nu^2 = -2\nu$ resolves the problem; the extension is a cyclic group Z_{2^f} generated by ν.

We have now done all that is needed to determine the additive and multiplicative structures of our groups; it remains to calculate the operations Ψ^k_O.

According to Theorem 5.1 (iii) we have $\Psi^k_O\xi = \xi^k$ for a line-bundle ξ. As remarked above, the line-bundle $\pi^*\eta$ over RP^n satisfies $(\pi^*\eta)^2 = 1$. Therefore

$$\Psi^k_O(\pi^*\eta) = \begin{cases} 1 & (k \text{ even}) \\ \pi^*\eta & (k \text{ odd}) . \end{cases}$$

That is,

$$\Psi^k_O(1 + \nu) = \begin{cases} 1 & (k \text{ even}) \\ 1 + \nu & (k \text{ odd}) . \end{cases}$$

Therefore

$$\Psi^k_O(\nu) = \begin{cases} 0 & (k \text{ even}) \\ \nu & (k \text{ odd}) , \end{cases}$$

and

$$\Psi^k_O(\nu^s) = \begin{cases} 0 & (k \text{ even}) \\ \nu^s & (k \text{ odd}) . \end{cases}$$

Since $\widetilde{K}_O(RP^n/RP^{2t})$ is monomorphically imbedded in $K_O(RP^n)$, the result about $\Psi^k_O\nu^{(t+1)}$ follows.

We necessarily have

$$\Psi^k_O\bar{\nu}^{(t)} = a\bar{\nu}^{(t)} + b\nu^{(t+1)}$$

for some coefficients a, b; our problem is to determine them. By using the injection $RP^{2t}/RP^{2t-1} \rightarrow RP^n/RP^{2t-1}$ and Corollary 5.2 for $RP^{2t}/RP^{2t-1} = S^{2t}$, we see that $a = k^t$. Now project into RP^n/RP^{2t-2}; $\bar{\nu}^{(t)}$ maps into $\nu^{(t)}$ and $\nu^{(t+1)}$ into $-2\nu^{(t)}$, and we see that

$$\Psi^k_O\nu^{(t)} \equiv a\nu^{(t)} - 2b\nu^{(t)} \qquad \bmod 2^{f+1} .$$

Therefore

$$b \equiv \tfrac{1}{2}(k^t - \varepsilon) \qquad \bmod 2^f ,$$

where $\varepsilon = 0$ or 1 according as k is even or odd. This completes the proof.

REMARK. An alternative method for obtaining the last formula is as follows. According to Theorem 7.2, we have in $\tilde{K}_C(CP^N/CP^{t-1})$ the formula

$$\Psi^k_C \mu^{(t)} = k^t \mu^{(t)} + \Sigma ,$$

where Σ denotes a sum of higher terms. Applying the projection $\varpi: RP^{2N+1}/SP^{2t-1} \to CP^N/CP^{t-1}$, we find

$$\Psi^k_C \bar{\nu}^{(t)} = k^t \bar{\nu}^{(t)} + \varpi^* \Sigma .$$

It is therefore only necessary to evaluate $\varpi^*\Sigma$, which leads to the same result.

In order to state our next theorem, we define $\varphi(n, m)$ to be the number of integers s such that $m < s \leqq n$ and $s \equiv 0, 1, 2$ or $4 \bmod 8$.

THEOREM 7.4. *Assume $m \not\equiv -1 \bmod 4$. Then we have $\tilde{K}_R(RP^n/RP^m) = Z_{2^f}$, where $f = \varphi(n, m)$. If $m = 0$, then $K_R(RP^n)$ may be described by the generator λ and the two relations*

$$\lambda^2 = -2\lambda , \qquad \lambda^{f+1} = 0$$

(so that $2^f\lambda = 0$). Otherwise the projection $RP^n \to RP^n/RP^m$ maps $\tilde{K}_R(RP^n/RP^m)$ isomorphically onto the subgroup of $\tilde{K}_R(RP^n)$ generated by λ^{g+1}, where $g = \varphi(m, 0)$. We write $\lambda^{(g+1)}$ for the element in $\tilde{K}_R(RP^n/RP^m)$ which maps into λ^{g+1}.

In the case $m \equiv -1 \bmod 4$ we have

$$\tilde{K}_R(RP^n/RP^{4t-1}) = Z + \tilde{K}_R(RP^n/RP^{4t}) .$$

Here the second summand is imbedded by the projection $RP^n/RP^{4t-1} \to RP^n/RP^{4t}$, and the first is generated by an element $\bar{\lambda}^{(g)}$ which will be defined below. (We have written g for $\varphi(4t, 0)$.)

The operations are given by the following formulae.

$$\text{(i)} \quad \Psi^k_R \lambda^{(g+1)} = \begin{cases} 0 & (k \text{ even}) \\ \lambda^{(g+1)} & (k \text{ odd}) , \end{cases}$$

$$\text{(ii)} \quad \Psi^k_R \bar{\lambda}^{(g)} = k^{2t} \bar{\lambda}^{(g)} + \begin{cases} \tfrac{1}{2} k^{2t} \lambda^{(g+1)} & (k \text{ even}) \\ \tfrac{1}{2}(k^{2t} - 1)\lambda^{(g+1)} & (k \text{ odd}) . \end{cases}$$

REMARK. So far as the additive structure of $K_R(RP^n)$ goes, the result is due to R. Bott and A. Shapiro (unmimeographed notes).

PROOF. We begin by applying the spectral sequence of § 6 to the space

$X = RP^n/RP^m$. Let us recall from § 6 that

$$K^{-q}(P) = \begin{cases} Z_2 & \text{if } q \equiv 1, 2 \bmod 8 \\ Z & \text{if } q \equiv 0, 4 \bmod 8 \\ 0 & \text{otherwise .} \end{cases}$$

The group $H^p(X, Z_2)$ is Z_2 for $m < p \leqq n$, otherwise zero. If n and m are even the group $H^p(X; Z)$ is Z_2 for even p such that $m < p \leqq n$, otherwise zero. However, if n is odd we obtain an extra group $H^n(X; Z) = Z$, and if m is odd we obtain $H^{m+1}(X; Z) = Z$ instead of Z_2. We can now enumerate the terms $E_2^{p,q}$ in our spectral sequence which have total degree zero. If $m + 1 \not\equiv 0 \bmod 4$ we find (apart from zero groups) just $\varphi(n, m)$ copies of Z_2. If $m + 1 \equiv 0 \bmod 4$ we find $\varphi(n, m)$ groups, of which one is Z and the remainder Z_2.

LEMMA 7.5. *If* $n \equiv 6, 7$ *or* $8 \bmod 8$ *then*

$$c: \widetilde{K}_R(RP^n) \longrightarrow \widetilde{K}_C(RP^n)$$

is an isomorphism.

PROOF. The homomorphism c is always an epimorphism, because $\widetilde{K}_C(RP^n)$ is generated by ν (Theorem 7.3) and $\nu = c\lambda$ (Lemma 7.1). If $n - 8t = 6$ or 7, then $\varphi(n, 0) = 4t + 3$, so that $\widetilde{K}_R(RP^n)$ contains at most 2^{4t+3} elements. On the other hand, $\widetilde{K}_C(RP^n)$ contains exactly 2^{4t+3} elements, by Theorem 7.3. Therefore c is an isomorphism in this case. Similarly if $n = 8t + 8$ (with $4t + 3$ replaced $4t + 4$). This completes the proof.

It follows that $\widetilde{K}_R(RP^n)$ is generated by λ when $n \equiv 6, 7$ or $8 \bmod 8$.

Let us reconsider the spectral sequence for the space $X = RP^n$. We have found $\varphi(n, 0)$ copies of Z_2 with total degree zero in our E_2 term; we have shown that if $n \equiv 6, 7$ or $8 \bmod 8$ they all survive unchanged to E_∞. It follows that the same thing holds for smaller values of n. We conclude that for any n we have $\widetilde{K}_R(RP^n) = Z_{2^f}$, where $f = \varphi(n, 0)$; this group is generated by λ. We have already shown that $\lambda^2 = -2\lambda$ (see the proof of Theorem 7.3). The formula $\lambda^{f+1} = 0$ therefore follows from the fact that $2^f\lambda = 0$.

Let us now consider the exact sequence

$$\widetilde{K}_R(RP^m) \xleftarrow{i^*} \widetilde{K}_R(RP^n) \longleftarrow \widetilde{K}_R(RP^n/RP^m) .$$

The kernel of i^* has 2^f elements, where $f = \varphi(n, 0) - \varphi(m, 0) = \varphi(n, m)$. If $m \not\equiv -1 \bmod 4$ then $\widetilde{K}_R(RP^n/RP^m)$ has at most 2^f elements. It is now clear that $\widetilde{K}_R(RP^n/RP^m)$ maps isomorphically onto the subgroup of $\widetilde{K}_R(RP^n)$ generated by $\pm 2^g\lambda = \pm\lambda^{g+1}$, where $g = \varphi(m, 0)$. We write $\lambda^{(g+1)}$ for the element in $\widetilde{K}_R(RP^n/RP^m)$ which maps into λ^{g+1}. This completes our consideration of the case $m \not\equiv -1 \bmod 4$.

In the case $m \equiv -1 \bmod 4$, our first concern is to show that the following exact sequence splits.

$$Z = \tilde{K}_R(RP^{4t}/RP^{4t-1}) \xleftarrow{i} \tilde{K}_R(RP^n/RP^{4t-1}) \xleftarrow{j} \tilde{K}_R(RP^n/RP^{4t}) .$$

It is clear that j is a monomorphism, since we have just shown that the composite

$$\tilde{K}_R(RP^n/RP^{4t}) \xrightarrow{j} \tilde{K}_R(RP^n/RP^{4t-1}) \longrightarrow \tilde{K}_R(RP^n)$$

is monomorphic.

LEMMA 7.6. *The map i is an epimorphism.*

PROOF. Inspect the following commutative diagram, in which the row and columns are exact.

$$\begin{array}{ccccc}
 & & \tilde{K}_R(RP^{4t-1}/RP^{4t-2}) & = & \tilde{K}_R(RP^{4t-1}/RP^{4t-2}) \\
 & & \uparrow & & \uparrow \\
 & & \tilde{K}_R(RP^{4t}/RP^{4t-2}) & \xleftarrow{i_1} & \tilde{K}_R(RP^n/RP^{4t-2}) \\
 & & j_1\uparrow & & \uparrow j_2 \\
K_R^1(RP^n/RP^{4t}) & \xleftarrow{\delta} & \tilde{K}_R(RP^{4t}/RP^{4t-1}) & \xleftarrow{i} & \tilde{K}_R(RP^n/RP^{4t-1}) .
\end{array}$$

We have $RP^{4t-1}/RP^{4t-2} = S^{4t-1}$ and $RP^{4t}/RP^{4t-1} = S^{4t}$; thus $\tilde{K}_R(RP^{4t-1}/RP^{4t-2}) = 0$, and the maps j_1, j_2 are epimorphic. We have also calculated that $\tilde{K}_R(RP^{4t}/RP^{4t-2}) = Z_2$ and i_1 is epimorphic. Hence $j_1 i$ is epimorphic. But j_1 is an epimorphism from Z to Z_2; hence $\operatorname{Im} i$ consists of the multiples of some odd number ω, and $\operatorname{Im}\delta = Z_\omega$. But it is clear from the spectral sequence that $K_R^1(RP^n/RP^{4t})$ contains no elements of odd order (except zero). Hence $\omega = 1$ and i is epimorphic. This completes the proof.

We wish next to specify a generator $\bar{\lambda}^{(g)}$. For this purpose we consider the map

$$c: \tilde{K}_R(RP^n/RP^{4t-1}) \longrightarrow \tilde{K}_C(RP^n/RP^{4t-1}) .$$

LEMMA 7.7. *If $n \equiv 6, 7$ or $8 \bmod 8$ then c is an isomorphism for t even, a monomorphism for t odd.*

PROOF. Inspect the following commutative diagram, in which each row is a split exact sequence.

$$\begin{array}{ccccc}
Z = \tilde{K}_R(RP^{4t}/RP^{4t-1}) & \xleftarrow{i} & \tilde{K}_R(RP^n/RP^{4t-1}) & \xleftarrow{j} & \tilde{K}_R(RP^n/RP^{4t}) \\
\downarrow c_1 & & \downarrow c & & \downarrow c_3 \\
Z = \tilde{K}_C(RP^{4t}/RP^{4s-1}) & \longleftarrow & \tilde{K}_C(RP^n/RP^{4t-1}) & \longleftarrow & \tilde{K}_C(RP^n/RP^{4t}) .
\end{array}$$

We will establish the nature of c_3. Suppose that $t = 2u$ and $n - 8v = 6, 7$ or 8. Then $\widetilde{K}_R(RP^n/RP^{4t}) = Z_{2^f}$, where

$$f - 4(v - u) = \begin{cases} 3 & (n - 8v = 6 \text{ or } 7) \\ 4 & (n - 8v = \quad 8 \quad) . \end{cases}$$

This group is generated by $\lambda^{(4u+1)}$. We also have $\widetilde{K}_O(RP^n/RP^{4t}) = Z_{2^f}$, generated by $\nu^{(4u+1)}$. We have $c_3\lambda^{(4u+1)} = \nu^{(4u+1)}$, so c_3 is an isomorphism if t is even. Next suppose that $t = 2u + 1$ and $n - 8v = 6, 7$ or 8. Then $\widetilde{K}_R(RP^n/RP^{4t}) = Z_{2^f}$, where

$$f - 4(v - u) = \begin{cases} 0 & (n - 8v = 6 \text{ or } 7) \\ 1 & (n - 8v = \quad 8 \quad) . \end{cases}$$

This group is generated by $\lambda^{(4u+4)}$. We also have $\widetilde{K}_O(RP^n/RP^{4t}) = Z_{2^{f+1}}$, generated by $\nu^{(4u+3)}$. We have $c_3\lambda^{(4u+4)} = -2\nu^{(4u+3)}$, so c_3 is a monomorphism if t is odd.

According to the results of Bott (as explained during the proof of Corollary 5.2), the map c_1 is an isomorphism for t even and a monomorphism for t odd. The result now follows by the Five Lemma.

We next explain how to choose the generator $\bar{\lambda}^{(g)}$, assuming that $n \equiv 6, 7$ or 8. If $t = 2u$ we take $\bar{\lambda}^{(4u)}$ to be the unique element in $\widetilde{K}_R(RP^n/RP^{4t-1})$ such that $c\bar{\lambda}^{(4u)} = \bar{\nu}^{(4u)}$. If $t = 2u + 1$ we define $\bar{\lambda}^{(4u+3)} = -r\bar{\nu}^{(4u+2)}$; then

$$\begin{aligned} c\bar{\lambda}^{(4u+3)} &= -cr\bar{\nu}^{(4u+2)} \\ &= -(1 + \Psi_O^{-1})\bar{\nu}^{(4u+2)} && \text{(Lemma 3.9)} \\ &= -2\bar{\nu}^{(4u+2)} && \text{(Theorem 7.3)} . \end{aligned}$$

Since $\operatorname{Im} c_1 = 2Z$ in this case, $i\bar{\lambda}^{(4u+3)}$ is a generator, and we may take $\bar{\lambda}^{(4u+3)}$ as our generator for the summand Z in $\widetilde{K}_R(RP^n/RP^{4t-1})$.

So far we have only defined $\lambda^{(g)}$ for $n \equiv 6, 7$ or $8 \bmod 8$. However, by naturality we obtain for smaller values of n an image element, also written $\lambda^{(g)}$, with the same properties. This procedure is clearly self-consistent if we reduce n from n_1 to n_2, where both n_1 and n_2 are congruent to 6, 7, or 8 mod 8.

Whether t is odd or even, one verifies that the image of $\lambda^{(g)}$ in $\widetilde{K}_O(RP^n)$ is ν^g. Therefore the image of $\lambda^{(g)}$ in $\widetilde{K}_R(RP^n)$ is λ^g. This explains the notation.

We now turn to the operations Ψ_R^k. Their values may be obtained by either of the following methods.

(i) The argument given in proving Theorem 7.3 goes over immediately to the case $\Lambda = R$, using the fact that $\lambda = \xi - 1$ and ξ is a line-bundle.

(ii) The operations in $\tilde{K}_C(RP^n/RP^m)$ are known by Theorem 7.3. The map

$$c\colon \tilde{K}_R(RP^n/RP^m) \longrightarrow \tilde{K}_C(RP^n/RP^m)$$

is known, and commutes with the operations. We can deduce the values of the operations Ψ_R^k in $\tilde{K}_R(RP^n/RP^m)$ if $n \equiv 6, 7$ or $8 \bmod 8$, because c is then a monomorphism (this follows from Lemmas 7.5, 7.7). The results follow for smaller values of n by naturality.

REMARK. It is also possible to compute the groups $\tilde{K}_R(RP^n/RP^m)$ directly from the spectral sequence of § 6; and this was, of course, the author's original approach. The group extensions are determined by computing

$$K_R^*(RP^n/RP^m; Z_2) = K_R^*((RP^n/RP^m) \# RP^2) \;,$$

and examining the universal coefficient sequence. It is necessary to know the expression of certain differentials in the spectral sequence in terms of Steenrod squares; it is easy to compute these squares in $(RP^n/RP^m) \# RP^2$, using the Cartan formula. No details will be given, but the earnest student may reconstruct them.

8. Proof of Theorem 1.2

In this section we complete the proof of Theorem 1.2. Let us suppose given, then, a map

$$f\colon RP^{m+\rho(m)}/RP^{m-1} \longrightarrow S^m$$

such that the composite

$$S^m = RP^m/RP^{m-1} \xrightarrow{\ i\ } RP^{m+\rho(m)}/RP^{m-1} \xrightarrow{\ f\ } S^m$$

has degree 1. (Here, as in § 1, we have $m = (2a+1)2^b$, $b = c + 4d$ and $\rho(m) = 2^c + 8d$.)

We first remark that if $d = 0$, then the Steenrod squares suffice to contradict the existence of f (cf. [15]). In what follows, then, we may certainly assume that $m \equiv 0 \bmod 8$. This ensures that

$$\varphi(m + \rho(m), m) = b + 1 \;,$$

where $\varphi(m, n)$ is the function introduced in § 7. According to Theorem 7.4, then, we have

$$\tilde{K}_R(RP^{m+\rho(m)}/RP^{m-1}) = Z + Z_{2^{b+1}} \;;$$

here the summands are generated by $\bar{\lambda}^{(g)}$ and $\lambda^{(g+1)}$, where $g = \frac{1}{2}m$. We know that $i^*\lambda^{(g+1)} = 0$ and $i^*\bar{\lambda}^{(g)}$ is a generator γ of $\tilde{K}_R(S^m) = Z$. If we had a map f, we would have

$$f^*\gamma = \bar{\lambda}^{(g)} + N\lambda^{(g+1)}$$

for some integer N. From the equation

$$f^*\Psi_R^k\gamma = \Psi_R^k f^*\gamma$$

we obtain (using Corollary 5.2)

$$f^*(k^{m/2}\gamma) = \Psi_R^k(\bar{\lambda}^{(g)} + N\lambda^{(g+1)}) \ ;$$

that is (using Theorem 7.4),

$$k^{m/2}\bar{\lambda}^{(g)} + k^{m/2}N\lambda^{(g+1)} = k^{m/2}\bar{\lambda}^{(g)} + \tfrac{1}{2}(k^{m/2} - \varepsilon)\lambda^{(g+1)} + \varepsilon N\lambda^{(g+1)} \ ,$$

where $\varepsilon = 0$ or 1 according as k is even or odd. That is,

$$(N - \tfrac{1}{2})(k^{m/2} - \varepsilon)\lambda^{(g+1)} = 0 \ ,$$

or equivalently,

$$(N - \tfrac{1}{2})(k^{m/2} - \varepsilon) \equiv 0 \qquad \bmod 2^{b+1} \ .$$

It remains only to prove that for a suitable choice of k, we have

$$k^{m/2} - \varepsilon \equiv 2^{b+1} \qquad \bmod 2^{b+2} \ ;$$

this will establish the required contradiction. We take $k = 3$.

LEMMA 8.1. *If* $n = (2a + 1)2^f$, *then* $3^n - 1 \equiv 2^{f+2} \bmod 2^{f+3}$.

(Note that since $n = \tfrac{1}{2}m$ or $m = 2n$, we have $b = f + 1$.)

PROOF. We first note that since $3^2 \equiv 1 \bmod 8$, we have $3^{2n} \equiv 1 \bmod 8$ and $3^{2n} + 1 \equiv 2 \bmod 8$. We now prove by induction over f that

$$3^{(2^f)} - 1 \equiv 2^{f+2} \qquad \bmod 2^{f+4} \qquad (\text{for} \quad f \geqq 1) \ .$$

For $f = 1$ the result is true, since $3^2 - 1 = 8$. Suppose the result true for some value of f. Then we have

$$\begin{aligned} 3^{(2^{f+1})} - 1 &= (3^{2^f} - 1)(3^{2^f} + 1) \\ &= (2^{f+2} + x2^{f+4})(2 + y2^3) \\ &\equiv 2^{f+3} \qquad \bmod 2^{f+5} \ . \end{aligned}$$

This completes the induction.

We now note that, since

$$3^{2^{f+1}} \equiv 1 \qquad \bmod 2^{f+3} \ ,$$

we have

$$3^{(2a)2^f} \equiv 1 \qquad \bmod 2^{f+3} \ ,$$

and

$$3^{(2a+1)2^f} - 1 \equiv 3^{2^f} - 1 \quad \bmod 2^{f+3}$$
$$\equiv 2^{f+2} \quad \bmod 2^{f+3} .$$

This establishes Lemma 8.1; the proof of Theorem 1.2 is thus completed.

9. Proofs of Theorem 1.1 and Corollary 1.3

We begin with Theorem 1.1.

Suppose, for a contradiction, that there were some n for which S^{n-1} admits $\rho(n)$ linearly independent vector fields. Then it is not hard to see that for each integer p, the sphere S^{pn-1} admits at least $\rho(n)$ linearly independent vector fields; this and more is proved by James [13, Corollary 1.4]. If p is sufficiently large then the appropriate Stiefel manifold $V_{pn,\rho(n)+1}$ may be approximated by a truncated projective space, which in James's notation is called $Q_{pn,\rho(n)+1}$ [12]. From the cross-section in the Stiefel manifold, we deduce that the complex $Q_{pn,\rho(n)+1}$ is reducible, at least for $pn \geqq 2(\rho(n) + 1)$ [12, Theorem 8.2, p. 131]. According to Atiyah, $Q_{pn,\rho(n)+1}$ is S-dual to $P_{\rho(n)+1-pn,\rho(n)+1}$, and therefore the latter object is S-co-reducible (see [1, p. 299 and Theorem 6.1, p. 307]). The latter object, however, is somewhat fictitious if $\rho(n) + 1 - pn$ is negative (which is generally so); one has to interpret it as $P_{\rho(n)+1-pn+qr,\rho(n)+1}$, where r is an integer arising from Atiyah's work but not explicitly determined by him, and q is an integer sufficient to make $qr - pn$ positive (see [1, p. 307, second footnote]). In our notation $P_{\rho(n)+1-pn+qr,\rho(n)+1}$ becomes

$$RP^{qr-pn+\rho(n)}/RP^{qr-pn-1} = X , \quad \text{say} .$$

If q is chosen large enough (the precise condition being $qr \geqq pn + \rho(n) + 3$) we enter the domain of stable homotopy theory, and the complex X is S-co-reducible if and only if it is co-reducible. It remains to show how this contradicts Theorem 1.2. We may suppose that p is odd, and (by choice of q if necessary) that qr is divisible by $2n$. If we set $m = qr - pn$, we see that m is an odd multiple of n, so that $\rho(m) = \rho(n)$. We now have

$$X = RP^{m+\rho(m)}/RP^{m-1}$$

with X co-reducible. This contradiction establishes Theorem 1.1.

We turn now to the proof of Corollary 1.3. The affirmative part of the result is due to James [13, Theorem 3.1, p. 819]. Given Theorem 1.1, the negative result follows from the same theorem of James, provided we have $n - 1 > 2\rho(n)$. The only possible exceptions to this are $n = 1, 2, 3, 4, 8$ and 16. In the first five cases the result is trivially true; and in the case $n = 16$, it follows from the work of Toda, as has been remarked by

James [13, p. 819]. This completes the proof.

THE INSTITUTE FOR ADVANCED STUDY

REFERENCES

1. M. F. ATIYAH, *Thom complexes*, Proc. London Math. Soc., 11 (1961), 291-310.
2. ——— and F. HIRZEBRUCH, *Riemann-Roch theorems for differentiable manifolds*, Bull. Amer. Math. Soc., 65 (1959), 276-281.
3. ———, *Vector bundles and homogeneous spaces*, Proc. of Symposia in Pure Mathematics 3, Differential Geometry, Amer. Math. Soc., 1961, pp. 7-38.
4. ——— and J. A. TODD, *On complex Stiefel manifolds*, Proc. Cambridge Philos. Soc., 56 (1960), 342-353.
5. A. BOREL, *Sur la cohomologie des espaces fibrés principaux et des espaces homogènes de groupes de Lie compacts*, Ann. of Math., 57 (1953), 115-207.
6. R. BOTT, *The stable homotopy of the classical groups*, Ann. of Math., 70 (1959), 313-337.
7. ———, *Quelques remarques sur les théorèmes de périodicité*, Bull. Soc. Math. de France, 87 (1959), 293-310.
8. B. ECKMANN, *Gruppentheoretischer Beweis des Satzes von Hurwitz-Radon über die Komposition quadratischer Formen*, Comment. Math. Helv., 15 (1942) 358-366.
9. S. EILENBERG and N. E. STEENROD, Foundations of Algebraic Topology, Princeton University Press, 1952.
10. I. M. JAMES, *The intrinsic join: a study of the homotopy groups of Stiefel manifolds*, Proc. London Math. Soc., (3) 8 (1958), 507-535.
11. ———, *Cross-sections of Stiefel manifolds*, Proc. London Math. Soc., (3) 8 (1958), 536-547.
12. ———, *Spaces associated with Stiefel manifolds*, Proc. London Math. Soc., (3) 9 (1959), 115-140.
13. ———, *Whitehead products and vector-fields on spheres*, Proc. Cambridge Philos. Soc., 53 (1957), 817-820.
14. N. E. STEENROD, The Topology of Fibre Bundles, Princeton University Press, 1951.
15. ——— and J. H. C. WHITEHEAD, *Vector fields on the n-sphere*, Proc. Nat. Acad. Sci. USA, 37 (1951), 58-63.
16. H. TODA, *Vector fields on spheres*, Bull. Amer. Math. Soc., 67 (1961), 408-412.

Proc. Camb. Phil. Soc. (1965), **61**, 81

On complex Stiefel manifolds

BY J. F. ADAMS AND G. WALKER
University of Manchester

Communicated by C. T. C. WALL

(*Received* 1 *June* 1964)

1. *Introduction and results.* We shall study the following complex Stiefel fibring:

$$U(n)/U(n-k) \to U(n)/U(n-1) = S^{2n-1}. \qquad (1\cdot 1)$$

In particular we shall study the problem: *for what values of n and k does the fibring* (1·1) *admit a cross-section*? A necessary condition for the existence of a cross-section has been found by Atiyah and Todd (8). We shall show (Theorem 1·1 below) that the condition of Atiyah and Todd is sufficient (as well as necessary) for the existence of a cross-section. The problem stated above is therefore completely solved.

We shall begin by discussing the condition found by Atiyah and Todd, for they state it in various equivalent forms. First we have three forms which are purely algebraic, but are still more or less close transcriptions of topological conditions; we refer to conditions (A_k), (B_k) and (C_k) on page 343 of (8). Secondly, Atiyah and Todd reduce their condition to an explicit number-theoretical condition on n and k.

We quote condition (C_k) from (8), page 343, with minor changes of notation. By definition, the integer n satisfies condition (C_k) if and only if the coefficient of z^q in

$$\left(\frac{\operatorname{Log}(1+z)}{z}\right)^n$$

is an integer for $0 \leqslant q \leqslant k-1$.

It is clear that the integers n which satisfy condition (C_k) are precisely the multiples of a certain integer M_k. The number M_k is explicitly determined by Atiyah and Todd ((8), page 343), but we shall not need its explicit determination until section 6.

We can now state our first result.

THEOREM 1·1. *The complex Stiefel fibring* (1·1) *has a cross-section if and only if n is divisible by* M_k.

This theorem has already been announced in (4) and (5).

James (15) has proved that for any given k there exist integers n such that (1·1) has a cross-section, and that the set of all such n is exactly the set of multiples of the least such n, which he has called b_k. We can therefore reformulate Theorem 1·1 as follows.

THEOREM 1·2. *The James number* b_k *is equal to the Atiyah–Todd number* M_k.

This theorem establishes the main conjecture of Atiyah and Todd ((8), page 344). The theorem of James has been reproved by Atiyah (6). At this point we require-some notation. We write $K_\Lambda(X)$ for the Grothendieck–Atiyah–Hirzebruch group: see (7) and (12). We write $J(X)$ for the group introduced by Atiyah in (6), and

$$J: K_R(X) \to J(X)$$

for the canonical epimorphism, as in (6). We remark that our group $J(X)$ is written $\tilde{J}(X)$ in the more systematic notation of (1) and (2); however, we shall avoid encumbering the present paper with these tildes. We write CP^l for complex projective space of l complex dimensions, and η for the canonical complex line bundle over CP^l. Then Atiyah (6) has reduced the problem of the cross-sections of (1·1) to the problem of computing $J(CP^{k-1})$. Using James's language, we may paraphrase Theorem 6·5 on page 308 of (6) as follows.

LEMMA 1·3. *The James number b_k is the order of the element $J(\eta)$ in the group $J(CP^{k-1})$.*

In section 6 we shall prove

LEMMA 1·4. *The order of $J(\eta)$ in $J(CP^{k-1})$ is M_k.*

It is clear that given Lemma 1·3, Theorem 1·2 is equivalent to Lemma 1·4.

The cases $k = 1, 2, 3, 4$ of Theorem 1·2 are due to Atiyah and Todd ((8), page 344, line 2). In what follows we can therefore restrict our attention to CP^l for $l > 1$ and ignore CP^1, which would behave as an exception to some of our statements.

Our method may be outlined as follows. We shall compute the group $J(CP^{k-1})$ by sandwiching it between a 'lower bound' $J'(CP^{k-1})$ (defined in section 4) and an 'upper bound' $J''(CP^{k-1})$ (defined in section 5). We emphasize that while the group $J(CP^{k-1})$ is 'unknown', the groups $J'(CP^{k-1})$ and $J''(CP^{k-1})$ are 'known' (at least in principle) and serve as computable bounds for $J(CP^{k-1})$. For further details of this method, we refer the reader to (1). However, we remark that the group $J''(X)$ of this paper is written $\tilde{J}''_R(X)$ in (1) and (2). Also the lower bound J' which we shall use in this paper does not coincide with the J' (or the $\tilde{J}'$) of (1), (2), the latter being historically later, and in general more powerful. The present J' arises from a re-interpretation of the method of Atiyah and Todd, due primarily to Atiyah. As such, it is closely related to the condition (C_k) of Atiyah and Todd, and this of course is essential for our purposes. Extensive explicit calculations of the groups $J'(CP^l)$ (in the sense of the present paper) have been made by Atiyah and Todd (unpublished).

When we speak of a 'lower bound' and an 'upper bound', we mean that we shall construct a commutative diagram of the following form, in which all the maps will be epimorphisms.

$$\begin{array}{ccc} & \overset{J''}{\nearrow} & J''(CP^l) \\ & & \downarrow{\scriptstyle \theta''} \\ K_R(CP^l) & \xrightarrow{J} & J(CP^l) \\ & \underset{J'}{\searrow} & \downarrow{\scriptstyle \theta'} \\ & & J'(CP^l) \end{array}$$

Actually, we shall define two functors $J'_C(X)$, $J'_R(X)$; but it will follow from Lemma 4·3 below that we can identify $J'_C(CP^l)$ with $J'_R(CP^l)$ for $l > 1$; therefore we shall often write $J'(CP^l)$ for either.

This method results in the following theorem.

THEOREM 1·5. *The map*

$$\theta'': J''(CP^l) \to J(CP^l)$$

is an isomorphism for each l. The map

$$\theta': J(CP^l) \to J'(CP^l)$$

is an isomorphism if $l \not\equiv 1 \bmod 4$. *If* $l \equiv 1 \bmod 4, l > 1$, *then the kernel of* θ' *is* Z_2, *generated by* $Jr\mu^l$.

Here the homomorphism

$$r: K_C(X) \to K_R(X)$$

is as in (3) and (10), and $\mu = \eta - 1$.

The proof will be given in section 6.

The remainder of this paper is organized as follows. The proofs of Theorem 1·5 and Lemma 1·4 are in section 6. The essential point is to prove that

$$\theta'\theta'': J''(CP^{2v}) \to J'(CP^{2v})$$

is an isomorphism (Lemma 6·1). We do this by induction over v, using the cofibring

$$CP^{2v-2} \xrightarrow{i} CP^{2v} \xrightarrow{j} CP^{2v}/CP^{2v-2}.$$

For this purpose we require general exactness properties of J' and J'', and computations of $J'(CP^{2v}/CP^{2v-2})$ and $J''(CP^{2v}/CP^{2v-2})$. These are supplied by earlier sections (sections 4, 5). The results for CP^{2v+1} are then deduced from the results for CP^{2v} and CP^{2v+2} (Lemmas 6·2, 6·4).

The earlier sections are as follows. In section 2 we compute the groups $K_R(CP^l/CP^m)$. In section 3 we prove a lemma on $J(CP^l)$ which needs no machinery of ours. The functor J' will be set up in section 4, and the functor J'' in section 5.

2. *The groups* $K_R(CP^l/CP^m)$. In Lemma 2·1 we shall recall the structure of the groups $K_C(CP^l/CP^m)$ from (3) and (8) and in the rest of this section we shall establish all the facts we need about the groups $K_R(CP^l/CP^m)$.

We first fix certain notation. Throughout this paper, the letters X and Y will stand for finite connected CW-complexes. We shall write

$$c: K_R(X) \to K_C(X),$$
$$r: K_C(X) \to K_R(X),$$
$$t: K_C(X) \to K_C(X)$$

for the homomorphisms induced by complexification, 'realification', and complex conjugation. We refer the reader to (3) and (10) for properties of these homomorphisms. CP^l will mean complex projective space of l complex dimensions, and η will stand for the canonical complex line bundle over CP^l. We shall set

$$\mu = \eta - 1 \in \tilde{K}_C(CP^l),$$
$$\omega = r\mu \in \tilde{K}_R(CP^l).$$

All these symbols will be kept for these uses alone.

As we have remarked above, the following is a known result: see (8).

LEMMA 2·1. *The ring $K_C(\mathrm{CP}^l)$ may be presented by the generator $\mu = \eta - 1$ and the relation $\mu^{l+1} = 0$. The operations Ψ_C^k of* (3) *are given by*

$$\Psi_C^k \mu = (1+\mu)^k - 1 \quad ((3), \textit{Theorem } 7\cdot 2).$$

The cofibring $\mathrm{CP}^v/\mathrm{CP}^w \xrightarrow{i} \mathrm{CP}^u/\mathrm{CP}^w \xrightarrow{j} \mathrm{CP}^u/\mathrm{CP}^v$

leads to an exact sequence

$$0 \to \tilde{K}_C(\mathrm{CP}^u/\mathrm{CP}^v) \xrightarrow{j^*} \tilde{K}_C(\mathrm{CP}^u/\mathrm{CP}^w) \xrightarrow{i^*} \tilde{K}_C(\mathrm{CP}^v/\mathrm{CP}^w) \to 0.$$

The Chern character ((7), (8))

$$\mathrm{ch}: K_C(\mathrm{CP}^l/\mathrm{CP}^m) \to H^*(\mathrm{CP}^l/\mathrm{CP}^m; Q)$$

is a monomorphism.

Consider the cofibring

$$\mathrm{CP}^m \xrightarrow{i} \mathrm{CP}^l \xrightarrow{j} \mathrm{CP}^l/\mathrm{CP}^m.$$

By abuse of language, we shall use the notation μ^s for the (unique) element $(j^*)^{-1}\mu^s$ in $\tilde{K}_C(\mathrm{CP}^l/\mathrm{CP}^m)$, provided of course that $i^*\mu^s = 0$. We may therefore say, for example, that the group $\tilde{K}_C(\mathrm{CP}^l/\mathrm{CP}^m)$ is free Abelian, with a base consisting of the elements μ^s for $m < s \leqslant l$.

The next result contains the facts we need about the groups $K_R(\mathrm{CP}^l/\mathrm{CP}^m)$. In order to state it, we define $T_k(z)$ to be the polynomial such that

$$T_k(z - 2 + z^{-1}) = z^k - 2 + z^{-k}.$$

There is such a polynomial for each integer k; it is unique, and it has integral coefficients.

THEOREM 2·2 (i). *The ring $K_R(\mathrm{CP}^l)$ may be presented by the generator $\omega = r\mu$ and the following relations*:

$$\begin{aligned} \omega^{2w+1} &= 0 \quad \textit{if} \quad l = 4w; \\ 2\omega^{2w+1} = 0, \quad \omega^{2w+2} &= 0 \quad \textit{if} \quad l = 4w+1; \\ \omega^{2w+2} &= 0 \quad \textit{if} \quad l = 4w+2 \quad \textit{or} \quad 4w+3. \end{aligned}$$

(ii) *The operations Ψ_R^k of* (3) *are given by*

$$\begin{aligned} \Psi_R^k(\omega) &= T_k(\omega) \\ &\equiv k^2\omega \quad (\textit{modulo higher powers of } \omega). \end{aligned}$$

(iii) *The cofibring* $\mathrm{CP}^v/\mathrm{CP}^w \xrightarrow{i} \mathrm{CP}^u/\mathrm{CP}^w \xrightarrow{j} \mathrm{CP}^u/\mathrm{CP}^v$

leads to an exact sequence

$$0 \to \tilde{K}_R(\mathrm{CP}^u/\mathrm{CP}^v) \xrightarrow{j^*} \tilde{K}_R(\mathrm{CP}^u/\mathrm{CP}^w) \xrightarrow{i^*} \tilde{K}_R(\mathrm{CP}^v/\mathrm{CP}^w) \to 0.$$

(iv) *The sequence* $0 \to \tilde{K}_R(\mathrm{CP}^l/\mathrm{CP}^m) \xrightarrow{c} \tilde{K}_C(\mathrm{CP}^l/\mathrm{CP}^m)$

is exact provided $l \not\equiv 1 \bmod 4$.

(v) *The sequence*

$$\tilde{K}_R(\mathrm{CP}^l/\mathrm{CP}^m) \xrightarrow{c} \tilde{K}_C(\mathrm{CP}^l/\mathrm{CP}^m) \xrightarrow{1-t} \tilde{K}_C(\mathrm{CP}^l/\mathrm{CP}^m)$$

is exact provided $m \not\equiv 1$ mod 4.

(vi) *The sequence*

$$\tilde{K}_C(\mathrm{CP}^l/\mathrm{CP}^m) \xrightarrow{1-t} \tilde{K}_C(\mathrm{CP}^l/\mathrm{CP}^m) \xrightarrow{r} \tilde{K}_R(\mathrm{CP}^l/\mathrm{CP}^m)$$

is exact provided $l \not\equiv 3 \bmod 4$.

(vii) *The sequence* $\tilde{K}_C(\mathrm{CP}^l/\mathrm{CP}^m) \xrightarrow{r} \tilde{K}_R(\mathrm{CP}^l/\mathrm{CP}^m) \to 0$

is exact provided $m \not\equiv 3 \bmod 4$.

Part (i) of this theorem was published as Theorem 3·9 of (16).

The proof of Theorem 2·2 will occupy the rest of this section. Once parts (i) and (iii) are proved, we may introduce an abuse of language similar to that mentioned above. Consider the cofibring

$$\mathrm{CP}^m \xrightarrow{i} \mathrm{CP}^l \xrightarrow{j} \mathrm{CP}^l/\mathrm{CP}^m.$$

We shall use the notation ω^s for the (unique) element $(j^*)^{-1}\omega^s$ in $\tilde{K}_R(\mathrm{CP}^l/\mathrm{CP}^m)$, provided of course that $i^*\omega^s = 0$. Similarly for the element $2\omega^s$, if $i^*(2\omega^s) = 0$.

We shall prove part (i) by considering the usual spectral sequence (cf. (3), section 6). We shall say that an element $\alpha \in \tilde{K}_\Lambda(X)$ 'has filtration $\geqslant a$' if α is the image of an element in $\tilde{K}_\Lambda(X/X^{a-1})$, where X^q denotes the q-skeleton of X. We begin with a well-known lemma about the filtration of a product (cf. (7), page 20, line 4).

LEMMA 2·3. *If* α, β *have filtrations* $\geqslant a, b$ *then* $\alpha\beta$ *has filtration* $\geqslant a+b$.

This lemma is easily proved by observing that $\alpha\beta$ lies in the image of the homomorphism induced by an obvious map

$$\Delta: X \to (X/X^{a-1}) \mathbin{⋇} (X/X^{b-1});$$

the map Δ can be deformed into one which factors through X/X^{a+b-1}.

The terms $E_1^{p,q}$ of total degree zero in our spectral sequence are as follows:

$$\tilde{K}_R(\mathrm{CP}^u/\mathrm{CP}^{u-1}) = \tilde{K}_R(S^{2u}) = \begin{cases} Z & \text{if } u \equiv 0 \text{ or } 2 \bmod 4, \\ Z_2 & \text{if } u \equiv 1 \bmod 4, \\ 0 & \text{if } u \equiv 3 \bmod 4. \end{cases}$$

If $u = 4w$ there is a unique element γ_{4w} in $\tilde{K}_R(\mathrm{CP}^u/\mathrm{CP}^{u-1})$ such that

$$c\gamma_{4w} = \mu^{4w},$$

and γ_{4w} is a generator. If $u = 4w+1$ there is no choice for the generator γ_{4w+1} in $\tilde{K}_R(\mathrm{CP}^u/\mathrm{CP}^{u-1})$. If $u = 4w+2$ there is a unique element γ_{4w+2} in $\tilde{K}_R(\mathrm{CP}^u/\mathrm{CP}^{u-1})$ such that

$$c\gamma_{4w+2} = 2\mu^{4w+2},$$

and γ_{4w+2} is a generator.

The following lemma shows that these generators in E_1 are cycles for all the differentials, and relates them to the powers of ω.

LEMMA 2·4. *One can choose elements*

$$\begin{aligned} \alpha_{4w} &\in \tilde{K}_R(\mathrm{CP}^l/\mathrm{CP}^{4w-1}) && (\textit{if } l > 4w-1), \\ \alpha_{4w+1} &\in \tilde{K}_R(\mathrm{CP}^l/\mathrm{CP}^{4w}) && (\textit{if } l > 4w), \\ \alpha_{4w+2} &\in \tilde{K}_R(\mathrm{CP}^l/\mathrm{CP}^{4w+1}) && (\textit{if } l > 4w+1) \end{aligned}$$

with the following properties.

(i) *The image of* α_u *in* $\tilde{K}_R(\mathrm{CP}^u/\mathrm{CP}^{u-1})$ *is the generator* γ_u.

(ii) *The images of* α_{4w}, α_{4w+1} *in* $\tilde{K}_R(\mathrm{CP}^l)$ *are* ω^{2w}, ω^{2w+1}.

(iii) *The image of* α_{4w+2} *in* $\tilde{K}_R(\mathrm{CP}^l/\mathrm{CP}^{4w})$ *is* $2\alpha_{4w+1}$ *(and therefore its image in* $\tilde{K}_R(\mathrm{CP}^l)$ *is* $2\omega^{2w+1}$*).*

Proof. We begin with some hypothetical remarks. Suppose as a hypothesis that ω^{2w} has filtration $\geqslant 8w$ in CP^l for each l. Then if $l > 4w-1$ there exists

$$\alpha_{4w} \in \tilde{K}_R(\mathrm{CP}^l/\mathrm{CP}^{4w-1})$$

whose image in $\tilde{K}_R(\mathrm{CP}^l)$ is ω^{2w}; we will show that the image of α_{4w} in $\tilde{K}_R(\mathrm{CP}^{4w}/\mathrm{CP}^{4w-1})$ is γ_{4w}. In fact, in CP^l we have

$$\begin{aligned} c\omega &= cr\mu \\ &= (1+t)\mu \\ &= (1+\Psi_C^{-1})\mu \\ &\equiv \mu^2 \quad \text{(modulo higher powers of } \mu). \end{aligned}$$

Therefore $$c\omega^v \equiv \mu^{2v} \quad \text{(modulo higher powers of } \mu).$$

Hence in $\mathrm{CP}^l/\mathrm{CP}^{4w-1}$ we have

$$c\alpha_{4w} \equiv \mu^{4w} \quad \text{(modulo higher powers of } \mu).$$

Therefore the image of α_{4w} in $\tilde{K}_R(\mathrm{CP}^{4w}/\mathrm{CP}^{4w-1})$ is γ_{4w}.

Similarly, suppose as a hypothesis that ω^{2w+1} has filtration $\geqslant 8w+2$ in CP^l for each l. Set $q = \max(l, 4w+2)$; then there exists $\alpha_{4w+1} \in \tilde{K}_R(\mathrm{CP}^q/\mathrm{CP}^{4w})$ whose image in $\tilde{K}_R(\mathrm{CP}^q)$ is ω^{2w+1}. Let us consider the following diagram.

$$\begin{array}{ccccc} & & \tilde{K}_R(\mathrm{CP}^q/\mathrm{CP}^{4w}) & \xleftarrow{j_1^*} & \tilde{K}_R(\mathrm{CP}^q/\mathrm{CP}^{4w+1}) \\ & \swarrow^{i_1^*} & \downarrow i_3^* & & \downarrow \\ Z_2 = \tilde{K}_R(\mathrm{CP}^{4w+1}/\mathrm{CP}^{4w}) & & & & \\ & \nwarrow^{i_2^*} & & & \\ & & \tilde{K}_R(\mathrm{CP}^{4w+2}/\mathrm{CP}^{4w}) & \xleftarrow[j^*]{} & \tilde{K}_R(\mathrm{CP}^{4w+2}/\mathrm{CP}^{4w+1}) \end{array}$$

Since $i_1^*(2\alpha_{4w+1}) = 0$, there exists $\alpha_{4w+2} \in \tilde{K}_R(\mathrm{CP}^q/\mathrm{CP}^{4w+1})$ such that

$$j_1^* \alpha_{4w+2} = 2\alpha_{4w+1}.$$

Arguing as above, we have

$$c\alpha_{4w+2} \equiv 2\mu^{4w+2} \quad \text{(modulo higher powers of } \mu);$$

therefore the image of α_{4w+2} in $\tilde{K}_R(\mathrm{CP}^{4w+2}/\mathrm{CP}^{4w+1})$ is the generator γ_{4w+2}. Also the element $i_3^* \alpha_{4w+1}$ has infinite order in $\tilde{K}_R(\mathrm{CP}^{4w+2}/\mathrm{CP}^{4w})$, because $ci_3^* \alpha_{4w+1} = \mu^{4w+2}$; and the group $j_2^* \tilde{K}_R(\mathrm{CP}^{4w+2}/\mathrm{CP}^{4w+1})$ is generated by $2i_3^* \alpha_{4w+1}$, according to the work above; therefore $i_2^* i_3^* \alpha_{4w+1} \neq 0$ in $\tilde{K}_R(\mathrm{CP}^{4w+1}/\mathrm{CP}^{4w})$. That is, the image of α_{4w+1} in $\tilde{K}_R(\mathrm{CP}^{4w+1}/\mathrm{CP}^{4w})$ is the generator γ_{4w+1}.

The theory given above is applicable without any hypothesis in the special case $2w+1 = 1$; in fact, we can take

$$\alpha_1 = \omega \in \tilde{K}_R(\mathrm{CP}^l).$$

Thus α_2 exists. This provides us with generators for the groups in the following exact sequence.

$$0 = K_R^{-1}(\mathrm{CP}^1) \to \tilde{K}_R(\mathrm{CP}^2/\mathrm{CP}^1) \to \tilde{K}_R(\mathrm{CP}^2) \to \tilde{K}_R(\mathrm{CP}^1).$$

We see that $\tilde{K}_R(\mathrm{CP}^2) = Z$, generated by ω.

We have just shown that 2ω has filtration $\geqslant 4$; it follows by Lemma 2·3 that $4\omega^2$ has filtration $\geqslant 8$. Therefore $4\omega^2$ maps to zero in $\tilde{K}_R(\mathrm{CP}^2)$; but we have seen that $\tilde{K}_R(\mathrm{CP}^2) = Z$; so ω^2 maps to zero in $\tilde{K}_R(\mathrm{CP}^2)$, and ω^2 has filtration $\geqslant 6$. Since we have

$$\tilde{K}_R(\mathrm{CP}^3/\mathrm{CP}^2) = \tilde{K}_R(S^6) = 0,$$

we see that ω^2 has filtration $\geqslant 8$. Also ω has filtration $\geqslant 2$. Lemma 2·3 now shows that ω^{2w}, ω^{2w+1} have filtrations $\geqslant 8w$, $8w+2$ in CP^l (for any l). That is, the hypotheses of our 'hypothetical remarks' are true. This completes the proof of Lemma 2·4.

Proof of Theorem 2·2. It follows from the filtrations given in Lemma 2·4 that the powers of ω satisfy the relations given in Theorem 2·2 (i). It also follows from Lemma 2·4 that in our spectral sequence all differentials vanish on the groups of total degree zero. At this stage we have shown that the additive group $\tilde{K}_R(\mathrm{CP}^l/\mathrm{CP}^m)$ is generated by the images of the elements α_{4w} (for $m < 4w \leqslant l$), α_{4w+1} (for $m < 4w+1 \leqslant l$) and α_{4w+2} (if $m = 4w+1$).

We will now consider the map

$$c\colon \tilde{K}_R(\mathrm{CP}^l/\mathrm{CP}^m) \to \tilde{K}_C(\mathrm{CP}^l/\mathrm{CP}^m).$$

It is given by

$$\begin{aligned} c\alpha_{4w} &\equiv \mu^{4w} && \text{(modulo higher powers of } \mu),\\ c\alpha_{4w+1} &\equiv \mu^{4w+2} && \text{(modulo higher powers of } \mu),\\ c\alpha_{4w+2} &\equiv 2\mu^{4w+2} && \text{(modulo higher powers of } \mu); \end{aligned}$$

and, we repeat, the value $c\alpha_{4w+2}$ is needed only if $m = 4w+1$. We draw two conclusions. First, if $l \not\equiv 1 \bmod 4$ the map c is monomorphic; this proves part (iv) of the theorem. Secondly, if $l \not\equiv 1 \bmod 4$ then the elements which we have given as generators of $\tilde{K}_R(\mathrm{CP}^l/\mathrm{CP}^m)$ are linearly independent; therefore if $l \not\equiv 1 \bmod 4$ the terms of total degree zero in the spectral sequence for $\mathrm{CP}^l/\mathrm{CP}^m$ contain no boundaries, and survive unchanged to E_∞. It follows that the same thing happens for smaller values of l (i.e. for all l). (Alternatively, if $l = 4w+1$ we can deal with α_{4w+1} by the method of Lemma 3·1 below.) Therefore the spectral sequence is trivial, so far as the terms of total degree zero are concerned. At this stage we have proved parts (i) and (iii) of the theorem.

We turn to part (ii). We have

$$c\omega = cr\mu = (1+t)\mu = \eta - 2 + \eta^{-1}$$

and

$$\begin{aligned} c\Psi^k_R\omega &= \Psi^k_c c\omega \quad ((3), \text{ Theorem } 4{\cdot}1\text{ (iv)})\\ &= \Psi^k_c(\eta - 2 + \eta^{-1})\\ &= \eta^k - 2 + \eta^{-k} \quad ((3), \text{ Theorem } 5{\cdot}1)\\ &= T_k(\eta - 2 + \eta^{-1})\\ &= T_k(c\omega)\\ &= cT_k(\omega) \quad (10). \end{aligned}$$

This shows that $\Psi^k_R\omega = T_k(\omega)$ if $l \not\equiv 1 \bmod 4$, because c is then a monomorphism; this implies the same formula for smaller values of l by naturality.

To show that
$$T_k(\omega) \equiv k^2\omega \quad \text{(modulo higher powers of } \omega),$$
we proceed as follows. Take the formula
$$T_k(z-2+z^{-1}) = z^k - 2 + z^{-k}$$
and substitute $z = 1$; we find $$T_k(0) = 0.$$
By differentiating, we find
$$T'_k(z-2+z^{-1}) = k(z^{k-1}+z^{k-3}+\dots+z^{-k+3}+z^{-k+1}).$$
Substituting $z = 1$, we find $$T'_k(0) = k^2.$$
This completes the proof of part (ii).

In each of parts (v), (vi), (vii) it is clear that the composite of the two maps is zero. We shall prove the remaining three assertions
$$\operatorname{im} c \supset \ker(1-t),$$
$$\operatorname{im}(1-t) \supset \ker r,$$
$$\operatorname{im} r \supset \ker 0$$
by a common method, which consists of an inductive process. We treat part (v) as an example.

Consider the cofibring
$$CP^v/CP^w \xrightarrow{i} CP^u/CP^w \xrightarrow{j} CP^u/CP^v.$$
Using Lemma 2·1 and Theorem 2·2 (iii) we obtain the following commutative diagram, in which the rows are exact.
$$\begin{array}{ccccccc}
0 \leftarrow \tilde{K}_R(CP^v/CP^w) & \xleftarrow{i^*} & \tilde{K}_R(CP^u/CP^w) & \xleftarrow{j^*} & \tilde{K}_R(CP^u/CP^v) \leftarrow 0 \\
\downarrow c & & \downarrow c & & \downarrow c \\
0 \leftarrow \tilde{K}_C(CP^v/CP^w) & \xleftarrow{i^*} & \tilde{K}_C(CP^u/CP^w) & \xleftarrow{j^*} & \tilde{K}_C(CP^u/CP^v) \leftarrow 0 \\
\downarrow 1-t & & \downarrow 1-t & & \downarrow 1-t \\
0 \leftarrow \tilde{K}_C(CP^v/CP^w) & \xleftarrow{i^*} & \tilde{K}_C(CP^u/CP^w) & \xleftarrow{j^*} & \tilde{K}_C(CP^u/CP^v) \leftarrow 0
\end{array}$$
We see by elementary diagram-chasing that if the two outside columns are exact, then so is the middle one. In other words, if the assertion of part (v) holds for CP^u/CP^v and for CP^v/CP^w, then it holds for CP^u/CP^w also. This will allow us to prove part (v) by an inductive process, starting from a small number of special cases. Similarly for parts (vi) and (vii).

In fact, part (v) will follow by induction from the following special cases:
$$CP^u/CP^{u-1} \quad \text{for} \quad u \not\equiv 2 \mod 4,$$
$$CP^{2v}/CP^{2v-2}.$$
Part (vi) will follow by induction from the following special cases:
$$CP^u/CP^{u-1} \quad \text{for} \quad u \not\equiv 3 \mod 4,$$
$$CP^{2v}/CP^{2v-2}.$$

Part (vii) will follow by induction from the following special cases:

$$\mathrm{CP}^u/\mathrm{CP}^{u-1} \quad \text{for} \quad u \not\equiv 0 \mod 4,$$
$$\mathrm{CP}^{2v}/\mathrm{CP}^{2v-2}.$$

The space $\mathrm{CP}^u/\mathrm{CP}^{u-1}$ is a sphere S^{2u}, for which we may suppose that the results required are known (9). It remains to consider the space $\mathrm{CP}^{2v}/\mathrm{CP}^{2v-2}$. We have $\tilde{K}_R(\mathrm{CP}^{2v}/\mathrm{CP}^{2v-2}) = Z$, generated by ω^v. Since

$$t\mu \equiv -\mu + \mu^2 \quad \text{(modulo higher powers of } \mu\text{)},$$

the values of t in $\tilde{K}_C(\mathrm{CP}^{2v}/\mathrm{CP}^{2v-2})$ are given by

$$\begin{aligned} t\mu^{2v-1} &= -\mu^{2v-1} + (2v-1)\mu^{2v},\\ t\mu^{2v} &= \mu^{2v}.\end{aligned}$$

It follows that the homomorphisms $cr = 1+t$ and $1-t$ are given by

$$\begin{aligned} cr\mu^{2v-1} &= (2v-1)\mu^{2v},\\ cr\mu^{2v} &= 2\mu^{2v},\\ (1-t)\mu^{2v-1} &= 2\mu^{2v-1} - (2v-1)\mu^{2v},\\ (1-t)\mu^{2v} &= 0.\end{aligned}$$

Since $c\omega^v = \mu^{2v}$ and c is monomorphic we have

$$\begin{aligned} r\mu^{2v-1} &= (2v-1)\omega^v,\\ r\mu^{2v} &= 2\omega^v.\end{aligned}$$

We can now check directly that

$$\begin{aligned} \operatorname{im} c &= \ker(1-t) \quad &&(\text{generated by } \mu^{2v}),\\ \operatorname{im}(1-t) &= \ker r &&(\text{generated by } 2\mu^{2v-1} - (2v-1)\mu^{2v}),\\ \operatorname{im} r &= \ker 0. \end{aligned}$$

This completes the proof of Theorem 2·2.

3. *A lemma on* $J(\mathrm{CP}^l)$. The following lemma will be used in the proof of Lemma 6·2.

LEMMA 3·1. *In* $J(\mathrm{CP}^{4w+1})$ *we have* $Jr\mu^{4w+1} \neq 0$.

Proof. Let RP^u denote real projective space of u dimensions. It has been observed by Atiyah, and also by Bott (11), that the map

$$J: \tilde{K}_R(\mathrm{RP}^u) \to J(\mathrm{RP}^u)$$

is an isomorphism. We refer the reader to (2), Example 6·3, for a proof.

We can now use a simple naturality argument. Let $f: \mathrm{RP}^{8w+2} \to \mathrm{CP}^{4w+1}$ be the usual projection, arranged so that $f(\mathrm{RP}^{8w+1}) \subset \mathrm{CP}^{4w}$. Let $j: \mathrm{CP}^{4w+1} \to \mathrm{CP}^{4w+1}/\mathrm{CP}^{4w}$ be the quotient map. Since $\mathrm{CP}^{4w+1}/\mathrm{CP}^{4w} = S^{8w+2}$, we have (9) $\tilde{K}_C(\mathrm{CP}^{4w+1}/\mathrm{CP}^{4w}) = Z$ (generated by γ, say) and $\tilde{K}_R(\mathrm{CP}^{4w+1}/\mathrm{CP}^{4w}) = Z_2$ (generated by $r\gamma$). With μ as above, we have $\mu^{4w+1} = j^*\gamma$. Since f has degree 1, $f^*r\gamma$ is a generator of $\tilde{K}_R(\mathrm{RP}^{8w+2}/\mathrm{RP}^{8w+1})$. It is shown in (3), Theorem 7·4, that the image of this generator in $\tilde{K}_R(\mathrm{RP}^{8w+2})$ is non-zero; that is, $j^*f^*r\gamma \neq 0$, i.e. $f^*r\mu^{4w+1} \neq 0$. Since J is an isomorphism on RP^{8w+2}, we have $Jf^*r\mu^{4w+1} \neq 0$, i.e. $f^*Jr\mu^{4w+1} \neq 0$. Therefore $Jr\mu^{4w+1} \neq 0$. This completes the proof.

4. *The functor* J'. In this section we shall define and study the groups $J'_\Lambda(X)$. We want $J'_\Lambda(CP^l)$ to serve as a lower bound for $J(CP^l)$, in the sense explained in section 1, and so we define it by introducing invariants of stable fibre homotopy type. For these invariants we turn to the theory of characteristic classes.

We first recall something of Hirzebruch's theory of multiplicative sequences (13). By $1+\sum_{s>0} H^{2s}(X;Q)$ we shall mean the subset of $\sum_{s\geqslant 0} H^{2s}(X;Q)$ consisting of elements of the form $1+x$, with $x \in \sum_{s>0} H^{2s}(X;Q)$. This subset is a multiplicative group. Similar notations will occur later, and their meaning will be obvious. For example, in the case of an infinite complex such as BU, we have to consider $1+\prod_{s>0} H^{2s}(BU;Q)$.

Suppose given an (inhomogeneous) characteristic class

$$\theta \in 1+\prod_{s>0} H^{2s}(BU;Q).$$

We may say that θ is a 'multiplicative sequence' if

$$\theta(\xi \oplus \xi') = \theta(\xi)\,\theta(\xi').$$

(Actually we would prefer the term 'exponential' to 'multiplicative' since θ is homomorphic from addition to multiplication.)

If we apply θ to the universal bundle η over CP^∞ we obtain

$$\theta(\eta) = 1+\sum_{s>0} a_s y^s,$$

where y is the generator in $H^2(CP^\infty;Z)$ and the a_s are rationals. This establishes a (1-1) correspondence between multiplicative sequences and formal power series. We define bh to be the characteristic class corresponding to the formal power series

$$\frac{e^y-1}{y}.$$

The notation is intended to suggest 'Bernoulli'.

A precisely similar account applies to multiplicative sequences

$$\theta \in 1+\prod_{s>0} H^{4s}(BO;Q),$$

except that we consider $\theta(r\eta)$ instead of $\theta(\eta)$ and use formal power series in y^2. We define sh to be the characteristic class corresponding to the formal power series

$$\frac{e^{\frac{1}{2}y}-e^{-\frac{1}{2}y}}{y}.$$

The notation is intended to suggest 'sinh'.

The functions bh, sh are the same as those considered in (2), according to (2), Theorem 5·1. They yield homomorphisms

$$\text{bh}: K_C(X) \to 1+\sum_{s>0} H^{2s}(X;Q),$$

$$\text{sh}: K_R(X) \to 1+\sum_{s>0} H^{4s}(X;Q)$$

from the additive groups $K_C(X)$, $K_R(X)$ to the multiplicative groups written on the right.

The connexion between bh and sh is

$$\mathrm{bh}\,(\xi) = e^{\frac{1}{2}c_1(\xi)}\,\mathrm{sh}\,(r\xi).$$

(Since both sides are exponential, it is sufficient by Hirzebruch's theory to check this for the case $\xi = \eta$.)

The following lemma is useful in calculations, and would serve as an alternative definition of bh and sh. We quote it from (2), Corollary 5·2.

LEMMA 4·1. *Let $\xi \in K_C(X)$, $\xi' \in K_R(X)$. Then we have*

$$\mathrm{Log\,bh}\,\xi = \sum_{t=1}^{\infty} \alpha_t\,\mathrm{ch}_t\,\xi,$$

$$\mathrm{Log\,sh}\,\xi' = \sum_{s=1}^{\infty} \tfrac{1}{2}\alpha_{2s}\,\mathrm{ch}_{2s}\,c\xi'.$$

In these formulae, $\mathrm{Log}\,(1+x)$ is defined for $x \in \sum_{t>0} H^{2t}(X;Q)$ by means of the usual power series expansion. The coefficients α_t are as in (2), section 2, and ch_t means the component of ch in dimension $2t$.

The groups $J'_\Lambda(X)$ will be defined in terms of bh, sh. First, we define $V_C(X)$ to be the set of elements $\alpha \in K_C(X)$ such that

$$\mathrm{bh}\,\alpha = \mathrm{ch}\,(1+\beta)$$

for some $\beta \in \tilde{K}_C(X)$. Since

$$\mathrm{ch}: 1+\tilde{K}_C(X) \to 1 + \sum_{s>0} H^{2s}(X;Q)$$

is a homomorphism of multiplicative groups, it is clear that $V_C(X)$ is a subgroup.

Similarly, we define $V_R(X)$ to be the set of elements $\alpha \in K_R(X)$ such that

$$\mathrm{sh}\,\alpha = \mathrm{ch}\,c(1+\beta)$$

for some $\beta \in \tilde{K}_R(X)$. Since

$$\mathrm{ch}\,c: 1+\tilde{K}_R(X) \to 1 + \sum_{s>0} H^{4s}(X;Q)$$

is a homomorphism of multiplicative groups, it is clear that $V_R(X)$ is a subgroup.

In either case, we define $$J'_\Lambda(X) = K_\Lambda(X)/V_\Lambda(X),$$
and write $$J'_\Lambda: K_\Lambda(X) \to J'_\Lambda(X)$$
for the quotient map.

We now wish to obtain the following commutative diagrams of epimorphisms.

$$\begin{array}{ccc} K_R(\mathrm{CP}^l) & \xrightarrow{J} & J(\mathrm{CP}^l) \\ & \searrow^{J'_R} & \downarrow \theta'_R \\ & & J'_R(\mathrm{CP}^l) \end{array}$$

$$\begin{array}{ccc} K_C(\mathrm{CP}^l) & \xrightarrow{J_r} & J(\mathrm{CP}^l) \\ & \searrow^{J'_C} & \downarrow \theta'_C \\ & & J'_C(\mathrm{CP}^l) \end{array}$$

Since $r: \tilde{K}_C(\mathrm{CP}^l) \to \tilde{K}_R(\mathrm{CP}^l)$ is epimorphic by Theorem 2·2 (vii), it is sufficient to prove the following lemma.

LEMMA 4·2. *For any space X we have*

$$\ker J \subset \ker J'_R,$$
$$\ker Jr \subset \ker J'_C.$$

This lemma is due to Atiyah.

Proof. An element in $\ker J$ can be written in the form $\xi - n$, where n is an integer and ξ is a real vector bundle of dimension divisible by 8 whose associated sphere bundle is fibre homotopy trivial. Since the Stiefel–Whitney classes are fibre homotopy invariants ((17), Théorème IV. 4), it follows that $w_1(\xi)$ and $w_2(\xi)$ are zero; thus ξ is a spinor bundle. By (2), Corollary 5·8 we have

$$\mathrm{sh}\,\xi = \mathrm{ch}\,c(1+x)$$

for some $x \in \tilde{K}_R(X)$. Thus ξ (and so $\xi - n$) is an element of $\ker J'_R$.

Similarly, an element in $\ker Jr$ can be written in the form $\xi - n$, where n is an integer and ξ is a unitary bundle whose associated sphere bundle is fibre homotopy trivial. By (2), Corollary 5·8 we have

$$\mathrm{bh}\,\xi = \mathrm{ch}\,(1+x)$$

for some $x \in \tilde{K}_C(X)$. Thus ξ (and so $\xi - n$) is an element of $\ker J'_C$. This completes the proof.

The following lemma will justify us in omitting the subscript Λ from $J'_\Lambda(\mathrm{CP}^l)$ if $l > 1$.

LEMMA 4·3. *If m is even and $l > 1$, then the homomorphism*

$$r: K_C(\mathrm{CP}^l/\mathrm{CP}^m) \to K_R(\mathrm{CP}^l/\mathrm{CP}^m)$$

induces an isomorphism from $J'_C(\mathrm{CP}^l/\mathrm{CP}^m)$ to $J'_R(\mathrm{CP}^l/\mathrm{CP}^m)$.

The proof will require subsidiary lemmas.

LEMMA 4·4. *Assume $l > 1$, $\alpha \in K_C(\mathrm{CP}^l)$ and $r\alpha \in V_R(\mathrm{CP}^l)$. Then $c_1(\alpha) = 2h$ for some $h \in H^2(\mathrm{CP}^l; Z)$.*

Proof. Using the injection $i: \mathrm{CP}^2 \to \mathrm{CP}^l$, we see that it is sufficient to consider the case $l = 2$. Let us assume

$$\mathrm{sh}\,r\alpha = \mathrm{ch}\,c(1+\beta),$$

where

$$\alpha = a_0 + a_1\mu + a_2\mu^2$$

and

$$\beta = b\omega$$

for some integers a_0, a_1, a_2, b. (See Theorems 2·1, 2·2 (i).) We carry out the calculations using Lemma 4·1, and the assumptions yield

$$\alpha_2(\tfrac{1}{2}a_1 + a_2) = b.$$

Since $\alpha_2 = \frac{1}{12}$, it follows that a_1 is even. Now $c_1\alpha = a_1 y$, so the result follows.

LEMMA 4·5. *Let* $\alpha \in K_C(X)$ *satisfy the conditions*

(i) $r\alpha \in V_R(X)$

(ii) $c_1(\alpha) = 2h$ *for some* $h \in H^2(X; Z)$.

Then $\alpha \in V_C(X)$.

Proof. By condition (i),

$$\text{sh}\, r\alpha = \text{ch}\, c(1+\beta)$$

for some $\beta \in \tilde{K}_R(X)$. Since c_1 defines a (1-1) correspondence between complex line bundles over X and elements of $H^2(X; Z)$, by condition (ii) there exists a complex line bundle γ over X such that $c_1\gamma = h$. Thus

$$\text{ch}\, \gamma = e^{\frac{1}{2}c_1(\alpha)}.$$

Now we have

$$\begin{aligned} \text{bh}\, \alpha &= e^{\frac{1}{2}c_1(\alpha)}\, \text{sh}\, r\alpha \\ &= \text{ch}\, \gamma\, \text{ch}\, (1+c\beta) \\ &= \text{ch}\, \gamma\, (1+c\beta) \\ &= \text{ch}\, (1+\delta), \end{aligned}$$

where $\delta \in \tilde{K}_C(X)$. Thus $\alpha \in V_C(X)$, as claimed.

LEMMA 4·6. *If* $\alpha \in V_C(\text{CP}^l/\text{CP}^m)$, *where* $m \not\equiv 1 \bmod 4$, *then* $r\alpha \in V_R(\text{CP}^l/\text{CP}^m)$.

Proof. If $\alpha \in V_C(\text{CP}^l/\text{CP}^m)$, then

$$\text{bh}\, \alpha = \text{ch}\, (1+\beta)$$

for some $\beta \in \tilde{K}_C(\text{CP}^l/\text{CP}^m)$. That is,

$$e^{\frac{1}{2}c_1(\alpha)}\, \text{sh}\, r\alpha = \text{ch}\, (1+\beta).$$

Taking the component of this equation in $H^2(\text{CP}^l/\text{CP}^m; Q)$, we find

$$\tfrac{1}{2}c_1(\alpha) = c_1(\beta).$$

Hence we can find a complex line bundle γ so that $c_1\gamma = \frac{1}{2}c_1\alpha$, and so

$$\text{sh}\, r\alpha = \text{ch}\, (1+\delta)$$

for some $\delta \in \tilde{K}_C(\text{CP}^l/\text{CP}^m)$, as in the proof of Lemma 4·5.

We now apply the map

$$f: \text{CP}^l, \text{CP}^m \to \text{CP}^l, \text{CP}^m$$

defined by complex conjugation on each coordinate. We find

$$\text{sh}\, (rf^*\alpha) = \text{ch}\, (1+f^*\delta).$$

But in $\tilde{K}_C(\text{CP}^l/\text{CP}^m)$ we have $f^* = t$. (It is sufficient to prove this result in CP^l; here it is obvious that $f^*\eta = t\eta$, and every element of $K_C(\text{CP}^l)$ can be written as a polynomial in η. Alternatively, one can apply ch and so reduce matters to the effect of f on cohomology.) We now have

$$\text{sh}\, (rt\alpha) = \text{ch}\, (1+t\delta).$$

Since $rt\alpha = r\alpha$, we have

$$\text{ch}\, (1+t\delta) = \text{ch}\, (1+\delta).$$

Since ch is monomorphic, we have

$$t\delta = \delta.$$

By Theorem 2·2 (v), if $m \not\equiv 1 \bmod 4$ we have $\delta = c\beta$ for some $\beta \in \tilde{K}_R(\mathrm{CP}^l/\mathrm{CP}^m)$. Thus

$$\mathrm{sh}\, r\alpha = \mathrm{ch}\,(1 + c\beta),$$

and so

$$r\alpha \in V_R(\mathrm{CP}^l/\mathrm{CP}^m).$$

Proof of Lemma 4·3. If $m \not\equiv 3 \bmod 4$, then

$$r: \tilde{K}_C(\mathrm{CP}^l/\mathrm{CP}^m) \to \tilde{K}_R(\mathrm{CP}^l/\mathrm{CP}^m)$$

is an epimorphism, by Theorem 2·2 (vii). We therefore require to prove that

$$r^{-1}V_R = V_C.$$

Suppose therefore that $\alpha \in K_C(\mathrm{CP}^l/\mathrm{CP}^m)$ and $r\alpha \in V_R(\mathrm{CP}^l/\mathrm{CP}^m)$. Then $c_1(\alpha) = 0$ if $m > 0$, while $c_1(\alpha) = 2h$ if $m = 0$ by Lemma 4·4. Thus $\alpha \in V_C(\mathrm{CP}^l/\mathrm{CP}^m)$ by Lemma 4·5. Conversely, if $\alpha \in V_C(\mathrm{CP}^l/\mathrm{CP}^m)$, then $r\alpha \in V_R(\mathrm{CP}^l/\mathrm{CP}^m)$ by Lemma 4·6, since $m \not\equiv 1 \bmod 4$. This proves Lemma 4·3.

LEMMA 4·7. *$J'_C(n\eta) = 0$ in $J'_C(\mathrm{CP}^{k-1})$ if and only if n satisfies condition (C_k); in other words, the order of $J'_C(\eta)$ in $J'_C(\mathrm{CP}^{k-1})$ is M_k.*

This lemma states that Atiyah is correct in his re-interpretation of the methods of Atiyah and Todd—which will surprise nobody.

Proof of Lemma 4·7. We seek to re-express the condition that $J'_C(n\eta) = 0$ in $J'_C(\mathrm{CP}^{k-1})$. Since J'_C is a homomorphism, it is equivalent to say that

$$J'_C(-n\eta) = 0 \quad \text{in} \quad J'_C(\mathrm{CP}^{k-1}).$$

According to the definition, this condition means that

$$\mathrm{bh}\,(-n\eta) \in \mathrm{ch}\,(1 + \tilde{K}_C(\mathrm{CP}^{k-1})).$$

We have

$$\mathrm{ch}\,\mu^q = y^q \quad \text{(modulo higher powers of } y\text{)},$$

where y generates $H^2(\mathrm{CP}^{k-1}; Q)$. Thus the element $\mathrm{bh}\,(-n\eta)$ in $H^*(\mathrm{CP}^{k-1}; Q)$ has a unique expression in the form

$$\mathrm{bh}\,(-n\eta) = \sum_{0 \leqslant q \leqslant k-1} a_q\, \mathrm{ch}\,\mu^q,$$

where the coefficients a_q are rational and $a_0 = 1$. The condition

$$\mathrm{bh}\,(-n\eta) \in \mathrm{ch}\,(1 + \tilde{K}_C(\mathrm{CP}^{k-1}))$$

holds if and only if all the coefficients a_q are integers. Let us set

$$z = \mathrm{ch}\,\mu = e^y - 1;$$

then $y = \mathrm{Log}\,(1+z)$, where the function $\mathrm{Log}\,(1+z)$ is defined in $H^*(\mathrm{CP}^{k-1}; Q)$ by its usual power-series expansion. Also $\mathrm{ch}\,\mu^q = z^q$, so that a_q is the coefficient of z^q in

$$\mathrm{bh}\,(-n\eta) = \left(\frac{y}{e^y - 1}\right)^n = \left(\frac{\mathrm{Log}\,(1+z)}{z}\right)^n.$$

The condition $J'_C(n\eta) = 0$, then, is equivalent to the condition that the coefficient of z^q in

$$\left(\frac{\operatorname{Log}(1+z)}{z}\right)^n$$

should be integral for $0 \leqslant q \leqslant k-1$. This completes the proof.

LEMMA 4·8. *If $v > 0$, then the injection $i: CP^{2v} \to CP^{2v+1}$ induces an isomorphism from $J'(CP^{2v+1})$ onto $J'(CP^{2v})$.*

It has been observed by Atiyah that Lemmas 4·7 and 4·8 explain the fact ((8), page 344) that $M_{2k} = M_{2k-1}$ for $k > 1$.

Proof of Lemma 4·8. The map

$$i^*: K_R(CP^{2v+1}) \to K_R(CP^{2v})$$

is an epimorphism; it is therefore sufficient to prove that

$$(i^*)^{-1} V_R(CP^{2v}) \subset V_R(CP^{2v+1}).$$

Suppose then that $\alpha \in K_R(CP^{2v+1})$ is such that

$$i^* \operatorname{sh} \alpha = \operatorname{ch} c(1+\beta)$$

for some $\beta \in \tilde{K}_R(CP^{2v})$. Since i^* is epimorphic, we can write $\beta = i^*\gamma$, so that

$$i^* \operatorname{sh} \alpha = i^* \operatorname{ch} c(1+\gamma).$$

That is, $$\operatorname{sh} \alpha = \operatorname{ch} c(1+\gamma) \bmod H^{4v+2}(CP^{2v+1}; Q).$$

But $\operatorname{sh} \alpha$ and $\operatorname{ch} c(1+\gamma)$ have components zero in dimension $4v+2$. Therefore

$$\operatorname{sh} \alpha = \operatorname{ch} c(1+\gamma),$$

so $\alpha \in V_R(CP^{2v+1})$. This completes the proof.

We now proceed to build up results needed for our main proof.

LEMMA 4·9. *Suppose that*

$$X \xrightarrow{i} Y \xrightarrow{j} Y/X$$

is a cofibring such that $$\operatorname{ch}: K_C(X) \to H^*(X; Q)$$

and $$j^*: H^*(Y/X; Q) \to H^*(Y; Q)$$

are monomorphisms. Then $$j': J'_C(Y/X) \to J'_C(Y)$$

is a monomorphism.

(The results of section 2 show that we can apply this lemma in the case $X = CP^l$, $Y = CP^m$.)

Proof. Let $\alpha \in K_C(Y/X)$ be an element such that $J'_C j^*\alpha = 0$; that is,

$$\operatorname{bh} j^*\alpha = \operatorname{ch}(1+\beta),$$

where $\beta \in \tilde{K}_C(Y)$. Then we have

$$\begin{aligned} \operatorname{ch} i^*(1+\beta) &= i^* \operatorname{ch}(1+\beta) \\ &= i^* \operatorname{bh} j^*\alpha \\ &= \operatorname{bh} i^*j^*\alpha \\ &= 1. \end{aligned}$$

By assumption, this implies that

$$i^*(1+\beta) = 1.$$

By exactness, we have $$1+\beta = j^*(1+\gamma),$$

where $\gamma \in \tilde{K}_C(Y/X)$. It follows that

$$\begin{aligned} j^* \operatorname{bh} \alpha &= \operatorname{bh} j^* \alpha \\ &= \operatorname{ch}(1+\beta) \\ &= \operatorname{ch} j^*(1+\gamma) \\ &= j^* \operatorname{ch}(1+\gamma). \end{aligned}$$

By assumption, this implies that

$$\operatorname{bh} \alpha = \operatorname{ch}(1+\gamma).$$

Hence $J'_C \alpha = 0$. This completes the proof.

Our next object is to give the structure of $J'(\mathrm{CP}^{2v}/\mathrm{CP}^{2v-2})$. The answer involves the Bernoulli numbers, which enter via Lemma 4·1. We define $m(2v)$ to be the denominator of $\frac{1}{2}\alpha_{2v}$, when this fraction is written in its lowest terms (cf. (2), Theorem 2·6).

LEMMA 4·10. *The group* $J'(\mathrm{CP}^{2v}/\mathrm{CP}^{2v-2})$ *is cyclic of order* $m(2v)$, *generated by* $J'_R(\omega^v)$.

Proof. By Theorem 2·2 we have $\tilde{K}_R(\mathrm{CP}^{2v}/\mathrm{CP}^{2v-2}) = Z$, generated by ω^v. Let γ, δ be elements of $\tilde{K}_R(\mathrm{CP}^{2v}/\mathrm{CP}^{2v-2})$. By Lemma 4·1 we have

$$\operatorname{Log} \operatorname{sh} \gamma = \tfrac{1}{2}\alpha_{2v} \operatorname{ch}_{2v} c\gamma$$

while $$\operatorname{Log} \operatorname{ch} c(1+\delta) = \operatorname{ch}_{2v} c\delta,$$

since $H^q(\mathrm{CP}^{2v}/\mathrm{CP}^{2v-2}; Q) = 0$ except for $q = 0,\ 4v-2,\ 4v$. Also

$$\operatorname{ch}_{2v} c: \tilde{K}_R(\mathrm{CP}^{2v}/\mathrm{CP}^{2v-2}) \to H^{4v}(\mathrm{CP}^{2v}/\mathrm{CP}^{2v-2}; Q)$$

is monomorphic by section 2; actually an easy calculation shows that $\operatorname{ch}_{2v} c\omega^v = y^{2v}$. Thus $\operatorname{sh}\gamma$ can be written in the form $\operatorname{ch} c(1+\delta)$ for some $\delta \in \tilde{K}_R(\mathrm{CP}^{2v}/\mathrm{CP}^{2v-2})$ if and only if γ is divisible by the denominator of $\frac{1}{2}\alpha_{2v}$, i.e. by $m(2v)$. This completes the proof.

5. *The functor* J''. In this section we shall define and study the group $J''(X)$. We shall find that if $X = \mathrm{CP}^l$, $J''(X)$ is an upper bound for $J(X)$ in the sense of section 1. For a full account of the groups $J''(X)$ we refer the reader to (2), section 3; we remark once again that the J'' of this paper corresponds to the $\tilde{J}''$ of (1) and (2).

Let f be a function which assigns to each integer k a non-negative integer $f(k)$. Then we define $W(f, X)$ to be the subgroup of $K_R(X)$ generated by the elements

$$k^{f(k)}(\Psi^k_R - 1)\xi,$$

where k runs over the integers and ξ runs over $K_R(X)$. Here the operations Ψ^k_R are as in (3). That is, $W(f, X)$ is the subgroup of linear combinations

$$\sum_{k,\xi} a(k,\xi)\, k^{f(k)}(\Psi^k_R - 1)\xi,$$

where the $a(k,\xi)$ are integers and all but a finite number of them are zero.

If $f_1 \geqslant f_2$, then obviously $W(f_1, X) \subset W(f_2, X)$. We set

$$W(X) = \bigcap_f W(f, X),$$

where the intersection is taken over all such functions f. (Compare (2), Proposition 3·2.) Then $W(X)$ is a subgroup of $K_R(X)$; we set

$$J''(X) = K_R(X)/(Z + W(X)),$$

and write J'' for the quotient map.

It is easy to check that a map $f: X \to Y$ induces a homomorphism

$$f'': J''(Y) \to J''(X);$$

we refer to (2), section 3, for the details.

The following lemma states that $J''(\mathrm{CP}^l)$ is an upper bound for $J(\mathrm{CP}^l)$.

LEMMA 5·1. *The map*

$$J: K_R(\mathrm{CP}^l) \to J(\mathrm{CP}^l)$$

factors through J'', *giving the following commutative diagram of epimorphisms.*

$$\begin{array}{ccc} K_R(\mathrm{CP}^l) & \xrightarrow{J''} & J''(\mathrm{CP}^l) \\ & {}_{J}\searrow & \downarrow{\scriptstyle \theta''} \\ & & J(\mathrm{CP}^l) \end{array}$$

Proof. We shall first show that the additive group $K_R(\mathrm{CP}^l)$ is generated by $O(1)$-bundles and $O(2)$-bundles. In fact, the additive group $K_C(\mathrm{CP}^l)$ is generated by the elements μ^s for $0 \leqslant s \leqslant l$; and we can expand $\mu^s = (\eta - 1)^s$ in terms of η^q by the binomial theorem. Thus the additive group $K_C(\mathrm{CP}^l)$ is generated by the complex line bundles η^q for $0 \leqslant q \leqslant l$. By Theorem 2·2 (vii), the map

$$r: \tilde{K}_C(\mathrm{CP}^l) \to \tilde{K}_R(\mathrm{CP}^l)$$

is an epimorphism, and it follows that the additive group $K_R(\mathrm{CP}^l)$ is generated by the SO(1)-bundle 1 and the SO(2)-bundles $r(\eta^q)$ for $0 \leqslant q \leqslant l$.

The result now follows at once by (1), Theorem 1·3, and (2), Proposition 3·1. We will give the argument in detail. By (1), Theorem 1·3, if k is an integer and $\xi \in K_R(\mathrm{CP}^l)$, there is an exponent $e(k, \xi)$ such that

$$J k^{e(k,\xi)} (\Psi_R^k - 1)\xi = 0.$$

As in the proof of (2), Proposition 3·2, we can replace $e(k, \xi)$ by a function $f(k)$ independent of ξ; we let $\xi_1, \ldots, \xi_n$ be a finite set of generators for $K_R(\mathrm{CP}^l)$, and set

$$f(k) = \max_t e(k, \xi_t).$$

For this function f we have

$$W(f, X) \subset \ker J;$$

a fortiori, we have

$$W(X) \subset \ker J.$$

Also (on our definitions) $1 \in \ker J$; thus $\ker J'' \subset \ker J$, and J factors through J'', as required.

LEMMA 5·2. *Consider the cofibring*

$$CP^m \xrightarrow{i} CP^l \xrightarrow{j} CP^l/CP^m \quad (l \geqslant m).$$

Then the sequence $$J''(CP^l/CP^m) \xrightarrow{j''} J''(CP^l) \xrightarrow{i''} J''(CP^m) \to 0$$
is exact.

This follows at once from (2), Theorem 3·12.

In the following lemma, the number $m(2v)$ is the same as in Lemma 4·10 and (2), section 2.

LEMMA 5·3. *The group* $J''(CP^{2v}/CP^{2v-2})$ *is cyclic of order* $m(2v)$, *generated by* $J''(\omega^v)$.

Proof. By Theorem 2·2 we have $\tilde{K}_R(CP^{2v}/CP^{2v-2}) = Z$, generated by ω^v. For any element β in this group, we have

$$\Psi_R^k \beta = k^{2v}\beta.$$

For any function f assigning a non-negative integer $f(k)$ to each integer k, we therefore have

$$k^{f(k)}(\Psi_R^k - 1)\beta = k^{f(k)}(k^{2v} - 1)\beta.$$

Therefore the elements expressible in the form

$$\sum_{k,\beta} k^{f(k)}(\Psi_R^k - 1)\beta$$

are just the elements divisible by $h(f, 2v)$ where, as in (2), Theorem 2·7, $h(f, 2v)$ is the highest common factor of the numbers

$$k^{f(k)}(k^{2v} - 1)$$

as k varies. Now (2), Theorem 2·7, states that $h(f, 2v)$ divides $m(2v)$, and that, for a suitable choice of f, $h(f, 2v) = m(2v)$. Therefore the elements in $W(CP^{2v}/CP^{2v-2})$ are precisely those divisible by $m(2v)$. This completes the proof.

6. *Proof of the main theorems.*

LEMMA 6·1. *The map*

$$\theta'\theta'' : J''(CP^{2v}) \to J'(CP^{2v})$$

is an isomorphism.

Proof. The proof is by induction over v (starting from the case $v = 0$, which is trivial). Suppose the result true for CP^{2v-2}, and consider the cofibring

$$CP^{2v-2} \xrightarrow{i} CP^{2v} \xrightarrow{j} CP^{2v}/CP^{2v-2}.$$

We obtain the following commutative diagram.

$$\begin{array}{ccccccc}
0 \leftarrow J''(CP^{2v-2}) & \xleftarrow{i''} & J''(CP^{2v}) & \xleftarrow{j''} & J''(CP^{2v}/CP^{2v-2}) \\
\downarrow{\scriptstyle (\theta'\theta'')_{2v-2}} & & \downarrow{\scriptstyle (\theta'\theta'')_{2v}} & & \downarrow{\scriptstyle \cong} \\
0 \leftarrow J'(CP^{2v-2}) & \xleftarrow{i'} & J'(CP^{2v}) & \xleftarrow{j'} & J'(CP^{2v}/CP^{2v-2})
\end{array}$$

The insertion of the right-hand vertical isomorphism is justified by Lemmas 4·10 and 5·3. The upper row is exact by Lemma 5·2. The map j' is a monomorphism by Lemma

4·9. The central vertical map $(\theta'\theta'')_{2v}$ is known to be an epimorphism, and easy diagram-chasing shows that it is a monomorphism. This completes the induction, and proves the lemma.

Incidentally, we have shown that the two rows of the diagram are short exact sequences.

LEMMA 6·2. *Consider the epimorphisms*

$$J''(\mathrm{CP}^{2v+1}) \xrightarrow{\theta''} J(\mathrm{CP}^{2v+1}) \xrightarrow{\theta'} J'(\mathrm{CP}^{2v+1}).$$

Then (i) *if* $v = 2w+1, \theta'\theta''$ *is an isomorphism; and*

(ii) *if* $v = 2w$, $v \neq 0$, θ'' *is an isomorphism and* $\ker\theta' = Z_2$, *generated by* $Jr\mu^{4w+1}$.

Proof. Consider the following commutative diagram.

$$\begin{array}{ccccc} J''(\mathrm{CP}^{2v+1}/\mathrm{CP}^{2v}) & \xrightarrow{j''} & J''(\mathrm{CP}^{2v+1}) & \xrightarrow{i''} & J''(\mathrm{CP}^{2v}) \to 0 \\ & & \downarrow{\scriptstyle (\theta'\theta'')_{2v+1}} & & \downarrow{\scriptstyle (\theta'\theta'')_{2v}} \\ & & J'(\mathrm{CP}^{2v+1}) & \xrightarrow{i'} & J'(\mathrm{CP}^{2v}) \end{array}$$

The row is exact by Lemma 5·2, and the map $(\theta'\theta'')_{2v}$ is an isomorphism by Lemma 6·1. Hence $\ker(\theta'\theta'')_{2v+1} \subset \operatorname{im} j''$.

(i) If $v = 2w+1$, $\tilde{K}_R(\mathrm{CP}^{2v+1}/\mathrm{CP}^{2v}) = \tilde{K}_R(S^{8w+6}) = 0$. Hence $(\theta'\theta'')_{2v+1}$ is a monomorphism. It is known to be an epimorphism, so this proves (i).

(ii) If $v = 2w$, $\tilde{K}_R(\mathrm{CP}^{2v+1}/\mathrm{CP}^{2v}) = \tilde{K}_R(S^{8w+2}) = Z_2$, generated by $r\mu^{4w+1}$. Hence either $\ker(\theta'\theta'')_{2v+1}$ is 0 or it is Z_2 generated by $J''r\mu^{4w+1}$. In fact, $\theta''J''r\mu^{4w+1} = Jr\mu^{4w+1}$, and this is non-zero by Lemma 3·1. Hence θ'' is always an isomorphism.

It remains to calculate $J'_R(r\mu^{4w+1})$. We have

$$\operatorname{sh} r\mu^{4w+1} = 1,$$

so $J'_R(r\mu^{4w+1}) = 0$. This proves (ii).

Proof of Theorem 1·5. This theorem follows immediately from Lemmas 6·1 and 6·2.

COROLLARY 6·3. *The cofibring*

$$\mathrm{CP}^m \xrightarrow{i} \mathrm{CP}^l \xrightarrow{j} \mathrm{CP}^l/\mathrm{CP}^m \quad (l \geq m)$$

gives rise to an exact sequence

$$J(\mathrm{CP}^l/\mathrm{CP}^m) \xrightarrow{j^*} J(\mathrm{CP}^l) \xrightarrow{i^*} J(\mathrm{CP}^m) \to 0.$$

This follows immediately from Theorem 1·5 and Lemma 5·2. It establishes a conjecture of Atiyah and Todd.

LEMMA 6·4. *If* $w > 0$, *the order of* $Jr\eta$ *in* $J(\mathrm{CP}^{4w+1})$ *is* M_{4w+2}.

Proof. According to Lemmas 4·7 and 6·1, the orders of $Jr\eta$ in $J(\mathrm{CP}^{4w})$, $J(\mathrm{CP}^{4w+2})$ are M_{4w+1}, M_{4w+3}. Let us write

$$\alpha = Jr(M_{4w+1}\eta) \in J(\mathrm{CP}^{4w+2});$$

then the order of α is M_{4w+3}/M_{4w+1}.

Assuming $w > 0$, we shall show from the explicit definition of M_k given by Atiyah and Todd ((8), page 343) that M_{4w+3}/M_{4w+1} contains at most one power of 2. Following Atiyah and Todd, we define the function $\nu_p(n)$ so that the decomposition of n into prime powers is

$$n = 2^{\nu_2(n)} 3^{\nu_3(n)} 5^{\nu_5(n)} \ldots.$$

Then M_k is defined by

$$\nu_p(M_k) = \begin{cases} \max\,(r + \nu_p(r)), \quad 1 \leqslant r \leqslant \left[\dfrac{k-1}{p-1}\right] & \text{if } p \leqslant k \\ 0 & \text{if } p > k. \end{cases}$$

Thus we have

$$\nu_2(M_{4w+3}) = \max_{1 \leqslant s \leqslant 4w+2} (s + \nu_2(s)),$$

$$\nu_2(M_{4w+1}) = \max_{1 \leqslant s \leqslant 4w} (s + \nu_2(s)).$$

The second maximum contains a term with $s = 4w$, so

$$\nu_2(M_{4w+1}) \geqslant 4w + 2.$$

The only terms which enter into the first maximum but not into the second are those with $s = 4w+1$, $s = 4w+2$; these terms have

$$s + \nu_2(s) = 4w + 1,$$

$$s + \nu_2(s) = 4w + 3.$$

Therefore $\nu_2(M_{4w+3}) \leqslant \nu_2(M_{4w+1}) + 1$. We conclude that the order of α contains at most one power of 2.

The element α lies (by Corollary 6·3) in the subgroup $j^*J(CP^{4w+2}/CP^{4w})$. This subgroup of $J(CP^{4w+2})$ is cyclic of order $m(4w+2)$, by Lemmas 6·1, 4·9 and 4·10. The integer $m(4w+2)$ contains 2 to the power 3 exactly, by (2), section 2. Therefore α is divisible by 4 in the subgroup $j^*J(CP^{4w+2}/CP^{4w})$. Now apply the homomorphism induced by the inclusion

$$i : CP^{4w+1} \to CP^{4w+2};$$

we see that $Jr(M_{4w+1}\eta)$ is divisible by 4 in the subgroup $j^*J(CP^{4w+1}/CP^{4w})$ of $J(CP^{4w+1})$. But this subgroup is Z_2, according to our previous work. Therefore $Jr(M_{4w+1}\eta) = 0$ in $J(CP^{4w+1})$. Let the order of $Jr\eta$ in $J(CP^{4w+1})$ be M; then we have proved that M_{4w+1} is a multiple of M. Also M is a multiple of M_{4w+2}, since M_{4w+2} is the order of $J'_C\eta$ in $J'(CP^{4w+1})$; and M_{4w+2} is a multiple of M_{4w+1}, since M_{4w+1} is the order of $J'_C\eta$ in $J'(CP^{4w})$. Therefore $M = M_{4w+2} = M_{4w+1}$. This completes the proof.

Proof of Lemma 1·4. If $k-1 \not\equiv 1 \bmod 4$, the result follows immediately from Theorem 1·5 and Lemma 4·7. If $k-1 \equiv 1 \bmod 4$ and $k-1 > 1$, the result is given by Lemma 6·4.

With the proof of Lemma 1·4, the proof of Theorem 1·1 is complete.

APPENDIX

The treatment of the functor J'' given in section 5 is adequate for our needs in this paper. It is clear, however, that we can replace R by C in the definition of J'' and so obtain an analogous functor

$$J''_C(X) = K_C(X)/Z + W_C(X)$$

using the operations Ψ^k_C defined in (3). If we argue as for Lemma 5·1, we obtain the following lemma.

LEMMA A 1. *If $K_C(X)$ is generated additively by $U(1)$-bundles, then the homomorphism*

$$Jr: K_C(X) \rightarrow J(X)$$

factors to give the following commutative diagram.

$$\begin{array}{ccc} K_C(X) & \xrightarrow{J''_C} & J''_C(X) \\ {\scriptstyle r}\downarrow & & \downarrow{\scriptstyle \theta''_C} \\ K_R(X) & \xrightarrow{J} & J(X) \end{array}$$

This lemma states that for certain X, $J''_C(X)$ is an upper bound for $J(X)$ in an appropriate sense. By Lemma 2·1, it applies to $X = \mathrm{CP}^l$. We omit the details of the proof.

In this appendix we shall relate J''_C with J'' (which for consistency we now write as J''_R). In fact, we shall prove that $J''_C(\mathrm{CP}^l) = J''_R(\mathrm{CP}^l)$ for $l \not\equiv 3 \bmod 4$; it is for this reason that we did not need to introduce $J''_C(\mathrm{CP}^l)$ in the main paper. We shall also show that in general $J''_C(X)$ and $J''_R(X)$ differ only in their 2-primary components. (We recall from Theorem 3·11 of (2) that both groups are finite. Note that our group J''_Λ is written $\tilde{J}''_\Lambda$ in (2).)

Since $\Psi^k_C c = c\Psi^k_R$ ((3), Theorem 4·1 (iv)),

$$c: K_R(X) \rightarrow K_C(X)$$

induces a homomorphism $\qquad J''(c): J''_R(X) \rightarrow J''_C(X).$

The following lemma shows that

$$r: K_C(X) \rightarrow K_R(X)$$

induces a homomorphism $\qquad J''(r): J''_C(X) \rightarrow J''_R(X).$

LEMMA A 2. *The following diagram is commutative.*

$$\begin{array}{ccc} K_C(X) & \xrightarrow{r} & K_R(X) \\ {\scriptstyle \Psi^k_C}\downarrow & & \downarrow{\scriptstyle \Psi^k_R} \\ K_C(X) & \xrightarrow{r} & K_R(X) \end{array}$$

This lemma states that the operations Ψ^k_Λ commute with r (as well as with c). It should obviously have been published in (3): unfortunately the author of that paper omitted to think of it at the right time.

Proof of Lemma A 2. In the complex representation ring of $U(n)$ we have

$$\begin{aligned} cr\Psi^k_C &= (1+t)\Psi^k_C \\ &= (1+\Psi^{-1}_C)\Psi^k_C \\ &= \Psi^k_C + \Psi^{-k}_C \\ &= \Psi^k_C(1+\Psi^{-1}_C) \\ &= \Psi^k_C cr \\ &= c\Psi^k_R r, \end{aligned}$$

the steps being justified by (3), Theorems 5·1 and 4·1 (iv). But c is a monomorphism on the representation ring; therefore

$$r\Psi^k_C = \Psi^k_R r$$

as elements of the real representation ring.

If ξ is a $U(n)$-bundle over X, we therefore have in $K_R(X)$ the equation

$$\Psi^k_R r\xi = r\Psi^k_C \xi.$$

By linearity, the same equation holds if ξ is replaced by any element of $K_C(X)$. This completes the proof.

The following lemma shows that $J''_R(X)$ and $J''_C(X)$ coincide but for their 2-components.

LEMMA A 3. (i) *For all* X, $J''(r)$ *and* $J''(c)$ *are isomorphisms modulo the class of 2-primary finite Abelian groups.*

(ii) *If the sequence*

$$\tilde{K}_C(X) \xrightarrow{1-t} \tilde{K}_C(X) \xrightarrow{r} \tilde{K}_R(X) \to 0$$

is exact, then $J''(r)$ *is an isomorphism.*

Remark. According to Theorem 2·2 (vi), (vii), Part (ii) of this lemma applies to $X = CP^l/CP^m$ provided $l \not\equiv 3 \bmod 4$, $m \not\equiv 3 \bmod 4$.

Proof. The homomorphism

$$1-t: K_C(X) \to K_C(X)$$

commutes with the operations Ψ^k_C, as was shown in the course of the proof of Lemma A 2. Hence it induces a homomorphism

$$J''(1-t): J''_C(X) \to J''_C(X).$$

But for any element x in $K_C(X)$, $(1-t)x \in W_C(X)$. Hence $J''(1-t) = 0$.

To prove (i), we use the relations

$$rc = 2,$$
$$cr = 1+t = 2-(1-t).$$

Applying J'', we find

$$J''(r)J''(c) = 2,$$
$$J''(c)J''(r) = 2-J''(1-t).$$

But $J''(1-t) = 0$, so $J''(c)J''(r) = 2$. Thus the kernel and co-kernel of $J''(r)$ and $J''(c)$ are groups of exponent 2. This proves the assertion of (i) (and more).

To prove (ii), we observe that Lemma 3·8 of (2) can be applied to the exact sequence

$$\tilde{K}_C(X) \xrightarrow{1-t} \tilde{K}_C(X) \xrightarrow{r} \tilde{K}_R(X) \to 0.$$

Hence the sequence $$J''_C(X) \xrightarrow{J''(1-t)} J''_C(X) \xrightarrow{J''(r)} J''_R(X) \to 0$$

is also exact. Since $J''(1-t) = 0$, $J''(r)$ is an isomorphism. This completes the proof.

The second author wishes to thank the Carnegie Trust for the Universities of Scotland for a scholarship held during the period of this research.

REFERENCES

(1) ADAMS, J. F. On the groups $J(X)$. I. *Topology* **2** (1963), 181–195.

(2) ADAMS, J. F. On the groups $J(X)$. II. *Topology* (to appear).

(3) ADAMS, J. F. Vector fields on spheres. *Ann. of Math.* **75** (1962), 603–632.

(4) ADAMS, J. F. Applications of the Grothendieck–Atiyah–Hirzebruch functor $K(X)$, from *Colloquium on algebraic topology* (mimeographed notes; Aarhus, 1962).

(5) ADAMS, J. F. Applications of the Grothendieck–Atiyah–Hirzebruch functor $K(X)$, from *Proceedings of the international congress of mathematicians* (Stockholm, 1962).

(6) ATIYAH, M. F. Thom complexes. *Proc. London Math. Soc.* (3) **11** (1961), 291–310.

(7) ATIYAH, M. F. and HIRZEBRUCH, F. Vector bundles and homogeneous spaces. Proceedings of Symposia in Pure Mathematics 3, Differential Geometry. *Amer. Math. Soc.* (1961), pp. 7–38.

(8) ATIYAH, M. F. and TODD, J. A. On complex Stiefel manifolds. *Proc. Cambridge Philos. Soc.* **56** (1960), 342–353.

(9) BOTT, R. The stable homotopy of the classical groups. *Ann. of Math.* **70** (1959), 313–337.

(10) BOTT, R. Quelques remarques sur les théorèmes de périodicité. *Bull. Soc. Math. France* **87** (1959), 293–310.

(11) BOTT, R. A note on the *KO*-theory of sphere bundles. *Bull. Amer. Math. Soc.* **68** (1962), 395–400.

(12) BOTT, R. *Lectures on* $K(X)$ (mimeographed notes; Harvard, 1962).

(13) HIRZEBRUCH, F. *Neue topologische Methoden in der algebraischen Geometrie* (Springer: Berlin, 1956).

(14) JAMES, I. M. The intrinsic join: a study of the homotopy groups of Stiefel manifolds. *Proc. London Math. Soc.* (3) **8** (1958), 507–535.

(15) JAMES, I. M. Cross-sections of Stiefel manifolds. *Proc. London Math. Soc.* (3) **8** (1958), 536–547.

(16) SANDERSON, B. J. Immersions and embeddings of projective spaces. *Proc. London Math. Soc.* (3) **14** (1964), 137–153.

(17) THOM, R. Espaces fibrés en sphères et carrés de Steenrod. *Ann. Sci. École Norm. Sup.* **69** (1952), 109–182.

ON MATRICES WHOSE REAL LINEAR COMBINATIONS ARE NONSINGULAR

J. F. ADAMS, PETER D. LAX[1] AND RALPH S. PHILLIPS[2]

Let Λ be either the real field R, or the complex field C, or the skew field Q of quaternions. Let $A_1, A_2, \cdots, A_k$ be $n\times n$ matrices with entries from Λ. Consider a typical linear combination $\sum_{j=1}^{n} \lambda_j A_j$ with real coefficients λ_j; we shall say that the set $\{A_j\}$ "has the property P" if such a linear combination is nonsingular (invertible) except when all the coefficients λ_j are zero.

We shall write $\Lambda(n)$ for the maximum number of such matrices which form a set with the property P. We shall write $\Lambda_H(n)$ for the maximum number of Hermitian matrices which form a set with the property P. (Here, if $\Lambda=R$, the word "Hermitian" merely means "symmetric"; if $\Lambda=Q$ it is defined using the usual conjugation in Q.) Our aim is to determine the numbers $\Lambda(n)$, $\Lambda_H(n)$.

Of course, it is possible to word the problem more invariantly. Let W be a set of matrices which is a vector space of dimension k over R; we will say that W "has the property P" if every nonzero w in W is nonsingular (invertible). We now ask for the maximum possible dimension of such a space.

In [1], the first named author has proved that $R(n)$ equals the so-called Radon-Hurwitz function, defined below. In this note we determine $R_H(n)$, $C(n)$, $C_H(n)$, $Q(n)$ and $Q_H(n)$ by deriving inequalities between them and $R(n)$. The elementary constructions needed to prove these inequalities can also be used to give a simplified description of the Radon-Hurwitz matrices.

The study of sets of real symmetric matrices $\{A_j\}$ with the property P may be motivated as follows. For such a set, the system of partial differential equations

$$u_t = \sum_j A_j u_{x_j}$$

is a symmetric hyperbolic system in which the sound speeds are nonzero in every direction. For such systems the solution energy is propagated to infinity and a scattering theory can be developed.

To give our results, we require the Radon-Hurwitz numbers [2],

Received by the editors September 10, 1963.

[1] Sloan Fellow.

[2] Sponsored by the National Science Foundation, contract NSF-G 16434.

[3]. We set $n=(2a+1)2^b$ and $b=c+4d$, where a, b, c, d are integers with $0 \leqq c < 4$; then we define

$$\rho(n) = 2^c + 8d.$$

THEOREM 1. *We have*

$$R(n) = \rho(n), \qquad R_H(n) = \rho(\tfrac{1}{2}n) + 1,$$
$$C(n) = 2b + 2, \qquad C_H(n) = 2b + 1,$$
$$Q(n) = 2b + 4, \qquad Q_H(n) = 2b + 1.$$

The results for $\Lambda = Q$ are included so that topologists may avoid jumping to the conclusion that the subject is directly related to the Bott periodicity theorems. If this were so then it would be surprising to see the case $\Lambda = Q$ behaving like the case $\Lambda = C$.

The proof of Theorem 1 will be based on a number of simple constructions, which we record as lemmas.

LEMMA 1. $R_H(n) \leqq C_H(n) \leqq Q_H(n)$.

This is clear, since a matrix with entries from R may be regarded as a matrix with entries from C, and similarly for C and Q.

LEMMA 2. (a) $C(n) \leqq R(2n)$, (b) $Q(n) \leqq C(2n)$.

PROOF. We may regard our matrices as Λ-linear transformations of coordinate n-space Λ^n. Now by forgetting part of the structure of C^n it becomes a real vector space of dimension $2n$ over R, i.e., an R^{2n}. Thus any C-linear transformation of C^n gives an R-linear transformation of R^{2n}. Similarly for C and Q.

LEMMA 3. $\Lambda(n) + 1 \leqq \Lambda_H(2n)$.

PROOF. Let W be a k-dimensional space of $n \times n$ matrices with entries from Λ which has the property P. For each $A \in W$ and $\lambda \in R$, consider the following linear transformation from $\Lambda^n \oplus \Lambda^n$ to itself.

$$B(x, y) = (Ay + \lambda x, A^*x - \lambda y).$$

It is clear that its matrix is Hermitian, and that such B form a $(k+1)$-dimensional space. We claim that this set $\{B\}$ has property P. For suppose that some B is singular; then there exist x, y not both zero such that

$$Ay + \lambda x = 0, \qquad A^*x - \lambda y = 0.$$

Evaluating x^*Ay in two ways, we find

$$\lambda(x^*x + y^*y) = 0.$$

Hence $\lambda = 0$. Thus either A or A^* is singular; so A is singular and $A = 0$. This proves the lemma.

LEMMA 4. (a) $C_H(n)+1 \leqq C(n)$, (b) $Q_H(n)+3 \leqq Q(n)$.

PROOF. Let W be a k-dimensional space of $n \times n$ Hermitian matrices with entries from Λ which has the property P. Consider the matrices

$$A + \mu I,$$

where A runs over W and μ runs over the pure imaginary elements of Λ. We claim that they form a space with the property P and of dimension $k+1$ if $\Lambda = C$ or $k+3$ if $\Lambda = Q$. In fact, suppose that such a matrix is singular; then there is a nonzero x such that

$$Ax = -\mu x;$$

arguing as is usual for the complex case, we find

$$-\mu x^*x = x^*Ax = (-\mu x)^*x = \mu x^*x.$$

So μ is zero, A is singular, and thus A is zero. This proves the lemma.

LEMMA 5. $R_H(n)+7 \leqq R(8n)$.

PROOF. Let W be a k-dimensional space of real symmetric matrices with the property P. We require also the Cayley numbers K, which form an 8-dimensional algebra over R. We can thus form the real vector space

$$R^n \otimes_R K$$

of dimension $8n$. For each $A \in W$ and each pure imaginary $\mu \in K$ we consider the following linear transformation from $R^n \otimes_R K$ to itself:

$$B(x \otimes y) = Ax \otimes y + x \otimes \mu y.$$

We claim that the $(k+7)$-dimensional space formed by such B has property P. For suppose that some B is singular, and suppose, to begin with, that μ is nonzero. Then the elements 1, μ form an R-base for a sub-algebra of K which we may identify with C. Now every two elements of K generate an associative sub-algebra; in particular, K is a left vector space over C. Choose a C-base of K; this splits $R^n \otimes_R K$ as the direct sum of 4 copies of $R^n \otimes_R C$. Since B acts on each summand, it must be singular on at least one. That is, the real symmetric matrix A has a nonzero complex eigenvalue which is purely imaginary—a contradiction. Hence μ must be zero and $B=A\otimes 1$. Now choose an R-base of K; this splits $R^n\otimes_R K$ as the direct sum of 8 copies of R^n. Since B acts on each summand, it must be singular on at least one. That is, A must be singular; hence $A=0$. This completes the proof.

PROOF OF THEOREM 1. First we consider $R_H(n)$. If we use the fact that $R(n)=\rho(n)$, Lemmas 3 and 5 give

$$\rho(\tfrac{1}{2}n) + 1 \leqq R_H(n) \leqq \rho(8n) - 7.$$

But using the explicit definition of ρ, we have

$$\rho(8n) - 7 = \rho(\tfrac{1}{2}n) + 1.$$

This disposes of $R_H(n)$.

It follows from this argument that if we have a set of $\rho(n)$ $n\times n$ matrices with the property P, then by applying successively the constructions given in the proofs of Lemma 3 (taking $\Lambda=R$) and Lemma 5, we obtain a set of $\rho(2^4n)$ $2^4n\times 2^4n$ matrices with the property P. Now the set of 1, 2, 4, and 8 matrices which express the respective actions of R, C, Q, and K on R^m, C^m, Q^m, and K^m for $m=2a+1$ can be used to start the induction for the different cases $b\equiv 0$, 1, 2, and 3 (mod 4). This gives a slight variation of the construction of Hurwitz and Radon [2], [3]; the iterative procedures used by these authors require more steps and do not involve the Cayley numbers explicitly.

Next we consider $C(n)$. Lemmas 3 and 4(a) give

$$C(n) + 2 \leqq C(2n).$$

Now induction shows that

$$C(n) \geqq 2b + 2,$$

which gives us our inequality one way. Applying Lemma 2(a) directly only gives a good inequality for certain values of b, so we proceed as follows. Choose $e\geqq 0$ such that $b+e\equiv 0$ or 1 (mod 4). Then by induction we have

$$C(n) + 2e \leqq C(2^e n);$$

by Lemma 2(a) we have

$$C(2^e n) \leqq \rho(2^{e+1}n),$$

and by our choice of e we have

$$\rho(2^{e+1}n) = 2b + 2e + 2.$$

This gives

$$C(n) \leqq 2b + 2,$$

and proves the assertion made about $C(n)$.

Lemmas 3 and 4(a) now show that

$$C_H(n) = 2b + 1.$$

Again, Lemmas 1, 4(b) and 2(b) show that

$$Q_H(n) - C_H(n) \geqq 0,$$
$$Q(n) - Q_H(n) \geqq 3,$$
$$C(2n) - Q(n) \geqq 0.$$

But $C(2n) - C_H(n) = 3$, so all these inequalities are equalities. This completes the proof of Theorem 1.

References

1. J. F. Adams, *Vector fields on spheres*, Ann. of Math. (2) **75** (1962), 603–632.
2. A. Hurwitz, *Über die Komposition der quadratischen Formen*, Math. Ann. **88** (1923), 1–25.
3. J. Radon, *Lineare Scharen orthogonalen Matrizen*, Abh. Math. Sem. Univ. Hamburg **1** (1922), 1–14.

MANCHESTER UNIVERSITY, MANCHESTER, ENGLAND,
NEW YORK UNIVERSITY, AND
STANFORD UNIVERSITY

CORRECTION TO "ON MATRICES WHOSE REAL LINEAR COMBINATIONS ARE NONSINGULAR"

J. F. ADAMS, PETER D. LAX AND RALPH S. PHILLIPS

We are grateful to Professor B. Eckmann for pointing out an error in the proof of Lemma 4(b) of our paper [1]. This error invalidates Lemma 4(b) and that part of Theorem 1 which states the values of $Q(n)$, $Q_H(n)$. The error occurs immediately after the words "arguing as is usual for the complex case, we find"; it consists in manipulating as if the ground field Λ were commutative.

The proof of Lemma 4(b) can be repaired, as will be shown below, but it leads to a different conclusion from that given. Our paper should therefore be corrected as follows.

(i) In Theorem 1, the values of $Q(n)$ and $Q_H(n)$ should read

$$\text{"}Q(n) = \rho(\tfrac{1}{2}n) + 4, \qquad Q_H(n) = \rho(\tfrac{1}{4}n) + 5.\text{"}$$

The two sentence paragraph following Theorem 1 should be deleted. It remains interesting to ask what topological phenomena (if any) can be related to our algebraic results.

(ii) In Lemma 4, part (b) should read

$$\text{"}Q_H(n) + 3 \leqq R(4n).\text{"}$$

The proof is as follows.

Let W be a k-dimensional space of $n \times n$ Hermitian matrices with entries from Q which has the property P. The space Q^n is a real vector space of dimension $4n$. For each $A \in W$ and each pure imaginary $\mu \in Q$ we consider the following real-linear transformation from Q^n to itself:

Received by the editors December 20, 1965.

$$B(x) = Ax + x\mu.$$

We claim that the $(k+3)$-dimensional space formed by such B has the property P. For suppose that such a B is singular; then there is a nonzero x such that

$$Ax = -x\mu;$$

then we have

$$x^*(Ax) = -x^*x\mu,$$
$$(x^*A)x = (-x\mu)^*x = \mu x^*x.$$

Since x^*x is real and nonzero, we have $\mu=0$; hence A is singular and $A=0$. This completes the proof.

(iii) In Lemma 5, there should be added a second part, reading

"(b) $R_H(n) + 3 \leqq Q(n)$."

PROOF. Let W be a k-dimensional space of $n\times n$ real symmetric matrices which has the property P. Consider the matrices

$$A+\mu I,$$

where A runs over W and μ runs over the pure imaginary elements of Q. We claim that they form a space of dimension $k+3$ with the property P. In fact, suppose that such a matrix is singular; and suppose to begin with, that μ is nonzero. Then the elements $1, \mu$ form an R-base for a subalgebra of Q which we may identify with C. Choose a C-base of Q; this splits Q^n as the direct sum of two copies of C^n. Since the matrix $A+\mu I$ acts on each summand, it must be singular on at least one. That is, the real symmetric matrix A has a nonzero complex eigenvalue which is purely imaginary, a contradiction. Hence μ must be zero and $B=A$. Now choose an R-base of Q; this splits Q^n as the direct sum of 4 copies of R^n. Sinc A acts on each summand, it must be singular on at least one. That is, A must be singular; hence $A=0$. This completes the proof.

(iv) The final paragraph of the paper should be deleted, and replaced by the following proof.

"Finally, Lemmas 5(b), 3 and 4(b) show that

$$Q(n) - R_H(n) \geqq 3,$$
$$Q_H(2n) - Q(n) \geqq 1,$$
$$R(8n) - Q_H(2n) \geqq 3.$$

But we have already shown that

$$R(8n) - R_H(n) = 7,$$

so all these inequalities are equalities. This completes the proof of Theorem 1."

We note that this method provides an alternative proof of Lemma 5 ($R(8n)-R_H(n) \geqq 7$), without using the Cayley numbers.

Bibliography

1. J. F. Adams, Peter D. Lax and Ralph S. Phillips, *On matrices whose real linear combinations are nonsingular*, Proc. Amer. Math. Soc. **16** (1965), 318–322.

Manchester University, Manchester, England
New York University, and
Stanford University

Topology Vol. 2, pp. 181–195. Pergamon Press, 1963.

ON THE GROUPS $J(X)$—I

J. F. ADAMS

(*Received* 29 *May* 1963)

§1. INTRODUCTION

ATIYAH [6] has defined certain groups, which he has called $J(X)$. For our purposes, we shall define the groups $J(X)$ as follows. Let X be a good space, for example, a finite-dimensional CW-complex. Let $K_R(X)$ be the Grothendieck–Atiyah–Hirzebruch group [7, 8, 1] defined in terms of real vector bundles over X. Let $T(X)$ be the subgroup of $K_R(X)$ generated by elements of the form $\{\xi\} - \{\eta\}$, where ξ and η are orthogonal bundles whose associated sphere-bundles are fibre homotopy equivalent. (We think of $T(X)$ as the subgroup of fibre-homotopy-trivial virtual bundles.) We define

$$J(X) = K_R(X)/T(X).$$

If X is connected we have

$$K_R(X) = Z + \tilde{K}_R(X),$$

where $\tilde{K}_R(X)$ denotes the subgroup of virtual bundles whose virtual dimension is zero. We have $T(X) \subset \tilde{K}_R(X)$, so we may define

$$\tilde{J}(X) = \tilde{K}_R(X)/T(X).$$

We then have

$$J(X) = Z + \tilde{J}(X).$$

It is not hard to see that the group which we call $\tilde{J}(X)$ is isomorphic to that which Atiyah originally introduced and called $J(X)$ [6]. It was natural for Atiyah to concentrate on $\tilde{J}(X)$, since the summand Z is not interesting, and since Atiyah's theorem that $\tilde{J}(X)$ is finite [6, Proposition (1.5)] would not be true for $J(X)$.

Atiyah has also shown that the groups $J(X)$ have useful applications. If we take X to be a projective space (either real, complex or quaternionic) then the resulting group $J(X)$ holds the answer to classical questions about the existence of cross-sections of appropriate Stiefel fiberings [6, Theorem (6.5)]. If we take X to be a sphere, then the resulting group $\tilde{J}(X)$ is (up to isomorphism) the image of the classical J-homomorphism in an appropriate dimension [6, Proposition (1.4)]. It would therefore be amply worth-while to give means for computing the groups $J(X)$. The present series of four papers represents a start in this direction.

We shall attempt to compute the group $J(X)$ by introducing two further groups $J'(X)$, $J''(X)$. For the moment we need only emphasise three points about these groups.

(i) These groups are defined as quotients of $K_R(X)$; that is, we shall give definitions of the form

$$J'(X) = K_R(X)/V(X)$$

$$J''(X) = K_R(X)/W(X).$$

(ii) The groups $J'(X)$, $J''(X)$ are computable.

(iii) The group $J'(X)$ is intended to serve as a lower bound for $J(X)$, and the group $J''(X)$ is intended to serve as an upper bound for $J(X)$, in a sense which we will now explain.

We shall say that "$J''(X)$ is a lower bound for $J(X)$" if $T(X) \subset V(X)$, so that the quotient map $K_R(X) \to J'(X)$ factors through an epimorphism $J(X) \to J'(X)$. We shall prove that this is so for all X.

We shall say that "$J''(X)$ is an upper bound for $J(X)$" if $W(X) \subset T(X)$, so that the quotient map $K_R(X) \to J(X)$ factors through an epimorphism $J''(X) \to J(X)$. It is plausible to conjecture that this is so for all X; but so far we can prove this only in favourable cases, for example, $X = RP^n$ (real projective space), $X = CP^n$ (complex projective space), and $X = S^m$ with $m \not\equiv 0 \bmod 8$.

In such favourable cases we can proceed to compute the groups $J'(X)$, $J''(X)$; and if we find that the quotient map $J''(X) \to J'(X)$ is an isomorphism, then the group $J(X)$ is completely determined, being isomorphic to both $J'(X)$ and $J''(X)$.

We will now try to explain that the groups $J'(X)$, $J''(X)$ merely formalise two reasonable methods of attacking our problem. Let us start with the first. We shall sometimes wish to show that two bundles ξ, η represent different elements of $J(X)$. This is the sort of problem which one usually attacks by introducing suitable invariants. For example, the theory of characteristic classes sometimes allows one to prove that two bundles ξ, η are not fibre homotopy equivalent. This method has been pressed further by Atiyah (private communications; cf. [6] p. 291, lines 14, 15; p. 309, lines 6, 7) and Bott [9, 10]. Instead of characteristic classes with values in the ordinary cohomology groups $H^*(X; G)$, they use characteristic classes with values in the extraordinary cohomology groups $K_\Lambda(X)$. By using the best techniques available in this direction, one defines the group $J'(X)$; if two bundles ξ, η have different images in $J'(X)$, then they represent different elements of $J(X)$.

The group $J'(X)$, then, is essentially due to Atiyah and Bott. In particular, the notation $J'(X)$ is taken from unpublished work of Atiyah; it originally stood for a somewhat cruder lower-bound group. We adopt the notation $J''(X)$ by analogy with $J'(X)$.

Let us now turn to the second method of attack. We shall sometimes wish to show that two bundles ξ, η represent the same element of $J(X)$, although they represent different elements of $K_R(X)$. This is the sort of problem which one usually attacks by giving geometrical constructions. In this direction we offer Theorem (1.1) below.

Let ξ, η be sphere-bundles over a finite CW-complex X, with total spaces E_ξ, E_η and projections p_ξ, p_η. By a 'fibrewise map $f: E_\xi \to E_\eta$', we shall mean a map f such that the following diagram is commutative:

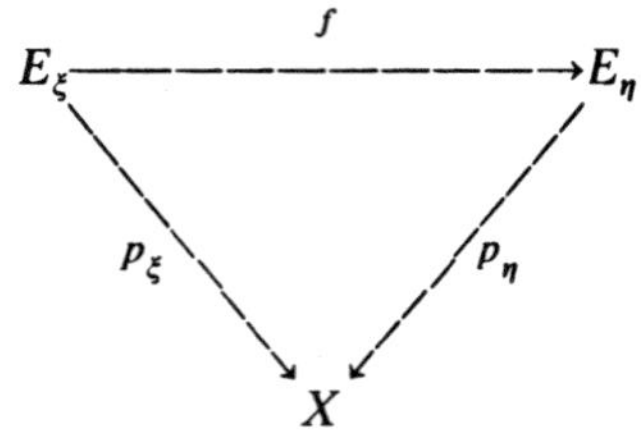

Let k be a positive integer.

THEOREM (1.1). *If there is a fibrewise map $f: E_\xi \to E_\eta$ of degree $\pm k$ on each fibre, then there exists a non-negative integer e such that the Whitney multiples $k^e\xi$, $k^e\eta$ are fibre homotopy equivalent.*

If we put $k = 1$ this is a theorem of Dold [11]. Therefore one may regard this theorem as a 'mod k' analogue of Dold's theorem.

By using Theorem (1.1), one can prove certain cases of the following conjecture (in which the operation Ψ^k is as in [1]).

CONJECTURE (1.2). *If k is an integer, X is a finite CW-complex and $y \in K_R(X)$, then there exists a non-negative integer $e = e(k, y)$ such that $k^e(\Psi^k - 1)y$ maps to zero in $J(X)$.*

More precisely, we shall prove the following cases of Conjecture (1.2).

THEOREM (1.3). *Assume that X is a finite CW-complex and that y is a linear combination of* O(1) *and* O(2) *bundles. Then there exists $e = e(k, y)$ such that $k^e(\Psi^k - 1)y$ maps to zero in $J(X)$.*

THEOREM (1.4). *Assume that X is a sphere S^{2n} and that y lies in the image of*

$$r: K_C(S^{2n}) \to K_R(S^{2n}).$$

Then there exists $e = e(k, y)$ such that $k^e(\Psi^k - 1)y$ maps to zero in $J(X)$.

In Part II we shall see that Theorem (1.4) leads to the result on the J-homomorphism which was announced in [2, Theorem (3); 3, Theorem (3)].

The definition of the group $J''(X)$ will be arranged so that if Conjecture (1.2) is true for all k and all y in $K_R(X)$, then $J''(X)$ is an upper bound for $J(X)$.

The arrangement of the present series of papers is as follows. The main object of Part I is to prove Theorem (1.1), the 'mod k Dold theorem'. We shall also prove Theorems (1.3) and (1.4). Parts II and III are devoted to a systematic account of the groups $J'(X)$ and $J''(X)$. In Part IV we shall apply the methods of K-theory to study the homotopy groups of spheres. Here we are concerned not only with the image of the J-homomorphism; we apply the methods of K-theory to give invariants defined for every homotopy class in the appropriate homotopy group.

A separate paper with G. Walker [5] will study the case $X = CP^n$. This paper depends essentially on Parts I and II of this series, but is independent of Parts III and IV.

The results of Parts I, II, III and [5] have been summarised elsewhere [2, 3, 4]. The results of Part IV have so far appeared only in lectures and mimeographed form, but there is some overlap with work of E. Dyer [13].

The present paper is arranged as follows. Theorem (1.1) is proved in §3; it depends on lemmas proved in §2. Theorems (1.3) and (1.4) are proved in §4.

§2. FUNCTION SPACES

Dold's theorem is proved [11] by using the topological monoid $H(n)$ of homotopy equivalences from S^{n-1} to S^{n-1}; the key idea is to take $H(n)$ seriously as a 'structural group'. This idea was developed further by Dold and Lashof [12].

We shall prove our 'mod k' analogue of Dold's theorem by using the space $G(n)$ of all maps from S^{n-1} to S^{n-1} (of whatever degree.) We write $G(n, k)$ for the component of $G(n)$ which consists of maps of degree k. These spaces are to be given the compact-open topology.

Various maps can be defined on the spaces $G(n)$, and in §3 we shall need to quote lemmas about the effect of these maps on the homotopy groups of $G(n)$. It is the object of this section to supply these results, which are stated as Lemmas (2.1) and (2.4).

We recall that if $r \leq n - 3$, the homotopy group $\pi_r(G(n, k))$ can be identified with the stable homotopy group π_r^S of the r-stem. We will give details of the identification below; it is hoped that these details will remove any doubt about the sign-conventions employed; we follow 'homology' conventions. The conventions appear more natural if one writes maps of spaces on the right of their arguments, and we therefore do so throughout this section.

If we are given a sphere map $\alpha : S^r \to G(n, k)$ or a homotopy class $\beta \in \pi_r(G(n, k))$, we shall write $[\alpha]$, or $[\beta]$, for the corresponding element in π_r^S (assuming always that $r \leq n - 3$).

Our identification of $\pi_r(G(n, k))$ with π_r^S passes through various intermediate groups, and we make the same convention for these groups; if γ is an element of one of these intermediate groups, we shall write $[\gamma]$ for the corresponding element in π_r^S.

We now discuss the various maps defined on $G(n)$.

If $g \in G(n)$, we shall define $\bar{g} : G(n) \to G(n)$ by composition with g, so that

$$(x)((f)\bar{g}) = ((x)f)g \qquad (x \in S^{n-1}, f \in G(n).)$$

If $g \in G(n, t)$, then $\bar{g}$ maps $G(n, s)$ into $G(n, st)$. Our first lemma describes the induced homomorphism $\bar{g}_*$ of homotopy groups.

LEMMA (2.1). *If $\alpha \in \pi_r(G(n, s))$, $g \in G(n, t)$ and $r \leq n - 3$, then*

$$[\bar{g}_*\alpha] = t[\alpha].$$

The join product

$$j : G(n) \times G(m) \to G(n + m)$$

is defined by $j(f, g) = f * g$, where S^{n+m-1} is regarded as the join $S^{n-1} * S^{m-1}$ of S^{n-1} and

S^{m-1}. The product j maps $G(n, s) \times G(m, t)$ into $G(n+m, st)$. Our second lemma describes the induced homomorphism j_*. For this purpose we identify $\pi_r(G(n, s) \times G(m, t))$ with the direct sum $\pi_r(G(n, s)) + \pi_r(G(m, t))$, as usual.

LEMMA (2.2). *If $\alpha \in \pi_r(G(n, s))$, $\beta \in \pi_r(G(m, t))$ and $r \leq min(n-3, m-3)$, then*

$$[j_*(\alpha + \beta)] = t[\alpha] + s[\beta].$$

The iterated join product

$$j^{(v)}: G(n_1) \times G(n_2) \times \ldots \times G(n_v) \to G\left(\sum_{i=1}^{v} n_i\right)$$

is defined by

$$j^{(v)}(f_1, f_2, \ldots, f_v) = f_1 * f_2 * \cdots * f_v.$$

Our third lemma describes the induced homomorphism $j^{(v)}{}_*$. For this purpose we identify $\pi_r(G(n_1, s_1) \times G(n_2, s_2) \times \ldots \times G(n_v, s_v))$ with the direct sum

$$\sum_{i=1}^{v} \pi_r(G(n_i, s_i)),$$

as usual.

LEMMA (2.3). *If $\alpha_i \in \pi_r(G(n_i, s_i))$ for $1 \leq i \leq v$ and $r \leq \mathrm{Min}(n_i - 3)$, then*

$$\left[j^{(v)}{}_*\left(\sum_{i=1}^{v} \alpha_i\right)\right] = \sum_{i=1}^{v} \frac{s_1 s_2 \ldots s_v}{s_i}[\alpha_i].$$

This follows immediately from Lemma (2.2), by induction over v.

LEMMA (2.4). *Let $\alpha : S^r \to G(n, s)$ be a sphere map, and define $\beta : S^r \to G(nv, s^v)$ by*

$$\beta(x) = \alpha(x) * \alpha(x) * \cdots * \alpha(x) \qquad (v \text{ factors}).$$

Then

$$[\beta] = vs^{v-1}[\alpha].$$

This follows immediately from Lemma (2.3). In fact, let

$$\gamma : S^r \to \mathop{X}_{i=1}^{v} G(n, s)$$

be the map all of whose components are α; then β is just the composite

$$S^r \xrightarrow{\gamma} \mathop{X}_{i=1}^{v} G(m, s) \xrightarrow{j^{(v)}} G(nv, s^v).$$

The reader is now warned that the rest of this section consists of routine homotopy theory designed to establish Lemmas (2.1) and (2.2); if these lemmas are found credible, the rest of this section may be omitted.

We now proceed to give the identification of $\pi_r(G(n, k))$ with π_r^S. Following [15], we first define $F(n)$ to be the subspace of $G(n)$ which consists of maps leaving the base-point fixed. Similarly, we define $F(n, k) = F(n) \cap G(n, k)$. We have an obvious fibering

$$F(n, k) \longrightarrow G(n, k) \longrightarrow S^{n-1},$$

so

$$i_* : \pi_r(F(n, k)) \longrightarrow \pi_r(G(n, k))$$

is an isomorphism for $r \leq n - 3$.

The space $F(n)$ is an H-space, under the following multiplication. We choose a fixed map $\phi : S^{n-1} \to S^{n-1} \vee S^{n-1}$ of type (1, 1) (preserving the base-points); we use this to define the product $\phi(f \vee g)$ of any two maps f, g in $F(n)$. Since the map ϕ is determined up to a homotopy (if $n \geq 3$), the product map $\phi^* : F(n) \times F(n) \to F(n)$ is determined up to a homotopy. The product ϕ^* is homotopy-associative and homotopy-commutative (assuming $n \geq 3$), and has a homotopy-unit.

The product ϕ^* maps $F(n, s) \times F(n, t)$ into $F(n, s + t)$. Thus the arcwise-components of $F(n)$ form a group under the product ϕ^* (namely the group Z). It follows that the arcwise-components $F(n, s)$ are r-simple for each r (so that the choice of base-points for their homotopy groups is immaterial); moreover, the homotopy groups of the various arcwise-components may be identified, using left or right translations. Since the product is homotopy-commutative (assuming $n \geq 3$) it is immaterial whether we use left translations or right translations.

We may identify the space $F(n, 0)$ with $\Omega^{n-1}(S^{n-1})$, and so identify $\pi_r(F(n, 0))$ with $\pi_{n-1+r}(S^{n-1})$. We give this identification explicitly. Let $S^p \times S^q$ denote the reduced product $S^p \times S^q / S^p \vee S^q$. Suppose given a map

$$h : S^r, e \longrightarrow F(n, 0), \omega$$

where e is the base-point in S^r and ω is the constant map at the base-point. Then we define the corresponding map

$$h' : S^{n-1} \times S^r \longrightarrow S^{n-1}$$

by the following formula:

$$(x, y)h' = (x)((y)h) \qquad (x \in S^{n-1}, y \in S^r).$$

If $r \leq n - 3$, we may identify $\pi_{n-1+r}(S^{n-1})$ with π_r^S. For this purpose it only remains to indicate our sign-convention for suspension. We define the suspension of $g : S^p \to S^q$ to be $1 \times g : S^1 \times S^p \to S^1 \times S^q$.

We now return to the proof of Lemmas (2.1), (2.2). We begin by replacing the spaces $G(n, s)$ by spaces $F(n, s)$. In fact, if we alter g inside $G(n, t)$, then we alter $\bar{g}$ by a homotopy, and do not alter $\bar{g}_*$; we may therefore suppose $g \in F(n, t)$, so that $\bar{g}$ maps $F(n)$ into $F(n)$. Similarly, the join product j maps $F(n, s) \times F(m, t)$ into $F(n + m, st)$, provided that we take the base-point in $S^{n-1} * S^{m-1}$ somewhere on the segment joining the base-points in S^{n-1} and S^{m-1}. Lemmas (2.1) and (2.2) will therefore follow from the following results.

LEMMA (2.5). *If* $\alpha \in \pi_r(F(n, s))$, $g \in F(n, t)$ *and* $r \leq n - 3$, *then*

$$[\bar{g}_* \alpha] = t[\alpha].$$

LEMMA (2.6). *If* $\alpha \in \pi_r(F(n, s))$, $\beta \in \pi_r(F(m, t))$ *and* $r \leq \mathrm{Min}(n - 3, m - 3)$, *then*

$$[j_*(\alpha + \beta)] = t[\alpha] + s[\beta].$$

We begin with Lemma (2.5). This will evidently follow from the following result.

LEMMA (2.7). (i) *If $g \in F(n, t)$ we have a commutative diagram of the following form, in which i_s, i_{st} are the identifications made earlier in this section*:

$$\begin{array}{ccc} \pi_r(F(n,0)) & \xrightarrow{i_s} & \pi_r(F(n,s)) \\ \bar{g}_* \downarrow & & \downarrow \bar{g}_* \\ \pi_r(F(n,0)) & \xrightarrow{i_{st}} & \pi_r(F(n,st)). \end{array}$$

(ii) *Lemma* (2.5) *is true if* $s = 0$.

Proof. We begin with part (i). The homomorphism $\bar{g}_* i_s$ is induced by a map of spaces which sends $f \in F(n, 0)$ into the following composite:

$$S^{n-1} \xrightarrow{\phi} S^{n-1} \vee S^{n-1} \xrightarrow{f \vee h} S^{n-1} \xrightarrow{g} S^{n-1}$$

(Here h is a fixed map of degree s.) The homomorphism $i_{st}\bar{g}_*$ is induced by a map of spaces which sends $f \in F(n, 0)$ into the following composite:

$$S^{n-1} \xrightarrow{\phi} S^{n-1} \vee S^{n-1} \xrightarrow{fg \vee k} S^{n-1}.$$

(Here k is a fixed map of degree st.) If we take $k = hg$, the two maps of $F(n, 0)$ become equal. This proves part (i).

We turn to part (ii). Let

$$y \longrightarrow (y)h : S^r, e \longrightarrow F(n, 0), \omega$$

be a representative map for α. Then a representative map r_1 for $[\alpha]$ is

$$(x, y) \longrightarrow (x)((y)h) : S^{n-1} \times S^r \longrightarrow S^{n-1}.$$

Also a representative map for $\bar{g}_*\alpha$ is

$$y \longrightarrow ((y)h)\bar{g} : S^r, e \longrightarrow F(n, 0), \omega.$$

Therefore a representative map r_2 for $[\bar{g}_*\alpha]$ is

$$(x, y) \longrightarrow (x)[((y)h)\bar{g}] : S^{n-1} \times S^r \longrightarrow S^{n-1}.$$

Evidently $r_2 = r_1 g$. But since g is a map of degree t, in the stable homotopy group π_r^S we have $[r_2] = t[r_1]$; that is, $[\bar{g}_*\alpha] = t[\alpha]$. This proves part (ii).

We now turn to Lemma (2.6). We recall that in defining j we have regarded S^{n+m-1} as $S^{n-1} * S^{m-1}$. However, for our purposes it makes no difference if we now replace $S^{n-1} * S^{m-1}$ by the quotient

$$\frac{S^{n-1} * S^{m-1}}{(S^{n-1} * e') \cup (e * S^{m-1})}.$$

Here e, e' denote the base-points in S^{n-1}, S^{m-1}. Since $S^{n-1} * e'$ and $e * S^{m-1}$ are cells with only the segment $e * e'$ in common, the quotient map

$$S^{n-1} * S^{m-1} \longrightarrow \frac{S^{n-1} * S^{m-1}}{(S^{n-1} * e') \cup (e * S^{m-1})}$$

is a homotopy equivalence. We may interpret the quotient space as $S^{n-1} \times S^1 \times S^{m-1}$.

If f, g are maps of S^{n-1}, S^{m-1} which preserve base-points, then $f*g$ passes to the quotient, and may be interpreted as $f \times 1 \times g$. In what follows, then, the 'join' symbol $*$ will be interpreted as referring to these quotient spaces and maps.

We now remark that since j_* is a homomorphism, Lemma (2.6) will be proved if we can calculate $j_*(\alpha + 0)$ and $j_*(0 + \beta)$. This calculation is equivalent to calculating the homomorphisms induced by two 'translation' maps. In fact, let f, g be fixed maps in $F(n, s)$, $F(m, t)$; then we can define maps

$$f^L : F(m, t) \longrightarrow F(n + m, st)$$

$$g^R : F(n, s) \longrightarrow F(n + m, st)$$

by $(h)f^L = f * h$, $(h)g^R = h * g$. We have

$$j_*(\alpha + 0) = g^R_*(\alpha), \quad j_*(0 + \beta) = f^L_*(\beta).$$

It will thus be sufficient to prove the following results.

LEMMA (2.8). (i) *If $f \in F(n, s)$ we have a commutative diagram of the following form, in which i_t, i_{st} are the identifications made earlier in this section:*

$$\begin{array}{ccc} \pi_r(F(m, 0)) & \xrightarrow{i_t} & \pi_r(F(m, t)) \\ \downarrow f^L_* & & \downarrow f^L_* \\ \pi_r(F(n + m, 0)) & \xrightarrow{i_{st}} & \pi_r(F(n + m, st)). \end{array}$$

(ii) *Similarly for g^R_*.*

(iii) *If $g \in F(m, t)$, $\alpha \in \pi_r(F(n, 0))$ (so that $s = 0$), then*

$$[g^R_* \alpha] = t[\alpha].$$

(iv) *Similarly for $f^L_* \beta$.*

Proof. We begin with part (i). The homomorphism $f^L_* i_t$ is induced by a map of spaces which sends $h \in F(m, 0)$ into

$$f * (\phi(h \vee k)),$$

where k is a fixed map of degree t. Owing to the fact that we are using the 'quotient' join, we can identify $S^{n-1} * (S^{m-1} \vee S^{m-1})$ with $(S^{n-1} * S^{m-1}) \vee (S^{n-1} * S^{m-1})$, and so write $f * (\phi(h \vee k))$ in the form

$$(1 * \phi)((f * h) \vee (f * k)).$$

The homomorphism $i_{st} f^L_*$ is induced by a map of spaces which sends $h \in F(m, 0)$ into

$$\phi'((f * h) \vee k'),$$

where ϕ' is a map of type (1, 1) and k' is a fixed map of degree st. If we take $\phi' = 1 * \phi$, $k' = f * k$ the two maps of $F(m, 0)$ become equal. This proves part (i); part (ii) is closely similar.

We now turn to part (iii). Let

$$h : S^r, e \longrightarrow F(n, 0), \omega$$

be a representative map for α. Then a representative map r_1 for $[\alpha]$ is given by

$$(x, y) \longrightarrow (x)((y)h) : S^{n-1} \times S^r \longrightarrow S^{n-1}.$$

A representative map for $g^R_* \alpha$ is given by assigning to each point $y \in S^r$ the map

$$(x, u, v) \longrightarrow ((x)(y)h), u, (v)g) : S^{n-1} \times S^1 \times S^{m-1} \longrightarrow S^{n-1} \times S^1 \times S^{m-1}.$$

Therefore a representative map r_2 for $[g^R_* \alpha]$ is given by

$$(x, u, v, y) \longrightarrow ((x)((y)h), u, (v)g) : S^{n-1} \times S^1 \times S^{m-1} \times S^r \longrightarrow S^{n-1} \times S^1 \times S^{m-1}.$$

This map may be factored in the form

$$(x, u, v, y) \xrightarrow{\rho} (u, v, x, y) \xrightarrow{1 \times 1 \times r_1} (u, v, (x)((y)h))$$

$$\xrightarrow{\sigma} ((x)((y)h), u, v) \xrightarrow{1 \times 1 \times g} ((x)((y)h), u, (v)g).$$

According to our definition of suspension, the stable class $[1 \times 1 \times r_1]$ is equal to $[r_1]$. The permutation maps ρ and σ have the same degree $(-1)^{(n-1)m}$, and the map $1 \times 1 \times g$ has degree t. Therefore in the stable homotopy group π_r^S we have $[r_2] = t[r_1]$, that is, $[g^R_* \alpha] = t[\alpha]$. This proves part (iii); the proof of part (iv) is closely similar, except that we do not need any permutation maps in the last step.

This completes the proof of Lemma (2.8), and establishes all the results of this section.

§3. PROOF OF 'DOLD'S THEOREM MOD k'

In this section we shall prove Theorem (1.1).

We begin by fixing some notation. Let ξ, ξ' be sphere bundles over X; I shall allow myself to speak of 'a fibrewise map $f : \xi \to \xi'$'; this is an abuse of language, or not, according to one's precise definition of a sphere-bundle.

Let $f : \xi \to \xi'$ be a fibrewise map of sphere bundles over X, and let $g : \eta \to \eta'$ be another such. Then we can clearly construct their Whitney sum

$$f \oplus g : \xi \oplus \eta \to \xi' \oplus \eta'$$

by taking joins on each fibre; it is again a fibrewise map of sphere bundles over X. By iterating this procedure we can construct Whitney multiples

$$mf : m\xi \to m\xi',$$

where m is any non-negative integer.

We shall write $X \times S^{n-1}$ to indicate a product bundle over X.

Our first lemma contains the main part of the proof. We shall suppose given (i) an integer $k > 0$, (ii) an $(n-1)$-sphere bundle ξ over a finite CW-complex X such that $\dim(X) \leq n-3$, and (iii) a fibrewise map $f : \xi \to X \times S^{n-1}$ of degree $\pm k$ on each fibre. It clearly follows that we can orient ξ so that the fibrewise map f has degree k on each fibre, and we will suppose this done.

LEMMA (3.1). *There exists* (i) *an integer* $t \geq 0$, (ii) *a fibrewise map* $g : k^t\xi \to X \times S^{N-1}$ (*where* $N = nk^t$) *of degree* 1 *on each fibre, and* (iii) *a map* $h : S^{N-1} \to S^{N-1}$, *such that the following diagram of fibrewise maps is fibre homotopy commutative*:

$$\begin{array}{ccc} k^t\xi & \xrightarrow{k^t f} & X \times S^{N-1} \\ & {\scriptstyle g}\searrow \quad \nearrow{\scriptstyle 1 \times h} & \\ & X \times S^{N-1} & \end{array}$$

Remark. It is clear that the degree of h must be $k^{(k^t)}$.

Proof. CW-complexes may be constructed by an inductive process, in which one attaches cells to what has already been constructed. The present proof (like many proofs about CW-complexes) consists of a corresponding induction. If X consists solely of 0-cells, then the result is clearly true, with $t = 0$. Let us suppose that X is formed by attaching a cell E^r to the subcomplex Y, with characteristic map $c : E^r, S^{r-1} \to X, Y$; and let us suppose, as our inductive hypothesis, that the result is true for Y; that is, we can find a fibre homotopy commutative diagram of the following form:

$$\begin{array}{ccc} k^u(\xi | Y) & \xrightarrow{k^u(f|Y)} & Y \times S^{M-1} \\ & {\scriptstyle g'}\searrow \quad \nearrow{\scriptstyle 1 \times h'} & \\ & Y \times S^{M-1} & \end{array}$$

(Here g' is supposed to be a map of degree 1 on each fibre.) Consider the induced bundle $c^*(k^u\xi)$ over E^r; it can be represented as a product bundle $E^r \times S^{M-1}$; we now have the following fibre homotopy commutative diagram of fibrewise maps:

$$\begin{array}{ccccccc} S^{r-1} \times S^{M-1} & \xrightarrow{c'} & k^u(\xi|Y) & \xrightarrow{g'} & Y \times S^{M-1} & \xrightarrow{1 \times h'} & Y \times S^{M-1} \\ \downarrow{\scriptstyle i} & & \downarrow{\scriptstyle i} & & & & \downarrow{\scriptstyle i} \\ E^r \times S^{M-1} & \xrightarrow{c''} & k^u\xi & & \xrightarrow{k^u f} & & X \times S^{M-1} \end{array}$$

(Here c', c'' lie over c.) The map $g'c'$ is equivalent to a map $\theta : S^{r-1} \to G(M, 1)$, where $G(M, 1)$ denotes the space of all maps from S^{M-1} to S^{M-1} of degree 1 (as in §2). If $r = 1$ then θ can be extended over E^r, since $G(M, 1)$ is arcwise-connected; we proceed to examine the case $r > 1$. Let K be the degree of h', which is $k^{(k^u)}$, as remarked above; and let us define $\bar{h}' : G(m, 1) \to G(m, K)$ by composition with h', as in §2; then the diagram shows that $\bar{h}'\theta : S^{r-1} \to G(m, K)$ can be extended over E^r. Now by Lemma (2.1) we have $[\bar{h}'\theta] = K[\theta]$, so $K[\theta] = 0$.

If we take the Whitney sum of the diagram with itself m times, we evidently replace $\theta(x)$ by $\theta(x) * \theta(x) * \ldots * \theta(x)$ (m factors). According to Lemma (2.4) this replaces $[\theta]$ by

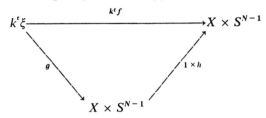

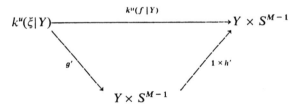

$m[\theta]$. If we take $m = K$, we replace $[\theta]$ by 0, and therefore we can find a fibrewise extension of $K(g'c')$ over $E^r \times S^{P-1}$ (where $P = KM$). This of course defines a fibrewise map

$$g'' : Kk^u\xi \longrightarrow X \times S^{P-1}$$

extending Kg'. (If $r = 1$ we arrive at the same conclusion with K replaced by 1.) At this stage we have the following diagram of fibrewise maps:

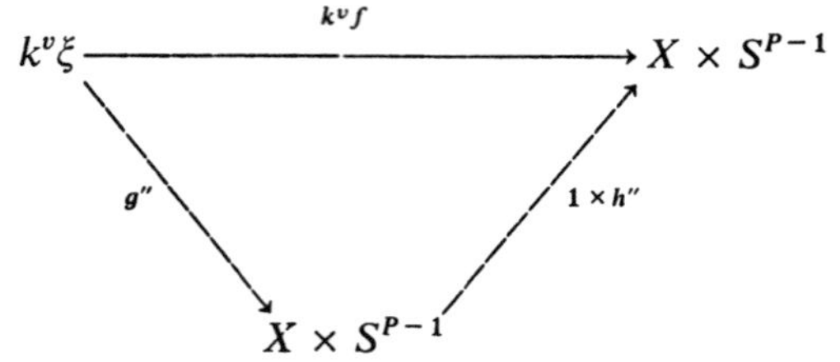

Here $k^v = Kk^u$; h'' is the join of K copies of h'; and the diagram is known to be fibre homotopy commutative on $k^v\xi | Y$.

The obstruction to extending a fibrewise homotopy over $I \times E^r$ is a map ϕ from the boundary of $I \times E^r$ to $G(P, L)$, where L is the present degree on the fibres (that is, $k^{(k^v)}$.) The map ϕ represents an element of $\pi_r(G(P, L))$. We can of course alter g'' by using any element α of $\pi_r(G(P, 1))$, and this alters $[\phi]$ by $[\bar{h}''\alpha] = L[\alpha]$ (Lemma (2.1)).

Let us now investigate the effect of taking the Whitney sum of this diagram with itself m times. We evidently replace $\phi(x)$ by $\phi(x) * \phi(x) * \ldots * \phi(x)$ (m factors). According to Lemma (2.4), this replaces $[\phi]$ by $mL^{m-1}[\phi]$. Since L is replaced by L^m, we can alter the obstruction $mL^{m-1}[\phi]$ by $L^m[\alpha]$. We now take $m = L$; the obstruction becomes $L^L[\phi]$ modulo $L^L[\alpha]$, that is, zero. We conclude that we can construct the following fibre homotopy commutative diagram of fibrewise maps:

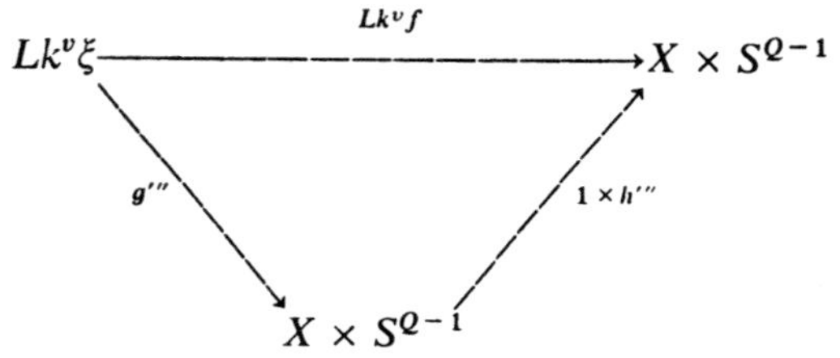

Here $Q = LP$, g''' has degree 1 on each fibre and h''' is the join of L copies of h''. Since Lk^v is a power of k, this completes the induction and proves Lemma (3.1).

COROLLARY (3.2). *Suppose given* (i) *an integer* $k > 0$, (ii) *an* $(n-1)$*-sphere bundle* ξ *over a finite CW-complex* X, *and* (iii) *a fibrewise map* $f : \xi \to X \times S^{n-1}$ *of degree* $\pm k$ *on each fibre. Then there exists an integer* t *such that the bundle* $k^t\xi$ *is fibre homotopy equivalent to a trivial bundle.*

Proof. The result is true for $k = 1$, by Dold's theorem [11]. We may thus suppose $k > 1$. For a suitable choice of s the bundle $k^s\xi$ and the fibrewise map $k^s f : k^s\xi \to X \times S^{N-1}$ (where $N = nk^s$) satisfy the dimensional restriction of Lemma (3.1). (The degree of $k^s f$ on each fibre is $k^{(k^s)}$). The conclusion of Lemma (3.1) provides a fibrewise map g of degree 1 on each fibre, and by Dold's theorem g must be a fibre homotopy equivalence. That is, there exists an integer t such that $k^{(tk^s)}\xi$ is fibre homotopy equivalent to a trivial bundle. This completes the proof.

We will now deduce Theorem (1.1) from Corollary (3.2).

Proof of Theorem (1.1). The result is true for $k = 1$, by Dold's theorem [11]. We may thus suppose $k > 1$. Suppose given a fibrewise map $f: \xi \to \eta$ of degree $\pm k$ on each fibre. There exists a sphere bundle ζ such that $\eta \oplus \zeta = \tau$, where τ is a trivial bundle. The map

$$f \oplus 1 : \xi \oplus \zeta \longrightarrow \eta \oplus \zeta = \tau$$

has degree $\pm k$ on each fibre. By Corollary (3.2), there exists an integer t such that $k^t\xi \oplus k^t \zeta$ is fibre homotopy equivalent to $k^t\tau$. Adding $k^t\eta$, we see that $k^t\xi \oplus k^t\tau$ is fibre homotopy equivalent to $k^t\eta \oplus k^t\tau$. That is, $k^t\xi$ and $k^t\eta$ are stably fibre homotopy equivalent. Now since $k > 1$, we can make the dimension of $k^t\xi$ as large as we please by increasing t; in particular we can make it so large that 'stable fibre homotopy equivalence' implies 'fibre homotopy equivalence' (cf. [6], pp. 293, 294). This completes the proof.

§4. APPLICATION OF 'DOLD'S THEOREM MOD k'

In this section we shall prove Theorems (1.3) and (1.4).

LEMMA 4.1. *Assume that X is a finite CW-complex and that $y \in K_R(X)$ is a linear combination of $O(1)$ bundles. Then there exists e (depending only on dim (X)) such that*

$$k^e(\Psi^k - 1)y = 0 \qquad in\ K_R(X).$$

Proof. Since $k^e(\Psi^k - 1)_y$ is linear in y, it is sufficient to consider the case in which y is an $O(1)$ bundle. In this case it is sufficient to consider the case in which y is the canonical real line bundle over RP^n, because any other $O(1)$ bundle can be induced from this by a map $f: X \to RP^n$, where $n = \dim(X)$. We now divide cases according to the parity of k. If k is odd, $\Psi^k(y) = y$ by [1, (5.1) (iii) or (7.4)(i)], and therefore $(\Psi^k - 1)y = 0$. If k is even, $(\Psi^k - 1)y = 1 - y$, and by [1, (7.4)] there exists e depending only on n such that

$$2^e(\Psi^k - 1)y = 0.$$

Since k is even,

$$k^e(\Psi^k - 1)y = 0.$$

This completes the proof.

Proof of Theorem (1.3). The result is trivial for $k = 0$; also by [1] we have $\Psi^{-k} = \Psi^k$, so we may assume $k > 0$.

Since $k^e(\Psi^k - 1)y$ is linear in y, it is sufficient to prove the result when y is an $O(1)$ bundle and when y is an $O(2)$ bundle. Lemma (4.1) deals with the case in which y is an $O(1)$ bundle, so we may suppose that y is an $O(2)$ bundle.

We will now recall something of the representation-theory of $O(2)$. $O(2)$ is a group of matrices acting on column vectors $(x_1, x_2)'$. Let us write

$$X_1 + iX_2 = (x_1 + ix_2)^r;$$

thus X_1 and X_2 are polynomials of degree r in x_1 and x_2. Each matrix in $O(2)$ induces an orthogonal transformation of $(X_1, X_2)'$; we have thus defined a representation $\mu_r : O(2) \to O(2)$. We also have λ_2, the determinant representation of $O(2)$, and λ_0, the trivial representation of degree 1.

By checking the characters we easily find:

$$\Psi^k = \begin{cases} \mu_k & (k \text{ odd}) \\ \mu_k - \lambda_2 + \lambda_0 & (k \text{ even}). \end{cases}$$

Now we have

$$(\lambda_2 - \lambda_0)y = (\lambda_2 y) - 1,$$

where $\lambda_2 y$ is an $O(1)$-bundle. By the argument of Lemma (4.1), if k is even there exists e such that

$$k^e(\lambda_2 - \lambda_0)y = 0 \qquad \text{in } K_R(X).$$

It remains then to prove that there exists e such that

$$k^e(\mu_k - 1)y$$

maps to zero in $J(X)$.

Consider the map $\phi : S^1 \to S^1$ defined by

$$\phi(x_1 + ix_2) = (x_1 + ix_2)^k.$$

By construction, ϕ is equivariant with respect to the homomorphism $\mu_k : O(2) \to O(2)$ of groups operating on S^1. Therefore it defines a map of bundles, say

$$f : y \longrightarrow \mu_k y.$$

The map f has degree $\pm k$ on each fibre. Therefore Theorem (1.1) applies; there is an integer e such that the multiples $k^e y$, $k^e \mu_k y$ are fibre homotopy equivalent. Thus $k^e(\mu_k y - y)$ maps to zero in $J(X)$. This completes the proof.

Proof of Theorem (1.4). We must recall some facts about the representability of our functors; the following details are taken from [6, pp. 293, 294]. Let $O(m)$ be the orthogonal group, and let $H(m)$ be the monoid of homotopy equivalences from S^{m-1} to S^{m-1}; then we have an inclusion map $i(m) : O(m) \to H(m)$. By passing to classifying spaces (in the sense of [12]) we obtain

$$Bi(m) : BO(m) \to BH(m).$$

Consider the induced function

$$(Bi(m))_* : \pi(X, BO(m)) \longrightarrow \pi(X, BH(m)),$$

where $\pi(X, Y)$ means the set of homotopy classes of maps from X to Y, and m is taken sufficiently large, depending on dim (X). Then there is a natural $(1-1)$ correspondence between $\tilde{J}(X)$ and $\mathrm{Im}(Bi(m))_*$.

Now let W be the Cartesian product of n copies of S^2; say $W = S^2 \times S^2 \times \ldots \times S^2$. We shall argue by considering the relation between S^{2n} and W. Let V be the set of points in W

which have at least one co-ordinate at the base-point in S^2; then we have $W = V \cup E^{2n}$. The attaching map $\alpha : S^{2n-1} \to V$ may be used to start a sequence of cofiberings, which up to homotopy type is the following:

$$S^{2n-1} \xrightarrow{\alpha} V \xrightarrow{i} W \xrightarrow{q} S^{2n} \xrightarrow{S\alpha} SV.$$

(Here SV and $S\alpha$ are the suspensions of V and α.) The cofibering

$$W \xrightarrow{q} S^{2n} \xrightarrow{S\alpha} SV$$

induces the following sequence of sets:

$$\pi(W, BH(m)) \xleftarrow{q^*} \pi(S^{2n}, BH(m)) \xleftarrow{(S\alpha)^*} \pi(SV, BH(m)).$$

This sequence is exact, in the sense that if $q^*x = 0$, then $x = (S\alpha)^*y$ for some y. But it is well known [14, Theorem (4.1)] that $S\alpha : S^{2n} \to SV$ is homotopic to the constant map; therefore $q^*x = 0$ implies $x = 0$. This shows that the map

$$q^* : \tilde{J}(S^{2n}) \longrightarrow \tilde{J}(W)$$

is monomorphic. By adding Z, we see that

$$q^* : J(S^{2n}) \longrightarrow J(W)$$

is monomorphic.

Now suppose given a class $y \in K_R(S^{2n})$ lying in the image of

$$r : K_C(S^{2n}) \longrightarrow K_R(S^{2n}),$$

so that $y = rz$, where $z \in K_C(S^{2n})$. Then in $K_R(W)$ we have $q^*y = rq^*z$. Now every element in $K_C(W)$ is a linear combination of complex line bundles; in particular, q^*z is such a linear combination. Therefore $q^*y = rq^*z$ is a linear combination of $SO(2)$ bundles. Theorem (1.3) thus applies to q^*y, and there exists $e = e(k,y)$ such that the element

$$k^e(\Psi^k - 1)q^*y = q^*k^e(\Psi^k - 1)y$$

maps to zero in $J(W)$. Since we have shown that

$$q^* : J(S^{2n}) \longrightarrow J(W)$$

is monomorphic, it follows that $k^e(\Psi^k - 1)y$ is zero in $J(S^{2n})$. This completes the proof.

REFERENCES

1. J. F. Adams: Vector fields on spheres, *Ann. Math., Princeton* **75** (1962), 603–632.
2. J. F. Adams: Applications of the Grothendieck–Atiyah–Hirzebruch functor $K(X)$, *Colloquium on Algebraic Topology*, pp. 104–113. Aarhus, 1962. (Mimeographed notes).
3. J. F. Adams: Applications of the Grothendieck–Atiyah–Hirzebruch functor $K(X)$, *Proceedings of the International Congress of Mathematicians* (Stockholm, 1962), (to be published).
4. J. F. Adams: On the groups $J(X)$, *Proceedings of a Symposium in Honour of M. Morse*, Princeton University Press (to be published).
5. J. F. Adams and G. Walker: On complex Stiefel manifolds (to be published).
6. M. F. Atiyah: Thom complexes, *Proc. Lond. Math. Soc.* **11** (1961), 291–310.
7. M. F. Atiyah and F. Hirzebruch: Riemann–Roch theorems for differentiable manifolds, *Bull. Amer. Math. Soc.* **65** (1959), 276–281.

8. M. F. Atiyah and F. Hirzebruch: Vector bundles and homogeneous spaces, *Proceedings of Symposia in Pure Mathematics* 3, *Differential Geometry*, 7–38. American Mathematical Society, 1961.
9. R. Bott: A note on the *KO*-theory of sphere-bundles, *Bull. Amer. Math. Soc.* **68** (1962), 395–400.
10. R. Bott: Lectures on $K(X)$ (mimeographed notes), Harvard University.
11. A. Dold: Uber fasernweise Homotopieaquivalenz von Faserraumen, *Math. Z.* **62** (1955), 111–136.
12. A. Dold and R. Lashof: Principal quasifibrations and fibre homotopy equivalence of bundles, *Illinois J. Math.* **3** (1959), 285–305.
13. E. Dyer: Chern characters of certain complexes, *Math. Z.* **80** (1963), 363–373.
14. I. M. James: Multiplication on spheres—II, *Trans. Amer. Math. Soc.* **84** (1957), 545–558.
15. G. W. Whitehead: On products in homotopy groups, *Ann. Math., Princeton* **47** (1946), 460–475.

Department of Mathematics,
University of Manchester.

Topology Vol. 3, pp. 137–171. Pergamon Press, 1965.

ON THE GROUPS $J(X)$—II

J. F. ADAMS

(*Received* 4 *September* 1963)

§1. INTRODUCTION

THE GENERAL object of this series of papers is to give means for computing the groups $J(X)$. A general introduction has been given at the beginning of Part I. The object of the present paper, Part II, is to set up the groups $J'(X)$ and $J''(X)$.

The arrangement of the present paper is as follows. We reach the group $J'(X)$ in §6. Its definition depends on the "cannibalistic characteristic class" ρ^k, which is treated in §5; and this in turn depends on the Thom isomorphism in K-theory, to which we devote §4. The group $J''(X)$ is treated in §3. Here we prove Theorem (3.12), which states a formal property of J'', and is required for use in [5]. §2 is devoted to necessary number-theory about the Bernoulli numbers.

§2. NUMBER-THEORY

The work of Milnor and Kervaire [15] shows the importance of the Bernoulli numbers in studying the J-homomorphism. In what follows, we shall need certain elementary number-theoretical results about Bernoulli numbers and related topics. These results are presumably known, but for completeness, they are collected and proved in the present section.

We begin by establishing some notation. The Bernoulli numbers enter algebraic topology in various ways. One of their bridgeheads is the power-series for the function.

$$\mathrm{Log}\left(\frac{\mathrm{Sinh}\,\frac{1}{2}x}{\frac{1}{2}x}\right).$$

Here the function

$$\frac{\mathrm{Sinh}\,\frac{1}{2}x}{\frac{1}{2}x}$$

is to be interpreted as 1 for $x = 0$; it is then analytic and non-zero for $|x| < 2\pi$, and is even. Thus we have

$$\mathrm{Log}\left(\frac{\mathrm{Sinh}\,\frac{1}{2}x}{\frac{1}{2}x}\right) = \sum_{s=1}^{\infty} \alpha_{2s} \frac{x^{2s}}{(2s)!} \tag{2.1}$$

(for $|x| < 2\pi$), where this expansion defines the coefficients α_{2s}. We have

$$\frac{e^x - 1}{x} = e^{\frac{1}{2}x} \cdot \frac{\operatorname{Sinh} \frac{1}{2}x}{\frac{1}{2}x};$$

therefore

$$\text{(2.2)} \qquad \operatorname{Log} \frac{e^x - 1}{x} = \sum_{t=1}^{\infty} \alpha_t \frac{x^t}{t!},$$

where we have defined α_t for odd t by setting

$$\alpha_1 = \tfrac{1}{2}, \qquad \alpha_{2s+1} = 0 \quad \text{for} \quad s > 0.$$

It is very easy to compare (2.2) with the expansion of $x/(e^x - 1)$. Following Hardy and Wright [12, p. 90] we set

$$\text{(2.3)} \qquad \frac{x}{e^x - 1} = \sum_{t=0}^{\infty} \beta_t \frac{x^t}{t!}.$$

LEMMA (2.4). *For $t > 1$ we have*

$$\alpha_t = \frac{\beta_t}{t}.$$

Proof. Differentiating (2.2), we have

$$\frac{x}{e^x - 1}\left(\frac{e^x}{x} - \frac{e^x - 1}{x^2}\right) = \sum_{t=1}^{\infty} t\alpha_t \frac{x^{t-1}}{t!}.$$

Rewriting this and using (2.3), we have

$$1 - \frac{1}{x} + \sum_{t=0}^{\infty} \beta_t \frac{x^{t-1}}{t!} = \sum_{t=1}^{\infty} t\alpha_t \frac{x^{t-1}}{t!}.$$

Equating coefficients, we find

$$\beta_t = t\alpha_t \qquad (t > 1).$$

This completes the proof.

The relation between the coefficients β_t and the classical Bernoulli numbers B_s is

$$\beta_{2s} = (-1)^{s-1} B_s \qquad (s > 0),$$

as on [12, p. 90].

The theorem of von Staudt [12, p. 91] determines the value of β_t mod 1. However, the numbers which arise in algebraic topology are not the numbers β_t themselves, but the numbers

$$\frac{\alpha_{2s}}{2} = \frac{\beta_{2s}}{4s} = (-1)^{s-1} \frac{B_s}{4s}.$$

We need to know the value of $\alpha_t/2 = \beta_t/2t$ as an element of the group of rationals mod 1. Since this group is a torsion group, it splits as the direct sum of its p-primary components. It will thus be sufficient if for each prime p we give the value mod Q'_p, where Q'_p is the additive groups of rationals with denominators prime to p.

In the next theorem, we suppose that t is even.

THEOREM (2.5). *If p is odd we have*

$$\frac{\alpha_t}{2} = \frac{\beta_t}{2t} \equiv \begin{cases} 0 & \text{mod } Q'_p \quad \text{if} \quad t \not\equiv 0 \text{ mod}(p-1) \\ \dfrac{1}{2pu} & \text{mod } Q'_p \quad \text{if} \quad t = (p-1)u. \end{cases}$$

For $p = 2$ we have

$$\frac{\alpha_t}{2} = \frac{\beta_t}{2t} \equiv \begin{cases} \frac{3}{8} & \text{mod } Q'_2 \quad \text{if} \quad t = 2 \\ \frac{1}{16} & \text{mod } Q'_2 \quad \text{if} \quad t = 4 \\ \dfrac{1}{2} + \dfrac{1}{4t} & \text{mod } Q'_2 \quad \text{if} \quad t \geqslant 6. \end{cases}$$

We defer the proof.

We now require some more notation. We write $\nu_p(n)$ for the exponent to which the prime p occurs in the decomposition of n into prime powers, so that

$$n = 2^{\nu_2(n)} 3^{\nu_3(n)} 5^{\nu_5(n)} \dots .$$

We define an explicit number-theoretic function $m(t)$ as follows.

For p odd,

$$\nu_p(m(t)) = \begin{cases} 0 & \text{if} \quad t \not\equiv 0 \text{ mod}(p-1) \\ 1 + \nu_p(t) & \text{if} \quad t \equiv 0 \text{ mod}(p-1). \end{cases}$$

For $p = 2$,

$$\nu_2(m(t)) = \begin{cases} 1 & \text{if} \quad t \not\equiv 0 \text{ mod } 2 \\ 2 + \nu_2(t) & \text{if} \quad t \equiv 0 \text{ mod } 2. \end{cases}$$

(Note that $\nu_p(m(t)) = 0$ except for a finite number of p.)

For example, we have $m(2s + 1) = 2$.

THEOREM (2.6). *$m(2s)$ is the denominator of*

$$\frac{\alpha_{2s}}{2} = \frac{\beta_{2s}}{4s} = (-1)^{s-1} \frac{B_s}{4s}$$

when this fraction is expressed in its lowest terms.

This theorem is due to Milnor and Kervaire [15, Lemma (3)]. It is clear that it follows immediately from Theorem (2.5).

The function $m(t)$ also appears in a rather different situation, to which we turn next. Roughly speaking, we want to say that $m(t)$ 'is the highest common factor of the expressions

$$k^{\infty} (k^t - 1)$$

as k runs over all integers'. We proceed to make this precise.

Let f be a function which assigns to each integer k (positive, negative or zero) a non-negative integer $f(k)$. Given such a function f and a non-negative integer t, we define

$h(f, t)$ to be the highest common factor of the integers

$$k^{f(k)}(k^t - 1)$$

as k varies over all integers (positive, negative and zero).

THEOREM (2.7). *$h(f, t)$ divides $m(t)$. For each t there is a function $f(k)$ such that $h(f, t) = m(t)$.*

This result is involved in a proof by E. Dyer [11, pp. 365, 366] although it is not given a separate statement there. I owe to Dyer a suggestion for expressing the proof more elegantly.

We will now prove the results stated above. The proof of Theorem (2.5) follows the pattern of von Staudt's theorem, in that we compare a summation formula involving β_t with an independent estimate of the sum. The only difference is that the estimate holds modulo a high power of p, instead of modulo p. We begin with two well-known lemmas.

LEMMA (2.8). *If $t > 0$ we have*

$$\sum_{1 \leqslant y \leqslant q-1} y^t = \sum_{1 \leqslant v \leqslant t+1} \frac{t!}{v!(t-v+1)!} \beta_{t-v+1} q^v.$$

This follows from the identity

$$1 + e^x + e^{2x} + \dots + e^{(q-1)x} = \frac{x}{e^x - 1} \cdot \frac{e^{qx} - 1}{x}$$

by expanding in powers of x and equating coefficients; see [12, p. 90].

We now introduce the ring J_m of residue classes mod m, and the multiplicative group G_m of units in J_m (that is, the group of residue classes of integers prime to m).

LEMMA (2.9). *If $m = p^f$ with p odd and $f \geqslant 1$ then G_m is cyclic of order $(p-1)p^{f-1}$. If $m = 2^f$ and $f \geqslant 2$ then G_m is the direct sum of the subgroup consisting of ± 1 and the subgroup of residue classes congruent to 1 mod 4; the latter subgroup is cyclic of order 2^{f-2}.*

This lemma is well-known. See [17, pp. 145, 146].

We next observe that if x lies in a given residue class mod p^a, where $a \geqslant 1$, then x^p lies in a well-determined residue class mod p^{a+1}; this follows immediately from the binomial theorem. By induction over b we see that if x lies in a given residue class mod p^a, then x^{p^b} lies in a well-determined residue class mod p^{a+b}.

For the next two lemmas, we write t in the form $t = fp^b$ with f prime to p. We also write $G_{2^a}^+$ for the subset of G_{2^a} consisting of the residue classes x for $0 < x < 2^{a-1}$.

LEMMA (2.10). *If p is odd we have*

$$\sum_{x \in G_{p^a}} x^t \equiv \begin{cases} 0 & \text{mod } p^{a+b} \quad \textit{if} \quad t \not\equiv 0 \text{ mod } (p-1) \\ (p-1)p^{a-1} & \text{mod } p^{a+b} \quad \textit{if} \quad t \equiv 0 \text{ mod } (p-1). \end{cases}$$

If $p = 2$, $a \geqslant 3$, $b \geqslant 1$ we have

$$\sum_{x \in G_{2^a}^+} x^t \equiv 2^{a-2} + 2^{a+b-1} \qquad \text{mod } 2^{a+b}.$$

Proof. We begin with the case $p = 2$. Consider the homomorphism

$$\theta : G_{2^{a+b}} \longrightarrow G_{2^{a+b}}$$

defined by $\theta(x) = x' = x^{f2^b}$. As remarked above, the map θ factors through G_{2^a}. Since we are assuming $a \geqslant 2$, $b \geqslant 1$, Lemma (2.9) shows that each element in Im θ is the image of just two elements in G_{2^a}, namely $\pm x$ for some x. That is, the elements in Im θ are the residue classes x' for $x \in G^+_{2^a}$. Thus we have

$$\sum_{x \in G^+_{2^a}} x' = \sum_{y \in \operatorname{Im}\theta} y \quad \text{in} \quad J_{2^{a+b}}.$$

By Lemma (2.9), Im θ is precisely the kernel of the obvious projection

$$G_{2^{a+b}} \longrightarrow G_{2^{b+2}}.$$

That is, Im θ consists of the elements $1 + j2^{b+2}$ (mod 2^{a+b}) for $1 \leqslant j \leqslant 2^{a-2}$. Thus we find

$$\begin{aligned}\sum_{y \in \operatorname{Im}\theta} y &\equiv \sum_{1 \leqslant j \leqslant 2^{a-2}} (1 + j2^{b+2}) \bmod 2^{a+b} \\ &\equiv 2^{a-2} + \tfrac{1}{2}.2^{a-2}(2^{a-2}+1)2^{b+2} \bmod 2^{a+b} \\ &\equiv 2^{a-2} + 2^{a+b-1} \bmod 2^{a+b}\end{aligned}$$

(since we have assumed $a \geqslant 3$). This completes the proof for $p = 2$.

In the case p odd we consider the homomorphism

$$\theta : G_{p^{a+b}} \longrightarrow G_{p^{a+b}}$$

defined by $\theta(x) = x' = x^{fp^b}$. The map θ factors through

$$\bar{\theta} : G_{p^a} \longrightarrow G_{p^{a+b}};$$

thus we have

$$\sum_{x \in G_{p^a}} x' = n \sum_{y \in \operatorname{Im}\theta} y \quad \text{in} \quad J_{p^{a+b}},$$

where n is the number of elements in Ker $\bar{\theta}$. The case $t \equiv 0 \bmod (p-1)$ is now similar to the case $p = 2$.

Let us therefore suppose that $t \not\equiv 0 \bmod (p-1)$. Consider the projection

$$G_{p^{a+b}} \longrightarrow G_p;$$

using Lemma (2.9), we see that there is an element z in Im θ such that $z \not\equiv 1 \bmod p$. As y runs over Im θ, so does zy; thus in $J_{p^{a+b}}$ we have

$$z \sum_{y \in \operatorname{Im}\theta} y = \sum_{y \in \operatorname{Im}\theta} y;$$

that is,

$$(z-1) \sum_{y \in \operatorname{Im}\theta} y \equiv 0 \qquad \bmod p^{a+b}.$$

Since $z \not\equiv 1 \bmod p$, we have

$$\sum_{y \in \operatorname{Im}\theta} y \equiv 0 \qquad \bmod p^{a+b}.$$

This completes the proof.

LEMMA (2.11). *If p is odd we have*

$$\sum_{0<x<p^a} x^t \equiv \begin{cases} 0 & \text{mod } p^{a+b} \quad \text{if} \quad t \not\equiv 0 \bmod (p-1) \\ (p-1)p^{a-1} & \text{mod } p^{a+b} \quad \text{if} \quad t \equiv 0 \bmod (p-1). \end{cases}$$

If $p = 2$, $a \geqslant 3$, $b \geqslant 1$, $t \geqslant 6$ then

$$\sum_{0<x<2^{a-1}} x^t \equiv 2^{a+2} + 2^{a+b-1} \bmod 2^{a+b}.$$

Proof. Consider the case $p = 2$. We argue by induction over a; let us assume that either (i) $a = 3$, or (ii) $a > 3$ and the result is true for $a - 1$. Then we have

$$\sum_{0<x<2^{a-1}} x^t = \sum_{x \in G^+_{2^a}} x^t + 2^t \sum_{0<x<2^{a-2}} x^t.$$

Using Lemma (2.10) and, if $a > 3$, the inductive hypothesis, we have

$$\sum_{0<x<2^{a-1}} x^t \equiv 2^{a-2} + 2^{a+b-1}$$

modulo 2^{a+b} and 2^{t+a-3}. It follows from the assumption $t \geqslant 6$ that $t \geqslant b + 3$; thus the congruence holds modulo 2^{a+b}. This completes the induction.

The case in which p is odd is proved similarly, starting the induction from $a = 1$.

Proof of Theorem (2.5). We consider the case $p = 2$. Since we can evidently compute α_2 and α_4 from (2.1), we shall suppose that $t \geqslant 6$. Write $t = f2^b$ with f odd. By Lemma (2.8), with $q = 2^{a-1}$, we have

$$\sum_{0<y<2^{a-1}} y^t = \sum_{1 \leqslant v \leqslant t+1} \frac{t!}{v!(t-v+1)!} \beta_{t-v+1} 2^{v(a-1)}$$

In the terms

$$\frac{t!}{v!(t-v+1)!} \beta_{t-v+1} 2^{v(a-1)},$$

the part

$$\frac{t!}{v!(t-v+1)!} \beta_{t-v+1}$$

does not depend on a; by choosing a large enough, we can ensure that all the terms

$$\frac{t!}{v!(t-v+1)!} \beta_{t-v+1} 2^{v(a-1)}$$

with $v \geqslant 2$ are divisible by 2^{a+b}. Using Lemma (2.11), we have

$$\frac{t!}{1!t!} \beta_t 2^{a-1} \equiv 2^{a-2} + 2^{a+b-1} \bmod 2^{a+b} Q'_2.$$

(Here $2^{a+b}Q'_2$ means the additive group of rationals $2^{a+b}r$, where $r \in Q'_2$). Dividing by $2^a t = f2^{a+b}$, we find

$$\frac{\alpha_t}{2} = \frac{\beta_t}{2t} \equiv \frac{1}{4t} + \frac{1}{2f} \bmod Q'_2.$$

Since f is odd, we have

$$\frac{1}{2f} \equiv \frac{1}{2} \bmod Q_2'.$$

This completes the proof for $p = 2$. The proof for p odd is similar, by substituting $q = p^a$ in Lemma (2.8).

We now turn to the proof of Theorem (2.7). We record the essential point of the proof as a lemma, for use in Part III.

LEMMA (2.12). *For each k prime to p we have*

$$\nu_p(k^t - 1) \geqslant \nu_p(m(t)).$$

Moreover, we have

$$\nu_p(k^t - 1) = \nu_p(m(t))$$

in the following cases.

(i) *p is odd and k is a generator of G_{p^2}.*

(ii) *$p = 2$, t is even and k is a generator of $G_8/\{\pm 1\}$.*

(iii) *$p = 2$, t is odd and k is a generator of G_4.*

Proof. Consider the case $p = 2$. If t odd then $\nu_2(m(t)) = 1$ and the result is trivial; for if k is odd, then $k^t - 1$ is divisible by 2, i.e. $\nu_2(k^t - 1) \geqslant 1$; and if $k \equiv -1 \bmod 4$, then $k^t - 1 \equiv -2 \bmod 4$, i.e. $\nu_2(k^t - 1) = 1$.

We may therefore suppose that $t = q\,2^{\nu-2}$, where q is odd and $\nu = \nu_2(m(t)) \geqslant 3$. By Lemma (2.9) we have

$$k^t \equiv 1 \bmod 2^\nu;$$

thus

$$\nu_2(k^t - 1) \geqslant \nu.$$

Now assume that k is a generator of $G_8/\{\pm 1\}$. Then k is a generator of $G_{2^{\nu+1}}/\{\pm 1\}$, and by Lemma (2.9) we have

$$k^t \not\equiv 1 \bmod 2^{\nu+1}.$$

Thus

$$\nu_2(k^t - 1) = \nu.$$

The proof for p odd is similar.

Proof of Theorem (2.7). Suppose given a function $f(k)$. Let p^v be the highest power of p dividing the integers

$$k^{f(k)}(k^t - 1)$$

for all k prime to p. Thus we shall certainly have

$$\nu_p(h(f, t)) \leqslant \nu;$$

but by Lemma (2.12), we have

$$\nu = \nu_p(m(t)).$$

Since this true for each prime p, $h(f, t)$ divides $m(t)$.

Given t, we may choose f so that

$$f(k) \geqslant \operatorname*{Max}_{p|k} v_p(m(t)).$$

Then the numbers p^v considered above will also divide the integers

$$k^{f(k)}(k^t - 1)$$

when k is divisible by p. In this case therefore we shall have

$$h(f, t) = \prod_p p^v = m(t).$$

This completes the proof of Theorem (2.7).

§3. THE GROUP $J''(X)$

In this section we shall introduce the group $J''(X)$, which will serve, in favourable cases, as an upper bound for $J(X)$. After giving the definition, elementary properties and examples, we come to the result on the groups $\tilde{J}(S^{4n})$ which was announced in [2, Theorem (3); 3, Theorem (3)]; see Theorem (3.7). Finally, we establish formal properties of the groups $J''(X)$; see especially Theorem (3.12).

In what follows, a Ψ-group will mean an abelian group Y together with given endomorphisms Ψ^k: $Y \to Y$ for each $k \in Z$, that is, for each integer k (positive, negative or zero). We impose no axioms on the endomorphisms Ψ^k. A Ψ-map between Ψ-groups will mean a homomorphism which commutes with the operations Ψ^k. If we speak of a Ψ-subgroup (or Ψ-quotient group) we shall mean that the injection (or projection) map is a Ψ-map.

The groups $K_A(X)$ are thus Ψ-groups, and of course this is the example of most interest to us. However, for technical reasons we sometimes have to consider other Ψ-groups, for example, Ψ-subgroups and Ψ-quotient-groups of groups $K_A(X)$.

Let Y be a Ψ-group, and let e be a function which assigns to each pair $k \in Z$, $y \in Y$ a non-negative integer $e(k, y)$. Then we define Y_e to be the subgroup of Y generated by the elements

$$k^{e(k,y)}(\Psi^k - 1)y.$$

That is, Y_e is the subgroup of linear combinations

$$\sum_{k,y} a(k, y)k^{e(k,y)}(\Psi^k - 1)y\,;$$

here the coefficients $a(k, y)$ are integers, and are zero except for a finite number of pairs (k, y). If $e_1 \geqslant e_2$, then $Y_{e_1} \subset Y_{e_2}$. We now define

$$J''(Y) = Y / \bigcap_e Y_e,$$

where the intersection runs over all functions e.

It is clear that a Ψ-map $f\colon Y_1 \to Y_2$ induces a map from $J''(Y_1)$ to $J''(Y_2)$. In fact, suppose given a function $e_2(k, y_2)$ on $Z \times Y_2$; then one defines a corresponding function e_1 by

$$e_1(k, y_1) = e_2(k, fy_1);$$

then we have

$$f(Y_1)_{e_1} \subset (Y_2)_{e_2};$$

hence

$$f \bigcap_e (Y_1)_e \subset \bigcap_{e_2} (Y_2)_{e_2}.$$

If X is a space, we define

$$J''_\Lambda(X) = J''(K_\Lambda(X)).$$

The case of most interest to us is, of course, the case $\Lambda = R$; in this case we write

$$J''(X) = J''_R(X) = J''(K_R(X)).$$

This construction is suggested, of course, by the results of Part I [4]. Let us recall conjecture 1.2 of Part I.

Conjecture (1.2) *of Part I.* If k is an integer, X is a finite CW-complex and $y \in K_R(X)$, then there exists a non-negative integer $e = e(k, y)$ such that $k^e(\Psi^k - 1)y$ maps to zero in $J(X)$.

PROPOSITION (3.1). *Suppose that for some X, Conjecture* (1.2) *of Part I holds for all k and y. Then $J''(X)$ is an upper bound for $J(X)$, in the sense of Part I.*

Proof. Take $Y = K_R(X)$, and let $T(X)$ be the kernel of the quotient map from $K_R(X)$ to $J(X)$, as in Part I. Then Conjecture 1.2 of Part I states that there is a function $e(k, y)$ such that $Y_e \subset T(X)$; *a fortiori;* $\bigcap_e Y_e \subset T(x)$. This completes the proof.

An alternative definition of $J''(Y)$, in which the functions $e(k, y)$ are replaced by functions of one variable, can be given when the abelian group Y is finitely generated (which is of course the case in our applications). In fact, we let f run over the functions $e(k, y)$ which are independent of y, so that $f(k, y) = f(k)$.

PROPOSITION (3.2). *If Y is finitely-generated then*

$$\bigcap_e Y_e = \bigcap_f Y_f,$$

so that we can write

$$J''(Y) = Y / \bigcap_f Y_f.$$

Proof. It is clear that $\bigcap_e Y_e \subset \bigcap_f Y_f$; we wish to prove the converse. Let $y_1, y_2, \ldots, y_n$ generate y; for any function $e(k, y)$, define a corresponding function $f(k)$ by

$$f(k) = \operatorname*{Max}_{1 \leqslant r \leqslant n} e(k, y_r).$$

It is easy to check that $Y_f \subset Y_e$; hence $\bigcap_f Y_f \subset \bigcap_e Y_e$. This completes the proof.

In what follows we will always assume that our Ψ-groups are finitely-generated, so that Proposition (3.2) applies. Several of the results which we prove with this assumption can be proved without it, though the proofs become slightly more complicated.

PROPOSITION (3.3)(a). *Let Y_1, Y_2 be finitely-generated Ψ-groups; then*

$$J''(Y_1 \oplus Y_2) = J''(Y_1) \oplus J''(Y_2).$$

(b) *Let P be a point; then*

$$J''(P) = Z.$$

(c) *Let X be a finite connected CW-complex; then*

$$J''(X) = Z + \tilde{J}''(X),$$

where

$$\tilde{J}''(X) = J''(\tilde{K}_R(X)).$$

Proofs. (a). We have

$$(Y_1 \oplus Y_2)_f = (Y_1)_f \oplus (Y_2)_f,$$

so

$$\bigcap_f (Y_1 \oplus Y_2)_f = \bigcap_f (Y_1)_f \oplus \bigcap_f (Y_2)_f.$$

(b). $K_R(P) = Z$, and the operations are given by $(\Psi^k - 1)\, y = 0$ for all k, y.

Part (c) follows by applying (a) and (b) to the decomposition

$$K_R(X) = K_R(P) + \tilde{K}_R(X).$$

This completes the proof.

We will now present some illustrative samples.

EXAMPLE (3.4). Take X to be real projective space RP^n; then the quotient map

$$K_R(RP^n) \to J''(RP^n)$$

is an isomorphism.

Proof. By [1, Theorem (7.4)], $\tilde{K}_R(RP^n)$ is cyclic of order 2^g, say. Let us choose f so that $f(k) \geqslant g$ for k even. Then $k^{f(k)}(\Psi^k - 1)\, y$ will be zero for k even. But for k odd $\Psi^k y = y$ in $K_R(RP^n)$ [1, Theorem (7.4)], so that $k^{f(k)}(\Psi^k - 1)\, y = 0$. Thus we have $Y_f = 0$ for this function f, and hence $\bigcap_f Y_f = 0$. This completes the proof.

EXAMPLE (3.5). Take X to be the sphere S^n with $n \equiv 1$ or $2 \bmod 8$; then the quotient map

$$K_R(S^n) \to J''(S^n)$$

is an isomorphism.

Proof. Let $f: RP^n \to S^n$ be a map of degree 1; then we have the following commutative diagram.

$$\begin{array}{ccc} K_R(S^n) & \xrightarrow{f^*} & K_R(RP^n) \\ \downarrow & & \downarrow \\ J''(S^n) & \longrightarrow & J''(RP^n) \end{array}$$

The map f^* is monomorphic, by the proof of [1, Theorem (7.4)] The right-hand column is monomorphic by Example (3.4). Therefore the left-hand column is monomorphic. This completes the proof.

EXAMPLE (3.6). Take X to be the sphere S^{4n}; then the group $\tilde{J}''(S^{4n})$ is cyclic of order $m(2n)$, where the function $m(2n)$ is in §2.

Proof. If $y \in \tilde{K}_R(S^{4n})$, we have

$$k^{f(k)}(\Psi^k - 1)\, y = k^{f(k)}(k^{2n} - 1)\, y \qquad [1, \text{Corollary } (5.2)].$$

Thus the subgroup Y_f of $\tilde{K}_R(S^{4n}) = Z$ consists of the multiples of $h(f, 2n)$, where $h(f, 2n)$ is the highest common factor of the integers

$$k^{f(k)}(k^{2n} - 1) \qquad (k \in Z).$$

The result now follows from Theorem (2.7).

THEOREM (3.7). *The image $J(\pi_{4n-1}(SO))$ of the stable J-homonomorphism—or equivalently, the group $\tilde{J}(S^{4n})$—is cyclic of order*

(i) $m(2n)$ *if* $4n \equiv 4 \bmod 8$

(ii) *either* $m(2n)$ or $2m(2n)$ *if* $4n \equiv 0 \bmod 8$.

This result was announced in [2, Theorem (3); 3, Theorem (3)].

Proof. The fact that the order of $\tilde{J}(S^{4n})$ is a multiple of $m(2n)$ is the result of Milnor and Kervaire [15] as improved by Atiyah and Hirzebruch [6]. We wish to argue in the opposite direction.

Suppose that $4n \equiv 4 \bmod 8$. Then the map

$$r : \tilde{K}_C(S^{4n}) \longrightarrow \tilde{K}_R(S^{4n})$$

is epimorphic; Theorem (1.4) of Part I [4] shows that Conjecture (1.2) of Part I is true for $X = S^{4n}$; the results (3.1) and (3.6) now show that the order of $\tilde{J}(S^{4n})$ divides $m(2n)$. This completes the proof in this case.

In case $4n \equiv 0 \bmod 8$ the proof is similar; we lose a factor of 2 because the image of

$$r : \tilde{K}_C(S^{4n}) \longrightarrow \tilde{K}_R(S^{4n})$$

consists of the elements divisible by 2.

We now seek to obtain formal properties of the group J''.

LEMMA (3.8). *Suppose that*

$$A \xrightarrow{i} B \xrightarrow{j} C \longrightarrow 0$$

is an exact sequence of finitely-generated Ψ-groups such that $J''(A)$ is finite. Then the sequence

$$J''(A) \xrightarrow{i_*} J''(B) \xrightarrow{j_*} J''(C) \longrightarrow 0$$

is exact.

Proof. Since $J''(B)$, $J''(C)$ are quotients of B, C it is clear that j_* is an epimorphism; it is also clear that $j_* i_* = 0$. It remains to prove that $\operatorname{Ker} j_* \subset \operatorname{Im} i_*$. In what follows, then, we suppose given $b \in B$ such that

$$jb \in \bigcap_f C_f.$$

It is given that $J''(A)$ is finite; choose a set of representatives $\alpha_1, \alpha_2, \ldots, \alpha_q$ in A for the elements of $J''(A)$. As a first step, we will show that for each f we can find α_r such that

$$b - i\alpha_r \in B_f.$$

In fact, suppose given a function $f(k)$. Since $jb \in C_f$, we have

$$jb = \sum_k k^{f(k)}(\Psi^k - 1)c_k$$

for a suitable set of elements c_k in C, of which all but a finite number are zero. Since j is epimorphic, we can find b_k in B (of which all but a finite number are zero) such that $c_k = jb_k$. Then we have

$$j(b - \sum_k k^{f(k)}(\Psi^k - 1)b_k) = 0,$$

so by exactness there is an a in A such that

$$b = ia + \sum_k k^{f(k)}(\Psi^k - 1)b_k.$$

If α_r is the representative for the class of a in $J''(A)$ we have

$$a - \alpha_r \in A_f;$$

that is,

$$a = \alpha_r + \sum_k k^{f(k)}(\Psi^k - 1)a_k$$

for a suitable set of elements a_k in A. Hence

$$b = i\alpha_r + \sum_k k^{f(k)}(\Psi^k - 1)(b_k + ia_k);$$

that is, $b - i\alpha_r \in B_f$. This completes the first step.

We have shown that for each f there exists α_r such that $b - i\alpha_r \in B_f$. We will now show that there exists α_r such that $b - i\alpha_r \in B_f$ for all f. Suppose the contrary; then for each α_r there exists f_r such that $b - i\alpha_r \notin B_{f_r}$. Define a function f by

$$f(k) = \operatorname*{Max}_{1 \leqslant r \leqslant q} f_r(k);$$

then for each r we have $b - i\alpha_r \notin B_f$, contradicting the first step.

We have thus shown that for some α_r, $b - i\alpha_r \in \bigcap_f B_f$. That is, in $J''(B)$ we have $\{b\} = i_*\{\alpha_r\}$. This completes the proof.

LEMMA (3.9). *Suppose that a finitely-generated Ψ-group Y admits a filtration*

$$Y = Y_1 \supset Y_2 \supset \ldots \supset Y_n = 0$$

by Ψ-subgroups Y_q such that $J''(Y_q/Y_{q+1})$ is finite for each q. Then $J''(Y)$ is finite.

This is easily proved by induction over n, using Lemma (3.8) to make the inductive step.

LEMMA (3.10). *Let $\bigvee S^q$ be a finite wedge-sum of q-spheres. Let Y be a Ψ-quotient of a Ψ-subgroup of $\tilde{K}_\Lambda(\bigvee S^q)$. Then $J''(Y)$ is finite.*

Proof. If $\tilde{K}_\Lambda(S^q)$ is finite, then $\tilde{K}_\Lambda(\bigvee S^q)$ is finite, Y is finite and $J''(Y)$ is finite. It is therefore only necessary to consider the following cases:

$$\Lambda = R,\ q \equiv 0 \bmod 4\ ;\ \Lambda = C,\ q \equiv 0 \bmod 2.$$

Let us assume that $q = 2n$; then the operations Ψ^k in Y are given by

$$\Psi^k y = k^n y.$$

Arguing as in Example (3.6), we see that for each $y \in Y$ the multiple $m(n)y$ maps to zero in $J''(Y)$ (where $m(n)$ is as in §2). Since Y is finitely-generated, $J''(Y)$ must be finite.

THEOREM (3.11). *If X is a finite connected CW-complex, then $\tilde{J}''_\Lambda(X)$ is finite.*

Proof. Filter $Y = \tilde{K}_\Lambda(X)$ by taking Y_q to be the image of the map

$$j^* : K_\Lambda(X, X^{q-1}) \longrightarrow \tilde{K}_\Lambda(X),$$

where X^n is the n-skeleton of X. Then Y_q/Y_{q+1} is a Ψ-quotient of a Ψ-subgroup of $\tilde{K}_\Lambda(\bigvee S_q)$. Thus $J''(Y_q/Y_{q+1})$ is finite by Lemma (3.10) and $J''(Y)$ is finite by Lemma (3.9).

THEOREM (3.12). *Let $X \to Y \to Z$ be a cofibering of finite connected CW-complexes such that the sequence*

$$\tilde{K}_\Lambda(Z) \xrightarrow{j^*} \tilde{K}_\Lambda(Y) \xrightarrow{i^*} \tilde{K}_\Lambda(X) \longrightarrow 0$$

is exact. Then the sequence

$$\tilde{J}''_\Lambda(Z) \xrightarrow{j^*} \tilde{J}''_\Lambda(Y) \xrightarrow{i^*} \tilde{J}''_\Lambda(X) \longrightarrow 0$$

is exact.

This follows immediately from Theorem (3.11) and Lemma (3.8).

THEOREM (3.13). *Let X be a finite connected CW-complex, and let $Y = \tilde{K}_\Lambda(X)$. Then there exists a function $F(k)$ such that*

$$\bigcap_f Y_f = Y_F.$$

This theorem shows that although the definition of $J''(X)$ involves a limit over functions f (or e), the limit is actually attained.

Proof. By Theorem (3.11), $\tilde{J}''_\Lambda(X)$ is finite. Let $y_1, y_2, \ldots, y_n$ be representatives in $Y = \tilde{K}_\Lambda(X)$ for the non-zero elements of $\tilde{J}''_\Lambda(X)$. Since y_q is not in $\bigcap_f Y_f$, there is a function f_q such that y_q is not in Y_{f_q}. Define

$$F(k) = \operatorname*{Max}_{1 \leqslant q \leqslant n} f_q(k).$$

We have

$$\bigcap_f Y_f \subset Y_F,$$

so that we have a quotient map

$$\theta : Y/\bigcap_f Y_f \longrightarrow Y/Y_F.$$

By construction, y_q is not in Y_F, and this holds for each q; therefore θ is monomorphic. This proves the result.

§4. THE THOM ISOMORPHISM

In setting up the groups $J'(X)$, one should begin with a treatment of the "Thom isomorphism" in extraordinary cohomology. It is generally known that such an isomorphism can be set up. (As a matter of history, the relevant construction appears in the very sketchy sketch proof at the end of [6].) However, we have been waiting for an account which sets up this isomorphism in the best possible way, and proves that it enjoys the good properties one requires. Such an account has now been provided by Atiyah, Bott and Shapiro [18; see especially Theorem (12.3)].

In this section, I shall simply quote the result of Atiyah, Bott and Shapiro. In an earlier draft I included (for completeness and for my own security) a treatment of the Thom isomorphism, on which I based *ad hoc* proofs of certain results, especially Theorem (5.1) and (5.9) of the present paper. This treatment and these proofs are now omitted, at the referee's suggestion.

Let ξ be a vector bundle, with structural group $SO(n)$, over the finite connected CW-complex B. By the 'Thom pair $\bar{E}$, E of B', we shall mean either one of the following constructs.

(a) $\bar{E}$ is the associated bundle whose fibres are unit n-cells; E is the boundary of $\bar{E}$, so that E is the associated bundle whose fibres are unit $(n\text{-}1)$-spheres.

(b) $\bar{E}$ is the total space of the vector-bundle ξ; E is the complement of the zero cross-section in $\bar{E}$.

For cohomological purposes these two constructions are equivalent.

We recall that in ordinary cohomology we have a 'Thom isomorphism' [16]

$$\phi : H^q(B; G) \longrightarrow H^{n+q}(\bar{E}, E; G).$$

This is usually constructed as follows. We first construct a generator $u \in H^n(\bar{E}, E; Z)$. We then define

$$\phi(h) = u\,.(p^*h);$$

here $p\colon \bar{E} \to B$ is the projection map, so that p^*h lies in $H^q(\bar{E}; G)$ and the cup-product $u.(p^*h)$ lies in $H^{n+q}(\bar{E}, E; G)$.

In thinking about the Thom isomorphism in extraordinary cohomology, one follows the obvious analogy, replacing H^* by K_Λ^*.

The group $K_\Lambda^*(X)$ is conveniently defined for a finite-dimensional CW-complex X by using vector-bundles over X. It will be useful to generalise the definition to more general spaces X, in order to avoid having to discuss whether our Thom pairs $\bar{E}, E$ can be given the structure of CW-pairs.

We may replace X by a CW-complex Y which is weakly equivalent to X (for example, the total singular complex of X). We may now define

$$K_\Lambda^n(X) = \operatorname*{Inv\,Lim}_{q \to \infty} K_\Lambda^n(Y^q),$$

where Y^q is the q-skeleton of Y. We make the obvious definitions for pairs, maps etc.

The operations Ψ_Λ^k of [1] are defined in $K_\Lambda^n(Y^q)$; they pass to the inverse limit, and define operations Ψ_Λ^k in $K_\Lambda^n(X)$.

This use of the inverse limit is of course due to Atiyah and Hirzebruch [7] (except that they often restrict themselves to finite complexes when they could equally well allow finite-dimensional ones.)

The use of inverse limits has the disadvantage that it sacrifices exactness. However, if $H_*(X)$ is finitely-generated (which is of course the case for our Thom pairs) then the inverse limit is more apparent than real. In fact, in this case the double suspension $S^2 Y$ is simply-connected and has $H_*(S^2 Y)$ finitely-generated; thus $S^2 Y$ is equivalent to a finite CW-complex Z; and we have

$$\begin{aligned} \operatorname*{Inv\,Lim}_{q \to \infty} K_\Lambda^n(Y^q) &= \operatorname*{Inv\,Lim}_{q \to \infty} K_\Lambda^{n+2}(S^2 Y^q) \\ &= K_\Lambda^{n+2}(Z). \end{aligned}$$

For such spaces X, then, we do not lose exactness.

We can now describe the two cases which will concern us of the Thom isomorphism in extraordinary cohomology. In the first case, we suppose given a real vector bundle ξ over B with structural group Spin($8n$), and we obtain an isomorphism

$$\phi : K_R^*(B) \longrightarrow K_R^*(\bar{E}, E).$$

In the second case, we suppose given a complex vector bundle ξ with structural group $U(n)$, and we obtain an isomorphism

$$\phi : K_C^*(B) \longrightarrow K_C^*(\bar{E}, E).$$

In each case, φ is an isomorphism of modules over $K_\Lambda^*(B)$ (where $\Lambda = R$ or C, according to the case.) Moreover, φ is natural for bundle maps. For the definition of φ, we refer the reader to [18].

§5. THE CLASSES ρ^k.

In this section we shall study certain 'cannibalistic characteristic classes'. Following Atiyah (private communication dated 20 October 1961) we shall call them ρ^k; an independent account has been published by Bott, who calls them θ_k [8, 9].

We shall first define the class $\rho^k(\xi)$ for each bundle ξ of a suitable class, and establish certain formal properties. Then we shall give a result (Theorem (5.9)) which relates the operations ρ^k to representation-theory. After that we shall extend the definition of ρ^k from bundles ξ to virtual bundles. Finally we shall compute the values of ρ^k in RP^n and in S^n.

We begin by discussing the situation abstractly. Let K and H be extraordinary cohomology theories with products, and let $T:K \to H$ be a natural transformation (preserving products). Suppose given also some class of bundles ξ, for example, unitary bundles or Spin($8n$)-bundles ($n = 1, 2, \dots$.) For this class of bundles, we assume, there is given a Thom isomorphism

$$\phi_K : K^*(B) \longrightarrow K^*(\bar{E}, E);$$

this is a map of modules over $K^*(B)$, and is natural for maps of bundles. Similarly for

$$\phi_H : H^*(B) \longrightarrow H^*(\bar{E}, E).$$

Under these conditions the element

$$c(T, \xi) = \phi_H^{-1} T \phi_K(1) \in H^*(B)$$

may be considered as a 'characteristic class of ξ'; in particular, it is natural for bundle maps.

We have in mind the following special cases.

(i) Let us take $K = H = H^*(\quad; Z_2)$, $T = \sum_0^\infty Sq^i$. We obtain the (total) Stiefel–Whitney class of ξ [16].

(ii) Let us take $K = K_\Lambda$, $H = H^*(\quad; Q)$. Let us write $ch_C = ch$, $ch_R = ch.c$, so that we can take $T = ch_\Lambda : K_\Lambda \to H$. We obtain characteristic classes

$$\phi_H^{-1} ch_\Lambda \phi_K(1).$$

These classes are both classical and useful in calculations, and will be discussed below.

(iii) Let us take $K = H = K_\Lambda$, and take T to be the operation Ψ_Λ^k [1]. Then we obtain a chacteristic class which we call ρ_Λ^k:

$$\rho_\Lambda^k(\xi) = \phi_K^{-1} \Psi_\Lambda^k \phi_K(1) \in K_\Lambda(B).$$

Of course the characteristic class ρ_C^k is defined for unitary bundles and the class ρ_R^k is defined for Spin($8n$)-bundles ($n = 1, 2, \dots$).

The philosophy of characteristic classes $\phi_H^{-1} T \phi_K(1)$ has been expounded in [19, especially §§ 2.2, 2.15, 3.3; 22].

We will now discuss example (ii) above more fully. If we start from a Spin($8n$)-bundle ξ, then the classical expression for $\phi_H^{-1}\, ch\, c\phi_K 1$ is $(\hat{A}(\xi))^{-1}$, where $\hat{A}$ is as in [6; 21 § 23]. In fact, it is by now well known that this is the way $\hat{A}$ enters the theory of characteristic classes.

I will now indicate my objection to the notation $(\hat{A}(\xi))^{-1}$. In the theory of characteristic classes we should first do all we can for general bundles; only then should we apply the theory to the tangent and normal bundles of differentiable manifolds. (In historical terms, we should follow Whitney rather than Stiefel). From this point of view the characteristic class

$$\phi_H^{-1} ch\ c\ \phi_K 1$$

is clearly fundamental, and should have its own notation; in this paper I shall use the notation

$$sh(\xi) = \phi_H^{-1} ch\ c\ \phi_K 1.$$

(The choice of notation will be explained below). One now takes a differentiable manifold, with tangent bundle τ and normal bundle ν (for some embedding in R^n). One now encounters the class

$$\hat{A}(\tau) = sh(\nu).$$

That is to say, this class 'really' arises from the normal bundle; but one introduces $\hat{A}$ in order to write it in terms of the tangent bundle.

Similar remarks apply to unitary bundles, with $(\hat{A}(\xi))^{-1}$ replaced by

$$e^{c_1(\xi)}(T(\xi))^{-1}$$

where $T(\xi)$ is the Todd class [13, §§ 1.7, 10; 21 §22]. (This expression, like the previous one, depends on the precise choice of the Thom isomorphism ϕ_K).

For later use, we require explicit formulae for the characteristic classes $\phi_H^{-1}\ ch_\Lambda\ \phi_K 1$.

Following Borel and Hirzebruch, we consider in $U(n)$ the maximal torus T which consists of diagonal matrices. We have

$$BT = CP^\infty \times CP^\infty \times \ldots \times CP^\infty.$$

Let $x \in H^2(CP^\infty)$ be a generator; then the cohomology ring $H^*(BT; Q)$ is a polynomial ring on generators $x_1, x_2, \ldots, x_n$ corresponding to the factors. The embedding $i : T \to U(n)$ induces a monomorphism

$$(Bi)^* : H^*(BU(n); Q) \to H^*(BT; Q)$$

whose image is the subring of symmetric polynomials.

We write bh or bh_C for the characteristic class whose image under $(Bi)^*$ is

$$\prod_{1 \leq r \leq n} \frac{e^{x_r} - 1}{x_r}.$$

The notation bh is intended to suggest 'Bernoulli'.

By means of the usual embedding $U(n) \subset SO(2n)$ we obtain a maixmal torus T in $SO(2n)$. As before, the map

$$(Bi)^* : H^*(SO(2n); Q) \to H^*(BT; Q)$$

is a monomorphism. Its image is the subring of $H^*(BT; Q)$ additively generated by symmetric polynomials in which the exponents of the variables x_r are either all even, or all odd. Using the projection $\mathrm{Spin}(2n) \to SO(2n)$, we have

$$H^*(B\,\mathrm{Spin}(2n); Q) \cong H^*(BSO(2n); Q).$$

We write sh or bh_R for the characteristic class which corresponds to

$$\prod_{1\leq r\leq n} \frac{e^{\frac{1}{2}x_r} - e^{-\frac{1}{2}x_r}}{x_r}.$$

The notation sh is intended to suggest 'sinh'.

THEOREM (5.1). *We have*

$$\phi_H^{-1} ch_A \phi_K 1 = bh_A \xi.$$

This theorem was certainly known to previous authors; compare [19, foot of p. 149]. The proof which follows is due to Atiyah (private communication).

Proof. We shall proceed from the definition of ϕ_K given in [18], using the methods of Borel and Hirzebruch [20, 21]; compare [23, §5].

We first recall that Atiyah, Bott and Shapiro introduce a group $\text{Spin}^C(n)$, defined as a subset of a certain complex Clifford algebra [18]. We will begin by obtaining the result which corresponds to Theorem (5.1) when we consider bundles with structural group $\text{Spin}^C(2n)$ and take $K = K_C$. We have first to fix some notation.

Let S^1 be the subgroup of complex scalars of unit modulus in the complex Clifford algebra. Let T' be the maximal torus in $\text{Spin}^R(2n) \subset \text{Spin}^C(2n)$. Then $S^1 \cap T' = Z_2$, consisting of ± 1; and $S^1 \times_{Z_2} T'$ is a maximal torus T'' in $\text{Spin}^C(2n)$. This torus is a double cover of $(S^1/Z_2) \times T$, where T is the maximal torus in $SO(2n)$. We take the coordinate in S^1/Z_2 as x_0 mod 1; thus the coordinate in S^1 is $\frac{1}{2}x_0$ mod 1. Similarly, we write $x_1, \ldots, x_n$ for the coordinates in T.

We begin by considering the case $n = 1$. Let $\bar{E}, E$ be the universal Thom pair with structural group $\text{Spin}^C(2)$; and consider the induced homomorphisms

$$K_C(\bar{E}, E) \xrightarrow{j^*} K_C(\bar{E}) \underset{\cong}{\xleftarrow{p^*}} K_C(B).$$

Because of the 'difference bundle construction' employed by Atiyah, Bott and Shapiro, $j^*\phi_K 1$ can actually be written in the form $p^*\eta - p^*\zeta$, where η and ζ are bundles obtained from the universal bundle by known complex representations. We have to calculate the characters of these representations; it is sufficient to calculate their restrictions to S^1 and $\text{Spin}^R(2)$. The representations are one-dimensional, and the complex scalars in the Clifford algebra act as complex scalars; therefore the restriction of either character to S^1 is $e^{\frac{1}{2}x_0 \cdot 2\pi i}$. We turn to $\text{Spin}^R(2)$, which is the subset of elements

$$\text{Cos}\, \tfrac{1}{2}x_1 + e_1 e_2 \,\text{Sin}\, \tfrac{1}{2}x_1$$

in the Clifford algebra. By definition, the 'positive' basic representation is the one which represents $e_1 e_2$ as $+i$; the 'negative' basic representation is the one which represents $e_1 e_2$ as $-i$. Therefore the restriction of the characters to $\text{Spin}^R(2)$ are $e^{\frac{1}{2}x_1 \cdot 2\pi i}$, $e^{-\frac{1}{2}x_2 \cdot 2\pi i}$. Thus the characters are $e^{\frac{1}{2}(x_0 + x_1)\cdot 2\pi i}$ for the 'positive' representation, $e^{\frac{1}{2}(x_0 - x_1)\cdot 2\pi i}$ for the 'negative' representation. This leads immediately to the formula

$$\begin{aligned} ch(p^*)^{-1} j^* \phi_K 1 &= ch(\eta - \zeta) \\ &= e^{\frac{1}{2}x_0}(e^{\frac{1}{2}x_1} - e^{-\frac{1}{2}x_1}). \end{aligned}$$

Now by a standard result, we have

$$\phi_H^{-1} y = \frac{1}{x_1}(p^*)^{-1} j^* y.$$

This yields

$$\phi_H^{-1}\, ch\, \phi_k\, 1 = e^{\frac{1}{2}x_0}\left(\frac{e^{\frac{1}{2}x_1} - e^{-\frac{1}{2}x_1}}{x_1}\right).$$

This completes the calculation for $\mathrm{Spin}^C(2)$.

We now consider $\mathrm{Spin}^C(2n)$. We observe that our characteristic classes $c(T, \xi)$ are exponential, in the sense that

$$c(T, \xi \oplus \eta) = c(T, \xi) \, . \, c(T, \eta).$$

In fact, this follows from the "product formulae" for ϕ_H and ϕ_K in $\xi \oplus \eta$; that for ϕ_H is classical, while that for ϕ_K is one of the main results of Atiyah, Bott and Shapiro [18, Proposition (11.1)]. We also observe that the homomorphism

$$\mathrm{Spin}^C(2) \times \mathrm{Spin}^C(2) \times \ldots \times \mathrm{Spin}^C(2) \to \mathrm{Spin}^C(2n)$$

induces a monomorphism in rational cohomology of the classifying spaces. Therefore the result for $\mathrm{Spin}^C(2n)$ follows immediately from the result for $\mathrm{Spin}^C(2)$; we obtain the formula

$$e^{\frac{1}{2}x_0} \prod_{1 \leqslant r \leqslant n} \frac{e^{\frac{1}{2}x_r} - e^{-\frac{1}{2}x_r}}{x_r}.$$

Finally, we deduce the two parts of Theorem (5.1) by naturality. For a bundle with structural group $\mathrm{Spin}^R(8n)$ the constructions of [18] lead to

$$c\phi_R 1 = \phi_C 1$$

(with an obvious notation.) This yields the formula

$$\prod_{1 \leqslant r \leqslant 4n} \frac{e^{\frac{1}{2}x_r} - e^{-\frac{1}{2}x_r}}{x_r}$$

in $B\mathrm{Spin}^R(8n)$. For a $U(n)$-bundle one has to employ the homomorphism

$$U(n) \longrightarrow \mathrm{Spin}^C(2n)$$

given in [18, end of §3]. This homomorphism sends x_0 into $\sum_1^n x_r$, and sends x_r into x_r for $r \geq 1$. This yields the formula

$$\prod_{1 \leqslant r \leqslant n} \frac{e^{x_r} - 1}{x_r}$$

in $BU(n)$.

Alternatively, this last formula can be deduced from the construction in terms of exterior algebras, given in [18, Proposition (11.6)].

This completes the proof of Theorem (5.1).

PROPOSITION (5.2). *We have*

$$\operatorname{Log} bh\xi = \sum_{t=1}^{\infty} \alpha_t ch_t \xi$$

$$\operatorname{Log} sh\eta = \sum_{s=1}^{\infty} \tfrac{1}{2}\alpha_{2s} ch_{2s} c\eta.$$

In these formulae, we define Log $(1 + x)$ for $x \in \sum_{t>0} H^{2t}(X; Q)$ by means of the usual power-series expansion. The coefficients α_t are as in §2. We write ch_t for the component of ch in dimension $2t$.

This proposition follows from the definitions by standard methods and obvious manipulations.

We now return to the assumptions made at the beginning of this section, so that K, H are extraordinary cohomology theories and $T : K \to H$ is a natural transformation. We take up the study of the characteristic classes $\phi_H^{-1} T \phi_K(1)$.

LEMMA (5.3). *Suppose given the following commutative diagrams.*

$$\begin{array}{ccc} K^*(\bar{E}_1, E_1) & \xrightarrow{\ \alpha\ } & H^*(\bar{E}_2, E_2) \\ \uparrow{\scriptstyle \phi_1}\ \cong & & \uparrow{\scriptstyle \phi_2}\ \cong \\ K^*(B_1) & & H^*(B_1) \end{array}$$

$$\begin{array}{ccc} K^*(\bar{E}_1) & \xrightarrow{\ \beta\ } & H^*(\bar{E}_2) \\ \uparrow{\scriptstyle p_1^*} & & \uparrow{\scriptstyle p_2^*} \\ K^*(B_1) & \xrightarrow{\ \gamma\ } & H^*(B_2) \end{array}$$

Suppose that $\alpha(xy) = (\alpha x)(\beta y)$ for $x \in K^*(\bar{E}_1, E_1)$, $y \in K^*(\bar{E}_1)$. *Then we have*

$$\phi_2^{-1}\alpha\phi_1 x = (\phi_2^{-1}\alpha\phi_1 1)(Yx).$$

The proof is purely formal, and is obvious.

COROLLARY (5.4). *Taking* $\xi_1 = \xi_2 = \xi$, $\alpha = \beta = \gamma = T$ *we have*

$$\phi_H^{-1} T \phi_K(x) = c(T, \xi) . T(x).$$

In particular, we have

$$\phi_H^{-1} ch_A \phi_K(x) = bh_A(\xi) . ch_A(x)$$

$$\phi_K^{-1} \Psi_A^k \phi_K(x) = \rho_A^k(\xi) . \Psi_A^k(x).$$

PROPOSITION (5.5).

$$(\rho_A^k \xi) . (\Psi_A^k \rho_A^l \xi) = \rho_A^{kl} \xi.$$

This result has also been found by Bott [8, 9].

Proof. In [1] it is shown that $\Psi_\Lambda^k \Psi_\Lambda^l = \Psi_\Lambda^{kl}$. Take the equation

$$(\phi_K^{-1} \Psi_A^k \phi_K)(\phi_K^{-1} \Psi_A^l \phi_K)1 = (\phi_K^{-1} \Psi_A^{kl} \phi_K)1$$

and evaluate each side. We find

$$(\phi_K^{-1}\,\Psi_A^k\,\phi_K)(\rho_A^l\,\xi) = \rho_A^{kl}\,\xi,$$

or using Corollary (5.4),

$$(\rho_A^k\,\xi).(\Psi_A^k\,\rho_A^l\,\xi) = \rho_A^{kl}\,\xi.$$

This completes the proof.

For our next proposition, we define

$$\Psi_H^k : \sum_{s\geqslant 0} H^{2s}(X; Q) \longrightarrow \sum_{s\geqslant 0} H^{2s}(X; Q)$$

by

$$\Psi_H^k(x) = k^s x \qquad \text{if} \qquad x \in H^{2s}(X; Q).$$

The point of this definition is that

$$ch_A\,\Psi_A^k = \Psi_H^k\,ch_A;$$

see [1]. If ξ is a vector bundle whose dimension over the reals is $2n$, we have

$$\phi_H^{-1}\,\Psi_H^k\,\phi_H(x) = k^n\,\Psi_H^k(x).$$

PROPOSITION (5.6).

$$(bh_A\,\xi).(ch_A\,\rho_A^k\,\xi) = k^n\,(\Psi_H^k\,bh_A\,\xi).$$

Proof. Take the equation

$$(\phi_H^{-1}\,ch_A\,\phi_K)(\phi_K^{-1}\,\Psi_A^k\,\phi_K)1 = (\phi_H^{-1}\,\Psi_H^k\,\phi_H)(\phi_H^{-1}\,ch_A\,\phi_K)1$$

and evaluate both sides. We find

$$(\phi_H^{-1}\,ch_A\,\phi_K)\,(\rho_A^k\,\xi) = (\phi_H^{-1}\,\Psi_H^k\,\phi_H)\,(bh_A\,\xi);$$

using Corollary (5.4) and the remark above, we have

$$(bh_A\,\xi).(ch_A\,\rho_A^k\,\xi) = k^n\,(\Psi_H^k\,bh_A\,\xi).$$

This completes the proof.

We will next carry out the analogue, for our context, of the proof that Stiefel–Whitney classes are fibre-homotopy invariants. We suppose given a commutative diagram of the following form; it is not assumed that it arises from a bundle map.

$$\begin{array}{ccc} \bar{E}_1, E_1 & \xrightarrow{\;g\;} & \bar{E}_2, E_2 \\ \downarrow{\scriptstyle p_1} & & \downarrow{\scriptstyle p_2} \\ B_1 & \xrightarrow{\;f\;} & B_2 \end{array}$$

We define

$$k = \phi_{K,1}^{-1}\,g^*\,\phi_{K,2}(1) \in K^*(B_1)$$

$$h = \phi_{H,1}^{-1}\,g^*\,\phi_{H,2}(1) \in H^*(B_1).$$

PROPOSITION (5.7). *We have*

$$h.f^*c(T, \xi_2) = c(T, \xi_1).(Tk).$$

Proof. We have $g^*T = Tg^*$. Consider the equation

$$(\phi^{-1}g^*\phi)(\phi^{-1}T\phi)1 = (\phi^{-1}T\phi)(\phi^{-1}g^*\phi)1,$$

where $1 \in K^*(B_2)$, and the suffix for each φ can be determined from the context. The equation yields

$$(\phi^{-1}g^*\phi)\,c(T, \xi_2) = (\phi^{-1}T\phi)k,$$

or using Lemma (5.3),

$$h \cdot f^*c(T, \xi_2) = c(T, \xi_1) \cdot (Tk).$$

This completes the proof.

COROLLARY (5.8). *Let ξ_1, ξ_2 be unitary bundles over B in case $\Lambda = \mathbf{C}$, or Spin(8n)-bundles ($n = 1, 2, \ldots$) in case $\Lambda = \mathbf{R}$. If the sphere-bundles associated with ξ_1, ξ_2 are fibre-homotopy equivalent, then there exists an element $x_\Lambda \in \tilde{K}_\Lambda(B)$ such that*

$$bh_\Lambda \xi_2 = bh_\Lambda \xi_1 \cdot ch_\Lambda(1 + x_\Lambda)$$

$$\rho^l_\Lambda \xi_2 = \rho^l_\Lambda \xi_1 \cdot \frac{\Psi^l_\Lambda(1 + x_\Lambda)}{1 + x_\Lambda}.$$

Note that in the second equation x_Λ is independent of l.

Proof. If the sphere-bundles associated with ξ_1, ξ_2 are fibre-homotopy equivalent, then there is a diagram

$$\bar{E}_1, E_1 \xrightarrow{g} \bar{E}_2, E_2$$
$$\searrow_{p_1} \quad \swarrow_{p_2}$$
$$B$$

in which g has degree ± 1 on each fibre. We may therefore apply Proposition (5.7) with $f = 1$. The result involves $k = \phi_{K,1}^{-1} g^* \phi_{K,2}(1)$. We may determine the virtual dimension of k over each component of B by restricting on a single fibre; we find that this virtual dimension is ± 1 (according to the degree of g). Let ε be a trivial virtual bundle with the same virtual dimension as k on each component of B; then we have $\varepsilon k = 1 + x_\Lambda$ for some $x_\Lambda \in \tilde{K}_\Lambda(B)$.

Consider now the case of ρ^l_Λ. Since $h = k$, the result of Proposition (5.7) is

$$k \cdot (\rho^l_\Lambda \xi_2) = (\rho^l_\Lambda \xi_1) \cdot (\Psi^l_\Lambda k).$$

Multiplying by the equation $\varepsilon = \Psi^l_\Lambda$ or ε, we find

$$(1 + x_\Lambda) \cdot (\rho^l_\Lambda \xi_2) = (\rho^l_\Lambda \xi_1) \cdot \Psi^l_\Lambda(1 + x_\Lambda).$$

Now, $1 + x_\Lambda$ is invertible. (If B is finite this follows from the usual power-series for $(1 + x_\Lambda)^{-1}$; but in any case, if we take an equivalence γ inverse to g, we obtain an element $\phi_{K,2}^{-1} \gamma^* \phi_{K,1}(1)$ inverse to k.) We thus obtain

$$\rho^l_\Lambda \xi_2 = \rho^l_\Lambda \xi_1 \cdot \frac{\Psi^l_\Lambda(1 + x_\Lambda)}{1 + x_\Lambda}.$$

The case of bh_A is closely similar. We have $(ch_A\, \varepsilon)\ h = 1$ in $H^0(B)$. The result of Proposition (5.7) is

$$h \,.\, bh_A\, \xi_2 = bh_A\, \xi_1 \,.\, ch_A\, k.$$

Multiplying by $ch_A\, \varepsilon$, we find

$$bh_A\, \xi_2 = bh_A\, \xi_1 \,.\, ch_A(1 + x_A).$$

This completes the proof.

We now turn to a result useful for calculating ρ^k_Λ.

THEOREM (5.9). *If ξ is a $U(n)$-bundle then $\rho^k_C\, \xi$ is induced from ξ by the virtual representation whose character is*

$$\prod_{1 \leq r \leq n} \frac{z_r^k - 1}{z_r - 1} = \prod_{1 \leq r \leq n} (z_r^{k-1} + z_r^{k-2} + \ldots + z_r + 1).$$

If ξ is a Spin(8*n*)*-bundle then $\rho^k_R\, \xi$ is induced from ξ by the real virtual representation whose character is*

$$\prod_{1 \leq r \leq 4n} \frac{z_r^{\frac{1}{2}k} - z_r^{-\frac{1}{2}k}}{z_r^{\frac{1}{2}} - z_r^{-\frac{1}{2}}} = \prod_{1 \leq r \leq 4n} (z_r^{\frac{1}{2}(k-1)} + z_r^{\frac{1}{2}(k-3)} + \ldots + z_r^{-\frac{1}{2}(k-1)}).$$

This theorem was first published by Bott [8, 9]. It follows fairly easily from the definition of the Thom isomorphism ϕ used in [8, 9]. However, it is shown in [18] that this definition coincides with the definition given in [18].

In the above, we have defined $\rho^k_\Lambda(\xi)$ for suitable bundles ξ. Next we shall seek to extend the definition of ρ^h_Λ from bundles to virtual bundles.

We shall see that on bundles, ρ^k is 'homomorphic from addition to multiplication', or more shortly, 'exponential'. Also if τ is the trivial bundle of dimension $2n$ over the reals, we have $\rho^k(\tau) = k^n$. Therefore we are forced to define $\rho^k(-\tau) = k^{-n}$. This indicates that we can define ρ^k on virtual bundles only at the price of introducing denominators. We shall therefore define Q_k to be the additive group of fractions of the form p/k^q, where p and q are integers. If k is a virtual bundle over X, we shall seek to define $\rho^k(k)$ as an element of $K_A(X) \otimes Q_k$. More generally, we may be willing to consider $K_A(X) \otimes S$, where S is a suitable subring of C.

We face a similar situation if we try to define the composite $\rho^k\, \theta$, where θ is a virtual representation. (In Part III we shall be forced to consider such composites.) In this case we are forced, not only to introduce denominators, but also to introduce the completion of the representation ring. Let G be a compact connected Lie group, let $\Lambda = R$ or C, and let S be a subring of C. Let $K'_\Lambda(G)$ be the representation ring of G; then we can form $K'_\Lambda(G) \otimes S$. In $K'_\Lambda(G) \otimes S$ we take the ideal I consisting of elements of virtual dimension zero; these are the elements whose characters vanish at the identity of G. We can now form

$$\text{Comp}(K'_\Lambda(G) \otimes S) = \operatorname*{Inv\,Lim}_{m \to \infty} \frac{K'_\Lambda(G) \otimes S}{I^m}.$$

If θ is a virtual representation of G, we shall seek to define $\rho^k\theta$ as an element of $\mathrm{Comp}(K'_\Lambda(G) \otimes Q_k)$.

If X is a finite CW-complex, one may complete $K_\Lambda(X) \otimes S$ in a similar way. However, $\mathrm{Comp}(K_\Lambda(X) \otimes S)$ can be identified with $K_\Lambda(X) \otimes S$. For let I be the ideal of elements of virtual dimension zero in $K_\Lambda(X) \otimes S$, and let q be the dimension of X; then, as in [7], we have $(\tilde{K}_\Lambda(X))^{q+1} = 0$, so that $I^{q+1} = 0$, and

$$\frac{K_\Lambda(X) \otimes S}{I^m} = K_\Lambda(X) \otimes S$$

for $m \geqslant q + 1$.

We shall require the following lemma on completions. Here the letter K stands for an augmented ring, which in the applications becomes $K_\Lambda(X)$ or $K'_\Lambda(G)$.

LEMMA (5.10)(a). *An element of* $\mathrm{Comp}(K \otimes S)$ *whose virtual dimension is invertible in* S *is invertible in* $\mathrm{Comp}(K \otimes S)$.

(b) *An element* $\hat{c}$ *of* $\mathrm{Comp}(K \otimes S)$ *has in* $\mathrm{Comp}(K \otimes S)$ *a square root, unique up to sign, provided that the virtual dimension of* $\hat{c}$ *has a square root* s *in* S *and* $2s$ *is invertible in* S.

Proof of (b). Suppose given s, as in the data. Let c_m be the component of $\hat{c}$ in $(K \otimes S)/I^m$. In $(K \otimes S)/I^m$ we seek an element of the form $s + i_m$, where $i_m \in I/I^m$, such that

$$(s + i_m)^2 = c_m .$$

For $m = 1$ such an element exists and is unique. Suppose, as an inductive hypothesis, that such an element exists and is unique for some value of m. Choose in $(K \otimes S)/I^{m+1}$ a trial element $s + j_m$ mapping to $s + i_m$ in $(K \otimes S)/I^m$; then we have

$$(s + j_m)^2 = c_m - \varepsilon_m ,$$

where $\varepsilon_m \in I^m/I^{m+1}$. If $(K \otimes S)/I^{m+1}$ contains a square root $s + i_{m+1}$ for c_{m+1} at all, we can write this square root in the form $s + j_m + \delta_m$; and since the square root $s + i_m$ in $(K \otimes S)/I^m$ is unique, we must have $\delta_m \in I^m/I^{m+1}$. If we assume that $\delta_m \in I^m/I^{m+1}$, the equation

$$(s + j_m + \delta_m)^2 = c_{m+1} \quad \text{in} \quad (K \otimes S)/I^{m+1}$$

is equivalent to

$$2s\delta_m = \varepsilon_m \quad \text{in} \quad I^m/I^{m+1}.$$

By assumption, this equation has a unique solution for δ_m. This completes the induction. We have shown that for each m, c_m has a unique square root of virtual dimension s in $(K \otimes S)/I^m$. This yields the result stated.

Part (a) may be proved similarly, or by using the power-series for $(1 + x)^{-1}$.

The work to be done in defining $\rho^k(\kappa)$ is very similar to that in [1, pp. 606–609]. We follow [1] and adopt a convention. The letters f, g will denote maps of complexes such as X. The letters ξ, η will denote bundles; the letters κ, λ will denote elements of $K_\Lambda(X)$; the letters μ, ν will denote elements of $K_\Lambda(X) \otimes S$. The letters α, β will denote representations; the letters θ, ϕ will denote elements of $K'_\Lambda(G)$; the letters ψ, ρ will denote elements of $K'_\Lambda(G) \otimes S$; the letters $\hat{\psi}$, $\hat{\rho}$ will denote elements of $\mathrm{Comp}(K'_\Lambda(G) \otimes S)$. Initially, as on [1, p. 606], we have composites

$$\beta . \alpha, \qquad \alpha . \xi, \qquad \xi . f, \qquad f . g.$$

By linearising over the first factor, as on [1, p. 607], we obtain composites

$$\phi . \alpha, \qquad \theta . \xi, \qquad \kappa . f$$

By S-linearity over the first factor we obtain composites

$$\rho . \alpha, \qquad \psi . \xi, \qquad \mu . f$$

lying in appropriate groups $K \otimes S$. Since composition with a factor on the right preserves virtual dimensions and tensor-products, we obtain composites

$$\hat{\rho} . \alpha, \qquad \hat{\psi} . \xi.$$

These lie in the appropriate groups

$$\mathrm{Comp}(K'_\Lambda(G) \otimes S), \ \mathrm{Comp}(K_\Lambda(X) \otimes S) = K_\Lambda(X) \otimes S.$$

We have remarked earlier that ρ^k is 'homomorphic from addition to multiplication', that is, 'exponential'. We will now make this notion more precise. Let $G(n)$ be one of the series of groups $U(dn)$, $SO(dn)$ or $\mathrm{Spin}(dn)$ (for some integer d). Let

$$\pi : G(n) \times G(m) \longrightarrow G(n),$$
$$\varpi : G(n) \times G(m) \longrightarrow G(m)$$

be the projections of $G(n) \times G(m)$ onto its two factors; thus

$$\pi \oplus \varpi : G(n) \times G(m) \longrightarrow G(n+m)$$

is the 'universal Whitney sum map'. Here the universal Whitney sum map

$$\pi \oplus \varpi : \mathrm{Spin}(dn) \times \mathrm{Spin}(dm) \longrightarrow \mathrm{Spin}(d(n+m))$$

is constructed by lifting the map

$$\pi \oplus \varpi : SO(dn) \times SO(dm) \longrightarrow SO(d(n+m)).$$

For each n, let ρ_n be an element of $K'_\Lambda(G(n)) \otimes S$. We will say that the sequence $\rho = (\rho_n)$ is 'exponential' if we have

$$\rho_{n+m} . (\pi \oplus \varpi) = (\rho_n . \pi) \otimes (\rho_m . \varpi)$$

for all n, m. The sides of this equation lie in $K'_\Lambda(G(n) \times G(m)) \otimes S$; cf [1, p. 607, 609]. Similarly for a sequence

$$\hat{\rho}_n \in \mathrm{Comp}(K'_\Lambda(G(n)) \otimes S).$$

LEMMA (5.11). *If the sequence $\rho = (\rho_n)$ is exponential, then for any two representations $\alpha : H \to G(n)$, $\beta : H \to G(m)$ we have*

$$\rho_{n+m} . (\alpha \oplus \beta) = (\rho_n . \alpha) \otimes (\rho_m . \beta).$$

Moreover, for any two bundles ξ, η with groups $G(n)$, $G(m)$ we have

$$\rho_{n+m} . (\xi \oplus \eta) = (\rho_n . \xi) \otimes (\rho_n . \eta).$$

Similarly for a sequence

$$\hat{\rho} = (\hat{\rho}_n).$$

This lemma is strictly analogous to those of [1, pp. 607–609], and so is its proof.

For the next lemma, we introduce the Grothendieck groups $K_G(X)$, $K'_G(H)$. Here $K_G(X)$ is defined in the obvious fashion using bundles over X with group $G(n)$ for $n = 1, 2, \ldots$; similarly, $K'_G(H)$ is defined using representations $\alpha : H \to G(n)$ of the group H. If $G(n)$ is the sequence of groups $SO(dn)$ or $\mathrm{Spin}(dn)$, we write $K_{SO(d)}$ or $K_{\mathrm{Spin}(d)}$ for the resulting Grothendieck groups K_G. Thus (for example) $K_{\mathrm{Spin}(8)}(X)$ is defined in terms of bundles with structural group $\mathrm{Spin}(8n)$ for $n = 1, 2, \ldots$. If $d = 1$, we write $K_{SO}(X)$ for $K_{SO(1)}(X)$. The group $K_{SO}(X)$ is monomorphically embedded in $K_R(X)$ as the subgroup of classes κ such that $w_1(\kappa) = 0$. Under the decomposition

$$K_R(X) = Z + \tilde{K}_R(X),$$

we have

$$K_{SO(d)}(X) = dZ + \tilde{K}_{SO}(X).$$

We suppose given an exponential sequence $\hat{\rho} = (\hat{\rho}_n)$, where $\hat{\rho}_n \in \mathrm{Comp}(K'_\Lambda(G(n)) \otimes S)$. We assume that the virtual dimension of $\hat{\rho}_1$ is invertible in S.

LEMMA (5.12). *If $\theta \in K'_G(H)$, $\kappa \in K_G(X)$ it is possible to form composites*

$$\hat{\rho} \,.\, \theta \in \mathrm{Comp}(K'_\Lambda(H) \otimes S)$$

$$\hat{\rho} \,.\, \kappa \in \mathrm{Comp}(K_\Lambda(X) \otimes S) = K_\Lambda(X) \otimes S$$

so that these have the following properties.

(i) *$\hat{\rho}$ is exponential (in the obvious sense).*

(ii) *If we replace θ by α or κ by ξ, then these composites reduce to those considered above.*

Proof. If the virtual dimension of $\hat{\rho}_1$ is s, then the virtual dimension of $\hat{\rho}_n$ is s^n (since $\hat{\rho}$ is exponential). If s is invertible in S, so is s^n. Lemma (5.10)(a) now shows that every element $\hat{\rho}_n \cdot \alpha$ or $\hat{\rho}_n \cdot \xi$ is invertible. Therefore $\hat{\rho}$ can be defined, so as to be exponential, on the free abelian group F generated by the isomorphism classes of such α or ξ. It remains to show that $\hat{\rho}$ passes to the quotient, so that it is defined on $K'_G(H)$ or $K_G(X)$. This follows from the fact that $(\hat{\rho}_n)$ is exponential, using Lemma (5.11). This completes the proof.

Finally, we introduce one further generalised composite. Suppose that $\theta = (\theta_n)$ is an additive sequence of virtual representations with $\theta_n \in K'_\Lambda(G(n))$. Then by S-linearity over the second factor we can define composites

$$\theta \,.\, \mu, \qquad \theta \,.\, \rho$$

lying in the appropriate groups $K \otimes S$. If (moreover) θ is multiplicative and maps elements of virtual degree zero into elements of virtual degree zero, then we can define composites

$$\theta \,.\, \hat{\rho}$$

lying in the appropriate group $\mathrm{Comp}(K \otimes S)$. In practice this situation arises when θ is the sequence Ψ^k [1, §4].

As in [1, pp. 606–609], the appropriate associative laws continue to hold for all the composites we have discussed.

We shall now give some examples of exponential sequences.

EXAMPLE (5.13). Let ρ_n^k be the element in $K'_C(U(n))$ with character

$$\prod_{1 \leq r \leq n} \frac{z_r^k - 1}{z_r - 1}.$$

The sequence (ρ_n^k) is exponential, as one verifies immediately by checking characters. The virtual dimension of ρ_1^k is k (since $(z_1)^{k-1} + \ldots + z_1 + 1$ takes the value 1 at $z_1 = 1$). If $k = 0$, then $\rho_n^k = 0$; otherwise k is invertible in Q_k. The foregoing theory applies; if $\kappa \in K_C(X)$, we can define $\rho^k.\kappa$ as an element of $K_C(X) \otimes Q_k$. If κ is represented by a $U(n)$-bundle ξ, then $\rho^k.\kappa = \rho_C^k(\xi)$, according to Theorem (5.9).

We now turn to the "real" case. We have already defined $\rho_R^k(\xi)$ when ξ is a Spin($8n$)-bundle over X. It would therefore be plausible to define $\rho^k(\kappa)$ when $\kappa \in K_{\mathrm{Spin}(8)}(X)$. Actually we shall do more; we shall define $\rho^k(\kappa)$ when $\kappa \in K_{SO(2)}(X)$. (That is, κ may be a linear combination of $SO(2n)$-bundles for $n = 1, 2, \ldots$). For this purpose we need to distinguish the cases 'k odd' and 'k even'.

EXAMPLE (5.14). Assume that k is odd. Consider the formula

$$\prod_{1 \leq r \leq n} \frac{z_r^{\frac{1}{2}k} - z_r^{-\frac{1}{2}k}}{z_r^{\frac{1}{2}} - z_r^{-\frac{1}{2}}} = \prod_{1 \leq r \leq n} (z_r^{\frac{1}{2}(k-1)} + z_r^{\frac{1}{2}(k-3)} + \ldots + z_r^{-\frac{1}{2}(k-1)}).$$

It represents a polynomial in which the z_r occur to integral powers. It is also invariant under the Weyl group of $SO(2n)$; therefore it is the character of some virtual representation ρ_n^k of $SO(2n)$. We will show that this virtual representation is real. It must be real if n is even, because every virtual representation of $SO(4m)$ is real. It is also clear that the restriction of ρ_{2m}^k to $SO(4m\text{-}2)$ is $k\rho_{2m-1}^k$. In order to prove that ρ_{2m-1}^k is real, it is sufficient to recall the general fact that if $k\theta$ is real and k is odd, then θ is real. In fact, irreducible representations can be divided into those which coincide with their complex conjugate and those which do not; and the former can be divided into real and quaternionic representations. In order that a representation be real it is necessary and sufficient that it contain each quaternionic irreducible representation an even number of times, and each irreducible representation the same number of times as its complex conjugate when that is a distinct representation. If this condition holds for $k\theta$, it holds for θ.

Alternatively, assuming a little more representation-theory, we can argue that the given formula is invariant under the Weyl group of $O(2n)$; thus ρ_n^k can be written as a polynomial in the exterior powers, so it is real.

We have therefore established the existence of a sequence of virtual representations

$$\rho_n^k \in K'_R(SO(2n))$$

with the characters given. This sequence is exponential, as one verifies immediately by checking characters. The virtual dimension of ρ_1^k is k. The foregoing theory applies; if $\kappa \in K_{SO(2)}(X)$, we can define $\rho^k\kappa$ as an element of $K_R(X) \otimes Q_k$. If κ is represented by a

Spin($8n$)-bundle ξ, then $\rho^k_! \kappa = \rho^k_R(\xi)$, according to Theorem (5.9). This completes the case 'k odd'.

EXAMPLE (5.15). Assume that k is even. We will now construct a sequence of elements

$$\rho^k_n \in \mathrm{Comp}(K'_R(SO(2n)) \otimes Q_k).$$

First, let θ^k_n be the virtual representation of Spin($2n$) with character

$$\prod_{1 \leqslant r \leqslant n} \frac{z_r^{\frac{1}{2}k} - z_r^{-\frac{1}{2}k}}{z_r^{\frac{1}{2}} - z_r^{-\frac{1}{2}}}.$$

Since this character is real, it follows that $2\theta^k_n$ is a real virtual representation. Since k is even, $\frac{1}{2} \in Q_k$ and we have

$$\theta^k_n \in K'_R(\mathrm{Spin}(2n)) \otimes Q_k.$$

We now remark that if $S \subset C$, the character of an element of $K'_\Lambda(G) \otimes S$ is defined, and is a finite Laurent series in the z_r with coefficients in S. The elements of $K'_\Lambda(G) \otimes S$ are distinguished by their characters. Therefore we can prove that the sequence θ^k_n is exponential, by checking characters in the obvious way.

Next, consider the map

$$1 \oplus 1 : SO(2n) \longrightarrow SO(4n).$$

This sends a matrix M into

$$\begin{bmatrix} M & O \\ O & M \end{bmatrix}.$$

It can be lifted to a unique homomorphism

$$\tau : SO(2n) \longrightarrow \mathrm{Spin}(4n).$$

We will now define

$$\rho^k_n = (\theta^k_{2n} \cdot \tau)^{\frac{1}{2}} \in \mathrm{Comp}(K'_R(SO(2n)) \otimes Q_k).$$

Here the square root exists by Lemma (5.10)(b); we choose its sign so that the virtual dimension of ρ^k_n is k^n.

It is now easy to check that the sequence ρ^k_n is exponential, using the fact that θ^k_n is exponential and the fact that square roots are unique (Lemma (5.10)(b)).

The virtual dimension of ρ^k_1 is k. The foregoing theory applies; if $\kappa \in K_{SO(2)}(X)$, we can define $\rho^k_! \kappa$ as an element of $K_R(X) \otimes Q_k$.

Using Theorem (5.9), it is easy to check that if κ is represented by a Spin($8n$)-bundle ξ, then $\rho^k_! \kappa = \rho^k_R(\xi)$. This completes the case 'k even'.

We add one note. The character of an element

$$\hat{\rho} \in \mathrm{Comp}(K'_\Lambda(G) \otimes S)$$

can be defined, and is a formal power-series in the variables $\zeta_r = z_r - 1$ with coefficients in

S. Cf [7]; we shall give further details in Part III. One can then check that the character of ρ_n^k is

$$\prod_{1 \leqslant r \leqslant n} \frac{z_r^{\frac{1}{2}k} - z_r^{-\frac{1}{2}k}}{z_r^{\frac{1}{2}} - z_r^{-\frac{1}{2}}},$$

where this formula is interpreted using the expansion of $(z_r)^m$ as a binomial series in ζ_r.

We turn next to the calculation of the operations ρ_R^k for the space $X = RP^n$. We recall the structure of $\tilde{K}_R(RP^n)$ from [1, Theorem (7.4)]. Let ξ be the canonical line bundle over RP^n; then $\xi^2 = 1$, and $\lambda = \xi - 1$ is a generator in $K_R(RP^n)$, which is cyclic of order 2^f, where f is the number of integers s such that $0 < s \leqslant n$ and $s \equiv 0, 1, 2$ or $4 \bmod 8$. In particular, $\tilde{K}_R(RP^n) \otimes Q_k = 0$ if k is even. The only case of interest is therefore that in which k is odd. The operation ρ^k is defined on $K_{SO(2)}(X)$, so that ρ^k is defined on all multiples of 2λ in $\tilde{K}_R(RP^n)$. The value $\rho^k(2l\lambda)$ will lie in the multiplicative group $1 + \tilde{K}_R(RP^n) \otimes Q_k \cong 1 + \tilde{K}_R(RP^n)$ of elements of virtual dimension 1. In order to make the structure of this group more transparent, let $J_{2^{f+1}}$ be the ring of residue classes mod 2^{f+1}, and let $G_{2^{f+1}}$ be the multiplicative group of odd residue classes mod 2^{f+1}. Then we have $J_{2^{f+1}} \otimes Q_k \cong J_{2^{f+1}}$ for k odd; and we can define a ring homomorphism

$$\alpha : K_R(RP^n) \otimes Q_k \longrightarrow J_{2^{f+1}} \otimes Q_k$$

by setting $\alpha(\xi) = -1$, or equivalently $\alpha(\lambda) = -2$. The map α induces an isomorphism of $1 + \tilde{K}_R(RP^n) \otimes Q_k$ onto $G_{2^{f+1}}$. The subgroup $1 + \tilde{K}_{SO}(RP^n) \otimes Q_k$ (defined in term of orientable bundles) maps by α onto the group of residue classes congruent to 1 mod 4.

THEOREM (5.16). *The operations ρ^k on $\tilde{K}_{SO}(RP^n)$ are given by*

$$\rho^k(2l\lambda) = 1 + \frac{k^l - \varepsilon^l}{2k^l}\lambda$$

where $$\varepsilon = \begin{cases} 1 & \text{if } k \equiv 1 \bmod 4 \\ -1 & \text{if } k \equiv 3 \bmod 4. \end{cases}$$

Equivalently, they are given by

$$\alpha\rho^k(2l\lambda) = \left(\frac{\varepsilon}{k}\right)^l.$$

Remark (*a*). For l divisible by 4 this result has been found by Bott [8, 9]. Similar calculations have been made by Atiyah (private communication).

Remark (*b*). The values of ρ^k lie in $1 + \tilde{K}_{SO}(RP^n) \otimes Q_k$. This must necessarily happen, since all the representations we have used map into $SO(m)$.

Remark (*c*). One can choose an odd number k so that k (and hence ε/k) has the maximum possible order in $G_{2^{f+1}}$; then ε/k will have no square root in $G_{2^{f+1}}$; this proves that it is impossible to define ρ^k on λ so as to preserve the exponential property.

Proof. We have $2l\xi = i\xi$, where $i : O(1) \to SO(2l)$ maps ± 1 into ± 1. We will compute the representation $\rho_l^k i$ of $O(1)$, where ρ_l^k is as in (5.14). The value of the character of ρ_l^k at 1 is k^l. If we substitute $z_r = -1$, the value of

$$z_r^{\frac{1}{2}(k-1)} + z_r^{\frac{1}{2}(k-3)} + \dots + z_r^{-\frac{1}{2}(k-1)}$$

is ε, where

$$\varepsilon = \begin{cases} 1 & \text{if} \quad k \equiv 1 \bmod 4 \\ -1 & \text{if} \quad k \equiv 3 \bmod 4 \end{cases}$$

Therefore the value of the character at -1 is ε^l. Let λ^1 be the identity representation of $O(1)$; we conclude that

$$\rho_l^k\, i = a + b\lambda^1$$

where

$$a + b = k^l$$
$$a - b = \varepsilon^l.$$

Therefore

$$\rho_l^k\,(2l\xi) = \rho_l^k\, i\, \xi = a + b\xi,$$
$$\alpha\, \rho_l^k\,(2l\xi) = a - b = \varepsilon^l,$$
$$\alpha\, \rho_l^k\,(2l\lambda) = \frac{\varepsilon^l}{k^l}.$$

The same method yields the first part of the theorem. This completes the proof.

We will now consider the case $X = S^n$, where $n \equiv 1$ or $2 \bmod 8$. In this case the group $\tilde{K}_R(S^n)$ is Z_2, and we have $\tilde{K}_R(S^n) \otimes Q_k = 0$ for k even. The only case of interest is therefore again that in which k is odd. The operation ρ^k is defined on $\tilde{K}_R(S^n)$ for $n \geqslant 2$.

THEOREM (5.17). *If $n \equiv 1$ or $2 \bmod 8$ and $n \geqslant 2$, then the operations ρ^k on $\tilde{K}_R(S^n)$ are given by*

$$\rho^k x = \begin{cases} 1 & \text{if} \quad k \equiv \pm 1 \bmod 8 \\ 1 + x & \text{if} \quad k \equiv \pm 3 \bmod 8. \end{cases}$$

Proof. Consider a map $g : RP^n \to S^n$ of degree 1, as in 3.5. The map

$$g^* : \tilde{K}_R(S^n) \longrightarrow \tilde{K}_R(RP^n)$$

is monomorphic, by the proof of [1, Theorem (7.4)]. We can now compute $\rho^k x$ by naturality. If $x = 0$ the result is trivial, so we may assume $x \neq 0$; then $g^* x = 2^{f-1}\lambda$, where 2^f is the order of λ. By Lemma (2.9), the group $G_{2^{f+1}}$ is $Z_2 + Z_{2^{f-1}}$. The element ε/k (see Theorem (5.16)) has multiplicative order dividing 2^{f-2} if $k \equiv \pm 1 \bmod 8$, or 2^{f-1} if $k \equiv \pm 3 \bmod 8$. Thus

$$\alpha\, \rho^k\,(2^{f-1}\lambda) = (\varepsilon/k)^{2^{f-2}}$$

is equal to 1 if $k \equiv \pm 1 \bmod 8$, and otherwise not equal to 1. Thus $\rho^k x$ is equal to 1 or not according to the case; but if it is not 1, it can only be $1 + x$. This completes the proof.

It remains to consider the case $X = S^{4n}$.

THEOREM (5.18). *If $x \in \tilde{K}_R(S^{4n})$ then*

$$\rho^k(x) = 1 + \tfrac{1}{2}(k^{2n} - 1)\,\alpha_{2n}\, x,$$

where α_{2n} is as in §2.

Remark. The coefficient $\frac{1}{2}(k^{2n}-1)\alpha_{2n}$ lies in Q_k, according to Theorem (2.7).

Proof. We may suppose that x is a linear combination of Spin($8m$)-bundles for various values of m. By applying Proposition (5.6) to each bundle, we find

$$(shx).(ch\, c\, \rho^k x) = \Psi^k_H\, sh\, x.$$

By Corollary (5.2) we have

$$sh\, x = 1 + \tfrac{1}{2}\alpha_{2n}\, ch_{2n}\, cx,$$

so that

$$\Psi^k_H\, sh\, x = 1 + \tfrac{1}{2}k^{2n}\, \alpha_{2n}\, ch_{2n}\, cx$$

and

$$ch\, c\, \rho^k x = 1 + \tfrac{1}{2}(k^{2n}-1)\, \alpha_{2n}\, ch_{2n}\, cx.$$

Hence

$$\rho^k x = 1 + \tfrac{1}{2}(k^{2n}-1)\, \alpha_{2n}\, x.$$

This completes the proof.

§6. THE GROUP $J'(X)$.

In this section we shall introduce the group $J'(X)$. We shall prove (Theorem (6.1)) that $J'(X)$ is a lower bound for $J(X)$. We also compute the groups $J'(X)$ when $X = RP^n$ or S^n.

We will now give the definition of J'. First recall that $K_{SO(2)}(X)$ is monomorphically embedded in $K_R(X)$ as the subgroup of elements x such that (i) the first Stiefel–Whitney class $w_1(x)$ is zero, and (ii) the virtual dimension of x is even. We define $V(X)$ to be the subgroup of elements $x \in K_{SO(2)}(X)$ which satisfy the following condition: there exists $y \in \tilde{K}_R(X)$ such that

$$\rho^k x = \frac{\Psi^k(1+y)}{1+y} \quad \text{in} \quad K_R(X) \otimes Q_k$$

for all $k \neq 0$. We now define

$$J'(X) = K_R(X)/V(X).$$

It is necessary to check that $V(X)$ is a subgroup. We first note that any x which satisfies the condition given has virtual dimension zero. Let $1 + \tilde{K}_R(X) \otimes Q_k$ be the multiplicative group of elements of virtual dimension 1 in $K_R(X) \otimes Q_k$. Let Π be the multiplicative group

$$\prod_{k \neq 0} (1 + \tilde{K}_R(X) \otimes Q_k).$$

Let us define a function

$$\delta : 1 + \tilde{K}_R(X) \longrightarrow \prod$$

by

$$\delta(1+y) = \left\{\frac{\Psi^k(1+y)}{1+y}\right\};$$

then δ is a homomorphism, because Ψ^k is multiplicative for each k. Similarly, the function

$$\rho : \tilde{K}_{SO}(X) \longrightarrow \prod$$

defined by

$$\rho(x) = \{\rho^k(x)\}$$

is a homomorphism. Therefore the set

$$V(X) = \rho^{-1}\delta(1 + \tilde{K}_R(X))$$

is a subgroup.

THEOREM (6.1). *$J'(X)$ is a lower bound for $J(X)$, in the sense of Part I* [4].

We recall that in Part I we defined

$$J(X) = K_R/T(X);$$

here $T(X)$ is the subgroup of $K_R(X)$ generated by elements of the form $\{\xi\} - \{\eta\}$, where ξ and η are orthogonal bundles whose associated sphere-bundles are fibre homotopy equivalent. The theorem states that $T(X) \subset V(X)$, so that the quotient map $K_R(X) \to J'(X)$ factors through $J(X)$.

Proof. Suppose given a finite connected CW-complex X. I claim that $T(X)$ is generated by elements $\{\xi'\} - \{\eta'\}$, where ξ' and η' are orthogonal bundles whose associated sphere bundles are fibre homotopy equivalent, and η' is trivial of dimension divisible by 8. In fact, let ξ, η be orthogonal bundles over X whose associated sphere-bundles are fibre homotopy equivalent. Then the same is true for $\xi \oplus \zeta$ and $\eta \oplus \zeta$, whatever the bundle ζ. We have

$$\{\xi \oplus \zeta\} - \{\eta \oplus \zeta\} = \{\xi\} - \{\eta\}.$$

We can choose ζ so that $\eta \oplus \zeta$ is a trivial bundle of dimension divisible by 8.

Let ξ', η' be as above. Then the Stiefel–Whitney classes of ξ' are zero, since the Stiefel–Whitney classes are fibre homotopy invariants. Thus we can lift ξ' to a Spin($8n$)-bundle. Corollary (5.8) applies, and shows that there exists $y \in \tilde{K}_R(X)$ such that

$$\rho^k(\xi') = \rho^k(\eta') \, . \, \frac{\Psi^k(1+y)}{1+y}$$

for all k. That is,

$$\rho^k(\{\xi'\} - \{\eta'\}) = \frac{\Psi^k(1+y)}{1+y}$$

in $K_R(X) \otimes Q_k$. This shows that $\{\xi'\} - \{\eta'\} \in V(X)$; thus $T(X) \subset V(X)$. This completes the proof.

By way of illustration, we will now calculate the groups J' for the examples considered in §3.

EXAMPLE (6.3). Take $X = RP^n$. Then the quotient map

$$K_R(RP^n) \longrightarrow J'(RP^n)$$

is an isomorphism, and consequently the quotient map

$$K_R(RP^n) \longrightarrow J(RP^n)$$

is an isomorphism.

The fact that the results of [1] can be rephrased in this way is an observation of ATIYAH (private communication) and of BOTT [8, 9].

Proof. As observed above, the group $\tilde{K}_R(RP^n)$ is of order 2^f, and therefore $\tilde{K}_R(RP^n) \otimes Q_k$ is zero for k even. It is therefore sufficient to consider odd values of k, for which

$$\tilde{K}_R(RP^n) \otimes Q_k \cong \tilde{K}_R(RP^n).$$

According to [1, Theorem (7.4), p. 625], for k odd and $y \in \tilde{K}_R(RP^n)$ we have

$$\frac{\Psi^k(1+y)}{1+y} = 1.$$

The elements $v \in V(RP^n)$ have therefore to satisfy the conditions

(i) $w_1(v) = 0$

(ii) $\rho^k(v) = 1$ for k odd.

The first condition ensures that $v = 2l\lambda$, and disposes entirely of the low-dimensional case $n = 1$. By Theorem (5.16), if $k \equiv \pm 3 \bmod 8$ the element $\rho^k(2\lambda)$ has order 2^{f-1} in the multiplicative group $1 + \tilde{K}_R(RP^n)$. (Here the integer f is as in §5. The same application of Theorem (5.16) was made in the proof of Theorem (5.17).) Therefore the condition

$$\rho^k(2l\lambda) = 1, \qquad \text{all odd } k$$

implies that l is divisible by 2^{f-1}, i.e. that $2l\lambda = 0$ in $\tilde{K}_R(RP^n)$. Thus $V_R(RP^n) = 0$. This completes the proof.

EXAMPLE (6.4). Take $X = S^n$ with $n \equiv 1$ or $2 \bmod 8$. Then the quotient map

$$K_R(S^n) \longrightarrow J'(S^n)$$

is an isomorphism, and consequently the quotient map

$$K_R(S^n) \longrightarrow J(S^n)$$

is an isomorphism. Equivalently, the image $J(\pi_{n-1}(SO))$ of the stable J-homomorphism is Z_2 for $n \equiv 1$ or $2 \bmod 8$.

Proof. As in proving Theorem (5.17), we consider a map $f : RP^n \to S^n$ of degree 1, so that the induced map

$$f^* : \tilde{K}_R(S^n) \longrightarrow \tilde{K}_R(RP^n)$$

is monomorphic. Consider the following commutative diagram.

$$\begin{array}{ccc} K_R(S^n) & \xrightarrow{K(f)} & K_R(RP^n) \\ \downarrow{\scriptstyle q_S} & & \downarrow{\scriptstyle q_P} \\ J'(S^n) & \xrightarrow{J'(f)} & J'(RP^n) \end{array}$$

Since $K(f)$ is monomorphic and q_P is an isomorphism, q_S must be monomorphic. This completes the proof.

EXAMPLE (6.5). Take $X = S^{4n}$. Then $J'(S^{4n})$ is cyclic of order $m(2n)$.

This is essentially the theorem of MILNOR and KERVAIRE [15], as improved by ATIYAH and HIRZEBRUCH [6]; it states that the image $J(\pi_{4n-1}(SO))$ of the stable J-homomorphism has an order divisible by the denominator of $B_n/4n$.

Proof. Suppose $x \in V(S^{4n})$; that is, suppose that

$$\rho^k x = \frac{\Psi^k(1+y)}{1+y}$$

for all k. Using Theorem (5.18), this becomes

(6.6) $$1 + \tfrac{1}{2}(k^{2n} - 1)\alpha_{2n} x = 1 + (k^{2n} - 1)y$$

where α_{2n} is as in §2; by Theorem (2.6) we have

$$\tfrac{1}{2}\alpha_{2n} = \frac{d(2n)}{m(2n)},$$

where $d(2n)$ and $m(2n)$ are coprime. Equation (6.6) holds in $1 + \tilde{K}_R(S^{4n}) \otimes Q_k$; in $\tilde{K}_R(S^{4n})$ we have

(6.7) $$k^{f(k)}(k^{2n} - 1)\frac{d(2n)}{m(2n)}x = k^{f(k)}(k^{2n} - 1)y,$$

for some exponent $f(k)$. According to Theorem (2.7) the highest common factor of the integers $k^{f(k)}(k^{2n} - 1)$ divides $m(2n)$. Therefore by taking a linear combination of the equations (6.7), we can show

$$d(2n)x = m(2n)y.$$

That is, x is divisible by $m(2n)$ in $\tilde{K}_R(S^{4n})$.

Conversely, if x is divisible by $m(2n)$ in $\tilde{K}_R(S^{4n})$, then we can solve the equation

$$d(2n)x = m(2n)y$$

for y, and we can calculate that

$$\rho_x^k = \frac{\Psi^k(1+y)}{1+y}$$

for all k. Thus $x \in V(S^{4n})$.

This determines the subgroup $V(S^{4n})$, and proves the result stated.

REFERENCES

1. J. F. ADAMS: Vector fields on spheres, *Ann. Math., Princeton* **75** (1962), 603–632.
2. J. F. ADAMS: Applications of the Grothendieck-Atiyah-Hirzebruch functor $K(X)$, *Colloquium on Algebraic Topology, Aarhus*, pp. 104–113 (mimeographed notes), 1962.
3. J. F. ADAMS: Applications of the Grothendieck-Atiyah-Hirzebruch functor $K(X)$, *Proceedings of the International Congress of Mathematicians, Stockholm*, pp. 435–441, 1962.
4. J. F. ADAMS: On the groups $J(X)$. I, *Topology* 2 (1963), 181–195.
5. J. F. ADAMS and G. WALKER: On complex Stiefel manifolds, *Proc. Camb. Phil. Soc.* **61** (1965), 81–103.
6. M. F. ATIYAH and F. HIRZEBRUCH: Riemann–Roch theorems for differentiable manifolds, *Bull. Amer. Math. Soc.* **65** (1959), 276–281.
7. M. F. ATIYAH and F. HIRZEBRUCH: Vector bundles and homogeneous spaces, *Proceedings of Symposia in Pure Mathematics* 3, *Differential Geometry*, Amer. Math. Soc. pp. 7–38, 1961.
8. R. BOTT: A note on the *KO*-theory of sphere-bundles, *Bull. Amer. Math. Soc.* **68** (1962), 395–400.
9. R. BOTT: *Lectures on* $K(X)$ (mimeographed notes), Harvard University.
10. A. DOLD: Relations between ordinary and extraordinary homology, *Colloquium on Algebraic Topology, Aarhus*, pp. 2–9 (mimeographed notes), 1962.
11. E. DYER: Chern characters of certain complexes, *Math. Z.* 80 (1963), 363–373.
12. G. H. HARDY and E. M. WRIGHT: *An Introduction to the Theory of Numbers*, (4th ed.) Oxford University Press, London, 1960.
13. F. HIRZEBRUCH: *Neue Topologische Methoden in der Algebraischen Geometrie*, Springer, Berlin, 1956.
14. J. MILNOR: *Lectures on Characteristic Classes* (mimeographed notes), Princeton University.
15. J. MILNOR and M. A. KERVAIRE: Bernoulli numbers, homotopy groups, and a theorem of Rohlin, *Proceedings of the International Congress of Mathematicians, Edinburgh* 1958, pp. 454–458. Cambridge University Press, London, 1960.
16. R. THOM: Espaces fibrés en spheres et carrés de Steenrod, *Ann. Sci. Ec. Norm. Sup.* **69** (1952), 109–182.
17. H. ZASSENHAUS: *The Theory of Groups*, (2nd ed.) Chelsea (1958).
18. M. F. ATIYAH, R. BOTT and A. SHAPIRO: Clifford modules, *Topology* 3 (Suppl. 1) (1964), 3–38.
19. M. F. ATIYAH and F. HIRZEBRUCH: Cohomologie-Operationen und charakteristische Klassen, *Math. Z.* 77 (1961), 149–187.
20. A. BOREL and F. HIRZEBRUCH: Characteristic classes and homogeneous spaces I, *Amer. J. Math.* **80** (1958), 458–538.
21. A. BOREL and F. HIRZEBRUCH: Characteristic classes and homogeneous spaces II, *Amer. J. Math.* **81** (1959), 315–382.
22. E. DYER: Relations between cohomology theories, *Colloquium on Algebraic Topology, Aarhus*, pp. 89–93 (mimeographed notes), 1962.
23. F. HIRZEBRUCH: A Riemann–Roch theorem for differentiable manifolds, *Sem. Bourbaki* **177** (1959).

Topology Vol. 3, pp. 193–222. Pergamon Press, 1965.

ON THE GROUPS $J(X)$—III

J. F. ADAMS

(*Received* 25 *November* 1963)

§1. INTRODUCTION

THE GENERAL OBJECT of this series of papers is to give means for computing the groups $J(X)$. A general introduction has been given at the beginning of Part I [3]. We recall that in Part II [4] we set up two further groups $J'(X)$ and $J''(X)$; here $J'(X)$ is a "lower bound" for $J(X)$, and we conjecture that $J''(X)$ is an "upper bound" for $J(X)$. The present paper, Part III, has two main objects; the first is to prove the following theorem.

THEOREM (1.1). *For each finite CW-complex X we have $J'(X) = J''(X)$.*

The precise sense in which the groups $J'(X)$ and $J''(X)$ are "equal" is the following. Both groups can be defined as quotients of $K_R(X)$, say

$$J'(X) = K_R(X)/V(X)$$
$$J''(X) = K_R(X)/W(X).$$

We shall prove that the subgroups $V(X)$ and $W(X)$ of $K_R(X)$, although differently defined, are in fact the same. Therefore the corresponding quotient groups $J'(X)$ and $J''(X)$ are (as a matter of logic) identical; that is, they are one and the same object.

Theorem (1.1) will be proved in §4. The proof is completely dependent on the existence of a certain commutative diagram (Diagram 3.1), which is established in §3. This, in turn, depends on an extension of a result of Atiyah and Hirzebruch [5], which is stated as Theorem (2.2) and proved in §2.

The proof given in §§3 and 4 may also be found in [2]. However, I would like the present account to be regarded as more complete and final; in particular, Theorem (2.2) of this paper is to be regarded as superseding Lemma (2.1) of [2].

The second main object of this paper may be explained as follows. We shall show in Part IV that the image of the stable J-homomorphism

$$J : \pi_{8m+3}(SO) \longrightarrow \pi^S_{8m+3}$$

is a direct summand. (Here π^S_r is the stable r-stem.) In other words, for the case $X = S^{8m+4}$ the group $J(X)$ is a direct summand in something else. It is reasonable to ask whether this is a special case of a result true for some general class of spaces X. The answer appears to be "yes", modulo some doubt as to the best way of setting up the foundations of the subject. This will be explained in §7.

Needless to say, §7 depends on §6, and §6 depends on §5. In §6 we observe that some, but not all, elements of $\prod_k (K_R(X) \otimes Q_k)$ have the form $\{\rho^k(x)\}$, and we essentially give a characterisation of the elements which have this form (Theorem (6.2)). In §5 we establish a "periodicity" property of the operations Ψ^k (Theorem (5.1)). Besides being used in §6, this property will be used in Part IV of the present series.

§2. COMPLETIONS AND CHARACTERS

In the present paper we shall sometimes have to work in the completion of a representation ring. This completion has already appeared in [4, §5]; we shall recall the details below. One calculates in such a ring by using characters; in this section we shall explain this topic, following Atiyah and Hirzebruch [5, pp. 24–27]. We shall also set up certain results needed later. Corollary (2.9) states that an element of a completed representation ring is determined by its character. Corollary (2.10) is a technical result needed in §3. These corollaries follow at once from the main result, Theorem (2.2); this is a slight extension of a result of Atiyah and Hirzebruch. Much of the proof is obtained by following these authors; but we are forced to add extra arguments, since we are interested in the real representation ing as well as the complex one.

G will be a compact connected Lie group, in which we choose a maximal torus T, whose points are given by complex coordinates $z_1, z_2, \dots, z_n$ with $|z_r| = 1$ for each r. If $\Lambda = R$ or C, we can form the representation ring $K'_\Lambda(G)$; the notation is as in [1]. If $\theta \in K'_\Lambda(G)$, then the character $\chi(\theta)$ of θ is a finite Laurent series in the variables $z_1, z_2, \dots, z_n$, with integer coefficients. We may identify the ring of such Laurent series with $K'_C(T)$; this amounts to identifying χ with $i^* : K'_\Lambda(G) \longrightarrow K'_C(T)$. An element of $K'_\Lambda(G)$ is determined by its character; in other words, χ (or i^*) is a monomorphism.

Let S be a subring of C; then we can form $K'_\Lambda(G) \otimes S$. If $\theta \in K'_\Lambda(G) \otimes S$, then the character $\chi(\theta)$ of θ is defined by S-linearity, and is a finite Laurent series with coefficients in S. The ring of such Laurent series may be identified with $K'_C(T) \otimes S$; this amounts to identifying χ with

$$i^* \otimes 1 : K'_\Lambda(G) \otimes S \longrightarrow K'_C(T) \otimes S.$$

Since S is torsion-free, $i^* \otimes 1$ (or χ) is again a monomorphism.

In $K'_\Lambda(G) \otimes S$ we take the ideal $I = \tilde{K}'_\Lambda(G) \otimes S$; this consists of the elements whose characters vanish at the origin. We give $K'_\Lambda(G) \otimes S$ the I-adic uniform structure and complete it; that is, we define

$$\mathrm{Comp}(K'_\Lambda(G) \otimes S) = \operatorname*{Inv\,Lim}_{q \to \infty} \frac{K'_\Lambda(G) \otimes S}{I^q}.$$

We will now define the character of an element in Comp $(K'_\Lambda(G) \otimes S)$. By substituting $z_r = 1 + \zeta_r$, any finite Laurent series in $z_1, z_2, \dots, z_n$ yields a power-series in $\zeta_1, \zeta_2, \dots, \zeta_n$ (convergent in a neighbourhood of the origin). An element of I yields a power-series vanishing at the origin; an element of I^q yields a power series starting with terms of the qth order. Therefore an element of Comp $(K'_\Lambda(G) \otimes S)$ yields a formal power-series.

More formally, let $S[[\zeta_1, \zeta_2, \ldots, \zeta_n]]$ be the ring of formal power-series in $\zeta_1, \zeta_2, \ldots, \zeta_n$ with coefficients in S, and let J be the ideal consisting of power-series with zero constant term; we give $S[[\zeta_1, \zeta_2, \ldots, \zeta_n]]$ the J-adic uniform structure. The substitution $z_r = 1 + \zeta_r$ yields a map

$$\chi : K'_\Lambda(G) \otimes S \longrightarrow S[[\zeta_1, \zeta_2, \ldots, \zeta_n]],$$

which is uniformly continuous, since

$$I^q \subset \chi^{-1} J^q.$$

By completion we obtain a map

$$\chi : \mathrm{Comp}(K'_\Lambda(G) \otimes S) \longrightarrow S[[\zeta_1, \zeta_2, \ldots, \zeta_n]],$$

since $S[[\zeta_1, \zeta_2, \ldots, \zeta_n]]$ is its own completion.

This account applies, of course, to the Lie group T, and may be used to identify $S[[\zeta_1, \zeta_2, \ldots, \zeta_n]]$ with Comp $(K'_C(T) \otimes S)$; cf. [5, 4.3, pp. 26, 27]. The only significant step is to check the following proposition; we include an elementary proof for completeness, but the reader who prefers to refer to [5] may omit it.

PROPOSITION (2.1). *The I-adic uniform structure on $K'_C(T) \otimes S$ coincides with that induced from the J-adic uniform structure on $S[[\zeta_1, \zeta_2, \ldots, \zeta_n]]$. More precisely, we have*

$$\chi^{-1} J^q = I^q.$$

Proof. We have already seen that $I^q \subset \chi^{-1} J^q$; it remains to prove that $\chi^{-1} J^q \subset I^q$. Let

$$L = \sum_e s_e \, z_1^{e_1} z_2^{e_2} \ldots z_n^{e_n}$$

be a finite Laurent series such that $\chi(L) \in J^q$. Without loss of generality we may suppose that all the exponents e_r which occur are positive; for otherwise we can replace L by $L z_1^{f_1} z_2^{f_2} \ldots z_n^{f_n}$, since $z_1^{f_1} z_2^{f_2} \ldots z_n^{f_n}$ is invertible in $K'_C(T) \otimes S$. Assuming that the e_r are positive, we may substitute $z_r = 1 + \zeta_r$ and so write L as a finite sum

$$L = \sum_g s'_g \, \zeta_1^{g_1} \zeta_2^{g_2} \ldots \zeta_n^{g_n}.$$

By assumption, all the terms with $\sum_r g_r < q$ are zero. This displays L as an element of I^q.

The central result required for our applications is the following.

THEOREM (2.2). *If S is a subring of the ring Q of rational numbers, then the I-adic uniform structure on $K'_\Lambda(G) \otimes S$ coincides with that induced from the J-adic uniform structure on $S[[\zeta_1, \zeta_2, \ldots, \zeta_n]]$. More precisely, given q there exists $r = r(G, \Lambda, q)$ independent of S such that*

$$\chi^{-1} J^r \subset I^q \subset \chi^{-1} J^q.$$

The proof is based on an argument of Atiyah and Hirzebruch [5, pp. 24–27]; it will require several lemmas.

LEMMA (2.3). *Let A be a finitely-generated commutative ring; let W be a finite group of automorphisms of A; let B be the subring of elements of A invariant under W. Then A is a finitely-generated B-module and B is a finitely-generated ring.*

This lemma is given in [5, p. 24 (iii)]. (I have altered the statement slightly, but the proof is unchanged.)

COROLLARY (2.4). *$K'_\Lambda(G)$ is a finitely-generated ring, and $K'_C(T)$ is a finitely-generated module over it.*

The case $\Lambda = C$ is due to Atiyah and Hirzebruch [5, pp. 24–27].

Proof. Since the case $\Lambda = C$ has been proved by Atiyah and Hirzebruch, and since we are mainly interested in the case $\Lambda = R$, we will give the proof for the case $\Lambda = R$.

Let us recall that $K'_C(G)$ is a free abelian group, generated by the irreducible representations of G. These may be classified into three classes. Class (c) contains each irreducible representation ρ which is distinct from its complex conjugate $\bar{\rho}$. Class (r) contains the complex forms of real representations. Class (q) contains the complex forms of quaternionic representations.

Let us now apply Lemma (2.3), taking A to be $K'_C(T)$, and W to be the direct product $\Gamma \times Z_2$, where Γ is the Weyl group of G and Z_2 acts by complex conjugation. The resulting ring B consists of the elements of $K'_C(G)$ invariant under complex conjugation. These constitute a free abelian group generated by the following elements: the irreducible representations of classes (r) and (q), together with the elements $\rho + \bar{\rho}$, where ρ runs over class (c). The lemma shows that B is a finitely-generated ring, and that $K'_C(T)$ is a finitely-generated module over B. Since B is finitely-generated, we may choose a set of generators consisting of a finite number of representations in the classes (r) and (q), together with a finite number of representations $\rho + \bar{\rho}$. Let the generators in class (q) be $q_1, q_2, \dots, q_m$. Let D be the subring of B generated by the remaining generators, together with the products $q_i q_j$; thus D is a finitely-generated ring. We have $D \subset K'_R(G)$, since the representations $\rho + \bar{\rho}$ and $q_i q_j$ are real. Also B is a finitely-generated D-module, generated by $1, q_1, q_2, \dots, q_m$. Since $K'_C(T)$ is a finitely-generated B-module, it follows that $K'_C(T)$ is a finitely-generated D-module, *a fortiori* a finitely-generated module over $K'_R(G)$. Finally, D is Noetherian and $K'_R(G)$ is a D-submodule of the finitely-generated D-module B; therefore $K'_R(G)$ is a finitely-generated D-module, and hence a finitely-generated ring. This completes the proof.

We will require one further consequence of the way $K'_R(G)$ is embedded in $K'_C(G)$. As in the proof above, B will be the subring of elements in $K'_C(G)$ invariant under complex conjugation. We set

$$\tilde{B} = B \cap \tilde{K}'_C(G).$$

LEMMA (2.5). *The ideals $\tilde{B}$ and $\tilde{K}_R(G) \cdot B$ define the same uniform structure on B.*

Proof. We clearly have $\tilde{K}_R(G) \cdot B \subset \tilde{B}$; we wish to argue in the opposite direction. Let $q_1, q_2, \dots, q_m$ be as in the proof above; we will show that

$$(\tilde{B})^{m+1} \subset \tilde{K}_R(G) \cdot B.$$

In fact, $\tilde{B}$ is generated as a B-module by the elements $g_r - \gamma_r$, where g_r runs over the generators of B and $\gamma_r \in Z$ is the dimension of g_r. Therefore $(\tilde{B})^{m+1}$ is generated as a

B-module by the elements

$$\prod_{1 \leqslant i \leqslant m+1} (g_{r_i} - \gamma_{r_i}).$$

If any g_{r_i} is a generator in $K_R(G)$, then the product lies in $\tilde{K}_R(G) \cdot B$. It remains only to consider the case in which every g_{r_i} is a generator q_j. In this case at least one q_j must occur twice, yielding a factor

$$(q_j - \gamma)^2 = q_j^2 - 2q_j\gamma + \gamma^2.$$

Since q_j^2 and $2q_j$ are real representations, the factor $(q_j - \gamma)^2$ lies in $\tilde{K}_R(G)$. This completes the proof.

We now require further lemmas of Atiyah and Hirzebruch.

LEMMA (2.6). *Let A, W and B be as in Lemma* (2.3). *Let J be an ideal of A such that $wJ = J$ for each $w \in W$; set $I = J \cap B$. Then J and $I \cdot A$ define the same uniform structure on A; more precisely, there exists an integer m such that*

$$J^m \subset I \cdot A \subset J.$$

This lemma is quoted (with minor changes of notation) from [5, p. 25 (iv)].

LEMMA (2.7). *Let A be a ring and B a subring of A such that B is Noetherian and A is a finitely-generated module over B. Let I be an ideal of B. Then the I-adic uniform structure of B coincides with that induced from the $I \cdot A$-adic uniform structure of A.*

This follows directly from [5, p. 24 (i)] by taking the modules M, N mentioned there to be the rings A, B.

COROLLARY (2.8). *Theorem* (2.2) *is true in the special case $S = Z$. More explicitly, let $I = \tilde{K}_\Lambda'(G)$; then the I-adic uniform structure on $K_\Lambda'(G)$ coincides with that induced from the J-adic uniform structure on $Z[[\zeta_1, \zeta_2, \ldots, \zeta_n]]$.*

The case $\Lambda = C$ is due to Atiyah and Hirzebruch [5 pp. 24–27].

Proof. Since the case $\Lambda = C$ is due to Atiyah and Hirzebruch, and since we are mainly interested in the case $\Lambda = R$, we will give the proof for the case $\Lambda = R$.

It will ease the statement of the proof if we make one convention. Let A be a ring, B a subring and J an ideal in A; then the "J-adic uniform structure on B" will mean that induced by the J-adic uniform structure on A; that is, it consists of the ideals $J^m \cap B$.

Let $J \subset Z[[\zeta_1, \zeta_2, \ldots, \zeta_n]]$ be as in the corollary. Then according to Proposition (2.1), the J-adic structure on $K_C'(T)$ coincides with the $\tilde{K}_C'(T)$-adic structure. The same is therefore true for the structures induced on any subring of $K_C'(T)$; in particular, the J-adic and $\tilde{K}_C'(T)$-adic structures on $K_r'(G)$ coincide.

We now apply Lemma (2.6), taking A to be $K_C'(T)$ and W to be $\Gamma \times Z_2$, as in the proof of Corollary (2.4). We take J to be $\tilde{K}_C'(T)$; thus I becomes $\tilde{B}$. According to Lemmas (2.6) and (2.7), the $\tilde{K}_C'(T)$-adic and $\tilde{B}$-adic structures coincide on B, and therefore on any subring of B, in particular $K_R'(G)$.

We now apply Lemma (2.7) to the ring B and the subring $K'_R(G)$. According to Lemmas (2.5) and (2.7), the $\tilde{B}$-adic and $\tilde{K}'_R(G)$-adic structures on $K'_R(G)$ coincide. This completes the proof.

Proof of Theorem (2.2). This theorem follows from the special case $S = Z$. Let us use subscripts to distinguish the ideals which occur in the two cases; thus

$$J_Z \subset Z[[\zeta_1, \zeta_2, \dots, \zeta_n]]$$

$$J_s \subset S[[\zeta_1, \zeta_2, \dots, \zeta_n]].$$

Then we have

$$J_s^r = J_Z^r \cdot S$$

and we easily check that

$$\chi^{-1}(J_S^r) = (\chi^{-1}(J_Z^r)) \cdot S.$$

The result follows.

COROLLARY (2.9). *If* $S \subset Q$, *the map* $\chi : \mathrm{Comp}(K'_\Lambda(G) \otimes S) \longrightarrow S[[\zeta_1, \zeta_2, \dots, \zeta_n]]$ *is monomorphic.*

This follows immediately from Theorem (2.2). It is this corollary which allows us to handle elements of $\mathrm{Comp}(K'_\Lambda(G) \otimes S)$ by means of their characters. The case $S = Z$, $\Lambda = C$ is due to Atiyah and Hirzebruch [5, pp. 26, 27].

In order to state the next corollary, suppose given two subrings $S \subset T \subset Q$. Then we have ideals

$$I_S = \tilde{K}'_\Lambda(G) \otimes S \subset K'_\Lambda(G) \otimes S$$

$$I_T = \tilde{K}'_\Lambda(G) \otimes T \subset K'_\Lambda(G) \otimes T.$$

COROLLARY (2.10). *Given* q, *there exists* $r = r(G, \Lambda, q)$ *independent of* S, T *such that*

$$(K'_\Lambda(G) \otimes S) \cap (I_T)^r \subset (I_S)^q.$$

This follows immediately from Theorem (2.2). It is needed for the arguments in §3.

§3. AN IDENTITY BETWEEN VIRTUAL REPRESENTATIONS

The object of this section is to prove Theorem (3.2), which is vital to the proof of Theorem (1.1). The most important part of this theorem will state the commutativity of the following diagram.

(3.1)
$$\begin{array}{ccc} \tilde{K}_{SO}(X) & \xrightarrow{k^e(\Psi^k - 1)} & \tilde{K}_{SO}(X) \\ \downarrow{\scriptstyle \theta^k} & & \downarrow{\scriptstyle \rho^l} \\ 1 + \tilde{K}_{SO}(X) & \xrightarrow{\delta^l} & 1 + \tilde{K}_{SO}(X) \otimes Q_l \end{array}$$

Here the groups are as in [4, §5]; those in the top row are additive, and those on the bottom row are multiplicative. In particular, $1 + \tilde{K}_{SO}(X)$ is the multiplicative group of elements $1 + y$ in $K_R(X)$, where $y \in \tilde{K}_{SO}(X)$; similarly for $1 + \tilde{K}_{SO}(X) \otimes Q_l$, where Q_l is the ring of rational numbers of the form a/l^b. The additive homomorphism $k^e(\Psi^k - 1)$ is defined by

$$k^e(\Psi^k - 1)x = k^e((\Psi^k x) - x);$$

thus the 1 in $\Psi_k - 1$ means the identity function. The existence of the homomorphism θ^k will be asserted as part of the theorem. The homomorphism ρ^l is as in [4, §5]. The multiplicative homomorphism δ^l is defined by

$$\delta^l(1 + y) = \frac{\Psi^l(1 + y)}{1 + y};$$

cf. [4, §6].

THEOREM (3.2). *Given integers q, k and a sufficiently large integer e* (viz. $e \geqq e_0(q, k)$) *there exists a function*

$$\theta^k = \theta^k(q, e) : \tilde{K}_{SO}(X) \longrightarrow 1 + \tilde{K}_{SO}(X)$$

defined for CW-complexes of dimension $<q$ and having the following properties.

(1) *θ^k is homomorphic from addition to multiplication, that is, exponential.*

(2) *θ^k is natural for maps of X.*

(3) *The image of $\theta^k(x)$ in $1 + \tilde{K}_{SO}(X) \otimes Q_k$ is $\rho^k(k^e x)$.*

(4) *Diagram* 3.1 *is commutative.*

We will now give a heuristic plausibility argument for Theorem (3.2). For this purpose we will abandon the real K-theory and work instead with the complex K-theory, for simplicity. We now argue that in some suitable formal setting we may hope to have

$$(\rho^l\Psi^k) \otimes \rho^k = \rho^{kl} = (\Psi^l\rho^k) \otimes \rho^l \tag{3.3}$$

In fact, all three expressions are exponential; therefore it is presumably sufficient to check the result for a complex line bundle ξ. Here we have

$$\Psi^k\xi = \xi^k, \qquad \rho^k\xi = \frac{\xi^k - 1}{\xi - 1}.$$

Therefore

$$(\rho^l\Psi^k\xi) \otimes (\rho^k\xi) = \frac{\xi^{kl} - 1}{\xi^k - 1} \cdot \frac{\xi^k - 1}{\xi - 1},$$

$$\rho^{kl}\xi = \frac{\xi^{kl} - 1}{\xi - 1},$$

$$(\Psi^l\rho^k\xi) \otimes (\rho^l\xi) = \frac{\xi^{kl} - 1}{\xi^l - 1} \cdot \frac{\xi^l - 1}{\xi - 1}.$$

(The last line uses the fact that Ψ^l preserves both addition and multiplication.) The three results are equal. We may therefore agree to suspend disbelief in (3.3). Rewriting this equation, we obtain

$$\rho^l(\Psi^k - 1) = \frac{\Psi^l\rho^k}{\rho^k}.$$

That is, if $x \in \tilde{K}_C(X)$, then $1 + y = \rho^k x$ is a formal solution of the equation

$$\rho^l(\Psi^k - 1)x = \frac{\Psi^l(1+y)}{1+y}.$$

Now, this formal solution involves denominators, that is, coefficients in Q_k. However, we can remove these denominators, up to dimension q, by considering

$$1 + z = (1 + y)^{k^e},$$

where e is suitably large. Raising the equation

$$\rho^l(\Psi^k - 1)x = \frac{\Psi^l(1+y)}{1+y}$$

to the power k^e, we obtain

$$\rho^l k^e(\Psi^k - 1)x = \frac{\Psi^l(1+z)}{1+z}.$$

This completes the plausibility argument.

We turn now to a rigorous version of this argument for the real case. The "suitable formal setting" for equations such as (3.3) will be the completion of a representation ring. Formulae such as

$$\frac{\xi^{kl} - 1}{\xi - 1}$$

will appear when we calculate characters.

In order to state the first lemma, let

$$\alpha_n, \beta_n \in \mathrm{Comp}(K'_\Lambda(SO(2n)) \otimes S)$$

be two exponential sequences (see [4, §5], just before Lemma (5.9)). We assume that $S \subset Q$.

LEMMA (3.4). *If* $\chi(\alpha_1) = \chi(\beta_1)$, *then* $\alpha_n = \beta_n$ *for all* n.

Proof. Suppose that

$$\chi(\alpha_1) = \chi(\beta_1) = \phi(\zeta_1),$$

where $\phi(\zeta_1)$ is a formal power series in ζ_1. Then by the exponential law,

$$\chi(\alpha_n) = \phi(\zeta_1)\phi(\zeta_2) \dots \phi(\zeta_n) = \chi(\beta_n).$$

The result now follows from Corollary (2.9).

We will now apply this lemma. First we observe that (with an obvious notation)

$$K'_{SO}(SO(n)) = K'_R(SO(n)).$$

In fact, any representation $\gamma : SO(n) \longrightarrow O(m)$ must map into $SO(m)$, since $SO(n)$ is connected. In particular,

$$\Psi^k - 1 \in \tilde{K}'_{SO}(SO(2n)).$$

(Here, we emphasise, 1 means the identity map of $SO(2n)$.) Thus $\rho^l \cdot (\Psi^k - 1)$ is defined, and lies in

$$\mathrm{Comp}(K'_R(SO(2n)) \otimes Q_l).$$

Again (see [4, §5]) $\Psi^l \cdot \rho^k$ is defined, and lies in

$$\mathrm{Comp}(K_R'(SO(2n)) \otimes Q_k).$$

Since ρ^k is invertible in this ring, we can define

$$\delta^l \cdot \rho^k = \frac{\Psi^l \cdot \rho^k}{\rho^k}.$$

LEMMA (3.5). *Consider the sequences of elements*

$$\alpha_n = \rho^l \cdot (\Psi^k - 1)$$

$$\beta_n = \delta^l \cdot \rho^k$$

in $\mathrm{Comp}(K_R'(SO)(2n)) \otimes Q_{kl})$. *These sequences are exponential, and satisfy* $\chi(\alpha_1) = \chi(\beta_1)$.

Proof. It is easy to check that the sequences are exponential. In fact, α_n is exponential because $(\Psi^k - 1)$ is additive and ρ^l is exponential; β_n is exponential because ρ^k is exponential and δ^l is multiplicative.

It remains to check that $\chi(\alpha_1) = \chi(\beta_1)$. The virtual representation Ψ^k of $SO(2)$ is the representation $z \longrightarrow z^k$. By definition, $\rho^l \cdot (\Psi^k - 1)$ means the element $(\rho^l \cdot \Psi^k)/\rho^l$ of $\mathrm{Comp}(K_R'(SO(2)) \otimes Q_l)$. Its character is given by

$$\chi(\alpha_1) = \frac{(z_1)^{\frac{1}{2}kl} - (z_1)^{-\frac{1}{2}kl}}{(z_1)^{\frac{1}{2}k} - (z_1)^{-\frac{1}{2}k}} \cdot \frac{(z_1)^{\frac{1}{2}} - (z_1)^{-\frac{1}{2}}}{(z_1)^{\frac{1}{2}l} - (z_1)^{-\frac{1}{2}l}}.$$

Of course, this expression is interpreted as a formal power-series in ζ_1, where $z_1 = 1 + \zeta_1$; see [4, §5].

For any virtual representation θ of G, the character of $\Psi^l \cdot \theta$ is given by

$$\chi(\Psi^l \cdot \theta)g = \chi(\theta)g^l$$

[1, Theorem (4.1) (vi).] Evidently this equation remains true when θ is replaced by an element of $K_\Lambda'(G) \otimes S$ or $\mathrm{Comp}(K_\Lambda'(G) \otimes S)$. Therefore the character of $\delta^l \cdot \rho^k = (\Psi^l \cdot \rho^k)/\rho^k$ is given by

$$\chi(\beta_1) = \frac{(z_1)^{\frac{1}{2}kl} - (z_1)^{-\frac{1}{2}kl}}{(z_1)^{\frac{1}{2}l} - (z_1)^{-\frac{1}{2}l}} \cdot \frac{(z_1)^{\frac{1}{2}} - (z_1)^{-\frac{1}{2}}}{(z_1)^{\frac{1}{2}k} - (z_1)^{-\frac{1}{2}k}}.$$

Of course, this expression also is interpreted as a formal power-series in ζ_1. We have $\chi(\alpha_1) = \chi(\beta_1)$. This completes the proof of Lemma (3.5).

PROPOSITION (3.6). *In* $\mathrm{Comp}(K_R'(SO(2n)) \otimes Q_{kl})$ *we have*

$$\rho^l \cdot (\Psi^k - 1) = \delta^l \cdot \rho^k$$

and

$$\rho^l \cdot (\Psi^k - 1) = \frac{\Psi^l(k^{-n}\rho^k)}{k^{-n}\rho^k}.$$

Proof. The first assertion follows immediately from Lemmas (3.4, 3.5). The second follows by rewriting the first.

The element $k^{-n}\rho^k$ lies in $\mathrm{Comp}(K_R'(SO(2n)) \otimes Q_k)$, and has virtual dimension 1.

LEMMA (3.7). *Given integers n, k, r and sufficiently large e* (viz. $e \geqq e_0(n, k, r)$) *we have*

$$(k^{-n}\rho^k)^{k^e} = 1 + x \quad \text{in} \quad (K'_R(SO(2n)) \otimes Q_k)/I^r,$$

where $x \in \tilde{K}'_R(SO(2n))$.

Proof. It is clear that if the conclusion holds for one value of e, then it holds for all larger values of e. We now proceed by induction over r. Suppose that we have found e such that

$$(k^{-n}\rho^k)^{k^e} = 1 + x \quad \text{in} \quad (K'_R(SO(2n)) \otimes Q_k)/I^r,$$

where $x \in \tilde{K}'_R(SO(2n))$. (The induction starts with $r = 1$.) Then in

$$(K'_R(SO(2n)) \otimes Q_k)/I^{2r}$$

we can write

$$(k^{-n}\rho^k)^{k^e} = 1 + x + k^{-f}y,$$

where

$$y \in K'_R(SO(2n)) \cap I^r.$$

(Here we regard $K'_R(SO(2n))$ as embedded in $K'_R(SO(2n)) \otimes Q_k$.) Now we have

$$\begin{aligned}(k^{-n}\rho^k)^{k^{e+f}} &= (1 + x + k^{-f}y)^{k^f} \\ &= (1 + x)^{k^f} + y(1 + x)^{k^f - 1} \mod I^{2r} \\ &= 1 + z \mod I^{2r},\end{aligned}$$

where $z \in \tilde{K}'_R(SO(2n))$. This completes the induction.

Proof of Theorem (3.2). In what follows, X will always be a CW-complex of dimension $<q$. We can thus determine $n = n(q)$ so that there is a (1–1) correspondence between homotopy classes of maps $f: X \longrightarrow BSO$ and homotopy classes of maps $f: X \longrightarrow BSO(2n)$. That is, there is a $(1 - 1)$ correspondence between the isomorphism classes of $SO(2n)$-bundles ξ over X and the elements of $\tilde{K}_{SO}(X)$; the correspondence is given by $\xi \longrightarrow \{\xi\} - 2n$.

We will now invoke Lemma (3.7), and for this purpose we define an integer r depending only on q and n. (Thus r depends ultimately only on q.) With the notation of Corollary (2.10), we set

$$\begin{aligned}r_1 &= r(SO(2n), R, q) \\ r_2 &= r(SO(2n) \times SO(2n), R, q) \\ r &= \operatorname{Max}(q, r_1, r_2).\end{aligned}$$

We now employ Lemma (3.7) to choose elements

$$\begin{aligned}(\theta^k)_n &\in K'_R(SO(2n)) \\ (\theta^k)_{2n} &\in K'_R(SO(4n))\end{aligned}$$

such that the images of $(\theta^k)_n$, $(\theta^k)_{2n}$ in

$$(K'_R(SO(2n)) \otimes Q_k)/I^r,$$
$$(K'_R(SO(4n)) \otimes Q_k)/I^r$$

are

$$(k^{-n}\rho^k)^{k^e}, \qquad (k^{-2n}\rho^k)^{k^e}.$$

We can do this for all sufficiently large e (viz. for $e \geqq e_0(q, k)$).

We shall use the element $(\theta^k)_n$ to define the function

$$\theta^k : \tilde{K}_{SO}(X) \longrightarrow 1 + \tilde{K}_{SO}(X).$$

More precisely, we define

$$\theta^k(\{\xi\} - 2n) = (\theta^k)_n \cdot \{\xi\},$$

where ξ runs over the $SO(2n)$-bundles. It is clear that this does define a function, which is natural for maps of X. Since $r \geqq q$, it also follows that the image of $\theta^k(x)$ in $1 + \tilde{K}_{SO}(X) \otimes Q_k$ is $\rho^k(k^e x)$.

We shall use the element $(\theta^k)_{2n}$ to prove that the function θ^k is exponential. More precisely, let π, ϖ to be the projections of $SO(2n) \times SO(2n)$ onto its first and second factors. Then in

$$K_R'(SO(2n) \times SO(2n)) \otimes Q_k)/I^r$$

we have the following equations.

$$(\theta^k)_{2n} \cdot (\pi \oplus \varpi) = (k^{-2n} \rho^k)^{k^e} \cdot (\pi \oplus \varpi),$$

$$(k^{-2n} \rho^k)^{ke} \cdot (\pi \oplus \varpi) = (k^{-n} \rho^k)^{k^e} \cdot \pi \otimes (k^{-n} \rho^k)^{k^e} \cdot \varpi,$$

$$(k^{-n} \rho^k)^{k^e} \cdot \pi \otimes (k^{-n} \rho^k)^{k^e} \cdot \varpi = (\theta^k)_n \cdot \pi \otimes (\theta^k)_n \cdot \varpi.$$

We may now apply Corollary (2.10) with $G = SO(2n) \times SO(2n)$, $S = Z$, $T = Q_k$. The element

$$(\theta^k)_{2n} \cdot (\pi \oplus \varpi) - (\theta^k)_n \cdot \pi \otimes (\theta^k)_n \cdot \varpi$$

lies in

$$[K_R'(SO(2n) \times SO(2n)) \otimes S] \cap (I_T)^r.$$

By our choice of r, Corollary (2.10) applies and shows that this element lies in $(I_S)^q$. Therefore this element will annihilate any $SO(2n) \times SO(2n)$-bundle over X; for since X has dimension $<q$, we have $(\tilde{K}_R(X))^q = 0$. Let ξ, η be $SO(2n)$-bundles over X; we can apply the preceding remark to $\xi \times \eta$; we obtain the equation

$$(\theta^k)_{2n} \cdot (\xi \oplus \eta) = (\theta^k)_n \cdot \pi \otimes (\theta^k)_n \cdot \varpi$$

in $K_R(X)$. From this it follows that the function

$$\theta^k : \tilde{K}_{SO}(X) \longrightarrow 1 + \tilde{K}_{SO}(X)$$

is exponential.

We argue similarly to show that Diagram (3.1) is commutative. Proposition (3.6) states that

$$\rho^l \cdot (\Psi^k - 1) = \frac{\Psi^l \cdot (k^{-n} \rho^k)}{k^{-n} \rho^k}$$

in

$$\mathrm{Comp}(K_R'(SO(2n)) \otimes Q_{kl}).$$

Raising this equation to the power k^e, we have

$$\rho^l \cdot k^e (\Psi^k - 1) = \frac{\Psi^l (k^{-n} \rho^k)^{k^e}}{(k^{-n} \rho^k)^{k^e}}.$$

Thus

$$\rho^l \cdot k^e(\Psi^k - 1) = \frac{\Psi^l \cdot (\theta^k)_n}{(\theta^k)_n}$$

in

$$[K'_R(SO(2n)) \otimes Q_{kl}]/I^r.$$

We will now apply Corollary (2.10) with $G = SO(2n)$, $S = Q_l$, $T = Q_{kl}$. Let us take a representative

$$y \in K'_R(SO(2n)) \otimes Q_l$$

for the element

$$\rho^l \cdot k^e(\Psi^k - 1)$$

in

$$[K'_R(SO(2n)) \otimes Q_l]/(I_S)^r.$$

Then we have

$$y \in K'_R(SO(2n)) \otimes S,$$

$$\frac{\Psi^l \cdot (\theta^k)_n}{(\theta^k)_n} \in K'_R(SO(2n)),$$

$$\text{and} \quad y - \frac{\Psi^l \cdot (\theta^k)_n}{(\theta^k)_n} \in [K'_R(SO(2n)) \otimes S] \cap (I_T)^r.$$

By our choice of r, Corollary (2.10) applies and shows that this element lies in $(I_S)^q$. That is, we have

$$\rho^l \cdot k^e(\Psi^k - 1) = \frac{\Psi^l \cdot (\theta^k)_n}{(\theta^k)_n}$$

in

$$[K'_R(SO(2n)) \otimes Q_l]/(I_S)^q.$$

Arguing as above, it follows that Diagram (3.1) is commutative. This completes the proof of Theorem (3.2).

§4. PROOF OF THEOREM (1.1)

In proving Theorem (1.1), we shall have to work with square diagrams like Diagram (3.1). By a "square" S, we shall mean a commutative diagram of groups and homomorphisms which has the following form.

$$\begin{array}{ccc} A & \xrightarrow{f} & B \\ {\scriptstyle g}\downarrow & & \downarrow{\scriptstyle h} \\ C & \xrightarrow{i} & D \end{array}$$

We shall call a square "special" if it has the following property: given $b \in B$ and $c \in C$ such that $hb = ic$, there exists $a \in A$ such that $fa = b$ and $ga = c$. This is equivalent to demanding the exactness of the following sequence:

$$A \xrightarrow{(f,g)} B \oplus C \xrightarrow{(h,-i)} D.$$

By a "short exact sequence

$$0 \longrightarrow S' \longrightarrow S \longrightarrow S'' \longrightarrow 0\text{"}$$

of squares, we shall mean a commutative diagram composed of three squares S', S, S'' and four short exact sequences

$$\begin{array}{l} 0 \longrightarrow A' \longrightarrow A \longrightarrow A'' \longrightarrow 0 \\ 0 \longrightarrow B' \longrightarrow B \longrightarrow B'' \longrightarrow 0, \text{ etc.} \end{array}$$

LEMMA (4.1). *Suppose that*

$$0 \longrightarrow S' \longrightarrow S \longrightarrow S'' \longrightarrow 0$$

is a short exact sequence of squares in which S' and S'' are special; then S is special.

Proof. Each square determines a sequence

$$A \xrightarrow{(f,g)} B \oplus C \xrightarrow{(h,-i)} D$$

which we may regard as a chain complex. We now have a short exact sequence of chain complexes. This yields an exact homology sequence, which leads immediately to the required result.

Alternatively, one may give a direct proof by routine diagram-chasing.

Let X be a finite CW-complex, say of dimension $<q$. The main part of the proof of Theorem (1.1) will proceed by filtering X. Let us define F_r to be the image of $\tilde{K}_R(X/X^{r-1})$ in $K_R(X)$; then for $r \geqq 2$ and sufficiently large e, Theorem (3.2) provides us with the following commutative square.

$$\begin{array}{ccc} F_r & \xrightarrow{k^e(\Psi^k - 1)} & F_r \\ \Big\downarrow{\scriptstyle \theta^k} & & \Big\downarrow{\scriptstyle \rho^l} \\ 1 + F_r & \xrightarrow{\delta^l} & 1 + F_r \otimes Q_l \end{array}$$

(Since Q_l is torsion-free, $\otimes Q_l$ is an exact functor, and the image of $\tilde{K}_R(X/X^{r-1}) \otimes Q_l$ in $\tilde{K}_R(X) \otimes Q_l$ is $F_r \otimes Q_l$.)

If we pass to a (restricted) direct sum over k and an (unrestricted) direct product over l, we obtain the following commutative square S_r (for $r \geqq 2$).

$$(4.2)\qquad \begin{array}{ccc} \sum_k F_r & \xrightarrow{\sum_k k^{e(k)}(\Psi^k - 1)} & F_r \\ \Big\downarrow{\scriptstyle \sum_k \theta^k} & & \Big\downarrow{\scriptstyle \prod_l \rho^l} \\ 1 + F_r & \xrightarrow{\prod_l \delta^l} & \prod_l (1 + F_r \otimes Q_l) \end{array}$$

More precisely, we obtain this commutative square whenever the function $e(k)$ is sufficiently large, viz. for $e(k) \geqq e_0(q, k)$, where q is our fixed upper bound for the dimension of the complexes X considered.

THEOREM (4.3). *The square S_r displayed in Diagram* (4.2) *is special.*

Proof. We shall prove this result by downwards induction over r. For $r = q$ the result is trivial. Let us therefore assume as an inductive hypothesis that the square S_{r+1} is special, and prove that the square S_r is special. For this purpose, by Lemma (4.1), it is sufficient to construct a short exact sequence of squares

$$0 \longrightarrow S_{r+1} \longrightarrow S_r \longrightarrow S_r/S_{r+1} \longrightarrow 0$$

and prove that the square S_r/S_{r+1} is special.

Since the square S_{r+1} is embedded in S_r, it is clear that the required quotient square exists. In order to establish its structure, let us recall that $F_r \cdot F_s \subset F_{r+s}$; thus F_{r+1} is an ideal in F_r, and F_r/F_{r+1} is a ring in which the product is zero (assuming $r \geqq 1$). Besides the short exact sequence of additive groups

$$0 \longrightarrow F_{r+1} \longrightarrow F_r \longrightarrow F_r/F_{r+1} \longrightarrow 0,$$

we have also a short exact sequence of multiplicative groups:

$$1 \longrightarrow 1 + F_{r+1} \longrightarrow 1 + F_r \longrightarrow 1 + F_r/F_{r+1} \longrightarrow 1.$$

Here the product in the last group is given by

$$(1 + a)(1 + b) = 1 + (a + b).$$

Similar remarks hold for the following short exact sequence of multiplicative groups:

$$1 \longrightarrow 1 + F_{r+1} \otimes Q_l \longrightarrow 1 + F_r \otimes Q_l \longrightarrow 1 + (F_r/F_{r+1}) \otimes Q_l \longrightarrow 1$$

This shows that the square S_r/S_{r+1} has the following form.

(4.4)
$$\begin{array}{ccc}
\sum_k F_r/F_{r+1} & \xrightarrow{\sum_k k^{e(k)}(\Psi^k - 1)} & F_r/F_{r+1} \\
\Big\downarrow{\scriptstyle \sum_k \theta^k} & & \Big\downarrow{\scriptstyle \prod_l \rho^l} \\
1 + (F_r/F_{r+1}) & \xrightarrow{\prod_l \delta^l} & \prod_l (1 + (F_r/F_{r+1}) \otimes Q_l)
\end{array}$$

The maps of the square S_r/S_{r+1} are induced by those of the square S_r.

Theorem (4.3) will thus be proved once we have established the following lemma.

LEMMA (4.5). *The square displayed in Diagram* (4.4) *is special.*

Proof. The quotient group F_r/F_{r+1} of Diagram (4.4) is isomorphic to a quotient group of a subgroup of

$$\tilde{K}_R(X^r/X^{r-1}) = \tilde{K}_R(\vee S^r).$$

(Compare [4, proof of Theorem (3.11)].) Since this isomorphism is given by induced maps, the operations Ψ^k, ρ^k and θ^k on F_r/F_{r+1} are given by the same formulae that hold in S^r:

$$\Psi^k y = a(k, r)y$$
$$\rho^k y = 1 + b(k, r)y$$
$$\theta^k y = 1 + c(k, r, e)y.$$

Here the coefficients a, b and c do not depend on X or y. To give the coefficients, we have to divide the cases. Unless $r \equiv 0$, 1, 2 or 4 mod 8 the group F_r/F_{r+1} is zero and there is nothing to prove. Consider first the case $r \equiv 0$ or 4 mod 8; say $r = 4t$, $t > 0$. In this case we have

$$\Psi^k y = k^{2t} y$$
$$\rho^k y = 1 + \tfrac{1}{2}\alpha_{2t}(k^{2t} - 1)y$$
$$\theta^k y = 1 + \tfrac{1}{2}\alpha_{2t} k^{e(k)}(k^{2t} - 1)y.$$

Here the first result is quoted from [1, Corollary (5.2)], the second is quoted from [4, Theorem (5.18)] and the third is deduced using Theorem (3.2), part (3).

We can now check that the square (4.4) is special for $r = 4t$. Suppose given $u, v \in F_r/F_{r+1}$ such that

$$\rho^l u = \frac{\Psi^l(1 + v)}{1 + v} \qquad \text{for all } l.$$

This yields

$$\tfrac{1}{2}\alpha_{2t}(l^{2t} - 1)u = (l^{2t} - 1)v$$

in $(F_r/F_{r+1}) \otimes Q_l$ for all l. This means that for some exponent $f(l)$ we have

$$\tfrac{1}{2}\alpha_{2t} l^{f(l)}(l^{2t} - 1)u = l^{f(l)}(l^{2t} - 1)v$$

in F_r/F_{r+1} (for each l). We may write $\tfrac{1}{2}\alpha_{2t} = n(2t)/m(2t)$ where $n(2t)$, $m(2t)$ are coprime and $m(2t)$ is as in [4, §2]. According to [4, Theorem [2.7)], by taking a suitable linear combination of the equations we have just obtained, we find

$$n(2t)\, u = m(2t)\, v.$$

Since the numbers $n(2t)$, $m(2t)$ are coprime we can choose integers a, b so that

$$am(2t) + bn(2t) = 1;$$

now define

$$w = au + bv \in F_r/F_{r+1};$$

this ensures that

$$m(t)w = u, \qquad n(t)w = v.$$

Using [4, Theorem (2.7)] again, let us choose integers $c(k)$, zero except for a finite number of k, such that

$$\sum_k c(k)k^{e(k)}(k^{2t} - 1) = m(2t).$$

Let σ be the element of $\sum F_r/F_{r+1}$ with components $c(k)w$. Then

$$\sum_k k^{e(k)}(\Psi^k - 1)\sigma = \sum_k c(k)k^{e(k)}(k^{2t} - 1)w$$
$$= m(2t)w$$
$$= u.$$

Again,

$$\Big(\sum_k \theta^k\Big)\sigma = 1 + \sum_k \tfrac{1}{2}\alpha_{2t}c(k)k^{e(k)}(k^{2t}-1)w$$
$$= 1 + n(t)w$$
$$= 1 + v.$$

Therefore the square (4.4) is special if $r \equiv 0$ or $4 \bmod 8$.

We turn now to the case $r \equiv 1$ or $2 \bmod 8$, $r \geqq 2$. In this case every element of F_r/F_{r+1} has order 2, so that

$$(F_r/F_{r+1}) \otimes Q_l \cong \begin{cases} F_r/F_{r+1} & (l \text{ odd}) \\ 0 & (l \text{ even}). \end{cases}$$

We have:

$$\Psi^k y = \begin{cases} 0 & (k \text{ even}) \\ y & (k \text{ odd}) \end{cases}$$

$$\left.\begin{array}{l} \rho^k y = 1 + y \\ \theta^k y = 1 + y \end{array}\right\} \quad (k \equiv \pm 3 \bmod 8).$$

Here the second result is quoted from [4, Theorem (5.17)], and the third is deduced using Theorem (3.2), part (3). The first result is deduced from the corresponding result for RP^n[1, §7] by naturality, according to the pattern of [**4**, 3.5, 5.17, 6.4].

It follows from this that the map $\sum_k k^{e(k)}(\Psi^k - 1)$ of Diagram (4.4) is zero, at least for $e(k) \geqq 1$. The map $\prod_l \delta^l$ is also trivial (with image 1). The map $\sum_k \theta^k$ is epimorphic. The map $\prod_l \rho^l$ is monomorphic. It follows immediately that the square (4.4) is special.

This completes the proof of Lemma (4.5), and (therefore) of Theorem (4.3).

COROLLARY (4.6). *The following square is special (for sufficiently large functions e(k).)*

$$\begin{array}{ccc} \sum_k \tilde{K}_{SO}(X) & \xrightarrow{\sum_k k^{e(k)}(\Psi^k - 1)} & \tilde{K}_{SO}(X) \\ \Big\downarrow{\scriptstyle \sum_k \theta^k} & & \Big\downarrow{\scriptstyle \prod_l \rho^l} \\ 1 + \tilde{K}_{SO}(X) & \xrightarrow{\prod_l \delta^l} & \prod_l 1 + \tilde{K}_{SO}(X) \otimes Q_l \end{array} \tag{4.7}$$

This follows immediately from Theorem (4.3), by setting $r = 2$.

Proof of Theorem (1.1). In proving Theorem (1.1), we may assume without loss of generality that X is connected. With the notation of §1, we require to prove that $V(X) = W(X)$.

Here we have defined

$$W(X) = \bigcap_e W(e, X),$$

where $W(e, X)$ is the subgroup of $\tilde{K}_R(X)$ generated by elements of the form

$$k^{e(k)}(\Psi^k - 1)x$$

as k runs over all integers and x runs over $K_R(X)$. I claim that for $e(k)$ sufficiently large (viz. for $e(k) \geqq e_0(q)$) we can obtain the same subgroup $W(e, X)$ by letting x run over $\tilde{K}_{SO}(X)$. In fact, any element $x \in K_R(X)$ can be written as $x = y + z$, where $y \in \tilde{K}_{SO}(X)$ and z is a linear combination of real line bundles. A real line bundle ζ can be induced by a map $f: X \longrightarrow RP_q$; from this we see that

$$2^{eo(q)}(\Psi^k - 1)\zeta = 0 \qquad (k \text{ even})$$

$$(\Psi^k - 1)\zeta = 0 \qquad (k \text{ odd})$$

Thus

$$k^{e(k)}(\Psi^k - 1)z = 0$$

if $e(k) \geqq e_0(q)$, and hence

$$k^{e(k)}(\Psi^k - 1)x = k^{e(k)}(\Psi^k - 1)y.$$

We have thus shown that for each sufficiently large function $e(k)$ the group $W(e; X)$ is the image of the map $\sum_k k^{e(k)}(\Psi^k - 1)$ appearing in Diagram (4.7).

Again, we have

$$V(X) = \left(\prod_l \rho^l\right)^{-1}\left(\prod_l \delta^l\right)(1 + \tilde{K}_R(X));$$

see [4, §6]. I claim that we can obtain the same subgroup by taking

$$\left(\prod_l \rho^l\right)^{-1}\left(\prod_l \delta^l\right)(1 + \tilde{K}_{SO}(X)).$$

In fact, any element $1 + x \in 1 + \tilde{K}_R(X)$ can be written in the form $(1 + y)\zeta$, where $1 + y \in 1 + \tilde{K}_{SO}(X)$ and ζ is a real line bundle over X. By dividing the cases "l even" and "l odd", as above, we see that

$$\frac{\Psi^l \zeta}{\zeta} = 1 \qquad \text{in } 1 + \tilde{K}_R(X) \otimes Q_l$$

for all l. Hence

$$\left(\prod_l \delta^l\right)(1 + x) = \left(\prod_l \delta^l\right)(1 + y)$$

and

$$\left(\prod_l \delta^l\right)(1 + \tilde{K}_R(X)) = \left(\prod_l \delta^l\right)(1 + \tilde{K}_{SO}(X)).$$

We have thus shown that the group $V(X)$ is the group

$$\left(\prod_l \rho^l\right)^{-1}\left(\prod_l \delta^l\right)(1 + \tilde{K}_{SO}(X))$$

of Diagram (4.7).

The fact that Diagram (4.7) is commutative now shows that $W(e, X) \subset V(X)$ for sufficiently large functions $e(k)$. The fact that Diagram (4.7) is special shows that

$V(X) \subset W(e, X)$ for sufficiently large functions $e(k)$. Thus for sufficiently large functions $e(k)$ we have $W(e, X) = V(X)$; hence

$$W(X) = \bigcap_e W(e, X) = V(X).$$

This completes the proof of Theorem (1.1).

§5. A PERIODICITY THEOREM FOR THE OPERATIONS Ψ^k

In this section we shall prove Theorem (5.1), which is needed in §6 and in Part IV of the present series. Roughly speaking, it will assert that $\Psi^k(x)$ is a periodic function of k. More accurately, it will make this assertion "modulo m".

We shall suppose that X is a CW-complex such that $H_*(X)$ is finitely-generated. We make this assumption because we wish to apply the results to a Thom complex, using the devices explained in [4, §4].

THEOREM (5.1). *If $x \in K_\Lambda(X)$ and $m \in Z$, then the value of $\Psi^k_\Lambda(x)$ in $K_\Lambda(X)/mK_\Lambda(X)$ is periodic in k with period m^e. Here e depends on X and Λ, but is independent of x and m.*

In this theorem, and below, the statement "$f(k)$ is periodic in k with period m^e" means simply "$k_1 \equiv k_2 \bmod m^e$ implies $f(k_1) = f(k_2)$". It is not asserted that m^e is the smallest possible period. In particular, the theorem is true for $m = 0$ in a trivial way.

The proof will require three lemmas.

LEMMA (5.2). *Let s be a fixed positive integer. Then the binomial coefficient*

$$\frac{k(k-1)\dots(k-s+1)}{1 \cdot 2 \dots s+1},$$

when taken mod m, is periodic in k with period m^s.

This result is of course not "best possible", but it is sufficient for our purposes.

Proof. We proceed by induction over s. The result is certainly true for $s = 1$; assume it true for s. Consider the summation formula

$$\frac{k(k-1)\dots(k-s)}{1 \cdot 2 \dots s+1} = \sum_{0 \leqslant l \leqslant k-1} \frac{l(l-1)\dots(l-s+1)}{1 \cdot 2 \dots s}.$$

By the inductive hypothesis, the summand

$$\frac{l(l-1)\dots(l-s+1)}{1 \cdot 2 \dots s+1},$$

when taken mod m, is periodic in l with period m^s. Let the sum over m^s consecutive terms be σ; then the sum over m^{s+1} consecutive terms is $m\sigma$, that is, 0 mod m. Therefore the sum is periodic with period m^{s+1}. This completes the induction, and proves the lemma.

LEMMA (5.3). *Let ξ be a real line bundle over X. Then the value of $\Psi^k_R(\xi)$ in $K_R(X)/mK_R(X)$ is periodic in k with period m.*

Proof. For a real line bundle ξ we have

$$\Psi^k\xi = \begin{cases} 1 & (k \text{ even}) \\ \xi & (k \text{ odd}) \end{cases}.$$

Thus $\Psi^k\xi$ is periodic with period 2. If m is even, this is all that is required. On the other hand, $\xi - 1$ has order 2^t for some t, and so if m is odd $\xi - 1$ is divisible by m; in this case, therefore, the mod m value of $\Psi^k\xi$ is constant. This completes the proof.

LEMMA (5.4). *Let ξ be an $SO(2n)$-bundle (if $\Lambda = R$) or a $U(n)$-bundle (if $\Lambda = C$) over X. Then there exists an integer $e = e(X, \Lambda, n)$ such that the value of $\Psi^k_\Lambda\xi$ in $K_\Lambda(X)/mK_\Lambda(X)$ is periodic in k with period m^e.*

Proof. Since $H_*(X)$ is finitely generated, there exists $q = q(X, \Lambda)$ such that the filtration subgroup F_q of $K_\Lambda(X)$ is zero. We now apply Theorem (2.2), taking $S = Z$ and $G = SO(2n)$ or $G = U(n)$ according to the case. We obtain $r = r(n, \Lambda, q)$ such that

$$\chi^{-1}(J^r) \subset I^q.$$

That is, if a virtual representation θ of G is such that its character $\chi(\theta)$ is small of the rth order at the identity of G, then we shall have $\theta\xi = 0$ in $K_\Lambda(X)$.

We now introduce virtual representations Φ^k of G, for $k \geqq 0$, by the following equation.

$$\Phi^k_\Lambda = \Psi^k_\Lambda - k\Psi^{k-1}_\Lambda + \frac{k(k-1)}{1\cdot 2}\Psi^{k-2}_\Lambda - \cdots + (-1)^k\Psi^0_\Lambda.$$

In the case $G = U(n)$, $\Lambda = C$ the character of Ψ^k_C is $\sum_{1 \leqslant t \leqslant n} (z_t)^k$, and therefore the character of Φ^k_C is

$$\sum_{1 \leqslant t \leqslant n} (z_t - 1)^k = \sum_{1 \leqslant t \leqslant n} (\zeta_t)^k.$$

Thus

$$\Phi^k_C \in \chi^{-1}(J^k).$$

Since $c\Phi^k_R = \Phi^k_C c$, we also have

$$\Phi^k_R \in \chi^{-1}(J^k).$$

We can invert the definition and write

$$\Psi^k_\Lambda = \Phi^0_\Lambda + k\Phi^1_\Lambda + \frac{k(k-1)}{1\cdot 2}\Phi^2_\Lambda + \cdots.$$

Here we only have to take r terms if we wish to work modulo $\chi^{-1}(J^r)$. With this interpretation, the formula is true whether k is positive, negative or zero; in the case $\Lambda = C$ this can be checked at once by taking characters; the case $\Lambda = R$ follows since the Φ's and Ψ's commute with c.

We now have

$$\Psi^k_\Lambda\xi = \sum_{0 \leqslant s \leqslant r-1} \frac{k(k-1)\ldots(k-s+1)}{1\cdot 2\cdot \ldots s}\Phi^s_\Lambda\xi.$$

According to Lemma (5.2), the mod m value of each summand is periodic in k with period m^r. This proves the lemma.

Proof of Theorem (5.1). We recall that since X may be infinite-dimensional, $K_\Lambda(X)$ is defined as an inverse limit. Consider first the case $\Lambda = C$; then an element of this inverse limit may be represented by a map $X \longrightarrow Z \times BU$. Since X is cohomologically finite-dimensional and $BU(n)$ is simply-connected, this map can be compressed into $Z \times BU(n)$ for some $n = n(X)$. That is, any element of $K_C(X)$ can be represented in the form $h + \xi$, where $h \in H^0(X; Z)$ and ξ is a bundle with structural group $U(n)$. Since $\Psi^k(h) = h$, the result now follows from Lemma (5.4).

Similarly, in the case $\Lambda = R$, every element of $K_R(X)$ can be represented in the form $h + \xi + \eta$, where ξ is a real line bundle and η is an $SO(2n)$-bundle. The result now follows from Lemmas (5.3, 5.4). This completes the proof.

§6. CHARACTERISATION OF THE POSSIBLE VALUES $\{\rho^k\}$

Let X be a finite CW-complex, and let $x \in \tilde{K}_{\mathrm{Spin}}(X)$, $y \in \tilde{K}_R(X)$ be two elements. Then the equation

$$v^k = \rho^k(x)\,\frac{\Psi^k(1 + y)}{1 + y} \qquad \text{(all } k\text{)} \tag{6.1}$$

defines an element

$$\{v^k\} \in \prod_{k \neq 0} (1 + \tilde{K}_R(X) \otimes Q_k).$$

It is the object of this section to characterise the elements which arise in this way.

THEOREM (6.2). *An element*

$$\{v^k\} \in \prod_{k \neq 0} (1 + \tilde{K}_R(X) \otimes Q_k)$$

can be written in the form (6.1) *if and only if the following conditions hold.*

(a) *Let* $i : X^2 \longrightarrow X$ *be the inclusion map of the* 2-*skeleton; then* $i^* v^k \in Q_k$.

(b) $v^{-1} = 1$.

(c) $v^k \cdot \Psi^k v^l = v^{kl}$ in $1 + \tilde{K}_R(X) \otimes Q^{kl}$.

(d) *For each prime power* p^f *there exists* p^g *such that the mod* p^f *value of* v^k *is periodic in* k *with period* p^g.

These conditions call for a few comments. Condition (a) states in effect that v^k "is of filtration 3". For the purposes of this theorem we could equally well have written "$i^* v^k = 1$" instead of "$i^* v^k \in Q^k$"; the condition is written in the form given so that it can be applied to more general sequences v^k. Conditions (a) and (b) are of course fairly trivial. Condition (c) has been stressed by Bott, who calls it the "cocycle condition" [7]. Personally I do not see the point of stressing (c) without (d), since in the presence of torsion both are certainly needed to make any realistic algebraic model of the topological situation.

It remains to explain what we mean by "periodic" in condition (d); for on the face of it, the mod p^f value of v^k lies in a group which is dependent on k. More precisely, let us write K for $K_R(X)$; then the mod p^f value of v^k lies in the ring

$$(K \otimes Q_k)/p^f(K \otimes Q_k).$$

However, we have canonical isomorphisms

$$(K \otimes Q_k)/p^f(K \otimes Q_k) \cong (K/p^f K) \otimes Q_k$$
$$\cong \begin{cases} K/p^f K & (k \not\equiv 0 \bmod p) \\ 0 & (k \equiv 0 \bmod p). \end{cases}$$

This allows us to identify the rings

$$(K \otimes Q_k)/p^f(K \otimes Q_k)$$
$$(K \otimes Q_l)/p^f(K \otimes Q_l)$$

whenever $k \equiv l \bmod p$.

The proof of Theorem (6.2) will require three lemmas.

LEMMA (6.3). *In Theorem* (6.2), *the conditions* (a), (b), (c) *and* (d) *are necessary.*

This is the easy half of Theorem (6.2).

Proof. We note that if two sequences v^k, w^k satisfy these conditions, so does their product $v^k w^k$ and the inverse of v^k. It is now sufficient to check the conditions for $\rho^k x$ when x is a Spin($8n$)-bundle, and for

$$\frac{\Psi^k(1 + y)}{1 + y}.$$

Let x be a Spin($8n$)-bundle ξ; we check condition (a). Since Spin($8n$) is 1-connected, the bundle $i^*\xi$ over X^2 is trivial; thus

$$i^*\rho^k\xi = \rho^k i^*\xi = k^{4m}.$$

Condition (c) is given by [4, Proposition (5.5)]. Now consider the equation

$$\rho^k\xi = \phi^{-1}\Psi^k\phi 1.$$

This makes condition (b) obvious, since Ψ^{-1} is the identity. Also the mod p^f value of $\Psi^k\phi 1$ is periodic in k with period p^{ef}, by Theorem (5.1) applied in the Thom space X^ξ; thus $\rho^k\xi$ satisfies condition (d).

We turn to the sequence

$$\frac{\Psi^k(1 + y)}{1 + y}$$

and check condition (a). We note (from the spectral sequence) that $K_R(X^2)$ is generated by real line bundles and elements of filtration 2; for each of these we have $\Psi^k z = z$ so long as k is odd; and for k even we have

$$\tilde{K}_R(X^2) \otimes Q_k = 0,$$

since $\tilde{K}_R(X^2)$ is 2-primary. Thus

$$\frac{\Psi^k(1 + y)}{1 + y} = 1 \quad \text{in} \quad 1 + \tilde{K}_R(X^2) \otimes Q_k,$$

proving condition (a). It is easy to check that

$$\frac{\Psi^k(1 + y)}{1 + y}$$

satisfies conditions (b) and (c); condition (d) is given by Theorem (5.1). This completes the proof of Lemma (6.3).

The proof of Theorem (6.2) will be by filtering $K_R(X)$ and using induction over the dimension. To make the inductive step we shall require Lemma (6.4).

We shall assume that Y is a finitely-generated abelian group, on which operations Ψ^k and ρ^k are defined by the same formulae that hold in the particular case $Y = \tilde{K}_R(\vee S^r)$ (compare the proof of Lemma (4.5)). If $r \equiv 1$ or 2 mod 8, we assume that evey element of Y has order 2.

LEMMA (6.4). *Suppose that an element*

$$\{v^k\} \in \prod_{k \neq 0} (1 + Y \otimes Q_k)$$

satisfies conditions (b), (c) *and* (d) *of Theorem* (6.2). *Then it can be written in the form* (6.1) *for some* x, y *in* Y.

Proof. Since Y is a finitely-generated abelian group, we may express it as a direct sum of cyclic groups Z and Z_{p^f}. It is now easy to see that it is sufficient to prove the result for these summands. This reduces the proof to three cases.

We consider first the case $r \equiv 0$ or 4 mod 8 (say $r = 4t$) and $Y = Z$. Let us write

$$v^k = 1 + w^k,$$

where

$$w^k \in Y \otimes Q_k \cong Q_k.$$

In this case condition (c) yields

$$v^k \cdot \Psi^k v^l = v^l \cdot \Psi^l v^k,$$

which gives

$$w^k + k^{2t} w^l = w^l + l^{2t} w^k,$$

that is,

$$(k^{2t} - 1) w^l = (l^{2t} - 1) w^k.$$

Therefore there is a rational number c such that

$$w^k = (k^{2t} - 1)c.$$

Now there exists $f(k)$ such that $k^{f(k)} w^k$ is integral; and by [4, Theorem (2.7)], the highest common factor of the expressions $k^{f(k)}(k^{2t} - 1)$ divides $m(2t)$; therefore $m(2t)c$ is integral. That is, we may write

$$w^k = \frac{(k^{2t} - 1)d}{m(2t)}$$

where d is integral.

Now let us write $\frac{1}{2}\alpha_{2t} = n(2t)/m(2t)$, where $n(2t)$ and $m(2t)$ are coprime. (See [4, §2]; compare the proof of Lemma (4.5)). We may now set

$$am(2t) + bn(2t) = 1.$$

Setting $x = bd$, $y = ad$ we easily calculate that

$$\rho^k(x) \frac{\Psi^k(1 + y)}{1 + y} = v^k.$$

This completes this case.

Secondly, we consider the case $r \equiv 0$ or $4 \bmod 8$ (say $r = 4t$), and $Y = Z_{p^f}$. In this case

$$Y \otimes Q_k \cong \begin{cases} Y & \text{if } k \not\equiv 0 \bmod p \\ 0 & \text{if } k \equiv 0 \bmod p. \end{cases}$$

We may therefore restrict attention entirely to those values of k prime to p.

When we apply condition (d), we shall of course take p^f to be the order of Y. Thus condition (d) asserts that the actual value of v^k is periodic with period p^g. According to Lemma (6.3) (for the space $X = S^r$), we can suppose g chosen so large that if x is a generator of $\tilde{K}_R(S^r)$, then the mod p^f value of $\rho^k(x)$ is periodic in k, with period p^g, and similarly for

$$\frac{\Psi^k(1+y)}{1+y}.$$

(Of course this is also easy to check using the explicit formulae for ρ^k and Ψ^k in S^r.) Therefore the same periodicity statement will be true for every expression

$$\rho^k(x)\frac{\Psi^k(1+y)}{1+y}$$

with x, y in Y. In Y, of course, the periodicity statement asserts that the actual value of this expression is periodic with period p^g.

We now fix attention on a particular value of k. If p is odd, we choose k to be a generator for the multiplicative group G_{p^g} of residue classes prime to p modulo p^g. If $p = 2$, we choose k to be a generator of the quotient group $G_{2^g}/\{\pm 1\}$, which is again cyclic.

By [4, Lemma (2.12)], when the fraction

$$\frac{k^{2t}-1}{m(2t)}$$

is written in its lowest terms, both numerator and denominator are prime to p. Therefore we can solve the equation

$$v^k = 1 + \frac{k^{2t}-1}{m(2t)} z$$

for a solution $z \in Y \otimes Q_k \cong Y$. As before, we set $\frac{1}{2}\alpha_{2t} = n(2t)/m(2t)$, $am(2t) + bn(2t) = 1$, $x = bz$ and $y = az$; we easily calculate that

$$\rho^k(x)\frac{\Psi^k(1+y)}{1+y} = v^k$$

for this particular value of k.

Thus the two expressions

$$\rho^l(x)\frac{\Psi^l(1+y)}{1+y}, \qquad v^l$$

agree for $l = k$. But both expressions satisfy condition (c), which allows us to calculate v^{k^r} in terms of v^k; thus we see that the two expressions agree for $l = k^r$. In fact, using condition (b) also, we see that they agree for $l = \pm k^r$. But by the choice of k, the integers $\pm k^r$ give all residue classes prime to p modulo p^g; hence by periodicity, the two expressions agree for all l prime to p. This completes this case.

Finally, we consider the case $r \equiv 1$ or $2 \bmod 8$, $Y = Z_2$. In this case

$$Y \otimes Q_k \cong \begin{cases} Z_2 & \text{if k is odd} \\ 0 & \text{if } k \text{ is even.} \end{cases}$$

We may therefore restrict attention entirely to odd values of k. By condition (d), the value of v^k is periodic in k with period 2^g.

If k and l are odd, condition (c) gives

$$v^k \cdot v^l = v^{kl}.$$

Therefore the function v^k of k gives a homomorphism from the multiplicative group of odd residue classes mod 2^g to the multiplicative group $1 + Y \cong Z_2$. By condition (b), this homomorphism factors through $G_{2^g}/\{\pm 1\}$. But this group is cyclic; so there are only two possible homomorphisms. We must have

$$v^k = \begin{cases} 1 & \text{for } k \equiv \pm 1 \bmod 8 \\ 1 + x & \text{for } k \equiv \pm 3 \bmod 8 \end{cases}$$

where x is one of the two elements in Y. According to [4, Theorem (5.17)], this shows that

$$v^k = \rho^k x.$$

This completes the proof of Lemma (6.4).

We need one more lemma. As in §4, we define F_r to be the image of $\tilde{K}_R(X/X^{r-1})$ in $K = K_R(X)$.

LEMMA (6.5). *Suppose given a sequence*

$$\{v^k\} \in \prod_{k \neq 0} (1 + F_r \otimes Q_k)$$

and suppose that for each prime power p^f there exists p^g such that the value of v^k in

$$(K \otimes Q_k)/p^f(K \otimes Q_k)$$

is periodic with period p^g. Then for each prime power p^f there exists p^h such that the value of v^k in

$$1 + (F_r \otimes Q_k)/p^f(F_r \otimes Q_k)$$

is periodic with period p^h.

Proof. Suppose given p^f. Consider the subgroup S_t of elements x in $K_R(X)$ such that

$$p^{ft} x \in F_r.$$

This is an increasing sequence of Z-submodules in the finitely-generated Z-module $\tilde{K}_R(X)$, therefore convergent. That is, there is a t such that $x \in \tilde{K}_R(X)$ and $p^{f(t+1)}x \in F_r$ imply $p^{ft}x \in F_r$. According to the data, there is now an h such that the value of v^k in $K \otimes Q_k/p^{f(t+1)}(K \otimes Q_k)$ is periodic in k with period p^h. Suppose then that $k \equiv l \bmod p^h$ and k, l are prime to p. We have

$$v^k = 1 + k^{-a} w^k$$

$$v^l = 1 + l^{-b} w^l$$

for some $w^k, w^l \in F_r$. The periodicity statement gives

$$l^b w^k - k^a w^l = p^{f(t+1)} x$$

for some $x \in \tilde{K}_R(X)$ with $p^{f(t+1)} x \in F_r$. By our choice of t, this gives $p^{ft} x \in F_r$; that is,

$$l^b w^k - k^a w^l = p^f y$$

with $y \in F_r$. Since k and l are prime to p, this gives

$$k^{-a} w^k = l^{-b} w^l$$

in $F_r/p^f F_r$. This proves the lemma.

Proof of Theorem (6.2). We will prove by downwards induction over r that if an element

$$\{v^k\} \in \prod_{k \neq 0} (1 + F_r \otimes Q_k)$$

satisfies conditions (b), (c) and (d) of Theorem (6.2), then it can be written in the form (6.1), provided $r \geqq 3$. Here it does not matter whether the periodicity condition (d) is interpreted as an equation in $K_R(X)/p^f K_R(X)$ or as an equation in $F_r/p^f F_r$, according to Lemma (6.5). The inductive hypothesis is trivial for r greater than the dimension of X; we assume it true for $r + 1$, where $r \geqq 3$. Let

$$\{v^k\} \in \prod_k (1 + F_r \otimes Q_k)$$

be an element satisfying conditions (b), (c) and (d) of Theorem (6.2). Then the image of $\{v^k\}$ in

$$\prod_k (1 + (F_r/F_{r+1}) \otimes Q_k)$$

satisfies the conditions of Lemma (6.4). Therefore there are elements x, y in F_r such that

$$v^k \left[\rho^k(x) \frac{\Psi^k(1 + y)}{1 + y}\right]^{-1} \in 1 + F_{r+1} \otimes Q_k$$

for all k. (Note that $x \in \tilde{K}_{\mathrm{Spin}}(X)$, since $r \geqq 3$.) By Lemma (6.3) the element

$$\left\{\rho^k(x) \frac{\Psi^k(1 + y)}{1 + y}\right\}$$

satisfies the conditions (b), (c) and (d); therefore the inductive hypothesis applies to the sequence

$$v^k \left[\rho^k(x) \frac{\Psi^k(1 + y)}{1 + y}\right]^{-1},$$

which can accordingly be written

$$\rho^k(x') \frac{\Psi^k(1 + y')}{1 + y'}.$$

Thus v^k can be written in the form (6.1) (with x replaced by $x + x'$, and similarly for y). This completes the induction. Therefore the inductive hypothesis is true for $r = 3$; this proves that the conditions (a), (b), (c) and (d) are sufficient. The proof of Theorem (6.2) is complete.

§7. POSSIBLE GENERALISATIONS

The trend of this section can be explained by considering the special case $X = S^{8m+4}$. Let π_r^S be the stable r-stem. We shall show in Part IV that the image of the stable J-homomorphism

$$J : \pi_{8m+3}(SO) \longrightarrow \pi_{8n+3}^S$$

is a direct summand. The reason is essentially that the quotient map

$$\tilde{J}(S^{8m+4}) \longrightarrow \tilde{J}'(S^{8m+4})$$

can be extended over the whole of π_{8m+3}^S.

It is reasonable to seek for a generalisation of this phenomenon. The generalisation should state that for a suitable class of spaces X, $J(X)$ is a natural direct summand in $L(X)$, where L is a functor such that $\tilde{L}(S^n) = \pi_{n-1}^S$. Unfortunately, I do not feel certain as to the best way of arranging the details of the construction of the functor L. For this reason, the present section is written as a tentative and heuristic explanation of the phenomena involved.

The construction of the functor L should secure the following properties.

(i) $L(X)$ should be a Grothendieck group generated by equivalence classes of "fiberings", in some weak sense.

The senses of the word "fibering" which might be considered include (for example) the following.

(a) Fiberings which are locally fibre homotopy equivalent to products $B \times S^n$; see [8].

(b) Hurewicz fiberings in which the fibres are homotopy-equivalent to spheres; see [9, 10, 11].

(c) Suitable CSS-fiberings; see [6].

As stated above, I do not feel certain as to the best choice of details.

(ii) The fiberings considered should admit suitable "Whitney sums" and "induced fiberings" in order that Grothendieck's construction should apply.

(iii) Sphere-bundles should qualify as "fiberings" in the sense considered; moreover the Whitney sums and induced bundles which we have for sphere-bundles should qualify as "Whitney sums" and "induced bundles" in the generalised sense. We require this in order to obtain a natural transformation

$$K_R(X) \longrightarrow L(X).$$

(iv) The image of $K_R(X)$ in $L(X)$ should be $J(X)$, up to natural isomorphism.

For practical purposes, this means that the equivalence relation used in constructing $L(X)$ must be fibre homotopy equivalence.

(v) The functor $L(X)$ should be a representable contravarient functor; moreover, the representing space should be essentially that constructed as follows.

Let H_n be the space of homotopy-equivalences from S^n to S^n, considered as a monoid under composition. Let BH_n be its classifying space (under some interpretation of this phrase). By suspension we define an embedding $H_n \longrightarrow H_{n+1}$, whence a map $BH_n B \longrightarrow H_{n+i}$. Set $BH = \operatorname{Lim}_{n \to \infty} BH_n$ and take the space $Z \times BH$.

(vi) The "fiberings", "Whitney sums" and "induced fiberings" employed in the construction of $L(X)$ should retain sufficient of the cohomological properties of their classical counterparts, in respect both of ordinary and of extraordinary cohomology theories.

This is necessary in order that we should be able to define cohomological invariants of these fiberings, according to the pattern introduced in Part II.

Let us assume that with some arrangement of the details, we can secure what has been indicated above. The whole of the rest of this section will depend on this assumption, although I will not bother to write every sentence in a conditional form. However, I will not call the results "theorems", since the underlying assumptions have not been stated precisely enough. Of course it would not be hard to drag every assumption out into the open and give it the status of an axiom; but the result might be somewhat tedious to read.

We will now consider the set of "fiberings" whose Thom pairs can be oriented over the cohomology theory K_Λ.

EXAMPLE (6.1). *If $H^*(X; Z)$ is torsion-free, then every "fibering" ξ over X is orientable over the cohomology theory K_C.*

For the Thom space X^ξ of ξ will also be torsion-free; hence the spectral sequence

$$H^*(X^\xi ; K_C^*(P)) \Rightarrow K_C^*(X^\xi)$$

will be trivial; therefore the required orientation will exist.

A similar argument will work if $H^*(X; Z)$ is even-dimensional.

The condition of "orientability over K_Λ" is invariant under fibre homotopy equivalence, and even under stable fibre homotopy equivalence; it is inherited by induced fiberings; if two fiberings admit orientations, we can put the product orientation on their Whitney sum. (This assumes that the details chosen for "Whitney sums" and "induced fiberings" allow one to draw the usual diagrams. (See (vi) above.))

We now introduce the subgroup $L_{OR}(X)$ of $L(X)$, generated by fiberings orientable over K_R and of dimension divisible by 8. Our cohomological invariants, such as ρ^k, are defined on such fiberings. More precisely, given an orientation u of ξ over K_R, this determines a Thom isomorphism ϕ in K_R-cohomology. It thus determines

$$\rho^k(\xi) = \phi^{-1}\Psi^k\phi 1.$$

However, if $y \in \tilde{K}_R(X)$, then we can replace u by $\pm(p^*(1 + y))u$; this changes $\rho^k(\xi)$, multiplying it by

$$\frac{\Psi^k(1 + y)}{1 + y}.$$

The classes $\rho^k\xi$ satisfy formulae such as

$$\rho^k(\xi \oplus \eta) = \rho^k\xi \, . \, \rho^k\eta$$

$$\rho^k(f^*\xi) = f^*(\rho^k\xi)$$

up to the relevant indeterminacy

$$\frac{\Psi^k(1+y)}{1+y}.$$

We can therefore define our invariant $\rho = \prod_k \rho^k$ on $L_{OR}(X)$ so as to be exponential. More precisely, let $k^* + \tilde{K}_R(X) \otimes Q_k$ be the set of elements in $K_R(X) \otimes Q_k$ of the form $k^f + x$, where $f \in Z$ and $x \in \tilde{K}_R(X) \otimes Q^k$. This set is a multiplicative group. Let us define a map

$$\delta : 1 + \tilde{K}_R(X) \longrightarrow \prod_k (k^* + \tilde{K}_R(X) \otimes Q_k)$$

by

$$\delta(1+y) = \left\{\frac{\Psi^k(1+y)}{1+y}\right\},$$

as above. Then we can define ρ so that its values lie in the cokernel of δ.

We will next prove that the image of the map ρ is no larger than it would have been if we had used only classical bundles, instead of more general fiberings. The proof uses Theorem (6.2), which was introduced for this purpose. More precisely, suppose given an extraordinary fibering ξ of dimension $8n$ over X, and an orientation u of the Thom space X^ξ. We will show that the sequence

$$\rho^k \xi = \phi^{-1} \Psi^k \phi\, 1$$

satisfies conditions (a,), (b), (c) and (d) of Theorem (6.2).

In fact, the proofs that conditions (b), (c) and (d) are necessary (see Lemma (6.3) and [4, Proposition (5.5)]) are purely cohomological; therefore these proofs remain valid when we consider extraordinary fiberings. Thus the sequence $\rho^k\xi$ satisfies conditions (b), (c) and (d).

It remains to give a new argument for condition (a). We observe that if ξ has an orientation u over K_R, then this yields an orientation h over $H^*(\quad; Z)$ such that $Sq^1 h = 0$ and $Sq^2 h = 0$. (The second assertion follows by considering the spectral sequence

$$H^*(X^\xi; K_R^*(P)) \Rightarrow K_R^*(X^\xi),$$

since Sq^2 is a differential in this spectral sequence.) It follows that the Stiefel–Whitney classes $w_1(\xi)$ and $w_2(\xi)$ are zero. Now we observe that the embedding $BO \longrightarrow BH$ induces an isomorphism of π_1 and π_2 (both groups being Z_2). Therefore extraordinary fiberings over 2-dimensional complexes are classified by w_1 and w_2, as in the classical case. Thus if $i : X^2 \longrightarrow X$ is the inclusion map, $i^*\xi$ is fibre homotopy trivial. It follows that

$$\rho^k(i^*\xi) = k^{4n}\,\frac{\Psi^k(1+y)}{1+y}.$$

But in proving Lemma (6.3), we have shown that

$$\frac{\Psi^k(1+y)}{1+y} = 1 \qquad \text{in} \qquad 1 + \tilde{K}_R(X^2) \otimes Q_k.$$

Thus $i^*\rho^k\xi = k^{4n}$, which establishes condition (a).

Theorem (6.2) now shows that if $x \in \tilde{L}_{OR}(X)$, then there exists $x' \in \tilde{K}_{\mathrm{Spin}}(X)$ such that

$$\rho(x) = \rho(x').$$

This completes our argument about the image of ρ.

We will now show that the phenomenon which we have observed for S^{8m+4} generalises to any space X such that (i) $J(X) = J'(X)$, and (ii) every extraordinary fibering over X is orientable over K_R.

In fact, if condition (ii) holds we have the following commutative diagram.

$$\begin{array}{ccc} \tilde{K}_R(X) & \longrightarrow & \tilde{L}(X) \\ & \searrow \quad \swarrow & \\ & \tilde{J}'(X) = \mathrm{Im}\,\rho & \end{array}$$

The image of $\tilde{K}_R(X)$ in $\tilde{L}(X)$ is $\tilde{J}(X)$. If condition (i) holds, then $\tilde{J}(X)$ is a natural direct summand in $\tilde{L}(X)$.

This certainly succeeds in providing an acceptable generalisation of the case $X = S^{8m+4}$. However, it must be pointed out that there are many spaces X which do not satisfy condition (ii).

For suppose we confine attention to spaces X satisfying condition (i) and (ii). Then $J(X)$ is a natural direct summand in $L(X)$, and $L(X)$ is an exact functor; therefore $J(X)$ is exact. But this is not true for general spaces X, as may be shown by the following example.

Consider the cofibering

$$S^4 \xrightarrow{f} S^4 \longrightarrow S^4 \cup_f e^5,$$

where f is a map of degree 24. Applying the functors $\tilde{K}$, $\tilde{J}$ we obtain

$$Z \xleftarrow{24} Z \longleftarrow 0,$$

$$Z_{24} \xleftarrow{0} Z_{24} \longleftarrow 0.$$

The last sequence is not exact.

In this example, both spaces satisfy condition (i) and S^4 satisfies condition (ii); therefore $S^4 \cup_f e^5$ does not satisfy condition (ii). We note that for the space $X = S^4 \cup_f e^5$, $H^*(X; Z)$ is neither torsion-free nor even-dimensional (see (6.1)).

An alternative line of argument is to point out that extraordinary fiberings (such as must be used to construct $L(X)$) have greater freedom to be non-orientable over K_R than sphere-bundles have. This may be shown by the following example.

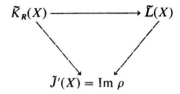

We know that every sphere-bundle over S^m is orientable over K_R, provided $m > 2$. If we translate this into terms of π^S_{m-1}, it states that every map $f: S^{n+m-1} \longrightarrow S^n$ which lies in the image of J induces the zero homomorphism of $\tilde{K}^*_R$ (provided $m > 2$). On the other hand, for $m = 3$ the map $\eta\eta$ (which is not in the image of J) induces a non-zero homomorphism of $\tilde{K}^*_R$; and we shall see in Part IV that the same thing is true whenever $m \equiv 2$ or 3 mod 8 and $m > 2$.

This second example tends to indicate that the question, "When is an extraordinary fibering non-orientable over K_R?" can sometimes be answered by quite simple invariants defined using K_R. Indeed there are some grounds for supposing it to be a less subtle question than those with which we have mainly been concerned.

In both examples it happens to be true that $J(X)$ is a direct summand in $L(X)$ (though not by a natural map or for a general reason). It would perhaps be interesting to look for an example in which $J(X)$ is not a direct summand in $L(X)$.

This completes our discussion of the extent to which we can hope to generalise the case $X = S^{8m+4}$.

Department of Mathematics,
University of Manchester

REFERENCES

1. J. F. ADAMS: Vector fields on spheres, *Ann. Math., Princeton* **75** (1962), 603–632.
2. J. F. ADAMS: On the groups $J(X)$, *Proceedings of a Symposium in Honour of Marston Morse*, Princeton University Press, to appear.
3. J. F. ADAMS: On the groups $J(X)$—I, *Topology* **2** (1963), 181–195.
4. J. F. ADAMS: On the groups $J(X)$—II, *Topology* **3** (1965), 137–172.
5. M. F. ATIYAH and F. HIRZEBRUCH: Vector bundles and homogeneous spaces, *Proceedings of Symposia in Pure Mathematics* **3**, 7–38. American Mathematical Society, 1961.
6. M. G. BARRATT, V. K. A. M. GUGENHEIM and J. C. MOORE: On semisimplicial fibre-bundles, *Amer. J. Math.* **81** (1959), 639–657.
7. R. BOTT: A note on the *KO*-theory of sphere-bundles, *Bull. Amer. Math. Soc.* **68** (1962), 395–400.
8. A. DOLD: Partitions of unity in the theory of fibrations, *Ann. Math., Princeton* **78** (1963), 223–255.
9. J. STASHEFF: A classification theorem for fibre spaces, to appear in *Topology*.
10. J. STASHEFF: Various classifications for fibre spaces, in preparation.
11. J. STASHEFF: Multiplications on classifying spaces, in preparation.

Topology Vol. 5, pp. 21–71. Pergamon Press, 1966.

ON THE GROUPS $J(X)$—IV

J. F. ADAMS

(*Received* 6 *July* 1965)

§1. INTRODUCTION

FROM ONE POINT of view, the present paper is mainly concerned with specialising the results on the groups $J(X)$, given in previous papers of this series [3, 4, 5], to the case $X = S^n$. It can, however, be read independently of the previous papers in this series; because from another point of view, it is concerned with the use of extraordinary cohomology theories to define invariants of homotopy classes of maps; and this machinery can be set up independently of the previous papers in this series. We refer to them only for certain key results.

From a third point of view, this paper represents a very belated attempt to honour the following two sentences in an earlier paper [2]. "However, it appears to the author that one can obtain much better results on the J-homomorphism by using the methods, rather than the results, of the present paper. On these grounds, it seems best to postpone discussion of the J-homomorphism to a subsequent paper." I offer topologists in general my sincere apologies for my long delay in writing up results which mostly date from 1961/62.

I will now summarise the results which relate to the homotopy groups of spheres. For this one needs some notation. The stable group $\operatorname{Lim}_{n\to\infty} \pi_{n+r}(S^n)$ will be written π_r^S. The stable J-homomorphism is thus a homomorphism

$$J: \pi_r(SO) \to \pi_r^S.$$

THEOREM 1.1. *If $r \equiv 0 \bmod 8$ and $r > 0$ (so that $\pi_r(SO) = Z_2$), then J is a monomorphism and its image is a direct summand in π_r^S.*

Before considering the case $r \equiv 1 \bmod 8$, we need a preliminary result. Suppose that $r \equiv 1$ or $2 \bmod 8$. Then any map $f: S^{q+r} \to S^q$ induces a homomorphism

$$f^*: \tilde{K}_R^q(S^q) \to \tilde{K}_R^q(S^{q+r}),$$

where the functor $\tilde{K}_R^*$ is that due to Grothendieck–Atiyah–Hirzebruch [10, 11, 2]. We have

$$\tilde{K}_R^q(S^q) = Z, \qquad \tilde{K}_R^q(S^{q+r}) = Z_2.$$

THEOREM 1.2. *Suppose that $r \equiv 1$ or $2 \bmod 8$ and $r > 0$. Then π_r^S contains an element μ_r, of order 2, such that any map $f: S^{q+r} \to S^q$ representing μ_r induces a non-zero homomorphism of $\tilde{K}_R^q$.*

The elements μ_r may be described more precisely than is done in this theorem. We have $\mu_1 = \eta$ and $\mu_2 = \eta\eta$, where η is (as usual) the generator of π_1^S. The elements u_r constitute a systematic family of elements, generalising η and $\eta\eta$; they have interesting properties, which I hope to discuss on another occasion. I am indebted to M. G. Barratt for ideas about systematic families of elements.

THEOREM 1.3. *Suppose that* $r \equiv 1 \bmod 8$ *and* $r > 1$ *(so that* $\pi_r(SO) = Z_2$*). Then J is a monomorphism and* π_r^S *contains a direct summand* $Z_2 + Z_2$, *one summand being generated by* μ_r *and the other being* Im J.

The case $r = 1$ is exceptional, in that the two summands coincide.

THEOREM 1.4. *Suppose that* $r \equiv 2 \bmod 8$ *and* $r > 0$. *Then* π_r^S *contains a direct summand* Z_2 *generated by* μ_r.

THEOREM 1.5. *Suppose* $r = 4s - 1 \equiv 3 \bmod 8$, *so that* $\pi_r(SO) = Z$. *Then the image of J is a cyclic group of order* $m(2s)$, *and is a direct summand in* π_r^S.

In this theorem, $m(t)$ is the numerical function discussed in [4, §2]. More explicitly, let B_s be the sth Bernoulli number; then $m(2s)$ is the denominator of $B_s/4s$, when this fraction is expressed in its lowest terms.

The direct sum splitting will be accomplished by defining (§7) a homomorphism

$$e'_R : \pi_r^S \to Z_{m(2s)}$$

such that

$$e'_R J : \pi_r(SO) \to Z_{m(2s)}$$

is an epimorphism.

THEOREM 1.6. *Suppose* $r = 4s - 1 \equiv 7 \bmod 8$, *so that* $\pi_r(SO) = Z$. *Then the image of J is a cyclic group of order either* $m(2s)$ *or* $2m(2s)$. *Moreover, there is a homomorphism*

$$e'_R : \pi_r^S \to Z_{m(2s)}$$

such that

$$e'_R J : \pi_r(SO) \to Z_{m(2s)}$$

is an epimorphism.

It follows that if the order of Im J is $m(2s)$, then Im J is a direct summand; this happens (for example) if $r = 7$ or 15. In any event, the subgroup of elements of odd order in Im J is a direct summand in π_r^S.

It will not be proved in this paper, but by more delicate arguments one can show that even for $r \equiv 7 \bmod 8$, the group π_r^S splits as (Ker e'_R) + $Z_{m(2s)}$; however, I do not know how the subgroup Im J lies with respect to this splitting.

The invariants (such as e'_R) which we shall introduce have convenient properties, and lend themselves to a variety of calculations; examples will be given in §§11, 12. They are not restricted to maps between spheres. The following result provides rather a striking example. We take p to be an odd prime, $g : S^{2q-1} \to S^{2q-1}$ to be a map of degree p^f, and Y to be the Moore space $S^{q-1} \cup_g e^{2q}$. Thus $\tilde{K}_C(Y) = Z_{p^f}$. $S^{2r}Y$ will mean the $2r$-fold suspension of Y; we take $r = (p-1)p^{f-1}$.

THEOREM 1.7. *For suitable q there is a map*

$$A : S^{2r}Y \to Y$$

which induces an isomorphism

$$A^* : \tilde{K}_C(Y) \to \tilde{K}_C(S^{2r}Y).$$

Therefore the composite

$$A . S^{2r}A . S^{4r}A . \dots S^{2r(s-1)}A : S^{2rs}Y \to Y$$

induces an isomorphism of $\tilde{K}_C$, *and is essential for every* s.

For $f = 1$ this result is related to Toda's sequence of elements $\alpha_s \in \pi^S_{2(p-1)s-1}$ [16,17], as will be explained in §12.

From the point of view of history or motivation, the sequence of ideas in this paper may be ordered as follows. Suppose given a map $f: X \to Y$. We may form the mapping cone $Y \cup_f CX$; by studying the group $K_C(Y \cup_f CX)$ and the homomorphism

$$ch: K_C(Y \cup_f CX) \to H^*(Y \cup_f CX; Q)$$

we may sometimes succeed in distinguishing $Y \cup_f CX$ from $Y \vee SX$; thus we may sometimes show that f is essential. This method was presumably known to Atiyah and Hirzebruch (*ca.* 1960/61); it is given in [6] (for the case in which X and Y are spheres) and was published by Dyer [13]. See also [19]. We touch on it in §7 of this paper.

One next realises that in the preceding construction, the possible Chern characters that can arise are severely limited by the fact that $K_C(Y \cup_f CX)$ admits operations Ψ^k. This observation leads to a proof of the non-existence of elements of Hopf invariant one (mod 2 and mod p); this proof was given in [6], and was first published by Dyer [13]. We touch on it in §8 of this paper. It should be said, however, that the most elegant proof by K-theory of the non-existence of elements of Hopf invariant one is somewhat different; see [8].

One next realises that the essential phenomenon we have to study is the short exact sequence

$$\tilde{K}_C(Y) \leftarrow \tilde{K}_C(Y \cup_f CX) \leftarrow \tilde{K}_C(SX)$$

of groups admitting operations Ψ^k. The class of this short exact sequence yields an element of a suitable group

$$\mathrm{Ext}^1(\tilde{K}_C(Y), \tilde{K}_C(SX)).$$

This element gives an invariant of f. If $K_C(Y \cup_f CX)$ is torsion-free this approach is equivalent to that using the Chern character; if $K_C(Y \cup_f CX)$ has torsion this approach is better than that using the Chern character. We therefore adopt this as our basic approach. It has been sketched in [7], and will be fully explained in §3.

In the above, we can of course use $\tilde{K}_R$ instead of $\tilde{K}_C$. The use of $\tilde{K}_R$ and the use of spaces with torsion gives the extra power needed to prove results such as Theorems 1.1, 1.3.

Once we realise that our invariants should take values in suitable Ext^1 groups, certain properties of the invariants become very plausible. Our invariants carry composition products (of homotopy classes) into composition products (in Ext) (§3); they carry Toda brackets (in homotopy) into Massey products (in Ext) (§§4,5). These products enable one to perform many calculations.

The arrangement of the paper is as follows. Since we make constant use of cofibre sequences

$$X \xrightarrow{f} Y \to Y \cup_f CX \to SX \dots ,$$

we devote §2 to them. In §3 we define our invariants and give their basic properties. §§4, 5 are devoted to their properties on Toda brackets, as indicated above. So far the work has been done for a quite general cohomology theory; in §§6, 7 we specialise to the case of $\tilde{K}_C$ and $\tilde{K}_R$. §7 contains the main theorem about the cases in which X and Y are spheres and $\tilde{K}$ is torsion-free. §8 contains the relationship between the invariants of §7 and the classical Hopf invariant in the sense of Steenrod. §9 considers the case needed for Theorems 1.1, 1.3, in which X and Y are spheres but $\tilde{K}$ is not torsion-free. In §10 we discuss the value of our invariants on the image of J. In §11 we work out the general theory of §§4, 5 (about Toda brackets) for the special cases which most concern us. In §§12 we prove Theorem 1.7 and discuss related matters; since the same machinery serves to discuss certain 2-primary phenomena, we also prove Theorem 1.2 there. In §12 we also give a number of examples and applications; the reader's attention is particularly directed to these, since they provide essential motivation.

Since drafting the body of this paper, I have become aware of Toda's paper [19], which has a considerable overlap with the present paper. I am very grateful to Toda for a letter about his results.

Toda defines an invariant

$$CH^{n+k} : \pi_{2n+2k-1}(S^{2n}) \to Q/Z$$

which is presumably the same as the invariant e_C discussed in this paper. He also defines an invariant CH_*^{4m+2h}, which is presumably the same (up to a certain constant factor) as the invariant e'_R discussed in this paper.

To give Toda proper credit for his priority, I offer the following concordance of results. Corollary 7.7 of this paper is to be found in Toda's paper, and is the essential step in the proof of his Theorems 6.3, 6.5(i) and (ii) which give restrictions on the values that can be taken by his invariants (compare 7.14, 7.15 of this paper). Proposition 7.20 of this paper is Theorem 6.5 (iii) of [19]. Corollary 8.3 of this paper is Theorem 6.7 of [19]. The case $\Lambda = C$ of Theorem 11.1 of this paper is Theorem 6.4 of [19]. Theorem 12.11 of this paper is contained in 6.8 of [19].

§2. COFIBERINGS

As explained in the introduction, this paper will make much use of sequences of cofiberings. We shall therefore devote this section to summarising some material about cofibre sequences, following [15]. We need only deal with "good" spaces; for the applications, it would be sufficient to consider finite CW-complexes.

Let $f: X \to Y$ be a map. We can construct from it a cofibering

$$X \xrightarrow{f} Y \xrightarrow{i} Y \cup_f CX.$$

Here i is an injection map; and $Y \cup_f CX$ is the space obtained from Y by attaching CX, the cone on X, using f as attaching map.

Iterating this construction, we can construct

$$Y \xrightarrow{i} (Y \cup_f CX) \xrightarrow{j} (Y \cup_f CX) \cup_i CY$$

and (setting $Z = Y \cup_f CX$)

$$Z \xrightarrow{j} (Z \cup_i CY) \xrightarrow{k} (Z \cup_i CY) \cup_j CZ.$$

Now the space $(Y \cup_f CX) \cup_i CY$ is homotopy-equivalent to the suspension SX; and similarly, the space $(Z \cup_i CY) \cup_j CZ$ is homotopy-equivalent to SY. In order to avoid errors of sign in what follows, it is desirable to use the "same" homotopy equivalence in the two cases. If we do this, then the map

$$k:(Y \cup_f CX) \cup_i CY \to (Z \cup_i CY) \cup_j CZ$$

corresponds to

$$-Sf: SX \to SY.$$

(This is easy to check; or see [15, p. 309, Satz 4].) We shall therefore take the following as our basic cofibre sequence.

$$X \xrightarrow{f} Y \xrightarrow{i} Y \cup_f CX \xrightarrow{j} SX \xrightarrow{-Sf} SY \ldots$$

This construction has various obvious properties, which we record for use later.

Proposition 2.1. *If $f \sim g$, then we can construct the following homotopy-commutative diagram, in which all the vertical arrows are homotopy equivalences.*

$$\begin{array}{ccccccccc}
X & \xrightarrow{f} & Y & \xrightarrow{i} & Y \cup_f CX & \xrightarrow{j} & SX & \xrightarrow{-Sf} & SY \\
\downarrow{\scriptstyle 1} & & \downarrow{\scriptstyle 1} & & \downarrow & & \downarrow{\scriptstyle 1} & & \downarrow{\scriptstyle 1} \\
X & \xrightarrow{g} & Y & \xrightarrow{i'} & Y \cup_g CX & \xrightarrow{j'} & SX & \xrightarrow{-Sg} & SY
\end{array}$$

Proposition 2.2. *Given a commutative diagram*

$$\begin{array}{ccc}
X & \xrightarrow{f} & Y \\
\downarrow{\scriptstyle h} & & \downarrow{\scriptstyle k} \\
X' & \xrightarrow{f'} & Y'
\end{array}$$

we can construct the following commutative diagram.

$$\begin{array}{ccccccccc}
X & \xrightarrow{f} & Y & \xrightarrow{i} & Y \cup_f CX & \xrightarrow{j} & SX & \xrightarrow{-Sf} & SY \\
\downarrow{\scriptstyle h} & & \downarrow{\scriptstyle k} & & \downarrow & & \downarrow{\scriptstyle Sh} & & \downarrow{\scriptstyle Sk} \\
X' & \xrightarrow{f'} & Y' & \xrightarrow{i'} & Y' \cup_{f'} CX' & \xrightarrow{j'} & SX' & \xrightarrow{-Sf'} & SY'
\end{array}$$

These obvious and elementary propositions are special cases of the more general results proved in [15, pp. 311–316].

Proposition 2.3. *Given*

$$X \xrightarrow{f} Y \xrightarrow{g} Z,$$

we can construct the following commutative diagram.

$$\begin{array}{ccccccccc}
X & \xrightarrow{f} & Y & \xrightarrow{i} & Y \cup_f CX & \xrightarrow{j} & SX & \xrightarrow{-Sf} & SY \\
\downarrow{\scriptstyle 1} & & \downarrow{\scriptstyle g} & & \downarrow & & \downarrow{\scriptstyle 1} & & \downarrow{\scriptstyle Sg} \\
X & \xrightarrow{gf} & Z & \xrightarrow{i'} & Z \cup_{gf} CX & \xrightarrow{j'} & SX & \xrightarrow{-S(gf)} & SZ \\
\downarrow{\scriptstyle f} & & \downarrow{\scriptstyle 1} & & \downarrow & & \downarrow{\scriptstyle Sf} & & \downarrow{\scriptstyle 1} \\
Y & \xrightarrow{g} & Z & \xrightarrow{i''} & Z \cup_g CY & \xrightarrow{j''} & SY & \xrightarrow{-Sg} & SZ
\end{array}$$

This follows from two applications of Proposition 2.2.

PROPOSITION 2.4. *For each r, we can construct the following homotopy-commutative diagram, in which all the vertical arrows are homotopy equivalences.*

$$\begin{array}{ccccc}
S^r Y & \xrightarrow{S^r i} & S^r(Y \cup_f CX) & \xrightarrow{S^r j} & S^{r+1}X \\
\downarrow{\scriptstyle 1} & & \downarrow & & \downarrow{\scriptstyle (-1)^r} \\
S^r Y & \xrightarrow{i'} & (S^r Y) \cup_{S^r} C(S^r X) & \xrightarrow{j'} & S^{r+1}X
\end{array}$$

This proposition is easy to check, provided we use the "reduced" cone and suspension. The map $(-1)^r$ of $S^{r+1}X$ arises as a permutation of the suspension coordinates.

§3. DEFINITION AND ELEMENTARY PROPERTIES OF THE INVARIANTS *d, e*

In this section we shall define our basic invariants d and e. We shall also establish the elementary properties of these invariants.

We shall suppose given a half-exact functor in the sense of [12]. For example, the functor may be one component of a (reduced) extraordinary cohomology theory. More precisely, k is to be a contravariant functor defined on (say) the category of finite CW-complexes and homotopy classes of maps, and taking values in some abelian category [14], say A. If

$$X \xrightarrow{i} Y \xrightarrow{j} Z$$

is a cofibre sequence, then

$$k(X) \xleftarrow{i^*} k(Y) \xleftarrow{j^*} k(Z)$$

is to be an exact sequence in the abelian category A. It follows that we may identify $k(X \vee Y)$ with the direct sum $k(X) \oplus k(Y)$ in the category A; see [12, p. 1].

Now suppose given a map $f: X \to Y$ between (say) finite connected CW-complexes. We can consider the induced homomorphism

$$f^*: k(Y) \to k(X).$$

If we take $X = Y = S^n$ and take k to be $H^n(\ ;Z)$, then the invariant f^* gives us the degree of f. We therefore regard

$$f^*: k(Y) \to k(X)$$

as "the degree of f, measured by k-theory". We define

$$d(f) = f^* \in \mathrm{Hom}(k(Y), k(X)).$$

Here $\mathrm{Hom}(M, N)$ means the set of maps from M to N in the abelian category A.

The invariant $e(f)$ will be defined when $d(f) = 0$ and $d(Sf) = 0$. In this case we use the map $f: X \to Y$ to start the following cofibre sequence.

$$X \xrightarrow{f} Y \xrightarrow{i} Y \cup_f CX \xrightarrow{j} SX \xrightarrow{-Sf} SY$$

Since we assume that $f^* = 0$ and $(Sf)^* = 0$, the functor k yields the following short exact sequence in the abelian category A.

$$0 \leftarrow k(Y) \xleftarrow{i^*} k(Y \cup_f CX) \xleftarrow{j^*} k(SX) \leftarrow 0$$

In an abelian category we can define Ext^1 by classifying short exact sequences; therefore the short exact sequence above yields an element of

$$\mathrm{Ext}^1(k(Y), k(SX)).$$

We call this element $e(f)$. The letter e stands for "extension", and goes well with d.

For example, let us consider the case in which $k = \tilde{H}^*(\quad; Z_2)$ and A is the category of graded modules over the mod 2 Steenrod algebra. Let us take $X = S^{m+n-1}$, $Y = S^m$. Given a map $f: S^{m+n-1} \to S^m$, we are led to consider the following short exact sequence.

$$0 \leftarrow \tilde{H}^*(S^m; Z_2) \leftarrow \tilde{H}^*(S^m \cup_f e^{m+n}; Z_2) \leftarrow \tilde{H}^*(S^{m+n}; Z_2) \leftarrow 0$$

As an extension of modules over the Steenrod algebra, this is completely determined by the Steenrod square

$$Sq^n: H^m(S^m \cup_f e^{m+n}; Z_2) \to H^{m+n}(S^m \cup_f e^{m+n}; Z_2).$$

We therefore recover Steenrod's approach to the mod 2 Hopf invariant.

The invariant $e(f)$ may thus be regarded as a "Steenrod–Hopf invariant" in which ordinary cohomology has been replaced by k-theory.

We have just defined

$$d(f) \in \mathrm{Ext}^0(k(Y), k(X))$$

(if we interpret $\mathrm{Ext}^0(M, N)$ as meaning $\mathrm{Hom}(M, N)$), and

$$e(f) \in \mathrm{Ext}^1(k(Y), k(SX)).$$

One would naturally hope to construct a third invariant, which should be defined when suitable d and e invariants vanish, and should take values in

$$\mathrm{Ext}^2(k(Y), k(S^2X)).$$

Similarly for a fourth invariant, and so on. However, we will not pursue this line of thought any further here.

In later sections we will give examples and applications of the invariants d and e, and develop the resources to do practical calculations with them. For the moment we consider the elementary properties of these invariants.

PROPOSITION 3.1 (a). *If $f \sim g$, then $d(f) = d(g)$.*

(b) *If $f \sim g$ and $e(f)$ is defined, the $e(g)$ is defined and $e(f) = e(g)$.*

Proof. Part (a) is obvious. Part (b) is proved by applying the functor k to the diagram given in Proposition 2.1.

We now consider the situation in which we have two maps

$$X \xrightarrow{f} Y \xrightarrow{g} Z.$$

We aim to show that the invariants d and e send composition products (in homotopy) into composition products, i.e. Yoneda products, in Ext groups.

PROPOSITION 3.2 (a). *We have*

$$d(gf) = d(f)d(g).$$

(b) *If $e(f)$ is defined then so is $e(gf)$, and we have*

$$e(gf) = e(f)d(g).$$

(c) *If $e(g)$ is defined then so is $e(gf)$, and we have*

$$e(gf) = d(Sf)e(g).$$

Here statements (b) and (c) use the pairing of Ext^0 and Ext^1 to Ext^1.

Proof. All the statements about invariants d are obvious. For the rest, we apply the functor k to the diagram given in Proposition 2.3, and we obtain the following commutative diagram.

$$\begin{array}{ccccc} k(Y) & \leftarrow & k(Y \cup_f CX) & \leftarrow & k(SX) \\ {\scriptstyle g^*}\uparrow & & \uparrow & & \uparrow{\scriptstyle 1} \\ k(Z) & \leftarrow & k(Z \cup_{gf} CX) & \leftarrow & k(SX) \\ {\scriptstyle 1}\uparrow & & \uparrow & & \uparrow{\scriptstyle (Sf)^*} \\ k(Z) & \leftarrow & k(Z \cup_g CY) & \leftarrow & k(SY) \end{array}$$

If $e(f)$ is defined, it is represented by the top row; similarly for $e(gf)$ and the middle row; similarly for $e(g)$ and the bottom row. By definition of the products in Ext, this shows that

$$e(gf) = e(f)\cdot g^*$$

in case (b), and

$$e(gf) = (Sf)^*\cdot e(g)$$

in case (c). This completes the proof.

For our next proposition, we assume that X is a co-H-space, for example, a suspension. That is, we are provided with a map

$$\Delta : X \to X \vee X$$

of type (1, 1). This allows us to define the sum of two (base-point-preserving) maps

$$f, g : X \to Y;$$

by definition, $f + g$ is the composite

$$X \xrightarrow{\Delta} X \vee X \xrightarrow{f \vee g} Y \vee Y \xrightarrow{\mu} Y,$$

where μ is a map of type (1, 1) in the dual sense.

PROPOSITION 3.3 (a). *We have*

$$d(f + g) = d(f) + d(g).$$

(b) *If $e(f)$ and $e(g)$ are defined then so is $e(f+g)$, and*

$$e(f+g) = e(f) + e(g).$$

In part (b), the sum occurring on the right-hand side is, of course, the Baer sum in Ext^1.

Proof. All the statements about invariants d are obvious. For the rest, we may identify $k(Y \vee Y)$ with the direct sum $k(Y) \oplus k(Y)$, and $k(S(X \vee X))$ with $k(SX) \oplus k(SX)$. In this way we can identify the sequence

$$k(Y \vee Y) \leftarrow k((Y \vee Y) \cup_{f \vee g} C(X \vee X)) \leftarrow k(S(X \vee X))$$

with the direct sum of the sequences

$$k(Y) \leftarrow k(Y \cup_f CX) \leftarrow k(SX)$$

$$k(Y) \leftarrow k(Y \cup_g CX) \leftarrow k(SX).$$

That is: if $e(f)$ and $e(g)$ are defined, so is $e(f \vee g)$, and it can be identified with the "external" sum $e(f) \oplus e(g)$. According to Proposition 3.2, we have

$$\begin{aligned} e(f+g) &= e(\mu(f \vee g)\Delta) \\ &= (S\Delta)^* e(f \vee g)\mu^* \\ &= (S\Delta)^*(e(f) \oplus e(g))\mu^*. \end{aligned}$$

But with our identifications,

$$(S\Delta)^* : k(SX) \oplus k(SX) \to k(SX)$$

is a map of type (1, 1) in the category A, and

$$\mu^* : k(Y) \to k(Y) \oplus k(Y)$$

is a map of type (1, 1) in the dual sense. Thus the element

$$(S\Delta)^*(e(f) \oplus e(g))\mu^*$$

is the Baer sum of $e(f)$ and $e(g)$. This completes the proof.

We will now discuss the behaviour of our invariants under suspension. For this purpose we shall suppose that for some integer r, $k(S^r X)$ is known as a function of $k(X)$. For example, when we take $k(X) = \tilde{K}_C(X)$ [10, 11, 2], we shall take $r = 2$; when we take $k(X) = \tilde{K}_R(X)$ we shall take $r = 8$. If we took $k(X) = \tilde{H}^*(X; Z_2)$ we could take $r = 1$. More formally, we shall suppose given a functor T, from the abelian category A to itself, which preserves exact sequences; and we shall suppose given an isomorphism

$$k(S^r X) \cong Tk(X)$$

natural for maps of X. We shall allow ourselves to identify $k(S^r X)$ and $Tk(X)$ under this isomorphism.

Since the functor T preserves exact sequences, it defines a function

$$T : \mathrm{Ext}^1(M, N) \to \mathrm{Ext}^1(TM, TN).$$

This function is actually a homomorphism.

PROPOSITION 3.4 (a). *We have*

$$d(S^r f) = Td(f).$$

(b) *If $e(f)$ is defined, then so is $e(S^r f)$, and we have*

$$e(S^r f) = (-1)^r Te(f).$$

Proof. All the statements about the invariant d are obvious. For the rest, we apply the functor k to the diagram given in Proposition 2.4 and use the fact that $kS^r = Tk$.

We now define stable track groups by

$$\mathrm{Map}_S(X, Y) = \operatorname*{Dir\,Lim}_{n \to \infty} \mathrm{Map}(S^{nr}X, S^{nr}Y).$$

We also define stabilised Hom groups in the abelian category A by iterating T and taking direct limits; thus,

$$\mathrm{Hom}_S(M, N) = \operatorname*{Dir\,Lim}_{n \to \infty} \mathrm{Hom}(T^n M, T^n N).$$

Similarly, we define stabilised Ext^1 groups by iterating the homomorphism $(-1)^r T$ and taking direct limits; thus,

$$\mathrm{Ext}^1_S(M, N) = \operatorname*{Dir\,Lim}_{n \to \infty} \mathrm{Ext}^1(T^n M, T^n N).$$

PROPOSITION 3.5 (a). *The invariant d defines a homomorphism from* $\mathrm{Map}_S(X, Y)$ *to* $\mathrm{Hom}_S(k(Y), k(X))$.

(b) *The invariant e defines a homomorphism from the subgroup* $\mathrm{Ker}\, d \cap \mathrm{Ker}(dS)$ *of* $\mathrm{Map}_S(X, Y)$ *to* $\mathrm{Ext}^1_S(k(Y), k(SX))$.

This follows immediately from Propositions 3.1, 3.3, 3.4.

The pairing of Ext groups used in Proposition 3.2 are evidently compatible with the operations T on Ext^0 and $(-1)^r T$ on Ext^1; therefore these pairings pass to the limit. With this interpretation, Proposition 3.2 continues to give the value of the invariants d, e on a composite gf of stable homotopy classes.

§4. MASSEY PRODUCTS IN HOMOLOGICAL ALGEBRA

In §3 we showed that the d and e invariants map composition products (in homotopy) into composition products (in homological algebra). In §5 we shall show that the d and e invariants map Toda brackets (in homotopy) into Massey products (in homological algebra). Of course it is necessary to begin by defining these Massey products, and that is the object of this section.

If we could work in a category containing sufficient projectives, so that we could use projective resolutions, the construction of Massey products would present no difficulty. Unfortunately, we have to work in a category which is not known to contain enough projectives. We have therefore to construct our Massey products without using projectives. In a work on homological algebra it would be desirable to show that if we accidentally have enough projectives, then the definitions which do not use projectives coincide (up to sign) with those which do use projectives. However, for present purposes we need not discuss this question; I hope that the definitions given below will commend themselves by their inherent plausibility and by the applications given in §5.

We shall suppose given four objects L, M, N and P of an abelian category, and three elements

$$\alpha \in \operatorname{Ext}^a(L, M)$$
$$\beta \in \operatorname{Ext}^b(M, N)$$
$$\gamma \in \operatorname{Ext}^c(N, P)$$

such that

$$\beta\alpha = 0 \qquad \text{in } \operatorname{Ext}^{a+b}(L, N)$$
$$\gamma\beta = 0 \qquad \text{in } \operatorname{Ext}^{b+c}(M, P).$$

Our object is to define the Massey product $\{\gamma, \beta, \alpha\}$, which should be an element of

$$\frac{\operatorname{Ext}^{a+b+c-1}(L, P)}{\gamma \operatorname{Ext}^{a+b-1}(L, N) + (\operatorname{Ext}^{b+c-1}(M, P))\alpha}.$$

Here the group $\operatorname{Ext}^{a+b-1}(L, N)$ is to be interpreted as zero if $a + b - 1 < 0$, and similarly for $\operatorname{Ext}^{b+c-1}(M, P)$. It is sufficient for us to consider the cases in which a, b, c and $a + b + c - 1$ are each either 0 or 1.

Case 1. $b = 1$. (Perhaps this should be counted as three cases.) In this case we can represent β by a short exact sequence, as follows.

$$0 \to N \xrightarrow{i} E \xrightarrow{j} M \to 0$$

This leads to the following exact sequences, in which the boundary maps coincide, up to sign, with multiplication by β.

$$\operatorname{Ext}^a(L, N) \xrightarrow{i} \operatorname{Ext}^a(L, E) \xrightarrow{j} \operatorname{Ext}^a(L, M) \xrightarrow{\delta} \operatorname{Ext}^{a+1}(L, N)$$
$$\operatorname{Ext}^c(M, P) \xrightarrow{j} \operatorname{Ext}^c(E, P) \xrightarrow{i} \operatorname{Ext}^c(N, P) \xrightarrow{\delta} \operatorname{Ext}^{c+1}(M, P)$$

Since $\beta\alpha = 0$ and $\gamma\beta = 0$, we can write α, γ in the form

$$\alpha = j\alpha', \qquad \gamma = \gamma' i$$

where

$$\alpha' \in \operatorname{Ext}^a(L, E), \qquad \gamma' \in \operatorname{Ext}^c(E, P).$$

We have only to take the element

$$\gamma'\alpha' \in \operatorname{Ext}^{a+c}(L, P).$$

It is easy to check that its indeterminacy is

$$\gamma \operatorname{Ext}^a(L, N) + (\operatorname{Ext}^c(M, P))\alpha,$$

as given above.

Case 2. $b = 0$, $a + c = 1$. Perhaps this should be counted as two cases. They are somewhat special, because they are low-dimensional. Suppose first that $a = 0$, $c = 1$. Let

$$0 \to P \xrightarrow{i} E \xrightarrow{j} N \to 0$$

be an extension representing γ. Then the fact that $\gamma\beta = 0$ allows us to factor $\beta : M \to N$ through j; using also the fact that $\beta\alpha = 0$, we obtain the following diagram.

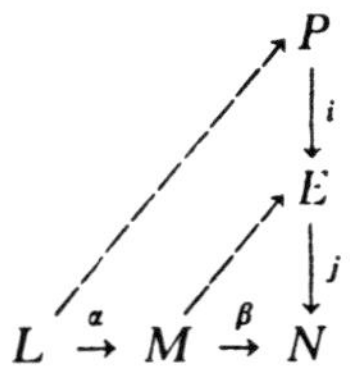

This yields an element of

$$\frac{\operatorname{Hom}(L, P)}{(\operatorname{Hom}(M, P))\alpha}.$$

The case $a = 1$, $c = 0$ is dual. Let

$$0 \to M \xrightarrow{i} E \xrightarrow{j} L \to 0$$

be an extension representing α; then we can construct the following diagram.

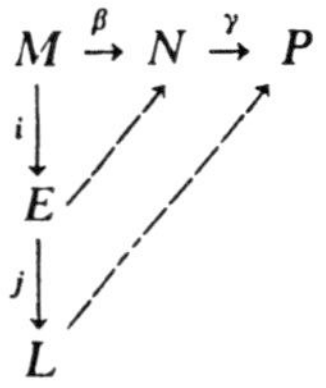

This yields an element of

$$\frac{\operatorname{Hom}(L, P)}{\gamma \operatorname{Hom}(L, N)}.$$

Case 3. $b = 0$, $a = c = 1$. Let

$$0 \to M \xrightarrow{i} E \xrightarrow{j} L \to 0$$
$$0 \to P \xrightarrow{i'} F \xrightarrow{j'} N \to 0$$

be extensions representing α, γ. The most convenient way to define an element of $\operatorname{Ext}^1(L, P)$ and check that it has the correct indeterminacy is to chase the element $\beta \in \operatorname{Ext}^0(M, N)$ back through the following diagram.

$$\begin{array}{ccccccc}
 & & \operatorname{Ext}^0(L, N) & \xrightarrow{j} & \operatorname{Ext}^0(E, N) & \xrightarrow{i} & \operatorname{Ext}^0(M, N) & \xrightarrow{\alpha} & \operatorname{Ext}^1(L, N) \\
 & & \downarrow{\scriptstyle\gamma} & & \downarrow{\scriptstyle\gamma} & & \downarrow{\scriptstyle\gamma} & & \\
\operatorname{Ext}^0(M, P) & \xrightarrow{\alpha} & \operatorname{Ext}^1(L, P) & \xrightarrow{j} & \operatorname{Ext}^1(E, P) & \xrightarrow{i} & \operatorname{Ext}^1(M, P) & &
\end{array}$$

The reader may wonder why we do not place an equal emphasis on the following dual diagram.

$$\begin{array}{ccccccc}
 & & \operatorname{Ext}^0(M, P) & \xrightarrow{i'} & \operatorname{Ext}^0(M, F) & \xrightarrow{j'} & \operatorname{Ext}^0(M, N) & \xrightarrow{\gamma} & \operatorname{Ext}^1(M, P) \\
 & & \downarrow{\scriptstyle\alpha} & & \downarrow{\scriptstyle\alpha} & & \downarrow{\scriptstyle\alpha} & & \\
\operatorname{Ext}^0(L, N) & \xrightarrow{\gamma} & \operatorname{Ext}^1(L, P) & \xrightarrow{i'} & \operatorname{Ext}^1(L, F) & \xrightarrow{j'} & \operatorname{Ext}^1(L, N) & &
\end{array}$$

The reason is that the element obtained from this diagram is the negative of that obtained from the first one. To prove this (and also for later use) it is convenient to give a direct construction of the required extension. Let us factor β in the form

$$\beta = j'\theta = \phi i;$$

we obtain the following commutative diagram.

$$\begin{array}{ccccccccc} 0 & \to & M & \xrightarrow{i} & E & \xrightarrow{j} & L & \to & 0 \\ & & & & \downarrow{\scriptstyle\theta} & & \downarrow{\scriptstyle\phi} & & \\ 0 & \to & P & \xrightarrow{i'} & F & \xrightarrow{j'} & N & \to & 0 \end{array}$$

We now form the maps

$$M \xrightarrow{(i,\theta)} E \oplus F \xrightarrow{(\phi,-j')} N$$

and define

$$G = \frac{\mathrm{Ker}(\phi, \, -j')}{\mathrm{Im}(i, \theta)}.$$

We check that we have an exact sequence

$$0 \to P \to G \to L \to 0,$$

yielding an element of $\mathrm{Ext}^1(L, P)$. By taking merely $\mathrm{Ker}(\phi, \, -j')$ or $\mathrm{Coker}(i, \theta)$, we obtain elements of $\mathrm{Ext}^1(E, P)$ and $\mathrm{Ext}^1(L, F)$. It is now easy to check that these are precisely the elements we want in chasing round the upper diagram, and their negatives are the elements we want in chasing round the lower diagram.

Finally, let us suppose given a functor T from our abelian category to itself, as in §3 above. Then it is clear that all the constructions above are compatible with T.

§5. TODA BRACKETS, I

In this section we shall show that the d and e invariants send Toda brackets (in homotopy) into Massey products (in homological algebra). For this purpose we shall generally suppose given four CW-complexes W, X, Y and Z, and three maps

$$W \xrightarrow{f} X \xrightarrow{g} Y \xrightarrow{h} Z$$

such that $hg \sim 0$, $gf \sim 0$.

First suppose given a specific homotopy

$$l : I \times W \to Y$$

such that $l(0, w)$ is constant and

$$l(1, w) = gfw.$$

Then we can define a map

$$G : X \cup_f CW \to Y$$

by

$$G(x) = g(x) \qquad (x \in X)$$
$$G(t, w) = l(t, w) \qquad (t \in I, \, w \in W).$$

Again, we can define a map

$$F : SW \to Y \cup_g CX$$

by

$$F(t, w) = \begin{cases} (2t, fw) & (0 \leqq t \leqq \frac{1}{2}) \\ l(2 - 2t, w) & (\frac{1}{2} \leqq t \leqq 1). \end{cases}$$

These maps figure in the following diagram.

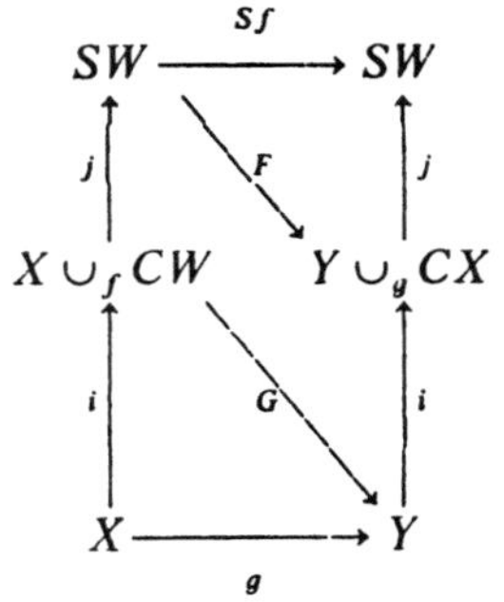

Here the two triangles are homotopy-commutative, and the parallelogram becomes homotopy-commutative if one inserts the map

$$-1: SW \to SW.$$

If we suppose given also a specific homotopy $hg \sim 0$, we can construct similarly the right-hand half of the following diagram.

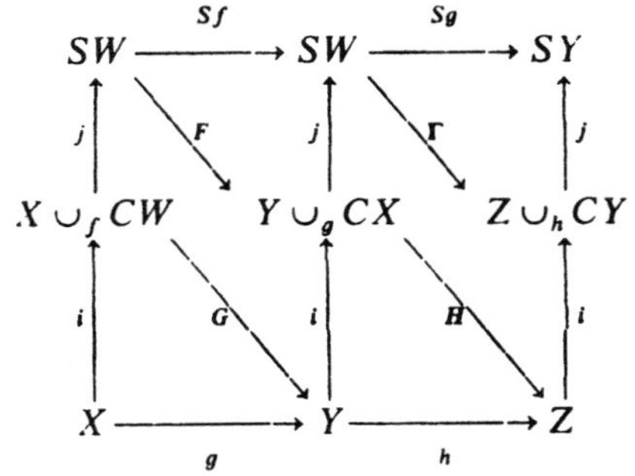

(5.1)

[The Toda bracket $\{h, g, f\}$ is the composite

$$HF: SW \to Z.$$

LEMMA 5.2 (a). *Suppose that $e(Sf)$ and $e(g)$ are defined. Then a homotopy $gf \sim 0$ is such that $e(F)$ is defined, if and only if it is such that $e(G)$ is defined.*

(b) *Suppose that $e(Sg)$ and $e(h)$ are defined. Then a homotopy $hg \sim 0$ is such that $e(\Gamma)$ is defined if and only if it is such that $e(H)$ is defined.*

Proof. We have the following diagram, in which the columns are exact and $j^* F^* = -G^* i^*$.

$$\begin{array}{c}
k(SX) \\
{\scriptstyle (-Sf)^*}\downarrow \\
k(SW) \xleftarrow{F^*} k(Y \cup_g CX) \\
{\scriptstyle j^*}\downarrow \quad \quad {\scriptstyle G^*} \quad \quad \downarrow{\scriptstyle i^*} \\
k(X \cup_f CW) \xleftarrow{} k(Y) \\
\quad \quad \quad \quad \quad \downarrow{\scriptstyle g^*} \\
\quad \quad \quad \quad \quad k(X)
\end{array}$$

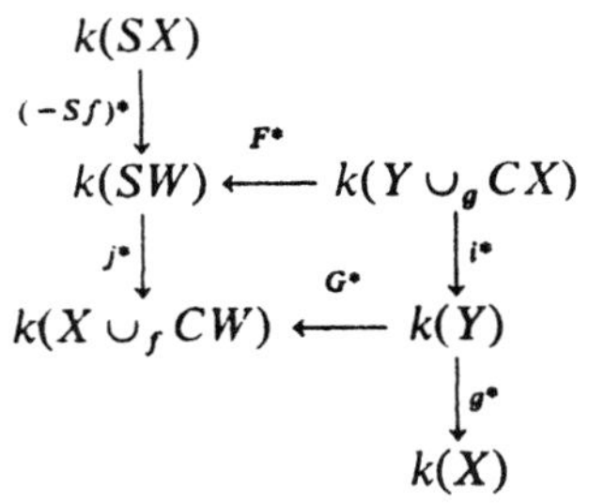

According to the data, j^* is mono and i^* is epi. Therefore $F^* = 0$ if and only if $G^* = 0$. Similarly for $(SF)^*$ and $(SG)^*$. This proves part (a); substituting $X \xrightarrow{g} Y \xrightarrow{h} Z$ for $W \xrightarrow{f} X \xrightarrow{g} Y$, we obtain part (b).

We can now state the main result of this section.

THEOREM 5.3 (i). *Suppose that $e(f)$ is defined. Then*

$$d\{h, g, f\} \subset -\{e(f), d(g), d(h)\}.$$

(ii) *Suppose that $e(g)$ is defined. Then*

$$d\{h, g, f\} \subset \{d(Sf), e(g), d(h)\}.$$

(iii) *Suppose that $e(h)$ is defined. Then*

$$d\{h, g, f\} \subset -\{d(Sf), d(Sg), e(h)\}.$$

(iv) *Suppose that $e(Sf)$ and $e(g)$ are defined, and that we only consider homotopies $gf \sim 0$ such that $e(F)$ is defined (or equivalently, by Lemma 5.2 (a), such that $e(G)$ is defined). Then $e\{h, g, f\}$ is defined and*

$$e\{h, g, f\} \subset \{e(Sf), e(g), d(h)\}.$$

(v) *Suppose that $e(Sf)$ and $e(h)$ are defined. Then $e\{h, g, f\}$ is defined and*

$$e\{h, q, f\} \subset -\{e(Sf), d(Sg), e(h)\}.$$

(vi) *Suppose that $e(Sg)$ and $e(h)$ are defined, and that we only use homotopies $hg \sim 0$ such that $e(H)$ is defined (or equivalently, by Lemma 5.2 (b), such that $e(\Gamma)$ is defined). Then $e\{h, g, f\}$ is defined and*

$$e\{h, g, f\} \subset -\{d(S^2 f), e(Sg), e(h)\}.$$

Proof. We tackle first the three cases in which the Massey product is defined by case (1) of §4, viz. the cases (ii), (iv) and (vi). For this purpose the objects L, M, E, N and P of §4 case (1) take the following values.

	L	M	E	N	P
Case (ii)	$k(Z)$	$k(Y)$	$k(Y \cup_g CX)$	$k(SX)$	$k(SW)$
Case (iv)	$k(Z)$	$k(Y)$	$k(Y \cup_g CX)$	$k(SX)$	$k(S^2W)$
Case (vi)	$k(Z)$	$k(SY)$	$k(S(Y \cup_g CX))$	$k(S^2X)$	$k(S^2W)$

It is to be noted that in case (vi), the invariant $e(Sg)$ is defined to be the short exact sequence

$$k(SY) \xleftarrow{i'^*} k(SY \cup_{Sg} CSX) \xleftarrow{j'^*} k(S^2X);$$

but by Proposition 2.4, this is the same as

$$k(SY) \xleftarrow{(Si)^*} k(S(Y \cup_g CX)) \xleftarrow{-(Sj)^*} k(S^2X).$$

We have now to construct

$$\alpha' \in \mathrm{Ext}^a(L, E), \qquad \gamma' \in \mathrm{Ext}^c(E, P)$$

as in §4 case (1). For this purpose we give the following values.

	α'	γ'
Case (ii)	$d(H)$	$d(F)$
Case (iv)	$d(H)$	$e(F)$
Case (vi)	$e(H)$	$-d(SF)$

The fact that these values have the required properties is proved by applying Propositions 3.1, 3.2 to the formulae

$$Hi \sim h, \qquad jF \sim Sf, \qquad (Sj)(SF) \sim S^2 f$$

where i, j are the maps appearing in

$$Y \xrightarrow{i} Y \cup_g CX \xrightarrow{j} SX.$$

Using Proposition 3.2 again for the composite HF, we find the following results. In case (ii), $d(HF)$ represents the Massey product. In case (iv) $e(HF)$ is defined, and represents the Massey product. In case (vi) $e(HF)$ is defined, and $-e(HF)$ represents the Massey product. This completes cases (ii), (iv) and (vi).

We tackle next the two cases in which the Massey product is defined by case (2) of §4, viz. the cases (i) and (iii). For this purpose the objects L, M, E, N and P of §4 case (2) take the following values.

	L	M	E	N	P
Case (i)	$k(Z)$	$k(Y)$	$k(X \cup_f CW)$	$k(X)$	$k(SW)$
Case (iii)	$k(Z)$	$k(SY)$	$k(Z \cup_h CY)$	$k(SX)$	$k(SW)$

We have now to construct diagrams as in §4 case (2). The appropriate diagrams are obtained from Diagram 5.1, and are as follows.

Case (i)

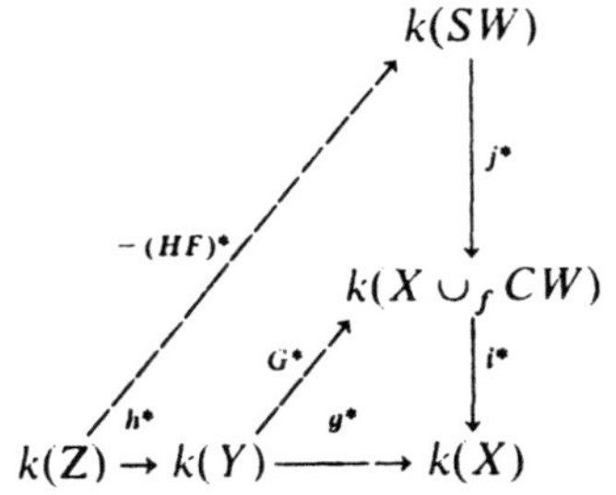

Case (ii)

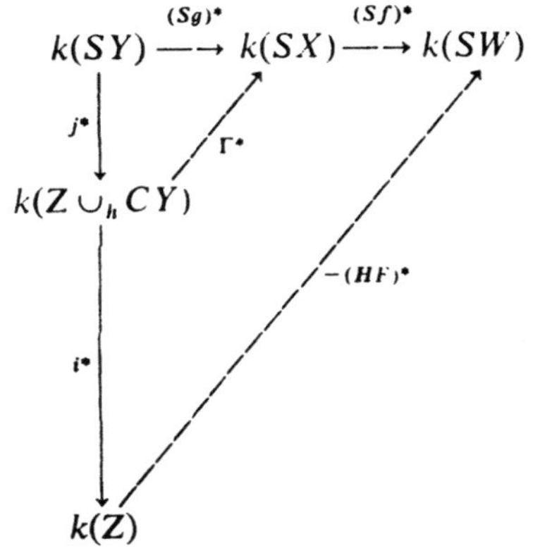

In both cases we see that $-d(HF)$ represents the Massey product. This completes cases (i) and (iii).

Finally, we tackle case (v). We first check that $e(HF)$ is defined. The fact that $(HF)^* = 0$, under the hypotheses given, follows by chasing round the following commutative diagram, in which the columns are exact.

$$\begin{array}{ccc} k(SX) & \xleftarrow{\Gamma^*} & k(Z \cup_h CY) \\ \big\downarrow{\scriptstyle (Sf)^*} & & \big\downarrow{\scriptstyle i^*} \\ k(SW) & \xleftarrow{-(HF)^*} & k(Z) \\ \big\downarrow{\scriptstyle j^*} & & \big\downarrow{\scriptstyle h^*} \\ k(X \cup_f CW) & \xleftarrow{G^*} & k(Y) \end{array}$$

Similarly for the fact that $(S(HF))^* = 0$.

We now recall that in case (v) the Massey product is defined by case (3) of §4. For this purpose the objects considered in §4 take the following values.

$$L = k(Z), \qquad M = k(SY), \qquad N = k(SX), \qquad P = k(S^2W),$$
$$E = k(Z \cup_h CY),$$
$$\alpha = e(h), \qquad \beta = (Sg)^*, \qquad \gamma = e(Sf).$$

We start with the element $-e(HF)$ in $\operatorname{Ext}^1(L, P)$. Its image in $\operatorname{Ext}^1(E, P)$ is $-e(HF)i^*$. By Proposition 3.2 this is

$$\begin{aligned} -e(iHF) &= e(\Gamma . Sf) \\ &= e(Sf).\Gamma^* \\ &= \gamma.\Gamma^*. \end{aligned}$$

But the element Γ^* in $\operatorname{Ext}^0(E, N)$ projects to $(Sg)^* = \beta$ in $\operatorname{Ext}^0(M, N)$. Therefore $-e(HF)$ qualifies as a representative for the Massey product. This proves case (v), and completes the proof of Theorem 5.3.

Perhaps it should be pointed out that Theorem 5.3 is consistent with the behaviour of d, e under suspension S^r (as in §3), because of the behaviour of Toda brackets under suspension:

$$S^r\{h, g, f\} \subset (-1)^r\{S^r h, S^r g, S^r f\}.$$

In our applications r will always be even, so the signs $(-1)^r$ can be forgotten.

§6. AN ABELIAN CATEGORY

The construction of §3 requires a half-exact functor k taking values in an abelian category A. In the applications we shall take $k(K)$ to be the Grothendieck–Atiyah–Hirzebruch group $\tilde{K}_\Lambda(X)$ [10, 11] equipped with its operations Ψ^k [2]. We shall therefore need to consider $\tilde{K}_\Lambda(X)$ as an object in a suitable abelian category A. Actually the category A will depend on Λ, where $\Lambda = R$ or C; but we shall not display the symbol Λ in the notation. It is the object of this section to define the category A.

By definition, an object of the category A is to be a finitely-generated abelian group M provided with endomorphisms

$$\Psi^k : M \to M$$

(one for each integer k) and satisfying the following axioms.

(6.1) $$\Psi^k . \Psi^l = \Psi^{kl}$$

(6.2) $$\Psi^0 = 0, \quad \Psi^1 = 1 \quad \text{and} \quad (\text{if } \Lambda = R)\ \Psi^{-1} = 1.$$

(6.3) For each $x \in M$ and $q \in Z$, the mod q value of $\Psi^k x$ is periodic in k with period q^e for some $e = e(x, q)$.

In this axiom, and below, the statement "$f(k)$ is periodic in k with period q^e" means simply "$k_1 \equiv k_2 \bmod q^e$ implies $f(k_1) = f(k_2)$". It is not asserted that q^e is the smallest possible period. In particular the condition is true for $q = 0$ in a trivial way.

By definition, a map in the category A is to be a homomorphism $\theta : M \to N$ of abelian groups which commutes with the operations Ψ^k.

EXAMPLE 6.4. *The functor $\tilde{K}_\Lambda$ associates to each finite connected CW-complex X an abelian group $\tilde{K}_\Lambda(X)$ provided with endomorphisms Ψ^k, and associates with each map $f : X \to Y$ an induced homomorphism*

$$f^* : \tilde{K}_\Lambda(Y) \to \tilde{K}_\Lambda(X).$$

The functor $\tilde{K}_\Lambda$ takes values in the category $A = A(\Lambda)$. In fact, axioms (6.1) and (6.2) are satisfied, according to [2 Theorem 5.1 (v), (vii)]; and axiom (6.3) is satisfied, according to [5 Theorem 5.1].

PROPOSITION 6.5. *The category A defined above is an abelian category, in the sense of* [14, Chapter IX].

The only point which requires detailed proof is the following.

LEMMA 6.6. *If M is an object in A, and N is a subgroup of M closed under the operations Ψ^k, then N satisfies axiom* (6.3).

Proof. This follows the lines of [5 Lemma 6.5]. Consider the subgroup S_t of elements x in M such that $q^t x \in N$. This an increasing sequence of Z-submodules in the finitely-generated Z-module M, therefore convergent. That is, there exists t such that $x \in M$, $q^{t+1}x \in N$ imply $q^t x \in N$. Now we use axiom (6.3) for M; given $y \in N$, there is an f such that the value of $\Psi^k y$ in $M/q^{t+1}M$ is periodic in k with period $q^{(t+1)f}$. That is, if $k \equiv l \bmod q^{(t+1)f}$, we have

$$\Psi^k y - \Psi^l y = q^{t+1}x$$

for some x in M. By our choice of t, this shows that

$$\Psi^k y - \Psi^l y \in qN.$$

We have only to take $e = (t + 1)f$. This completes the proof.

In §3, we assumed that $k(S^r X)$ could be calculated in terms of $k(X)$ by a functor T from A to A. It is clear what functor T we should take in the category A described above.

If M is an object in A, then the abelian group underlying TM is the same as that underlying M, but the operation Ψ^k in TM is $k^{\frac{1}{2}r}$ times that in M (where $r = 2$ if $\Lambda = C$ and $r = 8$ if $\Lambda = R$). It is clear that these new operations satisfy axioms (6.1), (6.2) and (6.3). Similarly, if $f: M \to N$ is a map in A, then Tf is to be the same homomorphism as f; this clearly commutes with the new operations.

It is now clear that we have an isomorphism

$$\tilde{K}_\Lambda(S^r X) \cong T\tilde{K}_\Lambda(X)$$

natural for maps of X; see [2 Corollary 5.3].

The theory given in §§3–5 can now be applied to the functors $k = \tilde{K}_R$ and $k = \tilde{K}_C$.

§7. AN INVARIANT DEFINED USING THE CHERN CHARACTER

We are now in a position to apply the theory given in §§3–6. To give applications, we shall begin by taking the spaces X and Y to be spheres of suitable dimension, so that we obtain information about stable homotopy groups of spheres. We shall write d_Λ, e_Λ for the invariants obtained by taking $k = \tilde{K}_\Lambda$, where $\Lambda = R$ or C.

We start with a preliminary discussion of the invariants d_Λ (7.1, 7.2). Next we show that the invariant e_C can be described in a more elementary way using the Chern character. As remarked in the introduction, there is considerable overlap at this point with work of Dyer [13]. We will discuss the relationship between e_C and the invariants d_R, e_R (7.14, 7.18). We will also give substantial information about the values taken by these invariants (e.g. 7.15, 7.16). There remain certain cases in which the invariant e_R is independent of e_C; we postpone these cases to §9.

We begin by considering the invariant d_Λ. Let θ be an element of π_r^S; choose a representative map

$$f: S^{q+r} \to S^q$$

for θ. *A priori*,

$$d_\Lambda(f) = f^*: \tilde{K}_\Lambda(S^q) \to \tilde{K}_\Lambda(S^{q+r})$$

depends on the residue class of q (mod 2 if $\Lambda = C$, mod 8 if $\Lambda = R$). I claim that it is sufficient to consider the case $q \equiv 0$ (mod 2 if $\Lambda = C$, mod 8 if $\Lambda = R$). In fact, suppose we know $d_\Lambda(f)$ in this case. Let P be a point; then $\tilde{K}_\Lambda^*(S^q)$ is a free module over $K_\Lambda^*(P)$, on one generator which lies in $\tilde{K}_\Lambda^0(S^q) = \tilde{K}_\Lambda(S^q)$. Therefore

$$d_\Lambda(f) = f^*: \tilde{K}_\Lambda^0(S^q) \to \tilde{K}_\Lambda^0(S^{q+r})$$

determines

$$f^*: \tilde{K}_\Lambda^{-t}(S^q) \to \tilde{K}_\Lambda^{-t}(S^{q+r}),$$

which is the same as

$$d_\Lambda(S^t f) = (S^t f)^*: \tilde{K}_\Lambda^0(S^{q+t}) \to \tilde{K}_\Lambda^0(S^{q+t+r}).$$

PROPOSITION 7.1. *d_Λ is zero on π_r^S for $r > 0$ unless $\Lambda = R$ and $r \equiv 1$ or 2 mod 8.*

First proof. By the above argument, d_Λ defines a homomorphism from π_r^S to G, where

$$G = \begin{cases} Z & \text{if } \Lambda = C \text{ and } r \equiv 0 \bmod 2 \\ & \text{or if } \Lambda = R \text{ and } r \equiv 0, 4 \bmod 8 \\ 0 & \text{if } \Lambda = C \text{ and } r \equiv 1 \bmod 2 \\ & \text{or if } \Lambda = R \text{ and } r \equiv 3, 5, 6, 7 \bmod 8. \end{cases}$$

Since π_r^S is a finite group, d_Λ must be zero.

Second proof. It is sufficient to consider the case of a map $f: S^{q+r} \to S^q$, where q, r are divisible by 2 if $\Lambda = C$, by 4 if $\Lambda = R$. Then the groups $\tilde{K}_\Lambda(S^q)$, $\tilde{K}_\Lambda(S^{q+r})$ are Z, and their operations Ψ^k are given by

$$\Psi^k x = k^{\frac{1}{2}q} x, \qquad \Psi^k x = k^{\frac{1}{2}(q+r)} x$$

respectively. If $r > 0$, the only homomorphism commuting with the operations is zero.

We now consider the case $\Lambda = R$, $r \equiv 1$ or $2 \bmod 8$. We take as our basic invariant the homomorphism

$$d_R : \pi_r^S \to Z_2$$

obtained by considering maps

$$f: S^{q+r} \to S^q$$

with $q \equiv 0 \bmod 8$. (It is understood, of course, that if we later wish to apply the theorems of §§3, 5 we shall still have to use the invariant d_R appropriate to spheres of the dimensions which actually arise).

THEOREM 7.2. *Assume $r \equiv 1$ or 2 mod 8 and $r > 0$. Then the invariant*

$$d_R : \pi_r^S \to Z_2$$

is an epimorphism; we have

$$\pi_r^S = Z_2 + \operatorname{Ker} d_R,$$

where the subgroup Z_2 is generated by μ_r.

This theorem includes Theorems 1.2 and 1.4. Its proof is deferred to §12.

We will now give an elementary construction, using the Chern character, for an invariant which we will later prove equivalent to e_C. This invariant has already been described in [6, 13]. See also [19].

Suppose given a map $f: S^{2n-1} \to S^{2q}$, where $n > q > 0$. If $\Lambda = R$ we assume that n and q are even. We use f to start the following cofibre sequence.

$$S^{2n-1} \xrightarrow{f} S^{2q} \xrightarrow{i} S^{2q} \cup_f e^{2n} \xrightarrow{j} S^{2n} \xrightarrow{-Sf} S^{2q+1}$$

Applying $\tilde{K}_\Lambda$, we obtain the following exact sequence.

$$0 \leftarrow \underset{\cong Z}{\tilde{K}_\Lambda(S^{2q})} \xleftarrow{i^*} \tilde{K}_\Lambda(S^{2q} \cup_f e^{2n}) \xleftarrow{j^*} \underset{\cong Z}{\tilde{K}_\Lambda(S^{2n})} \leftarrow 0$$

The group $\tilde{K}_\Lambda(S^{2q} \cup_f e^{2n})$ is therefore $Z + Z$; we can choose generators ξ, η so that ξ projects to the generator of $\tilde{K}_\Lambda(S^{2q})$, and η is the image of the generator in $\tilde{K}_\Lambda(S^{2n})$.

As in [4], we write ch_C for the Chern character

$$ch : K_C(X) \to H^*(X; Q),$$

and ch_R for the composite

$$K_R(X) \xrightarrow{c} K_C(X) \xrightarrow{ch} H^*(X; Q).$$

Let

$$h^{2q} \in H^{2q}(S^{2q} \cup_f e^{2n}; Z)$$
$$h^{2n} \in H^{2n}(S^{2q} \cup_f e^{2n}; Z)$$

be cohomology generators, corresponding under i^*, j^* to the generators in $H^{2q}(S^{2q}; Z)$, $H^{2n}(S^{2n}; Z)$. Then in $H^*(S^{2q} \cup_f e^{2n}; Q)$ we must have formulae of the following form.

$$(7.3) \qquad \begin{aligned} ch_\Lambda \xi &= a_{2q} h^{2q} + \lambda a_{2n} h^{2n} \\ ch_\Lambda \eta &= \qquad\qquad\quad a_{2n} h^{2n} \end{aligned}$$

Here we have

$$a_r = \begin{cases} 1 \text{ if } \Lambda = C \text{ and } r \equiv 0 \bmod 2 \\ 1 \text{ if } \Lambda = R \text{ and } r \equiv 0 \bmod 8 \\ 2 \text{ if } \Lambda = R \text{ and } r \equiv 4 \bmod 8. \end{cases}$$

(The coefficient a_{2n} is introduced into the term $\lambda a_{2n} h^{2n}$ for technical convenience). The coefficient $\lambda = \lambda(f)$ is some rational number. Of course, λ depends on the choice of ξ; we can replace ξ by $\xi + N\eta$, where N is any integer; this replaces λ by $\lambda + N$. To obtain an invariant of f we have therefore to consider the coset $\{\lambda(f)\}$ of $\lambda(f)$ in Q/Z, the rationals mod 1.

EXAMPLE 7.4. *Take $\Lambda = C$ and take f to be the Hopf map from S^3 to S^2. Then $S^2 \cup_f e^4$ is CP^2, the complex projective plane. We may take ξ to be the canonical line bundle minus the trivial line bundle. Then*

$$ch\xi = e^x - 1 = x + \tfrac{1}{2}x^2,$$

where x is the cohomology generator. Thus we have $\lambda = \frac{1}{2}$ and $\{\lambda(f)\} = \frac{1}{2} \bmod 1$.

It is easy to establish the properties of the invariant $\{\lambda(f)\}$ directly, by following the pattern of §3; but in fact this is not necessary, as we will establish that the invariant $\{\lambda(f)\}$ is equivalent to the invariant $e(f)$ introduced in §3 (see Proposition 7.8). We will first show that the invariant $\{\lambda(f)\}$ determines $e(f)$, by using the Chern character to compute the operations Ψ^k in $S^{2q} \cup_f e^{2n}$.

PROPOSITION 7.5. *With the notation introduced above, the operations Ψ^k in $\tilde{K}_\Lambda(S^{2q} \cup_f e^{2q})$ are given by the following formulae.*

$$(7.6) \qquad \begin{aligned} \Psi^k \xi &= k^q \xi + \lambda(k^n - k^q)\eta \\ \Psi^k \eta &= \qquad\qquad\quad k^n \eta. \end{aligned}$$

Proof. Since

$$ch_\Lambda : \tilde{K}_\Lambda(S^{2q} \cup_f e^{2n}) \to H^*(S^{2q} \cup_f e^{2n}; Q)$$

is monomorphic, the formulae can be checked by applying ch_Λ to both sides, using (7.3). To evaluate $ch_\Lambda \Psi^k \xi$ one uses [2 Theorem 5.1 (vi)].

COROLLARY 7.7. *The rational number λ has the form z/h, where $z \in Z$ and h is the highest common factor of the expressions $k^n - k^q$ as k runs over Z.*

This follows immediately, since the coefficients $\lambda(k^n - h^q)$ must be integers.

In order to discuss the invariant $e(f)$ we must now compute the appropriate Ext group. We write M, N for the objects $\tilde{K}_\Lambda(S^{2q})$, $\tilde{K}_\Lambda(S^{2n})$ of the abelian category A; thus the abelian group underlying M is Z and its operations are given by

$$\Psi^k x = k^q x;$$

similarly for N, in which

$$\Psi^k x = k^n x.$$

The following proposition computes $\mathrm{Ext}^1(M, N)$.

PROPOSITION 7.8. *There is a monomorphism*

$$\theta : \mathrm{Ext}^1(M, N) \to Q/Z$$

such that for any map

$$f : S^{2n-1} \to S^{2q}$$

we have

$$\theta(e(f)) = \{\lambda(f)\}.$$

The image of θ is the subgroup of cosets $\{z/h\}$, where z, h are as in Corollary 7.7.

The following proposition computes $\mathrm{Ext}^1_S(M, N)$.

PROPOSITION 7.9. *There is a monomorphism*

$$\theta_S : \mathrm{Ext}^1_S(M, N) \to Q/Z$$

such that for any map

$$f : S^{2n-1} \to S^{2q}$$

we have

$$\theta_S(e(f)) = \{\lambda(f)\}.$$

The image of θ_S is the subgroup of cosets $\{z/m(t)\}$, where $z \in Z$, $t = n - q$, and the numerical function $m(t)$ is as in [4 §2].

The explicit definitions of θ, θ_S will be given during the course of the proof. We begin by explaining the use of factor sets in studying our extensions.

Suppose given an extension

$$0 \leftarrow M \leftarrow E \leftarrow N \leftarrow 0$$

in the category A, where N, M are as above. Then we can choose generators ξ, η in E so that ξ projects to the generator in M and η is the image of the generator in N. The operations in E must be given by formulae of the following form.

$$\begin{aligned} \Psi^k \xi &= k^q \xi + c(k)\eta \\ \Psi^k \eta &= k^n \eta \end{aligned} \tag{7.10}$$

The integers $c(k)$ constitute a "factor set" describing the operations Ψ^k in the extension E.

LEMMA 7.11. *This factor set has the form*

$$c(k) = \lambda(k^n - k^q) \tag{7.12}$$

for some $\lambda \in Q$.

This lemma shows that the "abstract" algebraic extensions are described by the same formulae that we have already found in the "concrete" topological situation.

Proof of Lemma 7.11. By Axiom (6.1) we have in E the relation $\Psi^k\Psi^l = \Psi^{kl}$. This yields

$$c(kl) = c(k)l^q + c(l)k^n.$$

Interchanging k and l, we find

$$c(kl) = c(l)k^q + c(k)l^n.$$

Choosing l so that $l^n - l^q \neq 0$, we find

$$c(k) = \frac{c(l)(k^n - k^q)}{l^n - l^q}.$$

That is,

$$c(k) = \lambda(k^n - k^q)$$

for some rational λ. This proves the lemma.

If we replace ξ by $\xi + N\eta$, we replace the factor set $c(k)$ by $c(k) + N(k^n - k^q)$. This replaces λ by $\lambda + N$.

It is now clear how to define

$$\theta : \mathrm{Ext}^1(M, N) \to Q/Z;$$

by definition, the function θ will assign to any extension E the coset $\{\lambda\}$ in Q/Z given by formulae (7.10) and (7.12). The equation

$$\theta(e(f)) = \{\lambda(f)\}$$

follows immediately by comparing formulae (7.6), (7.10) and (7.12).

We have to remark that θ is a homomorphism; in fact, it is not hard to check that the Baer sum in $\mathrm{Ext}^1(M, N)$ corresponds to addition of factor sets, i.e. to addition in Q/Z. It is also clear that θ is a monomorphism.

It remains to discuss the image of θ. It is clear that in Lemma 7.11 the rational number λ has the form z/h, as in Corollary 7.7. We require the converse result.

LEMMA 7.13. *Each rational λ of the form z/h arises by formulae* (7.10), (7.12) *from some extension E and some choice of ξ.*

Proof. We use the formulae (7.10) and (7.12) to define operations Ψ^k on the free abelian group generated by ξ and η. We easily check that these operations satisfy axioms (6.1) to (6.3). This gives the extension E required.

This completes the proof of Proposition 7.8. It remains to check that our proceedings are compatible with suspension. We easily check from our formulae that if M and N are as above, then the following diagram is commutative.

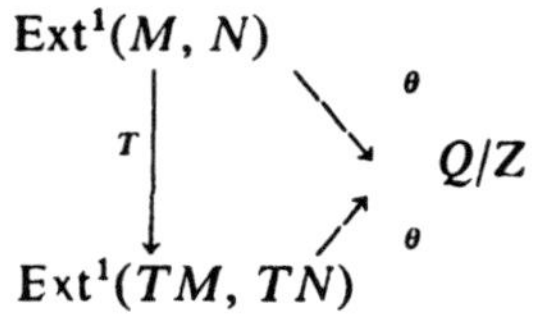

Therefore θ passes to the limit and defines a monomorphism

$$\theta_S : \mathrm{Ext}^1_S(M, N) \to Q/Z$$

such that $\theta_S(e(f)) = \{\lambda(f)\}$, as required. It remains to discuss the image of θ_S. Let n and q

tend to infinity so that their difference $n - q = t$ remains constant. Then according to [4 §2], the integer h increases, and ultimately attains a constant value, namely $m(t)$. This completes the proof of Proposition 7.9.

We shall now regard our invariant $e(f)$ as taking values in the rationals mod 1, in the case under discussion. We repeat that this is the case $X = S^{2n-1}$, $Y = S^{2q}$, where n and q are even if $\Lambda = R$.

At this point we possess a choice of invariants defined on the r-stem π_r^S for $r \equiv 3 \bmod 4$. In fact, by considering $e_R(f)$ for maps $f: S^{2q+r} \to S^{2q}$ with $2q \equiv 0 \bmod 8$ we obtain one invariant, say e'_R; by considering $e_R(f)$ for maps $f: S^{2q+r} \to S^{2q}$ with $2q \equiv 4 \bmod 8$ we obtain another invariant, say e''_R. We also have the invariant $e_C(f)$ for maps $f: S^{2q+r} \to S^{2q}$. We must discuss the relations between these invariants.

PROPOSITION 7.14. *If $r \equiv 7 \bmod 8$ then*

$$e_C = e'_R = e''_R : \pi_r^S \to Q/Z.$$

If $r \equiv 3 \bmod 8$ then

$$e_C = 2e'_R : \pi_r^S \to Q/Z$$

and

$$e''_R = 2e_C = 4e'_R : \pi_r^S \to Q/Z.$$

Proof. Consider the following diagram.

$$\begin{array}{ccccccccc} 0 & \leftarrow & \tilde{K}_R(S^{2q}) & \leftarrow & \tilde{K}_R(S^{2q} \cup_f e^{2n}) & \leftarrow & \tilde{K}_R(S^{2n}) & \leftarrow & 0 \\ & & \downarrow{\scriptstyle c'} & & \downarrow{\scriptstyle c} & & \downarrow{\scriptstyle c''} & & \\ 0 & \leftarrow & \tilde{K}_C(S^{2q}) & \leftarrow & \tilde{K}_C(S^{2q} \cup_f e^{2n}) & \leftarrow & \tilde{K}_C(S^{2n}) & \leftarrow & 0 \end{array}$$

Let us identify $\tilde{K}_\Lambda(S^{2q})$ with Z; then the map c' is multiplication by 1 if $2q \equiv 0 \bmod 8$, by 2 if $2q \equiv 4 \bmod 8$. Similarly for c''. So if $2q \equiv 0 \bmod 8$ we have

$$e_C(f) = c''.e_R(f);$$

if $2n \equiv 0 \bmod 8$ we have

$$e_R(f) = e_C(f).c'.$$

Similarly, consider the following diagram.

$$\begin{array}{ccccccccc} 0 & \leftarrow & \tilde{K}_C(S^{2q}) & \leftarrow & \tilde{K}_C(S^{2q} \cup_f e^{2n}) & \leftarrow & \tilde{K}_C(S^{2n}) & \leftarrow & 0 \\ & & \downarrow{\scriptstyle r'} & & \downarrow{\scriptstyle r} & & \downarrow{\scriptstyle r''} & & \\ 0 & \leftarrow & \tilde{K}_R(S^{2q}) & \leftarrow & \tilde{K}_R(S^{2q} \cup_f e^{2n}) & \leftarrow & \tilde{K}_R(S^{2n}) & \leftarrow & 0 \end{array}$$

This is a diagram in the category A, since r commutes with Ψ^k [9]. The map r' is multiplication by 2 if $2q \equiv 0 \bmod 8$, by 1 if $2q \equiv 4 \bmod 8$. Similarly for r''. So if $2q \equiv 4 \bmod 8$ we have

$$e_R(f) = r''e_c(f);$$

if $2n \equiv 4 \bmod 8$ we have

$$e_C(f) = e_R(f).r'.$$

This yields the results stated; actually it gives two proofs for each.

We will now describe the values taken by the invariants considered in Proposition 7.14.

THEOREM 7.15. *If* $r = 4s - 1$, *then the image of*

$$e'_R : \pi_r^S \to Q/Z$$

is precisely the subgroup of cosets $\{z/m(2s)\}$ $(z \in Z)$; *that is it is, a cyclic group of order* $m(2s)$.

It follows from Proposition 7.9 that the image of e'_R is contained in the subgroup indicated. In order to prove that the image of e'_R is the whole of this subgroup, we compute e'_R on the image of the J-homomorphism.

THEOREM 7.16. *If* $r = 4s - 1$, *then the value of the composite*

$$e'_R J : \pi_r(SO) \to Q/Z$$

on a suitable generator of $\pi_r(SO)$ *is* $\frac{1}{2}\alpha_{2s}$ *mod* 1.

In these theorems, the numerical function $m(t)$ and the rational number $\frac{1}{2}\alpha_{2s}$ are as in [4 §2]. That is,

$$\tfrac{1}{2}\alpha_{2s} = (-1)^{s-1}\frac{B_s}{4s},$$

where B_s is the sth Bernoulli number. Theorem 7.16 thus reproves the result of Milnor and Kervaire, as improved by Atiyah and Hirzebruch [10]. The value of $\frac{1}{2}\alpha_{2s}$ mod 1 is explicitly given by [4 Theorem 2.5], which was proved for this purpose. We recall from [4 Theorem 2.6] that the denominator of $\frac{1}{2}\alpha_{2s}$, when this fraction is expressed in its lowest terms, is precisely $m(2s)$. Theorem 7.15 will therefore follow from Theorem 7.16. The proof of Theorem 7.16 is deferred to §10.

I believe that Theorem 7.16 was known to earlier workers, for example, Atiyah (ca. 1960/61); see also Dyer [13, Theorem 1 and formulae on p.370].

Theorems 1.5, 1.6 will follow immediately from Theorems 7.15, 7.16 and [4 Theorem 3.7]. Suppose for example that $r = 4s - 1 \equiv 3 \bmod 8$. Then by Theorems 7.15, 7.16 we have the following diagram.

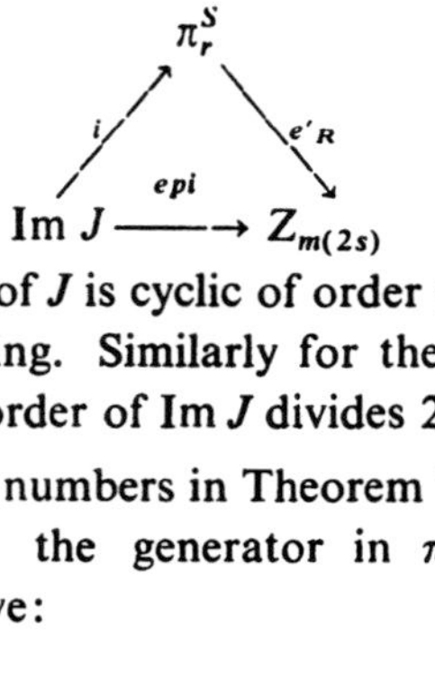

But by [4 Theorem 3.7] the image of J is cyclic of order dividing $m(2s)$. Therefore the diagram provides a direct sum splitting. Similarly for the case $r \equiv 7 \bmod 8$, except that [4 Theorem 3.7] only states that the order of Im J divides $2m(2s)$.

We will substitute a few small numbers in Theorem 7.16 in order to provide examples. For $r \equiv -1 \bmod 4$, let us take the generator in $\pi_r(SO)$, and let its image under $J : \pi_r(SO) \to \pi_r^S$ be j_r. Then we have:

EXAMPLE 7.17.

$$e'_R j_3 = \tfrac{1}{24}$$
$$e'_R j_7 = -\tfrac{1}{240}$$
$$e'_R j_{11} = \tfrac{1}{504}$$
$$e'_R j_{15} = -\tfrac{1}{480}$$
$$e'_R j_{19} = \tfrac{1}{264}$$

Inspecting Toda's tables [18, pp.186–188] we see that

$$e'_R : \pi_3^S \to Z_{24},$$

$$e'_R : \pi_7^S \to Z_{240}$$

and

$$e'_R : \pi_{11}^S \to Z_{504}$$

are isomorphisms, while

$$e'_R : \pi_{15}^S \to Z_{480}$$

and

$$e'_R : \pi_{19}^S \to Z_{264}$$

are epimorphisms with kernel Z_2. Toda gives the elements $\eta\kappa$ in π_{15}^S and $\bar{\sigma} \in \{\nu, \bar{\nu} + \varepsilon, \sigma\}$ in π_{19}^S as generating Z_2 summands; these elements are annihilated by e'_R, as we see using Proposition 3.2 and Theorem 5.3 (v).

We have still to describe the invariant e_C on the r-stem for $r \equiv 1 \bmod 4$. In this case the integer $m(t)$ occurring in Proposition 7.9 is 2, and so e_C gives a homomorphism from π_r^S to Z_2. We have already remarked that if $r \equiv 1 \bmod 8$ the invariant d_R gives a homomorphism from π_r^S to Z_2.

THEOREM 7.18. *If $r \equiv 1 \bmod 8$ we have*

$$e_C = d_R : \pi_r^S \to Z_2.$$

The behaviour of d_R has been described in Theorem 7.2. The proof of this theorem is deferred to §12.

For completeness we describe the value of this invariant on the image of the J-homomorphism.

PROPOSITION 7.19. *Suppose $r \equiv 1 \bmod 8$. Then the composite*

$$e_C J = d_R J : \pi_r(SO) \to Z_2$$

is an isomorphism for $r = 1$ and is zero for $r > 1$.

For $r = 1$ the J-homomorphism becomes an isomorphism from $\pi_1(SO) = Z_2$ to $\pi_1^S = Z_2$. The value of d_R on π_1^S is well known, and the value of e_C is given by Example 7.4. For $r > 1$ the proof of this proposition is deferred to §10.

PROPOSITION 7.20. *If $r \equiv 5 \bmod 8$ we have*

$$e_C = 0 : \pi_r^S \to Z_2.$$

This will follow immediately from Proposition 7.1, by using the following lemma.

LEMMA 7.21. *Suppose given $f : S^{2q+r} \to S^{2q}$ with $2q \equiv 0 \bmod 8$ and $r \equiv 1 \bmod 4$. If $d_R(f) \equiv 0$, then $e_C(f) = 0$.*

Proof. Consider the following diagram.

$$\begin{array}{ccccccc}
\tilde{K}_R(S^{2q+r}) & \xleftarrow{f^*} & \tilde{K}_R(S^{2q}) & \leftarrow & \tilde{K}_R(S^{2q} \cup_f e^{2q+r+1}) & \leftarrow & \tilde{K}_R(S^{2q+r+1}) = Z_2 \text{ or } 0 \\
\downarrow & & \downarrow \cong & & \downarrow & & \downarrow 0 \\
\tilde{K}_C(S^{2q+r}) & \leftarrow & \tilde{K}_C(S^{2q}) & \leftarrow & \tilde{K}_C(S^{2q} \cup_f e^{2q+r+1}) & \leftarrow & \tilde{K}_C(S^{2q+r+1}) = Z
\end{array}$$

If $f^* = 0$, the diagram provides a splitting of the extension $e_C(f)$.

§8. RELATION WITH THE HOPF INVARIANT

In this section we shall establish the relation between the invariant e_C discussed in §7 and the Hopf invariant (mod 2 or mod p) in the sense of Steenrod. As mentioned in the introduction, this leads to a proof, first published by Dyer [13], of the non-existence of elements of Hopf invariant one (mod 2 or mod p).

We first recall the definition of the Hopf invariant in the sense of Steenrod. As in §7, we take a map $f\colon S^{2n-1} \to S^{2q}$ and form $S^{2q} \cup_f e^{2n}$. Let p be a prime; and suppose that $n - q = k(p-1)$, where k is an integer. Then in $H^*(S^{2q} \cup_f e^{2n}; Z_p)$ we have a formula of the following form.

(8.1) $$P^k \rho h^{2q} = \mu \rho h^{2n}.$$

Here P^k is the Steenrod reduced power (interpreted as Sq^{2k} if $p = 2$); the homomorphism

$$\rho\colon H^*(X; Z) \to H^*(X; Z_p)$$

is induced by the quotient map $Z \to Z_p$ of coefficients; the classes h^{2q} and h^{2n} are generators in $H^*(S^{2a} \cup_f e^{2n}; Z)$, as in §7; and μ is some element of Z_p.

It is easy to see that μ is an invariant of f. We will now show that the value of μ is determined by $e_C(f)$. For this purpose we define Q'_p to be the additive group of rationals with denominators prime to p; then we have a unique homomorphism $\rho'\colon Q'_p \to Z_p$ extending the quotient map $Z \to Z_p$.

PROPOSITION 8.2. *We have*

$$p^k e_C(f) \in Q'_p$$
$$\mu = -\rho'(p^k e_C(f)).$$

Proof. Formula (7.3) states that

$$ch\xi = h^{2q} + \lambda h^{2n},$$

where $e_C(f) = \{\lambda\}$. We now appeal to [1, Theorems 1, 2]. The statements of this paper involve a further numerical function; we set

$$M(r) = \prod_p p^{[r/p-1]}.$$

(This function is written $m(r)$ in [1], but it is different from the function $m(t)$ of [4 §2].) In our application, we take the integer "r" of [1] to be $k(p-1)$. Theorem 1 of [1] now states that the class $M(r)\lambda h^{2n}$ is integral; that is, $M(r)\lambda \in Z$; thus $p^k\lambda \in Q'_p$. Moreover, in [1, Theorem 2], the class "$ch_{q,0}\xi$" must be h^{2q}, and the class "$ch_{q,r}\xi$" must be $M(r)\lambda h^{2n}$. Thus [1, Theorem 2 part (5)] gives

$$\rho(M(r)\lambda h^{2n}) = \frac{M(r)}{p^k}\chi(P^k)\rho h^{2q}.$$

Here χ means the canonical anti-automorphism of the Steenrod algebra. But in the complex $S^{2q} \cup_f e^{2n}$ decomposable Steenrod operations are zero; thus

$$\chi(P^k)\rho h^{2q} = -P^k \rho h^{2q}.$$

Since $M(r)/p^k$ is an integer prime to p, this leads at once to the result given.

COROLLARY 8.3. *The Hopf invariant in the sense of Steenrod is zero except in the following cases;*

(a) $p = 2$, $k = 1$, 2 *or* 4;

(b) p *is odd*, $k = 1$.

It is (of course) classical that non-zero values can occur in the exceptional cases given.

Proof. According to Proposition 7.9, we have $e_C(f) = \{z/m(t)\}$ where $z \in Z$ and $t = k(p-1)$. We have only to check that $m(t)$ contains the prime p to the power $(k-1)$ at most—except in the exceptional cases. This follows from the explicit definition of $m(t)$ given in [4, §2].

COROLLARY 8.4. *The stable group π^S_{2p-3} contains an element α with $p\alpha = 0$ and $e_C(\alpha) = -1/p$ mod 1.*

In fact, the p-component of π^S_{2p-3} is known to be Z_p; and it is known that we can choose a generator α whose Hopf invariant is 1 mod p.

The same argument shows that we can find elements in the 2-components of π^S_1, π^S_3 and π^S_7 whose e_C-invariants are $\frac{1}{2}$ mod 1, $\frac{1}{4}$ mod $\frac{1}{2}$ and $\frac{1}{16}$ mod $\frac{1}{8}$.

§9. THE INVARIANT e_R ON THE r-STEM FOR $r \equiv 0, 1$ *mod* 8

In this section we will add to the discussion of §7 by discussing the invariant e_R as it applied to maps $f: S^{2q+r} \to S^{2q}$ with $r \equiv 0$ or 1 mod 8 and $2q \equiv 0$ mod 8. The results are stated in Theorems 9.4, 9.5.

There are of course other possibilities for the dimensions of the spheres; one of them will actually arise in the proof of Proposition 12.17. The earnest student may consider the e_R-invariants of maps $f: S^{n-1} \to S^t$, where n, t run over the congruence classes 0, 1, 2 and 4 mod 8, so obtaining 16 cases. He will find that all the resulting invariants are determined by those we consider in this paper.

We will begin by computing the Ext groups which arise in our case. As before, let M be the object of the abelian category A in which the underlying group is Z and the operations are given by

$$\Psi^k x = k^q x.$$

Let N be the similar object in which the underlying group is Z and the operations are given by

$$\Psi^k x = k^n x.$$

Let N' be the quotient object $N/\nu N$, where ν is some positive integer; thus the abelian group underlying N' is Z_ν. We shall consider only the case $\Lambda = R$, and so we assume that q and n are even.

We have already computed $\text{Ext}^1_S(M, N)$, which is a cyclic group (Proposition 7.9). The next result computes $\text{Ext}^1_S(M, N')$.

Proposition 9.1. *The quotient map $N \to N'$ induces an isomorphism*

$$\mathrm{Ext}^1_S(M, N)/v\,\mathrm{Ext}^1_S(M, N) \xrightarrow{\cong} \mathrm{Ext}^1_S(M, N').$$

It follows that we may represent $\mathrm{Ext}^1_S(M, N')$ as the group of rationals $z/m(t)$ modulo 1 and $v/m(t)$, where $z \in Z$ and $t = n - q$.

Proof. The exact sequence

$$0 \to N \xrightarrow{v} N \to N' \to 0$$

induces an exact sequence

$$\mathrm{Ext}^1(M, N) \xrightarrow{v} \mathrm{Ext}^1(M, N) \to \mathrm{Ext}^1(M, N')$$

and so (passing to direct limits) an exact sequence

$$\mathrm{Ext}^1_S(M, N) \xrightarrow{v} \mathrm{Ext}^1_S(M, N) \to \mathrm{Ext}^1_S(M, N').$$

All that is required is to show that the map

$$\mathrm{Ext}^1_S(M, N) \to \mathrm{Ext}^1_S(M, N')$$

is epi. By splitting N' into p-components, we see that it is sufficient to consider the case $v = p^f$.

Suppose then that $v = p^f$, and suppose given an exact sequence

$$0 \leftarrow M \leftarrow E \leftarrow N' \leftarrow 0$$

in the category A. We may choose in E an element ξ projecting to the generator in M; we may write η for the image in E of the generator in N'. The operations Ψ^k in E must be given by formulae of the following form.

$$(9.2) \qquad \begin{aligned} \Psi^k\xi &= k^q\xi + c(k)\eta \\ \Psi^k\eta &= \qquad\; k^n\eta \end{aligned}$$

Here the coefficients $c(k)$ lie in Z_v, and constitute a "factor set".

We now invoke Axiom 6.3, which shows that the value of $c(k)$ modulo $v = p^f$ is periodic in k with period p^{ef} for some e. Now the multiplicative group G of residue classes prime to p, modulo p^{ef}, is cyclic if p is odd; let l be a generator for G, or for $G/\{\pm 1\}$ if $p = 2$. From the equation $\Psi^{kl} = \Psi^k\Psi^l$, we find

$$(9.3) \qquad c(kl) = l^q c(k) + k^n c(l) \qquad \bmod v.$$

(Compare the proof of Lemma 7.11.) By induction over r, we find that

$$c(l^r) = \frac{l^{rn} - l^{rq}}{l^n - l^q}\, c(l) \qquad \bmod v.$$

Since we have assumed we are in the case $\Lambda = R$, we have $\Psi^{-k} = \Psi^k$, and thus

$$c(-l^r) = \frac{l^{rn} - l^{rq}}{l^n - l^q}\, c(l) \qquad \bmod v.$$

(Recall now that n and q are even.) We have thus shown that

$$c(k) = (k^n - k^q)\mu \qquad \bmod v$$

for all k prime to p, where μ is the rational number $c(l)/(l^n - l^q)$. It is now easy to see that we have

$$c(k) = (k^n - k^q)\lambda \qquad \text{mod } \nu$$

for all k prime to p, where λ is a rational number whose denominator is a power of p.

Next recall that the class of E in $\text{Ext}_S^1(M, N')$ is not affected by applying the "eight-fold suspension operator" T (see §3). Suppose we do this t times; then the equation

$$c(k) = (k^n - k^q)\lambda \qquad \text{mod } \nu$$

(valid for k prime to p) becomes

$$k^{4t}c(k) = (k^{n+4t} - k^{q+4t})\lambda \qquad \text{mod } \nu$$

(for k prime to p). We can easily choose t large enough to satisfy the following two conditions.

(i) $k^{4t}c(k) = 0 \text{ mod } \nu$ wherever k is divisible by p.

(ii) $(k^{n+4t} - k^{q+4t})\lambda$ is integral and divisible by ν whenever k is divisible by p.

The equation

$$k^{4t}c(k) = (k^{n+4t} - k^{q+4t})\lambda \qquad \text{mod } \nu$$

will thus be true for all k. We have shown that the factor set $k^{4t}c(k)$ has the form considered in §7; thus E represents an element in the image of

$$\text{Ext}_S^1(M, N) \to \text{Ext}_S^1(M, N').$$

This completes the proof.

As a particular case of Proposition 9.1, we may put $\nu = 2$. Then the operations Ψ^k in N' are independent of n, being given by

$$\Psi^k x = \begin{cases} x & (k \text{ odd}) \\ 0 & (k \text{ even}). \end{cases}$$

We have

$$\text{Ext}_S^1(M, N') \cong Z_2.$$

In this case the proof given above specialises a little. Equation (9.3) shows that the factor set $c(k)$ gives a homomorphism from G, the multiplicative group of odd numbers modulo 2^{ef}, to the additive group Z_2. We arrive at two factor sets; the zero factor set, and that given by

$$c(k) = \begin{cases} 0 & \text{for } k \equiv \pm 1 \text{ mod } 8 \\ 1 & \text{for } k \equiv \pm 3 \text{ mod } 8. \end{cases}$$

The latter represents the non-zero element of $\text{Ext}_S^1(M, N')$.

Next, let $f: S^{2q+r} \to S^{2q}$ be a map with $r \equiv 0$ or $1 \text{ mod } 8$ and $2q \equiv 0 \text{ mod } 8$. Then we have

$$\tilde{K}_R(S^{2q}) = M, \qquad \tilde{K}_R(S^{2q+r+1}) = N'$$

and so

$$e_R(f) \in \text{Ext}_S^1(M, N') = Z_2.$$

Thus e_R gives a homomorphism from $\text{Ker } d_R \subset \pi_S$ to Z_2.

THEOREM 9.4. *If $r \equiv 0$ or 1 mod 8 and $r > 1$ then e_R maps* Ker d_R *onto Z_2, and* Ker e_R *is a direct summand in* Ker d_R.

We note that if $r \equiv 0 \bmod 8$ then $\operatorname{Ker} d_R = \pi_r^S$, by Proposition 7.1. If $r \equiv 1 \bmod 8$ then Ker d_R is a direct summand in π_r^S, by Theorem 7.17, and therefore Ker e_R is a direct summand in π_r^S.

Theorem 9.4 will follow immediately from the following result.

THEOREM 9.5. *If $r \equiv 0$ or 1 mod 8 and $r > 1$ then the composite*

$$e_R J : \pi_r(SO) \to Z_2$$

is an isomorphism.

(Note that e_R is defined on Im J, by Proposition 7.19.)

We see that Theorem 1.1 will follow immediately from Theorem 9.5; also Theorem 1.3 will follow immediately from Theorems 7.2 and 9.5. The proof of Theorem 9.5 will be given in §10.

§10. THE VALUES OF THE INVARIANTS ON THE IMAGE OF J

In §§7, 9 we have introduced certain invariants; in this section we shall compute the values which they take on the image of the stable J-homomorphism

$$J : \pi_r(SO) \to \pi_r^S.$$

Our main object, then, is to prove Theorems 7.16, 7.19 and 9.5.

We will first show that if we use an element in the image of the J-homomorphism as an attaching map, then the resulting two-cell complex is, in fact, a Thom complex. More precisely, suppose given a map $\varphi : S^r \to SO(q)$. We can apply the "Hopf construction" J to φ; we obtain the map

$$J\phi : S^{q+r} \to S^q$$

and the two-cell complex

$$X = S^q \cup_{J\phi} e^{q+r+1}.$$

On the other hand, we can use φ to define an E^q bundle over S^{r+1}, and so obtain a Thom complex, which actually has the form

$$Y = S^q \cup e^{q+r+1}.$$

LEMMA 10.1. *The complexes X and Y are homotopy-equivalent. With suitable choices of sign in the constructions given above, we can choose the equivalence to have degree* +1 *on both cells.*

I believe that this lemma was known to earlier workers, for example, Atiyah (ca. 1960); see also [13, p.370].

Proof. We first discuss the Thom complex Y. The E^q-bundle over S^{r+1} can be obtained from $E^q \cup (E^{r+1} \times E^q)$ by identifying each point (x, y) in $S^r \times E^q$ with the point $(\varphi x)y$ in E^q. (Here $SO(q)$ acts on E^q in the usual way.) We can now obtain the Thom complex by further identifying $S^{q-1} \cup (E^{r+1} \times S^{q-1})$ to a single point.

We now discuss the Hopf construction. To construct the map $J\varphi$, we realise S^{q+r} as the boundary of $E^{r+1} \times E^q$. We map $S^r \times S^{q-1}$ to S^{q-1} by

$$(J\phi)(x, y) = (\phi x)y \qquad (y \in S^{q-1});$$

we extend to a map from $S^r \times E^q$ to the upper hemisphere E^q_+ of S^q, say

$$(J\phi)(x, y) = (\phi x)y \qquad (y \in E^q);$$

we also extend it to a map from $E^{r+1} \times S^{q-1}$ to the lower hemisphere $E^q{}_-$ of S^q. (Actually this construction differs in sign from the one the author would usually prefer.)

The complex X is now

$$S^q \cup_{J\phi} (E^{r+1} \times S^q).$$

It will not alter its homotopy type if we identify $E^q{}_-$ to a point. By doing this we obtain precisely the description given above for Y. This completes the proof.

Proof of Theorem 7.16. We may start from a real bundle β over S^{r+1}, where $r = 4s$, such that β represents a generator of $\tilde{K}_R(S^{r+1})$. With the notation of §7, this is expressed by the equation

$$ch_{2s} c\beta = a_{4s} h^{4s}.$$

We may suppose that the structural group of β is Spin(q), where q is divisible by 8.

We now consider the Thom complex $S^q \cup e^{q+4s}$ corresponding to β, and we make use of the Thom isomorphism φ_K [4 §4]. In $\tilde{K}_R(S^q \cup e^{q+4s})$ we have the element $\varphi_K 1$; moreover, with the notation of [4 §§2, 5] we have

$$\varphi_H^{-1} ch_R \varphi_K 1 = 1 + \tfrac{1}{2}\alpha_{2s} a_{4s} h^{4s}$$

[4, Proposition 5.2]. That is, we have

$$ch_R \varphi_K 1 = h^q + \tfrac{1}{2}\alpha_{2s} a_{q+4s} h^{q+4s}.$$

We may take $\varphi_K 1$ for our generator ξ. This yields

$$e'_R(J\beta) = \tfrac{1}{2}\alpha_{2s} \qquad \text{mod } 1,$$

which proves Theorem 7.16.

Proof of Theorem 7.19, for the case $r > 1$. As in the previous proof, we may start with a real bundle β over S^{r+1}, with structural group Spin(q), where q is divisible by 8. As above, we obtain a generator $\varphi_K 1$ in $\tilde{K}_R(S^q \cup_{J\beta} e^{q+r+1})$, which restricts to the generator in $\tilde{K}_R(S^q)$. Therefore the generator in $\tilde{K}_R(S^q)$ is annihilated by $(J\beta)^*$; that is, $d_R(J\beta) = 0$. Lemma 7.21 now shows that $e_C J\beta = 0$.

First proof of Theorem 9.5. As in the two previous proofs, we may start from a real bundle β over S^{r+1} such that β represents a generator of $\tilde{K}_R(S^{r+1})$, and we may obtain a generator $\varphi_K 1$ in $\tilde{K}_R(S^q \cup_{J\beta} e^{q+r+1})$. We now wish to calculate $\Psi^k \varphi_K 1$ (at least for k odd). By [4 §5, especially Theorem 5.15], we have

$$\begin{aligned} \varphi_K^{-1} \Psi^k \varphi_K 1 &= \rho^k \beta \\ &= \begin{cases} 1 & \text{if } k \equiv \pm 1 \bmod 8 \\ 1 + \beta & \text{if } k \equiv \pm 3 \bmod 8. \end{cases} \end{aligned}$$

With the notation of §9, this gives

$$\Psi^k\xi = \begin{cases} \xi & \text{if } k \equiv \pm 1 \bmod 8 \\ \xi + \eta & \text{if } k \equiv \pm 3 \bmod 8. \end{cases}$$

If we recall the description of $\mathrm{Ext}^1_S(M, N)$ given in §9, this shows that $e_R J\beta$ is non-zero.

A second proof of Theorem 9.5 will be given in §12.

§11. TODA BRACKETS, II

The main purpose of this section is to show explicitly how the theorems of §5 apply to the invariants of §7. The spaces we shall deal with will thus be spheres; and we shall stay in those dimensions where the invariants e_Λ take values in Q/Z, the rationals mod 1.

We will begin by stating the main results, without proofs. The following result, which is typical, will be obtained by specialising Theorem 5.3 (v).

THEOREM 11.1. *Suppose given integers $a > b > c > 0$, which are even if $\Lambda = R$. Suppose given $f: S^{2a-2} \to S^{2b-1}$, $h: S^{2b-1} \to S^{2c}$ and $q \in Z$ such that $h(q\iota) \sim 0$ and $(q\iota)f \sim 0$. Then*

$$e_\Lambda\{h, q\iota, f\} = -qe\ (Sf)e_\Lambda(h) \qquad \textit{mod } 1.$$

We pause to check that both sides of this equation are well-defined as rationals mod 1. The indeterminacy of $\{h, q\iota, f\}$ is

$$h\pi_{2a-1}(S^{2b-1}) + \pi_{2b}(S^{2c})Sf,$$

and therefore (using (3.2) and (7.1)) $e_\Lambda\{h, q\iota, f\}$ is well-defined as a rational mod 1. If we change the fraction representing $e_\Lambda(Sf)$ by 1, we change $qe_\Lambda(Sf)\, e_\Lambda(h)$ by $qe_\Lambda(h)$, which is an integer since $h(q\iota) \sim 0$; similarly if we change the fraction representing $e_\Lambda(h)$ by 1. Thus $-qe_\Lambda(Sf)e_\Lambda(h)$ is well-defined mod 1.

In applying Theorem 11.1 in the case $\Lambda = R$, we have to distinguish when the invariant e_R means e'_R, and when it means e''_R, according to the dimensions of the spheres concerned.

Examples on Theorem 11.1. With the notation of Example 7.17, we have

$$\{j_3, 24, j_3\} = 40j_7$$

$$\{j_3, 24, 10j_7\} = 21j_{11}$$

$$\{j_3, 24, 21j_{11}\} = 80j_{15} \qquad \text{mod } \eta\kappa$$

$$\{j_7, 240, j_7\} = 2j_{15} \qquad \text{mod } \eta\kappa$$

etc.

In order to state the results obtained by specialising Theorem 5.3 (iv) and (vi) we need a little number theory. The numerical function $m(t)$ will be as in [4 §2]; as we shall need the explicit definition in our proofs, we recall it now. We write $v_p(n)$ for the exponent to which the prime p occurs in n, so that

$$n = 2^{v_2(n)}3^{v_3(n)}5^{v_5(n)}\ldots.$$

For odd primes p we set

$$v_p(m(t)) = \begin{cases} 0 & \text{if } t \not\equiv 0 \bmod (p-1) \\ 1 + v_p(t) & \text{if } t \equiv 0 \bmod (p-1). \end{cases}$$

For $p = 2$ we set

$$\nu_2(m(t)) = \begin{cases} 1 & \text{if } t \not\equiv 0 \bmod 2 \\ 2 + \nu_2(t) & \text{if } t \equiv 0 \bmod 2. \end{cases}$$

In order to avoid worrying about signs in what follows, we remark that this definition is equally valid if t is negative; only the case $t = 0$ need be excluded. Thus we have $m(-t) = m(t)$.

We shall suppose given two even integers u, v; the cases $u = 0$, $v = 0$ and $u = v$ are excluded.

LEMMA 11.2. *There exists a rational number $\delta(u, v)$ such that for sufficiently large t (depending on u and v), and for all $k \in Z$ we have*

$$(k^{t+u} - k^t) - \delta(u, v)(k^{t+v} - k^t) \equiv 0 \qquad \textit{mod } m(u)m(v-u).$$

The congruence is to be interpreted as meaning that the left-hand side is an integer multiple of $m(u)\, m(v-u)$.

We shall not only prove that $\delta(u, v)$ exists; we will give a definition for $\delta(u, v)$ which allows one to compute it easily. I am indebted to Dr. B. J. Birch for conversations about an earlier version of this lemma.

We recall from [4, §2] that for sufficiently large t, the highest common factor of the numbers $(k^{t+v} - k^t)$ (as k runs over Z) is $m(v)$. This shows that the property stated in Lemma 11.2 characterises $\delta(u, v)$ up to an integer multiple of $m(u)\, m(v-u)/m(v)$.

We shall need to refer to the following further properties of $\delta(u, v)$.

LEMMA 11.3.

(i) $\delta(-u, -v) \equiv \delta(u, v)$ *mod* $m(u)\, m(v-u)/m(v)$.

(ii) $\delta(u, v) + \delta(v-u, v) \equiv 1$ *mod* $m(u)\, m(v-u)/m(v)$.

(iii) $\delta(u, v) = \dfrac{\gamma m(u)}{m(v)}$

for some integer $\gamma = \gamma(u, v)$.

(iv) $\delta(u, v) = 1 + \dfrac{\gamma' m(v-u)}{m(v)}$

for some integer $\gamma' = \gamma'(u, v)$.

The following result may be obtained by specialising Theorem 5.3 (iv).

THEOREM 11.4. *Suppose given even integers $a > b > c > 0$. Suppose given $f\colon S^{2a-2} \to S^{2b-1}$, $g\colon S^{2b-1} \to S^{2c}$ and $q \in Z$ such that $(q\iota)g \sim 0$ and $gf \sim 0$. Then*

$$e_\Lambda\{q\iota, g, f\} = -q\delta e_\Lambda(Sf)e_\Lambda(g) \qquad \textit{mod } 1 \textit{ and } q/m(a-c)$$

where $\delta = \delta(a-b, a-c)$ (with the notation of Lemma 11.2).

As for Theorem 11.1, we have to check that both sides are well-defined modulo 1 and $q/m(a-c)$. For the left-hand side this is easy. Altering δ by $m(a-b)\, m(b-c)/m(a-c)$ alters the right-hand side by an integer multiple of $q/m(a-c)$, since $m(a-b)\, e_\Lambda(Sf)$ and

$m(b-c)\,e_\Lambda(g)$ are integers. Altering $e_\Lambda(Sf)$ by 1 alters the right-hand side by

$$q\left(1+\gamma'\frac{m(b-c)}{m(a-c)}\right)e_\Lambda(g)$$

(using Lemma 11.3 (iv)); since $qe_\Lambda(g)$ and $m(b-c)e_\Lambda(g)$ are integers this is zero mod 1 and $q/m(a-c)$. Altering $e_\Lambda(g)$ by 1 alters the right-hand side by

$$q\gamma\frac{m(a-b)}{m(a-c)}e_\Lambda(Sf)$$

(using Lemma 11.3 (iii)); since $m(a-b)\,e_\Lambda(Sf)$ is an integer, this is zero mod $q/(ma-c)$.

The following result may be obtained by specialising Theorem 5.3 (vi).

THEOREM 11.5. *Suppose given even integers $a>b>c>0$. Suppose given $g: S^{2a-2}\to S^{2b-1}$, $h: S^{2b-1}\to S^{2c}$ and $q\in Z$ such that $hg\sim 0$ and $g(q\iota)\sim 0$. Then*

$$e_\Lambda\{h, g, q\iota\} = -q\delta e_\Lambda(Sg)e_\Lambda(h) \qquad \textit{mod } 1 \textit{ and } q/m(a-c)$$

where $\delta=\delta(b-c, a-c)$ (with the notation of Lemma 11.2).

As before, we have to check that both sides are well-defined modulo 1 and $q/m(a-c)$. This is done exactly as for Theorem 11.4.

In applying Theorems 11.4 and 11.5, we have again to distinguish when the invariant e_R means e'_R, and when it means e''_R.

Examples on Theorem 11.5. With the notation of Example 7.17, we have

$$\begin{aligned}
\{j_3, 2j_3, 12\} &= 0 && \text{mod } 12j_7\\
\{j_7, j_3, 24\} &= -j_{11} && \text{mod } 24j_{11}\\
\{j_3, j_7, 240\} &= 7j_{11} && \text{mod } 24j_{11}\\
\{j_{11}, j_3, 24\} &= -4j_{15} && \text{mod } 24j_{15} \text{ and } \eta\kappa\\
\{j_7, 2j_7, 120\} &= j_{15} && \text{mod } 120j_{15} \text{ and } \eta\kappa\\
\{j_3, j_{11}, 504\} &= -4j_{15} && \text{mod } 24j_{15} \text{ and } \eta\kappa
\end{aligned}$$

etc.

The calculation of these examples requires a knowledge of the coefficients δ, which will be provided later in this section.

Theorems 11.4 and 11.5 are equivalent. In fact, if α and β belong to odd-dimensional stable groups, then we have

$$\{\alpha, \beta, q\iota\} = \{q\iota, \beta, \alpha\}$$

by a theorem of Toda [18, p.26 (3.4) (i) or p.33 (3.9) (i)]. The reader is warned not to suppose that this remark makes the equivalence completely obvious; in the case $\Lambda = R$ we still have to distinguish when e_R means the invariant e'_R, and when it means e''_R; we have then to use Proposition 7.14. However, these details lead to the required result. It will therefore be sufficient to prove one of these theorems and deduce the other. Similarly, we will state corollaries of only one of these theorems.

Other checks on our work are provided by the identities

$$\{\alpha, q\iota, \beta\} = \{\beta, q\iota, \alpha\}$$

and

$$\{\alpha, \beta, q\iota\} - \{\beta, q\iota, \alpha\} + \{q\iota, \alpha, \beta\} = 0$$

[18, p.26 (3.4) (ii) or p.33 (3.9) (ii)]. The first is consistent with Theorem 11.1; the second is consistent with Theorems 11.1, 11.4 and 11.5, as we see using Lemma 11.3 (ii).

We will now state two corollaries of Theorem 11.5 which are useful in dealing with p-components of stable homotopy groups. We retain the notation and assumptions of Theorem 11.5.

COROLLARY 11.6. *Let p be an odd prime such that $a - b$ and $b - c$ are divisible by $p - 1$. Then we have*

$$e_\Lambda\{h, g, q\iota\} = -q\,\frac{b - c}{a - c}\,e_\Lambda(Sg)e_\Lambda(h)$$

as an equation in the p-adic numbers modulo 1 and $q/m(a - c)$.

The case in which $a - b$ and $b - c$ are divisible by $(p - 1)$ is, of course, the only case of interest if we are studying p-components.

COROLLARY 11.7. *Let $p = 2$. Then we have*

$$e_\Lambda\{h, g, q\iota\} = -q\,\frac{b - c}{a - c}(1 + \omega 2^g)e_\Lambda(Sg)e_\Lambda(h)$$

(where ω is any odd number and $g = 1 + \nu_2(a - b)$) as an equation in the 2-adic numbers modulo 1 and $q/m(a - c)$.

It is no great surprise that the case $p = 2$ is exceptional.

In both corollaries, the phrase "modulo 1 and $q/m(a - c)$" refers to multiples of 1 and $q/m(a - c)$ by p-adic integers. The use of p-adic numbers is not essential, but it is convenient; it allows us to invert numbers prime to p, modulo a high power of p, without stating exactly which high power of p is required.

A further check on our work is now provided by the following observation. Suppose given a generator $\gamma \in \pi_{4r-1}(SO)$, a map $\theta : S^{4r+4s-2} \to S^{4r-1}$ and an integer q such that $\gamma\theta \sim 0$, $\theta(q\iota) \sim 0$. Then we can form in $\pi_{4r+4s-1}(SO)$ the Toda bracket $\{\gamma, \theta, q\iota\}$. Let $\gamma' \in \pi_{4r+4s-1}(SO)$ be a generator; then we have

$$\{\gamma, \theta, q\iota\} = -q e_R(S\theta)\gamma' \qquad \text{mod } q.$$

(Whether $e_R(S\theta)$ is an invariant e'_R or e''_R depends on the parity of r). In $\pi^S_{4r+4s-1}$ we shall have

$$J\{\gamma, \theta, q\} \subset \{J\gamma, \theta, q\},$$

that is,

$$-q e_R(S\theta)J\gamma' \subset \{J\gamma, \theta, q\}.$$

We may now apply e'_R to both sides, using Theorem 7.16 and [4, Theorem 2.5]. The results should agree modulo 1 and $q/m(2r+2s)$. Calculating in the p-adic numbers, both sides yield

$$-qe_R(S\theta)\frac{p-1}{4p(r+s)},$$

providing that $2r$ and $2s$ are divisible by $p-1$; otherwise 0. Calculating in the 2-adic numbers, both sides yield

$$-qe_R(S\theta)\left(\tfrac{1}{2}+\frac{1}{8(r+s)}\right)$$

provided $r \geqq 3$.

The remainder of this section is organised as follows. We begin with the number theory, leading up to the proofs of Lemmas 11.2 and 11.3. Then we prove Theorems 11.1, 11.4 and 11.5. Corollaries 11.6 and 11.7 then follow easily.

LEMMA 11.8. *Let p be an odd prime, let k be an integer prime to p, and let a, b, c be integers divisible by $(p-1)$. Then we have*

$$(a-b)k^c + (b-c)k^a + (c-a)k^b \equiv 0 \qquad mod\ p^{h+2},$$

where $h = v_p(a-b) + v_p(b-c) + v_p(c-a)$.

LEMMA 11.9. *Let $p = 2$, let k be an odd integer, and let a, b, c be even integers. Then we have*

$$(a-b)k^c + (b-c)k^a + (c-a)k^b \equiv \varepsilon 2^{h+3} \qquad mod\ 2^{h+4}$$

where

$$\varepsilon = \begin{cases} 0 & if\ k \equiv \pm 1\ mod\ 8 \\ 1 & if\ k \equiv \pm 3\ mod\ 8 \end{cases}$$

and $h = v_2(a-b) + v_2(b-c) + v_2(c-a)$.

We prove Lemma 11.9; the proof of Lemma 11.8 is similar but slightly simpler.

Without loss of generality we may assume that

$$v_2(a-b) = f$$
$$v_2(b-c) = f+g$$
$$v_2(c-a) = f$$

where $f \geqq 1$, $g \geqq 1$. Thus $h = 3f + g$. Set $d = 2^f$; by adding a constant to a, b and c we may assume they are all divisible by d. Set $K = k^d$; then $K \equiv 1 \bmod 2^{f+2}$. Hence

$$\frac{K^{a/d} - K^{b/d}}{K-1} \equiv (a-b)/d \qquad \text{mod } 2^{f+2}.$$

(Without loss of generality we may assume $a > b$; expand the left-hand side in powers of K.) Thus

$$k^a - k^b \equiv (K-1)(a-b)/d \qquad \text{mod } 2^{2f+4}$$

and

$$(b-c)(k^a - k^b) \equiv (K-1)(a-b)(b-c)/d \qquad \text{mod } 2^{3f+g+4}.$$

We now consider the sum of 2^g consecutive powers of K. I claim we have

$$K^{e+1} + K^{e+2} + \ldots + K^{e+2^g} \equiv 2^g + \varepsilon 2^{f+g+1} \qquad \text{mod } 2^{f+g+2}$$

where ε is as above. In fact, suppose that $K \equiv 1 \bmod 2^{\phi+2}$ but $K \not\equiv 1 \bmod 2^{\phi+3}$, where $\phi > f$ if $k \equiv \pm 1 \bmod 8$ and $\phi = f$ if $k \equiv \pm 3 \bmod 8$. Then the 2^g numbers

$$K^{e+1}, K^{e+2}, \ldots, K^{e+2^g}$$

give the 2^g residue classes $1 + q2^{\phi+2} \bmod 2^{\phi+g+2}$. Hence their sum is

$$2^g + \tfrac{1}{2}(2^g)(2^g+1)2^{\phi+2} \qquad \bmod 2^{\phi+g+2}.$$

This proves the assertion.

Arguing as above, we find

$$\begin{aligned} \frac{K^{b/d} - K^{c/d}}{K - 1} &\equiv (2^g + \varepsilon 2^{f+g+1}) \frac{b - c}{2^g d} \qquad &\bmod 2^{f+g+2} \\ &\equiv (b - c)/d + \varepsilon 2^{f+g+1} \qquad &\bmod 2^{f+g+2}. \end{aligned}$$

Thus

$$k^b - k^c \equiv (K - 1)(b - c)/d + \varepsilon 2^{2f+g+3} \qquad \bmod 2^{2f+g+4}$$

and

$$(a - b)(k^b - k^c) \equiv (K - 1)(a - b)(b - c)/d + \varepsilon 2^{3f+g+3} \qquad \bmod 2^{3f+g+4}.$$

Thus

$$(b - c)(k^a - k^b) - (a - b)(k^b - k^c) \equiv \varepsilon 2^{h+3} \qquad \bmod 2^{h+4},$$

which proves the lemma.

We now define $\delta(u, v)$. As above, let u, v be two even integers; the cases $u = 0$, $v = 0$ and $u = v$ are excluded. We propose to define the rational number $\delta(u, v)$ modulo $m(u)\, m(v - u)/m(v)$ by giving a finite number of congruences. Each congruence will be written as a congruence in the p-adic integers, holding mod p^f where

$$f = v_p m(u) + v_p m(v - u) - v_p m(v).$$

The primes p to be considered are those which divide $m(u)$, $m(v)$ or $m(v - u)$. We stipulate that the denominator of $\delta(u, v)$ is to contain no other primes; thus the definition given for $\delta(u, v)$ amounts to defining an integer (namely the numerator of $\delta(u, v)$) by a finite set of congruences modulo powers of different primes. This is always legitimate.

We now give the congruences.

Case (i). p is odd; $(p - 1)$ does not divide u or v, but divides $v - u$. Take

(11.10) $$\delta(u, v) \equiv 1 \qquad \bmod p^f.$$

Case (ii). p is odd; $(p - 1)$ divides just one of u, v and therefore does not divide $(v - u)$. Take

(11.11) $$\delta(u, v) \equiv 0 \qquad \bmod p^f.$$

Case (iii). p is odd; $(p - 1)$ divides both of u, v and therefore divides $v - u$. Take

(11.12) $$\delta(u, v) \equiv u/v \qquad \bmod p^f.$$

Case (iv). $p = 2$. Take

(11.13) $$\delta(u, v) = (1 + \omega 2^g)u/v \qquad \bmod 2^f$$

where ω is any odd number and $g = 1 + v_2(v - u)$. (Note that altering ω by 2 does not affect the result mod 2^f.)

Proof of Lemma 11.2. It is sufficient to verify the congruence in the p-adic numbers for a finite number of primes p, namely those mentioned above. For each prime p the congruence will be true for k divisible by p providing we choose t large enough; we may therefore restrict attention to the case $k \not\equiv 0 \bmod p$. In all cases we have given definitions of the form $\delta \equiv \delta' \bmod p^f$, where

$$f = \nu_p m(u) + \nu_p m(v-u) - \nu_p(v);$$

and we have

$$k^{t+v} - k^t \equiv 0 \qquad \bmod p^{\nu_p(v)}.$$

Thus we have

$$\delta(k^{t+v} - k^t) \equiv \delta'(k^{t+v} - k^t) \qquad \bmod p^h$$

where $h = \nu_p m(u) + \nu_p m(v-u)$. We may therefore replace δ by δ' in checking the congruence.

Case (i). p is odd; $(p-1)$ does not divide u or v, but divides $v-u$. We have

$$k^{t+v} - k^{t+u} \equiv 0 \qquad \bmod p^{\nu_p m(v-u)}$$

i.e.

$$k^{t+u} - k^t \equiv k^{t+v} - k^t \qquad \bmod p^{\nu_p m(v-u)}.$$

Since $\delta' = 1$ and $\nu_p m(u) = 0$ in this case, this is the result required.

Case (ii). p is odd; $(p-1)$ divides just one of u, v and therefore does not divide $(v-u)$. We have

$$k^{t+u} - k^t \equiv 0 \qquad \bmod p^{\nu_p m(u)}.$$

Since $\delta' = 0$ and $\nu_p m(v-u) = 0$ in this case, this is the result required.

Case (iii). p is odd; $(p-1)$ divides both of u, v and therefore divides $(v-u)$. Lemma 11.8 gives

$$v(k^{t+u} - k^t) \equiv u(k^{t+v} - k^t) \qquad \bmod p^{h+2},$$

where $h = \nu_p(u) + \nu_p(v) + \nu_p(v-u)$. This gives

$$k^{t+u} - k^t \equiv \frac{u}{v}(k^{t+v} - k^t) \qquad \bmod p^l,$$

where $l = \nu_p m(u) + \nu_p m(v-u)$. Since $\delta' = u/v$ in this case, this is the result required.

Case (iv). $p = 2$. Lemma 11.9 gives

$$v(k^{t+u} - k^t) - u(k^{t+v} - k^t) \equiv \varepsilon 2^{h+3} \qquad \bmod 2^{h+4}$$

where $h = \nu_2(u) + \nu_2(v) + \nu_2(v-u)$. We have

$$u(k^{t+v} - k^t) \equiv \varepsilon 2^r \qquad \bmod 2^{r+1}$$

where $r = \nu_2(u) + \nu_2(v) + 2$. Thus we have

$$v(k^{t+u} - k^t) - u(1 + \omega 2^g)(k^{t+v} - k^t) \equiv 0 \qquad \bmod 2^{h+4}$$

where ω is any odd number and $g = 1 + \nu_2(v-u)$. This gives

$$k^{t+u} - k^t \equiv \frac{u}{v}(1 + \omega 2^g)(k^{t+v} - k^t) \qquad \bmod 2^l$$

where $l = \nu_2(m(u)) + \nu_2(m(v-u))$. Since $\delta' = \frac{u}{v}(1 + \omega 2^g)$ in this case, this is the result required. This completes the proof.

Proof of Lemma 11.3.

(i) The congruence

$$\delta(-u, -v) \equiv \delta(u, v) \qquad \bmod m(u)m(v-u)/m(v)$$

follows immediately by inspecting the congruence (11.10) to (11.13).

(ii) For sufficiently large t we have

$$k^{t+u} - k^t \equiv \delta(u, v)(k^{t+v} - k^t)$$

$$k^{t+u} - k^{t+v} \equiv \delta(u-v, -v)(k^t - k^{t+v})$$

mod $m(u)\,m(v-u)$. Subtracting, we obtain

$$k^{t+v} - k^t \equiv (\delta(u, v) + \delta(u-v, -v))(k^{t+v} - k^t)$$

mod $m(u)\,m(v-u)$. Since the highest common factor of the expressions $(k^{t+v} - k^t)$ is $m(v)$, we find

$$\delta(u, v) + \delta(u-v, -v) \equiv 1 \qquad \bmod m(u)m(v-u)/m(v).$$

The result now follows by part (i).

Alternatively, we can check part (ii) from the congruences (11.10) to (11.13).

(iii) For sufficiently large t we have

$$(k^{t+u} - k^t) \equiv \delta(u, v)(k^{t+v} - k^t)$$

mod $m(v)\,m(v-u)$. For sufficiently large t, the highest common factor of the expressions $(k^{t+u} - k^t)$ is $m(u)$ and that of the expressions $(k^{t+v} - k^t)$ is $m(v)$. Taking linear combinations, we find

$$Nm(u) \equiv \delta(u, v)m(v)$$

mod $m(u)\,m(v-u)$, for some integer N. Hence the result.

Alternatively, we can check part (iii) from the congruences (11.10) to (11.13).

(iv) This follows immediately from (ii) and (iii).

This completes the proof of Lemma 11.3.

Proof of Theorem 11.1. We have to evaluate the Massey product $\{e_\Lambda(Sf), q, e_\Lambda(h)\}$ according to the definition of §4, Case 3. In that section we have objects L, M, N and P in our abelian category; in the present application they all have the underlying group Z, and they have operations given by $\Psi^k x = k^c x$, $\Psi^k x = k^b x$, $\Psi^k x = k^b x$ and $\Psi^k x = k^a x$ respectively. We write λ, μ, ν, π for their respective generators. We also have in mind two extensions

$$0 \leftarrow L \leftarrow E \leftarrow M \leftarrow 0$$

$$0 \leftarrow N \leftarrow F \leftarrow P \leftarrow 0$$

given by the following formulae.

$$\Psi^k \lambda' = k^c \lambda' + e'(k^b - k^c)\mu$$

$$\Psi^k \nu' = k^b \nu' + e''(k^a - k^b)\pi.$$

Here λ', ν' are elements lifting λ, ν and e', e'' are rationals representing $e_\Lambda(h)$, $e_\Lambda(Sf)$. According to §4, Case 3 we have to construct maps

$$\theta : M \to F, \qquad \phi : E \to N;$$

we do so by the following formulae.

$$\theta(\mu) = q\nu' - qe''\pi$$

$$\phi(\lambda') = qe'\nu$$

$$\phi(\mu) = q\nu.$$

(Note that qe' and qe'' are integers.) According to §4, Case 3 we have to consider an extension G; in it we construct a lifting λ'' of λ by

$$\lambda'' = (\lambda', qe'\nu').$$

We then compute in G the formula

$$\Psi \lambda'' = k^c \lambda'' + qe'e''(k^a - k^c)\pi.$$

We conclude that in this case the Massey product in Ext^1 is given by

$$\{e_\Lambda(Sf), q, e_\Lambda(h)\} = qe'e''.$$

Theorem 11.1 thus follows from Theorem 5.3 (v).

We have given this proof of Theorem 11.1 because it seems in keeping. However, it is possible to give an *ad hoc* proof using an intermediate space $S^{2b-1} \cup_q e^{2b}$, on the lines to be explained in §12 [cf. 6, 7. In Proposition 6 of 7 a minus sign has been left out by mistake.] If one defines e_Λ using the Chern character, it is not necessary to use the operations Ψ^k in proving Theorem 11.1. By contrast, in proving Theorems 11.4, 11.5 it seems essential to use the operations Ψ^k and number-theory. In fact, the number-theory we have given may be interpreted as an investigation of what limitations the Ψ^k impose on the Chern characters in a 3-cell complex

$$S^{2t} \cup e^{2(t+u)} \cup e^{2(t+v)}.$$

This gives a partial answer to questions raised by Dyer [13, second paragraph on p.371].

Proof of Theorem 11.5. We have first to verify the conditions of Theorem 5.3 (vi). With the notation of §5, we have to show that for any choice of homotopy $hg \sim 0$, the invariant $e_\Lambda(H)$ is defined. In our case we have

$$H : S^{2b-1} \cup_g e^{2a-1} \to S^{2c}.$$

Since a and b are even, the exact sequence

$$\tilde{K}_\Lambda(S^{2b-1}) \leftarrow \tilde{K}_\Lambda(S^{2b-1} \cup_g e^{2a-1}) \leftarrow \tilde{K}_\Lambda(S^{2a-1})$$

shows that

$$\tilde{K}_\Lambda(S^{2b-1} \cup_g e^{2a-1}) = 0.$$

Hence $d_\Lambda(H) = 0$. It follows that $d_\Lambda(SH) = 0$.

We have now to evaluate the Massey product $\{q, e_\Lambda(Sg), e_\Lambda(h)\}$ according to the definition of §4, Case 1. In that section we have objects L, M, N and P in our abelian category; in the present application they all have the underlying group Z, and they have operations given by $\Psi^k x = k^c x$, $\Psi^k x = k^b x$, $\Psi^k x = k^a x$ and $\Psi^k x = k^a x$ respectively. We write λ, μ, ν, π for their respective generators. We also have in mind two extensions $\alpha \in \operatorname{Ext}^1(L, M)$, $\beta \in \operatorname{Ext}^1(M, N)$ given by the following formulae.

$$\Psi^k \lambda' = k^c \lambda' + e'(k^b - k^c)\mu$$
$$\Psi^k \mu' = k^b \mu' + e''(k^a - k^b)\nu.$$

Here λ', μ' are elements lifting λ, μ and e', e'' are rationals representing $e_\Lambda(h)$, $e_\Lambda(Sg)$. We also have in mind a homomorphism $\gamma \in \operatorname{Ext}^0(N, P)$ given by

$$\gamma(\nu) = q\pi.$$

We have next to construct $\alpha' \in \operatorname{Ext}^1(L, E)$, $\gamma' \in \operatorname{Ext}^0(E, P)$ lifting α, γ (where E is the extension representing β.)

A suitable extension α' is defined by the following formula.

$$\Psi^k \lambda'' = k^c \lambda'' + e'(k^b - k^c)\mu' + e'e''(\delta(k^a - k^c) - (k^b - k^c))\nu.$$

Here λ'' is a lifting of λ, and $\delta = \delta(b - c, a - c)$ (with the notation of Lemma 11.2). Lemma 11.2 plays a crucial role; it shows that the coefficient of ν is an integer. (We may suppose that c is sufficiently large, because the result is not affected by suspension—provided of course that the number of suspensions is divisible by 2 or 8.) It is necessary, of course, to check that the formula satisfies Axioms (6.1), (6.2) and (6.3). The need to satisfy Axiom (6.1) accounts for the formula given.

A suitable homomorphism γ' is defined by the following formulae.

$$\gamma'(\mu') = qe''\pi$$
$$\gamma'(\nu) = q\pi.$$

(Note that qe'' is an integer). It is necessary, of course, to check that γ' commutes with Ψ^k. The need to do this accounts for the formula given for $\gamma'(\mu')$.

We have next to compute the extension

$$\gamma'\alpha' \in \operatorname{Ext}^1(L, P).$$

This is characterised by the following formula.

$$\Psi^k \lambda'' = k^c \lambda'' + qe'e''\delta(k^a - k^c)\pi.$$

We conclude that in this case the Massey product in Ext^1 is given by

$$\{q, e_\Lambda(Sg), e_\Lambda(h)\} = qe'e''\delta$$

modulo the indeterminacy of the Massey product; that is, modulo 1 and $q/m(a - c)$. Theorem 11.5 thus follows from Theorem 5.3 (vi).

Theorem 11.4 may be deduced from Theorem 11.5 (as remarked above), or proved similarly. For the convenience of any reader who wishes to do the latter, we record formulae for

$$\alpha', \in \operatorname{Ext}^0(L,E), \qquad \gamma' \in \operatorname{Ext}^1(E,P)$$

with a notation similar to that used above.

$$\alpha'\lambda = q\mu' - qe'\nu$$
$$\Psi^k\mu'' = k^c\mu'' + e'(k^b - k^c)\nu' + e'e''((k^a - k^b) - \delta(k^a - k^c))\pi$$
$$\Psi^k\nu' = k^b\nu' + e''(k^a - k^b)\pi.$$

Proof of Corollaries 11.6, 11.7. These follow from Theorem 11.5 by applying the homomorphism from the rationals to the p-adic numbers and using (11.12), (11.13).

§12. EXAMPLES

In this section we will give various examples and illustrations of our general methods, and prove certain results whose proof was deferred in earlier sections. To begin with, our work is directed towards proving Theorem 1.7.

We can actually make Theorem 1.7 a little more complete. As in §1, let p be an odd prime, let $g : S^{2q-1} \to S^{2q-1}$ be a map of degree p^f, and let Y be the Moore space $S^{2q-1} \cup_g e^{2q}$. Thus $\tilde{K}_C(Y) = Z_{p^f}$.

THEOREM 12.1. *There is a map*

$$A : S^{2r}Y \to Y$$

(*for suitable* q) *such that the image of*

$$A^* : \tilde{K}_C(Y) \to \tilde{K}_C(S^{2r}Y)$$

is Z_{p^t} (*where* $1 \leqq t \leqq f$), *if and only if* r *is divisible by* $(p-1)p^{t-1}$.

It is clear that this includes Theorem 1.7 (take $t = f$). We will show how to deduce Theorem 12.1 from Theorem 1.7.

First, suppose that there is a map $A : S^{2r}Y \to Y$ such that the image of A^* is Z_{p^t}. Then A^* commutes with the operations Ψ^k, which are given in Y and $S^{2r}Y$ by the formulae

$$\Psi^k x = k^q x, \qquad \Psi^k x = k^{q+r} x.$$

Therefore we have $k^{q+r} \equiv k^q \bmod p^t$; so r is divisible by $(p-1)p^{t-1}$.

Secondly, suppose that r is divisible by $(p-1)p^{t-1}$ and Theorem 1.7 is true. Set $Y' = S^{2q-1} \cup_h e^{2q}$, where h is a map of degree p^t. Then by Theorem 1.7 there is a map

$$A' : S^{2r}Y' \to Y'$$

inducing an isomorphism of $\tilde{K}_C$. We have only to take A to be the composite

$$S^{2r}Y \xrightarrow{S^{2r}i} S^{2r}Y' \xrightarrow{A'} Y' \xrightarrow{j} Y$$

where i, j are obvious maps such that $j^* : \tilde{K}_C(Y) \to \tilde{K}_C(Y')$ is an epimorphism and $i^* : \tilde{K}_C(Y') \to \tilde{K}_C(Y')$ is a monomorphism.

This completes the deduction of Theorem 12.1 from Theorem 1.7. We proceed with lemmas needed for the proof of Theorem 1.7. First we consider the cofibering

$$S^{2n-1} \xrightarrow{f} S^{2n-1} \xrightarrow{i} S^{2n-1} \cup_f e^{2n},$$

where f is a map of degree m. If $\Lambda = R$, we assume that n is even; thus we shall certainly have $d_R i = 0$, $d_R(Si) = 0$.

PROPOSITION 12.2. *$e_\Lambda i$ is the class of the extension*

$$0 \leftarrow Z_m \leftarrow Z \xleftarrow{-m} Z \leftarrow 0,$$

in which all the abelian groups have operations Ψ^k defined by

$$\Psi^k x = k^n x.$$

Proof. If we continue the cofibre sequence, it becomes

$$S^{2n-1} \cup_f e^{2n} \xrightarrow{j} S^{2n} \xrightarrow{-Sf} S^{2n};$$

we have only to apply $\tilde{K}_\Lambda$.

For the next proposition, we suppose given a diagram of the following form,

$$\begin{array}{ccc} & S^{2n-1} \cup_m e^{2n} & \\ {}^{i}\nearrow & & \searrow^{G} \\ S^{2n-1} & \xrightarrow{\quad g \quad} & S^{2q} \end{array}$$

(Here we have written $S^{2n-1} \cup_m e^{2n}$ instead of $S^{2n-1} \cup_f e^{2n}$, where f is a map of degree m.) If $\Lambda = R$, we assume that n and q are even. Thus $\tilde{K}_\Lambda(S^{2q}) = Z$ and $\tilde{K}_\Lambda(S^{2n-1} \cup_m e^{2n}) = Z_m$; we can regard $d_\Lambda(G)$ as an integer mod m. We can also regard $e_\Lambda(g)$ as a rational mod 1; since $mg \sim 0$, $me_\Lambda(g)$ is an integer mod m.

PROPOSITION 12.3. *We have*

$$d_\Lambda(G) = -me_\Lambda(g) \qquad \textit{mod } m$$

or equivalently

$$e_\Lambda(g) = -\frac{1}{m} d_\Lambda(G) \qquad \textit{mod } 1.$$

Proof. This proposition is a special case of Proposition 3.2 (b), which states that

$$e(Gi) = e(i)\, d(G).$$

The element $e(i)$ has been given in Proposition 12.2; one has only to compute the product $e(i)\, d(G)$, which is an easy exercise in homological algebra.

LEMMA 12.4. *Let p be an odd prime, $m = p^f$, and $r = (p-1)p^f$. Then there is an element $\alpha \in \pi_{2r-1}^S$ satisfying the following conditions.*

(i) $m\alpha = 0$.

(ii) $e_C \alpha = -\dfrac{1}{m}$.

(iii) *The Toda bracket $\{m, \alpha, m\}$ is zero mod $m\pi_{2r}^S$.*

Proof. For $f = 1$ the result is easy; we have only to take α to be an element of Hopf invariant one mod p in π_{2p-3}^S. Then (i), (ii) are given by Corollary 8.4 and (iii) follows from the fact that the p-component of π_{2p-2}^S is zero.

For any f we can take α to be a suitable element in Im J, using Theorem 1.5 or 1.6 to obtain (i), (ii). Condition (iii) follows from the fact that $\{m, \alpha, m\}$ is an element of order 2 [18, p.26 (2.4) (i), p.33 (3.9) (i)].

Lemma 12.4 supplies the data for the following lemma, which we shall also use with $m = 2$.

LEMMA 12.5. *Suppose given $\alpha \in \pi^S_{2r-1}$ and $m \in Z$ such that*

(i) $m\alpha = 0$,

(ii) $e_C\alpha = -\dfrac{1}{m}$,

(iii) $\{m, \alpha, m\} = 0 \bmod m\pi^S_{2r}$.

Then for suitably large q there exist maps A which make the following diagram homotopy-commutative; and for any such A we have $d_C(A) = 1$.

$$\begin{array}{ccc} S^{2q+2r-1} \cup_m e^{2q+2r} & \xrightarrow{A} & S^{2q-1} \cup_m e^{2q} \\ \uparrow i & & \downarrow j \\ S^{2q+2r-1} & \xrightarrow{\alpha} & S^{2q} \end{array}$$

Proof. Conditions (i), (iii) enable one to construct the diagram. By Proposition 12.3 and condition (ii) we have $d_C(jA) = 1$. Hence $d_C(A) = 1$.

Theorem 1.7 follows immediately from Lemmas 12.4, 12.5. Since A induces an isomorphism of $\tilde{K}_C$, so does the composite

$$A . S^{2r}A . S^{4r}A . \dots S^{2r(s-1)}A : S^{2rs}Y \to Y;$$

Indeed we have

$$d_C(A . S^{2r}A . S^{4r}A . \dots S^{2r(s-1)}A) = 1.$$

Therefore this composite is essential for every s.

Under the assumptions of Lemma 12.5, we construct a map

$$\alpha_s : S^{2q+2rs-1} \to S^{2q}$$

by the following diagram.

$$\begin{array}{ccc} S^{2q+2rs-1} \cup_m e^{2q+2rs} & \xrightarrow{A \cdot S^{2r}A \dots S^{2r(s-1)}A} & S^{2q-1} \cup_m e^{2q} \\ \uparrow i & & \downarrow j \\ S^{2q+2rs-1} & \xrightarrow{\alpha_s} & S^{2q} \end{array}$$

We have $\alpha_1 = \alpha$. The map α_s has order dividing m, since it can be extended over $S^{2q+2rs-1} \cup_m e^{2q+2rs}$. The maps α_s satisfy the equation

(12.6) $$\alpha_{s+t} \in \{\alpha_s, m, \alpha_t\}.$$

The case in which m is an odd prime p and $r = p - 1$ has been studied by Toda [16, 17].

PROPOSITION 12.7. *Under the assumptions of Lemma 12.5, the maps α_s are all essential; indeed we have*

$$e_C(\alpha_s) = -\frac{1}{m} \quad \bmod 1.$$

This improves and generalises a result of Toda [17]. Presumably the present proof is related to Toda's proof; however, it is hoped that the presentation given here may be found more conceptual.

Proofs. (i) Apply Proposition 12.3 to the diagram which defines α_s. (ii) Alternatively, apply Theorem 11.1 to equation (12.6) and use induction.

EXAMPLE 12.8. *We note that in* [16, 17] *Toda's elements* α_s *depend on the choice of* α_1, *which Toda does not fix; similarly, there is a choice for his element* α'_p. *However, we may take the choices so that*

$$e_C(\alpha_s) = -\frac{1}{p}, \qquad e_C(\alpha'_{rp}) = -\frac{1}{p^2}.$$

Then the coefficient δ *in Corollary* 11.6 *explains the coefficients which arise in Toda's formulae for*

$$\{\alpha_t, \alpha_s, p\} \quad \text{and} \quad \{\alpha'_{rp}, \alpha_s, p\}$$

[16, Theorem 4.17 (ii)].

We will now show how the invariant e_C applies to maps $f\colon S^{2r-1}Y \to Y$, where $Y = S^{2q-1} \cup_p e^{2q}$ for some odd prime p. We must first calculate the appropriate Ext groups. As in §9, let M be the object in A whose underlying group is Z and whose operations are given by $\Psi^k x = k^q x$; and let M' be the quotient object M/pM, whose underlying group is Z_p. Similarly for N', with q replaced by $q + r$.

PROPOSITION 12.9. *We have*

$$\mathrm{Ext}^1_S(M', N') = \begin{cases} Z_p + Z_p & \text{if } r \equiv 0 \bmod (p-1) \\ 0 & \text{if } r \not\equiv 0 \bmod (p-1). \end{cases}$$

Proof. The exact sequence

$$0 \to M \xrightarrow{p} M \to M' \to 0$$

induces the following exact sequence.

$$\mathrm{Hom}(M, N') \xrightarrow{p} \mathrm{Hom}(M, N') \to \mathrm{Ext}^1(M', N') \to \mathrm{Ext}^1(M, N') \xrightarrow{p} \mathrm{Ext}^1(M, N')$$

Passing to direct limits, we obtain the following exact sequence.

$$\mathrm{Hom}_S(M, N') \xrightarrow{p} \mathrm{Hom}_S(M, N') \to \mathrm{Ext}^1_S(M', N') \to \mathrm{Ext}^1_S(M, N') \xrightarrow{p} \mathrm{Ext}^1_S(M, N')$$

The group $\mathrm{Ext}^1_S(M, N')$ has been computed in Proposition 9.1; it is Z_p if $r \equiv 0 \bmod (p-1)$, 0 otherwise. (In §9 we assumed $\Lambda = R$; but this is not necessary if v is odd). The group $\mathrm{Hom}_S(M, N')$ is easy to compute; it is Z_p if $r \equiv 0 \bmod (p-1)$, 0 otherwise. This completes the proof in the case $r \not\equiv 0 \bmod (p-1)$. If $r \equiv 0 \bmod (p-1)$, we consider the functor from A to the category of abelian groups defined by forgetting the operations Ψ^k; this gives the following diagram.

$$\begin{array}{ccccccccc}
0 & \to & \mathrm{Hom}_S(M, N') & \to & \mathrm{Ext}^1_S(M', N') & \to & \mathrm{Ext}^1_S(M, N') & \to & 0 \\
 & & \downarrow{\cong} & & \downarrow & & \downarrow & & \\
0 & \to & \mathrm{Hom}(Z, Z_p) & \xrightarrow{\cong} & \mathrm{Ext}^1(Z_p, Z_p) & \longrightarrow & 0 & &
\end{array}$$

This shows that the exact sequence for $\mathrm{Ext}^1_S(M', N')$ splits.

COROLLARY 12.10. (of the proof). *If $r \equiv 0 \bmod (p-1)$, $\mathrm{Ext}^1_S(M', N')$ has a base consisting of the following two elements.*

(i) *An extension with underlying group Z_{p^2} and operations $\Psi^k x = k^q x$.*

(ii) *An extension with underlying group $Z_p + Z_p$ and operations*

$$\Psi^k \xi = k^q \xi + \lambda(k^{q+r} - k^q)\eta$$

for $\lambda = 1/m(r)$.

In fact, the element (i) represents a generator coming from $\mathrm{Hom}_S(M, N)'$, while the element (ii) maps to zero in $\mathrm{Ext}^1(Z_p, Z_p)$ and to a generator in $\mathrm{Ext}^1(M, N')$.

As above, let $Y = S^{2q-1} \cup_p e^{2q}$ for some odd prime p.

THEOREM 12.11. *If $r \equiv 0 \bmod (p-1)$ then the stable track group $Map^S(S^{2r-1}Y, Y)$ contains a direct summand $Z_p + Z_p$.*

Proof. Let $\beta : S^{-1}Y \to Y$ be the map which appears in the cofibre sequence

$$S^{-1}Y \xrightarrow{\beta} Y \to S^{2q-1} \cup_{p^2} e^{2q} \to Y.$$

Then $e_C(\beta)$ is the extension mentioned in Corollary 12.10 (i). Let $A : S^{2r}Y \to Y$ be a map with $d_C(A) = 1$, as above. Then by Proposition 3.2 (c) we have

$$e_C(\beta . S^{-1}A) = d_C(A) . e_C(\beta),$$

which is again the extension mentioned in Corollary 12.10 (i).

To construct the other generator, let $\gamma : S^{2t+2r+1} \to S^{2t}$ be an element in $\mathrm{Im}\, J$ such that

$$m(r)e_C(\gamma) \equiv 1 \qquad \mathrm{mod}\ p.$$

Then we can form the map

$$1 \wedge \gamma : Y \wedge S^{2t+2r-1} \to Y \wedge S^{2t},$$

where $A \wedge B$ is the "smash product" $A \times B / A \vee B$. If $e_C(\gamma)$ is represented by an extension E, then $e_C(1 \wedge \gamma)$ is represented (up to sign) by the extension E/pE; this is the extension mentioned in Corollary 12.10 (ii).

Since all elements of $\mathrm{Map}^S(S^{2r-1}Y, Y)$ have order dividing p, this proves Theorem 12.11.

Remark 12.12. In proving Theorem 12.11, we could have used $A . S^{2r}\beta$ instead of $\beta . S^{-1}A$. By Proposition 3.2 (b) we would then have

$$e_C(A . S^{2r}\beta) = e_C(S^{2r}\beta) . d_C(A),$$

giving an extension with underlying group Z_{p^2} and operations $\Psi^k x = k^{q+r}x$. Thus the invariant e_C serves to distinguish between $A . S^{2r}\beta$ and $\beta . S^{-1}A$ if $r \not\equiv 0 \bmod p(p-1)$, but not if $r \equiv 0 \bmod p(p-1)$. It might be interesting to know if these two elements are equal for $r \equiv 0 \bmod p(p-1)$. The groups $\mathrm{Map}^S_*(Y, Y)$ would perhaps repay study, since phenomena which in spheres appear as Toda brackets appear in $\mathrm{Map}^S_*(Y, Y)$ as compositions.

One could presumably obtain the analogue of Theorem 12.11 for Moore spaces $S^{2q-1} \cup_{p^f} e^{2q}$, or $S^{2q-1} \cup_2 e^{2q}$. In the latter case one would need to use $\tilde{K}_R$.

We now pass on to study 2-primary phenomena. To begin with we prove the following result.

THEOREM 12.13. *For each $s \geq 0$ there is an element μ_{8s+1} of order 2 in π^S_{8s+1} such that $e_C(\mu_{8s+1}) = \frac{1}{2}$ mod 1.*

Proof. Let α be the element of order 2 in π^S_7. Since $e'_R : \pi^S_7 \to Z_{240}$ is an isomorphism, we have $e_C(\alpha) = \frac{1}{2}$ mod 1. Also, by a delicate result of Toda [18, p.31 Corollary 3.7] we have

$$\begin{aligned} \{2, \alpha, 2\} &= \alpha\eta \qquad \text{mod } 2 \\ &= 0 \qquad \text{mod } 2, \end{aligned}$$

since α is divisible by 2 and $2\eta = 0$. Thus we can apply Lemma 12.5 to construct a map A. Now we have the following diagram.

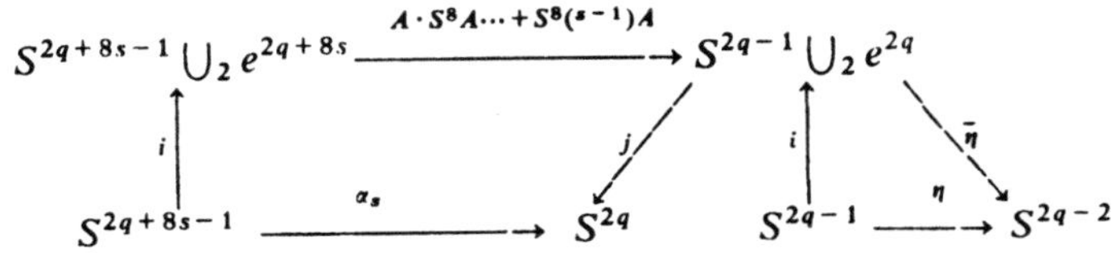

We define μ_{8s+1} to be the composite

$$\bar{\eta}.A.S^8A. \ldots S^{8(s-1)}A.i.$$

We have $\mu_1 = \eta$. The map μ_{8s+1} has order dividing 2, since it can be extended over $S^{2q+8s-1} \cup_2 e^{2q+8s}$. Since $e_C(\eta) = \frac{1}{2}$ mod 1, Proposition 12.3 shows that $d_C(\bar{\eta}) = 1$ mod 2. Hence

$$d_C(\bar{\eta}.A.S^8A. \ldots S^{8(s-1)}A) = 1 \qquad \text{mod } 2.$$

A second application of Proposition 12.3 now yields

$$e_C(\mu_{8s+1}) = \tfrac{1}{2} \qquad \text{mod } 1.$$

Alternatively, we can obtain the same result by applying Theorem 11.1 to the equation

$$\mu_{8s+1} \in \{\eta, 2, \alpha_s\},$$

in which $e_C(\alpha_s) = \frac{1}{2}$ mod 1 by Proposition 12.7.

Proof of Theorem 7.18. Suppose $r \equiv 1$ mod 8. Then by Theorem 12.13 the homomorphism

$$e_C : \pi^S_r \to Z_2$$

is an epimorphism. But we also have

$$d_R : \pi^S_r \to Z_2$$

and Ker $d_R \subset$ Ker e_C by Lemma 7.21. Therefore $d_R = e_C$. This proves Theorem 7.18.

We have just shown that

$$d_R \mu_{8s+1} \neq 0.$$

(It is possible to show this directly from the construction of μ_{8s+1}, but this is unnecessary.)

PROPOSITION 12.14. *If $r \equiv 1$ mod 8 and $s \equiv 1$ mod 8 then the composite $\mu_r\mu_s$ is non-zero; indeed*

$$d_R(\mu_r\mu_s) \neq 0.$$

This proposition generalises the behaviour of the composite $\eta\eta$. The proof is immediate.

Proof of Theorem 7.2. Let us define μ_{8s+2} to be one of the composites considered in Proposition 12.14, for example, $\eta\mu_{8s+1}$. Then we have shown that for $r \equiv 1, 2 \bmod 8$ and $r > 0$ we have $d_R\mu_r \neq 0$. Thus d_R is an epimorphism; and since μ_r is of order 2, π_r^S splits as a direct sum $Z_2 + \operatorname{Ker} d_R$, where the subgroup Z_2 is generated by μ_r.

EXAMPLE 12.15. *Suppose that $\theta \in \pi_{8t-1}^S$ is an element such that $m(4t)e_R(\theta)$ is odd. Then for $r \equiv 1, 2 \bmod 8$ the composite $\theta\mu_r$ is essential; indeed*

$$e_R(\theta\mu_r) \neq 0.$$

Proof. By Theorem 3.2 (c) we have

$$e_R(\theta\mu_r) = d_R(\mu_r)e_R(\theta).$$

Let us use the notation of §9; then $e_R(\theta)$ is a generator of the 2-component of $\operatorname{Ext}_S^1(M, N)$ and the homomorphism $d_R(\mu_r)$ may be identified with the quotient map $N \to N'$. So according to the discussion in §9, $d_R(\mu_r) . e_R(\theta)$ represents a generator of $\operatorname{Ext}_S^1(M, N')$.

This example provides a second proof for Theorem 9.5. In fact, let γ be a generator for $\pi_{8u-1}(SO)$ $(u > 0)$. Then the generators for $\pi_{8u}(SO)$, $\pi_{8u+1}(SO)$ can be written as composites $\gamma\eta$, $\gamma\eta\eta$; and we have

$$J(\gamma\eta) = J(\gamma)\eta$$
$$J(\gamma\eta\eta) = J(\gamma)\eta\eta.$$

Thus Theorem 9.5 follows from Example 12.15.

EXAMPLE 12.16. *If $r \equiv 1 \bmod 8$ then $\{2, \mu_r, 2\}$ is non-zero; indeed $d_R\{2, \mu_r, 2\} \neq 0$.*

This example generalises the behaviour of $\{2, \eta, 2\}$. The reader will find that it is an easy application of Theorem 5.3 (i). Alternatively, of course, one can quote [18, p.31 Corollary 3.7] to show that $\{2, \mu_r, 2\} = \mu_r\eta \bmod 2$ and use Proposition 12.14.

PROPOSITION 12.17. If $r \equiv 2 \bmod 8$ *and* $s \equiv 1 \bmod 8$ *then the composition $\mu_r\mu_s$ is non-zero; indeed*

$$e'_R(\mu_r\mu_s) = \tfrac{1}{2} \quad \bmod 1.$$

This proposition generalises the behaviour of the composite $\eta\eta\eta$.

Proof. Let

$$f : S^{2n-1} \to S^{2t}, \qquad g : S^{2t} \to S^{2q}$$

be maps representing μ_s, μ_r, where $2q \equiv 0 \bmod 8$, $2t \equiv 2 \bmod 8$, $2n - 1 \equiv 3 \bmod 8$. We have to consider the invariant $e_R(f)$. We have the following diagram.

$$\begin{array}{ccccccc}
Z_2 = \tilde{K}_R(S^{2t}) & \leftarrow & \tilde{K}_R(S^{2t} \cup_f e^{2n}) & \leftarrow & \tilde{K}_R(S^{2n}) = Z \\
\uparrow{\scriptstyle epi} & & \uparrow{\scriptstyle r} & & \uparrow{\scriptstyle iso} \\
Z = \tilde{K}_C(S^{2t}) & \leftarrow & \tilde{K}_C(S^{2t} \cup_f e^{2n}) & \leftarrow & \tilde{K}_C(S^{2n}) = Z
\end{array}$$

Let ξ, η be generators in $\tilde{K}_C(S^{2t} \cup_f e^{2n})$. Then since $e_C(f) = \frac{1}{2} \bmod 1$ we have (for a suitable choice of ξ)

$$\begin{aligned}\Psi^{-1}\xi &= (-1)^t\xi + \tfrac{1}{2}((-1)^n - (-1)^t)\eta \\ &= -\xi + \eta.\end{aligned}$$

Now in $\tilde{K}_R(S^{2t} \cup_f e^{2n})$ we have $r\Psi^{-1} = r$; thus we have $2r\xi = r\eta$. Thus $e_R(f)$ is the non-trivial extension

$$0 \leftarrow Z_2 \leftarrow Z \overset{2}{\leftarrow} Z \leftarrow 0$$

in which all the groups are given operations Ψ^k by the formula $\Psi^k x = k^n x$.

We must now compute the product $e_R(f)\, d_R(g)$, where

$$d_R(g) : \tilde{K}_R(S^{2q}) \to \tilde{K}_R(S^{2t})$$

is the epimorphism $Z \to Z_2$. We easily find that $e_R(f)\, d_R(g)$ is the extension corresponding to the rational $\frac{1}{2}$ mod 1.

PROPOSITION 12.18. *If $r \equiv 1$ mod 8 and $s \equiv 1$ mod 8 then any representative of the Toda bracket $\{\mu_r, 2, \mu_s\}$ is an element of order 4; indeed $e'_R\{\mu_r, 2, \mu_s\} = \frac{1}{4}$ mod $\frac{1}{2}$.*

This proposition generalises the behaviour of $\{\eta, 2, \eta\}$.

Proof. We have just shown that the indeterminacy of $\{\mu_r, 2, \mu_s\}$ consists at least of the integers $\frac{1}{2}$ mod 1. By Theorem 11.1 we have

$$\begin{aligned} e_C\{\mu_r, 2, \mu_s\} &= -\tfrac{1}{2}.2.\tfrac{1}{2} \qquad \text{mod } 1 \\ &= \tfrac{1}{2} \qquad \text{mod } 1. \end{aligned}$$

By Proposition 7.14 this is equivalent to

$$e'_R\{\mu_r, 2, \mu_s\} = \tfrac{1}{4} \qquad \text{mod } \tfrac{1}{2}.$$

On the other hand, we have

$$2\{\mu_r, 2, \mu_s\} = \{2, \mu_r, 2\}\mu_s \qquad \text{mod } 0.$$

This actually gives $\eta\mu_r\,\mu_s$; but at all events it is an element of order 2 at most, so $\{\mu_r, 2, \mu_s\}$ has order dividing 4. This completes the proof.

EXAMPLE 12.19. *Suppose given an even integer m and an element $\theta \in \pi_r^S$ (where $r \equiv -1$ mod 8) such that $m\theta = 0$ and $me_R(\theta)$ is odd. Then for $s \equiv 1$ or 2 mod 8 we have $\{\theta, m, \mu_s\} \neq 0$; indeed*

$$d_R\{\theta, m, \mu_s\} \neq 0.$$

Proof. If $s \equiv 1$ mod 8 we can make an easy calculation using Theorem 11.1:

$$\begin{aligned} e_C\{\theta, m, \mu_s\} &= -me_C(\theta)e_C(\mu_s) \\ &= \tfrac{1}{2} \qquad \text{mod } 1. \end{aligned}$$

If $s \equiv 2$ mod 8 then $d_R\{\theta, m, \mu_s\}$ depends only on $e_R(\theta)$, m and $d_R(\mu_s)$, by Theorem 5.3 (iii); so we may substitue $\mu_{s-1}\eta$ for μ_s, and then

$$\{\theta, m, \mu_s\} = \{\theta, m, \mu_{s-1}\}\eta.$$

So the result follows from the case $s \equiv 1$ mod 8.

Our final example is of interest in connection with certain rather technical manipulations with Toda brackets; this is perhaps not the place to explain the project from which these manipulations come, although the reader is assured that they are not without purpose.

We suppose given an element θ in π_r^S for $r = 2^{f-1} - 1$, such that $2^f\theta = 0$, $e_R(\theta) = 2^{-f}$ and $\theta\theta = 0$. For example, there is such an element if $f = 5$. We assume $f \geqq 4$, so that $r \equiv -1 \bmod 8$.

By [18, p.30, Theorem 3.6] there are elements ρ, φ in the 2-component of π_{2r-1} (where $2r - 1 = 2^f - 1$) such that

$$\rho \in \{\theta, \theta, 2^f\}$$

and

$$2\rho + 2^f\varphi \in \{\theta, 2^f, \theta\}.$$

EXAMPLE 12.20. *In the last equation the element* φ *cannot be zero; indeed we have*

$$e_R(\varphi) = 2^{-f-1} \qquad \textit{mod } 2^{-f}.$$

Proof. By Theorem 11.1 we have

$$e_R\{\theta, 2^f, \theta\} = -2^{-f}.$$

By Corollary 11.7 we have

$$\begin{aligned} e_R\{\theta, \theta, 2^f\} &= -2^f.\tfrac{1}{2}(1 + 2^{f-1})2^{-f}2^{-f} && \text{mod } \tfrac{1}{2} \\ &= -(2^{-f-1} + \tfrac{1}{4}) && \text{mod } \tfrac{1}{2}. \end{aligned}$$

Thus

$$e_R(2\rho) = -2^{-f} + \tfrac{1}{2} \qquad \text{mod } 1.$$

Hence

$$e_R(\varphi) = 2^{-f-1} \qquad \text{mod } 2^{-f}.$$

This completes the proof.

REFERENCES

1. J. F. ADAMS: On Chern characters and the structure of the unitary group, *Proc. Camb. Phil. Soc.* **57** (1961), 189–199.
2. J. F. ADAMS: Vector field on spheres, *Ann. Math.* **75** (1962), 603–632.
3. J. F. ADAMS: On the groups $J(X)$—I, *Topology* **2** (1963), 181–195.
4. J. F. ADAMS: On the groups $J(X)$—II, *Topology* **3** (1965), 137–171.
5. J. F. ADAMS: On the groups J(X)—III, *Topology* **3** (1965), 193–222.
6. J. F. ADAMS: Lectures on $K^*(X)$, (mimeographed notes, Manchester, 1962).
7. J. F. ADAMS: Cohomology operations (mimeographed notes, Seattle, 1963).
8. J. F. ADAMS and M. F. ATIYAH: *K*-theory and the Hopf invariant, *Quart. J. Math.*, to appear.
9. J. F. ADAMS and G. WALKER: Complex Stiefel Manifolds, *Proc. Camb. Phil. Soc.* **61** (1965), 81–103.
10. M. F. ATIYAH and F. HIRZEBRUCH: Riemann–Roch theorems for differentiable manifolds, *Bull. Amer. Math. Soc.* **65** (1959), 276–281.
11. M. F. ATIYAH and F. HIRZEBRUCH: Vector bundles and homogeneous spaces, *Proc. of Symposia in Pure Mathematics* 3, Differential Geometry, Amer. Math. Soc., 1961, pp. 7–38.
12. A. DOLD: Half exact functors and cohomology, (mimeographed notes, Seattle, 1963).
13. E. DYER: Chern characters of certain complexes, *Math. Z.* **80** (1963), 363–373.
14. S. MACLANE: *Homology*, Springer, Berlin, 1963.
15. D. PUPPE: Homotopiemengen und ihre induzierten Abbildungen. I *Math. Z.* **69** (1958), 299–344.
16. H. TODA: *p*-primary components of homotopy groups. IV. Compositions and toric constructions, *Mem. Coll. Sci. Kyoto*, Ser. **A 32** (1959), 297–332.
17. H. TODA: On unstable homotopy of spheres and classical groups, *Proc. Nat. Acad. Sci., Wash.* **46** (1960), 1102–1105.
18. H. TODA: Composition methods in homotopy groups of spheres, *Ann. Math. Stud. No.* **49**, Princeton 1962.
19. H. TODA: A survey of homotopy theory, *Sûgaku* **15** (1963/4), 141–155.

Manchester University

Topology Vol. 7, p. 331. Pergamon Press, 1968.

ON THE GROUPS $J(X)$—IV. CORRECTION

J. F. ADAMS

THE CORRECTION is to the first line of "*Examples on Theorem* 11.5", on p. 55 of my paper [1]. It appears that the numerical calculation is wrong, and should read

$$\{j_3, 2j_3, 12\} = 8j_7 \bmod 12j_7.$$

REFERENCE

1. J. F. ADAMS: On the groups $J(X)$—IV, *Topology* 5 (1966), 21–71.

University of Manchester

K-THEORY AND THE HOPF INVARIANT

By J. F. ADAMS (*Manchester*) *and* M. F. ATIYAH (*Oxford*)

[Received 29 December 1964]

Introduction

THE non-existence of elements of Hopf invariant one in $\pi_{2n-1}(S^n)$, for $n \neq 1, 2, 4$, or 8, was established in (**1**) by the use of secondary cohomology operations. The main purpose of this paper is to show how the use of primary operations in K-theory provides an extremely simple alternative proof of this result. In fact K-theory proofs have already been given in (**8**) and (**4**) but neither of these proofs is elementary: (**8**) uses results on complex cobordism, while (**4**) uses the connexion between the Chern character and the Steenrod squares established in (**3**) [see however (**6**) for a more elementary treatment of the results of (**3**)]. The simplicity and novelty of our present approach is that, unlike all previous attacks on the Hopf invariant problem, we consider not the stable but the *unstable* version of the problem: that is to say we shall prove directly

THEOREM A. *Let X be a 2-cell complex formed by attaching a $2n$-cell to an n-sphere, where $n \neq 1, 2, 4$, or 8. Then the cup-square*

$$H^n(X; Z_2) \to H^{2n}(X; Z_2)$$

is zero.

For other versions of this theorem and for the historical background of the problem we refer to (**1**).

Like the proofs of (**4**) and (**8**) our proof also extends to show the non-existence of elements of Hopf invariant one mod p, a result first proved by the use of secondary operations in (**9**), (**10**). In fact our methods yield a good deal more. In particular we shall establish the following new result suggested to us by James:

THEOREM B. *Let p be an odd prime and m a positive integer not dividing $p-1$. Let X be a finite complex such that*

(i) *$H^*(X; Z)$ has no p-torsion,*

(ii) *$H^{2k}(X; Q) = 0$ if $k \not\equiv 0 \bmod m$.*

Then the cup-p-th-power

$$H^{2m}(X; Z_p) \to H^{2mp}(X; Z_p)$$

is zero.

Quart. J. Math. Oxford (2), 17 (1966), 31–38.

Remark. Taking $m = 2$ and X to be the quaternionic projective space of dimension p we see that the condition $p \not\equiv 1 \bmod m$ cannot be dispensed with.

We begin in § 1 by presenting our proof of Theorem A. Our aim being to emphasize the simplicity of the proof, we refrain from any generalizations at this stage. The remainder of the paper is then devoted to extending the methods of § 1 to a more general context. In § 2 we make a short algebraic study of the operators ψ^k. Then in § 3 we prove Theorem B and finally in § 4 we show how Theorem B implies the non-existence of elements of Hopf invariant one mod p.

1. Non-existence of elements of Hopf invariant one

We assume here the basic results of K-theory for which we refer to (7). We shall also need the operations ψ^k introduced in (**2**). These are defined in terms of the exterior power operations λ^k, by the formula

$$\psi^k(x) = Q_k(\lambda^1(x), \ldots, \lambda^k(x)) \quad (x \in K(X)),$$

where Q_k is the polynomial which expresses the kth-power sum in terms of elementary symmetric functions. Their basic properties are:

$$\psi^k\colon K(X) \to K(X) \text{ is a ring homomorphism,} \tag{1.1}$$

$$\psi^k \text{ and } \psi^l \text{ commute,} \tag{1.2}$$

$$\text{if } p \text{ is a prime, } \psi^p(x) \equiv x^p \bmod p, \tag{1.3}$$

$$\text{if } u \in \tilde{K}(S^{2n}), \text{ then } \psi^k(u) = k^n u. \tag{1.4}$$

The proofs of (1.1), (1.2), and (1.4) are all elementary and can be found in (**2**) [§ 5], while (1.3) is an immediate consequence of the congruence

$$(\sum \alpha_i)^p \equiv \sum \alpha_i^p \mod p.$$

If we apply (1.4) to the wedge of spheres X^{2n}/X^{2n-1} (where X^q denotes the q-skeleton of X), we deduce at once that,

$$\text{if } \quad u \in K_{2n}(X), \quad \text{then} \quad \psi^k(u) \equiv k^n u \mod K_{2n+1}(X). \tag{1.5}$$

Here $K_q(X)$ denotes the qth filtration group of $K(X)$, i.e. it is the kernel of $K(X) \to K(X^{q-1})$.

We are now ready to give the proof of Theorem A. The result is trivial for n odd; in fact $2x^2 = 0$ for $x \in H^n(X; Z)$, while $H^{2n}(X; Z)$ is free (since $n \neq 1$). Thus we may suppose that $n = 2m$. Then $H^*(X; Z)$ is the associated graded ring of $K(X)$ (7) [§ 2], and so $\tilde{K}(X)$ is free on two generators $\quad a \in K_{2m}(X) \quad$ and $\quad b \in K_{4m}(X)$.

To prove the theorem we have to show that, if $m \neq 1$, 2, or 4, then

$$a^2 \equiv 0 \mod 2,$$

or equivalently by (1.3) that

$$\psi^2(a) \equiv 0 \mod 2.$$

Let us compute $\psi^2(a)$ and $\psi^3(a)$. By (1.5) these must be of the form

$$\psi^2(a) = 2^m a + \mu b, \qquad \psi^3(a) = 3^m a + \nu b,$$

for some integers μ, ν. Since, by (1.2), $\psi^2\psi^3 = \psi^3\psi^2$, we deduce, using (1.1) and (1.5), that

$$3^m(2^m a + \mu b) + \nu 2^{2m} b = 2^m(3^m a + \nu b) + \mu 3^{2m} b,$$

and so

$$3^m(3^m - 1)\mu = 2^m(2^m - 1)\nu. \tag{1.6}$$

But, by elementary number theory [cf. (**2**) Lemma 8.1], we have

if $m \neq 1$, 2, or 4, then 2^m does not divide $3^m - 1$. (1.7)

Thus (1.6) implies that μ is even and hence

$$\psi^2(a) \equiv 0 \mod 2$$

as required. This completes the proof of Theorem A.

2. Eigenspaces of ψ^k

We recall that the Chern character induces a ring homomorphism [(**7**) 1.10]

$$\text{ch}\colon K^*(X) \otimes Q \to H^*(X; Q)$$

and that, if $x \in K(X)$ with

$$\text{ch}\, x = \sum a_{2m}, \qquad a_{2m} \in H^{2m}(X; Q)$$

then [(**2**) Theorem 5.1 (vi)]

$$\text{ch}\, \psi^k(x) = \sum k^m a_{2m}. \tag{2.1}$$

Thus, if we use the Chern character to identify $K(X) \otimes Q$ with $\sum H^{2m}(X; Q)$, the subspace $H^{2m}(X; Q)$ becomes (for $k > 1$) just the eigenspace V_m of ψ^k corresponding to the eigenvalue k^m, which shows in particular that this eigenspace is independent of k. The dimension of V_m is just the $2m$th Betti number $B_{2m}(X)$. The following lemma is then a consequence of (2.1):

LEMMA 2.2. *Let X be a finite connected complex and assume that the Betti numbers $B_{2m}(X)$ are zero for $m \neq 0, m_1, m_2, \ldots, m_r$. Then, for any sequence of integers $k_1, \ldots, k_r$*

$$\prod_{i=1}^{r} (\psi^{k_i} - (k_i)^{m_i}) = 0 \quad \textit{in} \quad \tilde{K}(X) \otimes Q.$$

We now observe that this result is stated purely in terms of K-theory and makes no reference to cohomology or the Chern character, provided that we define the Betti numbers (as we may) by

$$B_{2m}(X) = \dim_Q\{K_{2m}(X)/K_{2m+1}(X) \otimes Q\}.$$

Moreover we can prove (2.2) purely in K-theory simply by using (1.5) and induction on the filtration. We propose therefore to take (2.2) as our starting point and to use only K-theory. Rational cohomology was mentioned only for motivation.

Let k, l denote integers greater than 1. From (2.2), taking $k_i = k$ for all i, we see that ψ^k is semi-simple and has eigenvalues k^{m_i} on $\tilde{K}(X) \otimes Q$. Let $V_{i,k}$ denote the eigenspace corresponding to k^{m_i}. Then

$$V_{i,k} = \operatorname{Im} \prod_{j \neq i} (\psi^k - k^{m_j}).$$

Applying (2.2) with $k_j = k$ for $j \neq i$ and $k_i = l$ we see that

$$V_{i,k} \subset V_{i,l}.$$

This being true for all k, $l > 1$ it follows that $V_{i,k} = V_{i,l}$ and so $V_{i,k}$ is independent of k (as was shown earlier by use of cohomology). We denote it therefore by V_i. Thus we have a decomposition

$$\tilde{K}(X) \otimes Q = \bigoplus_{i=1}^{r} V_i \tag{2.3}$$

invariant under all the ψ^k. Let π_i denote the projection operator corresponding to V_i. Then for any sequence of $r-1$ integers $k_1,\ldots, k_{i-1}, k_{i+1}, \ldots, k_r$ (all $k_i > 1$) we have the following expression for π_i:

$$\pi_i = \prod_{j \neq i} \left(\frac{\psi^{k_j} - k_j^{m_j}}{k_j^{m_i} - k_j^{m_j}} \right). \tag{2.4}$$

In fact π_i annihilates V_j for $j \neq 1$ and is the identity on V_i.

3. The pth power mod p

So far we have only considered the vector space $\tilde{K}(X) \otimes Q$. Now we turn our attention to the image of $\tilde{K}(X)$ in $\tilde{K}(X) \otimes Q$. An element of this image will be called an *integral element* of $\tilde{K}(X) \otimes Q$. If $x \in \tilde{K}(X) \otimes Q$, then there is a least positive integer d such that dx is integral. We call d the *denominator* of x. For convenience we shall now make the following definition. Given a sequence $m_1,\ldots, m_r$ of distinct positive integers and an integer i with $1 \leqslant i \leqslant r$ we define $d_i(m_1,\ldots, m_r)$ to be the highest common factor of all the products

$$\prod_{j \neq i} (k_j^{m_i} - k_j^{m_j}),$$

where $\{k_j\}$ $(j \neq i)$ runs over all sequences of $r-1$ integers > 1. With this notation (2.4) gives the following result.

Proposition (3.1). *Let X be as in (2.2) and let $x \in \tilde{K}(X) \otimes Q$ be integral. Then the denominator of $\pi_i x$ divides $d_i(m_1,\ldots, m_r)$.*

Now let p be a prime and let us compute $\psi^p(x)$ using the decomposition (2.3). Thus

$$x = \sum \pi_i x,$$

$$\psi^p(x) = \sum \psi^p(\pi_i x) = \sum p^{m_i} \pi_i x. \tag{3.2}$$

Suppose now that x is integral and that, for each i, p^{m_i} does not divide $d_i(m_1,\ldots, m_r)$. Then (3.1) and (3.2) show that in $\tilde{K}(X) \otimes Q$ we have

$$\psi^p(x) = \frac{py}{q},$$

where y is integral and q is prime to p. Transferring this result from $\tilde{K}(X) \otimes Q$ to $\tilde{K}(X)$ and using (1.3) we obtain

THEOREM C. *Let X be as in* (2.2), *let p be a prime and suppose for each i that p^{m_i} does not divide $d_i(m_1,\ldots, m_r)$. Then for any $x \in \tilde{K}(X)$ we have*

$$x^p \in p\tilde{K}(X) + \mathrm{Tors}\, \tilde{K}(X)$$

(*where* Tors *denotes the torsion subgroup*). *In particular, if $K(X)$ has no p-torsion, then $x^p \equiv 0 \bmod p$.*

This theorem, stated entirely in K-theory, is the most general result concerning the triviality of the pth power mod p given by our method. From it we shall now deduce a corollary about the pth power map in $H^*(X; Z_p)$:

COROLLARY. *Let X, p be as in Theorem C and assume further that $H^*(X; Z)$ has no p-torsion. Then, for any $m > 0$, the p-th power map*

$$H^{2m}(X; Z_p) \to H^{2mp}(X; Z_p)$$

is zero.

Proof. Let A_p denote Z localized at p, i.e. the ring of fractions m/n with n prime to p. Since the differentials of the spectral sequence $H^*(X; Z) \Rightarrow K^*(X)$ are all torsion operators [(7) 2.4] and since X has no p-torsion, it follows that the localized spectral sequence (i.e. the spectral sequence obtained by applying $\otimes A_p$) is trivial and that $K^*(X)$ has no p-torsion. Thus we have

$$H^{2m}(X; A_p) \cong H^{2m}(X; Z) \otimes A_p \cong K_{2m}(X)/K_{2m+1}(X) \otimes A_p. \tag{3.3}$$

Also, since X has no p-torsion,

$$H^{2m}(X; A_p) \to H^{2m}(X; Z_p) \tag{3.4}$$

is surjective. Hence, if $a \in H^{2m}(X; Z_p)$, we can find $x \in K_{2m}(X)$, $\alpha \in A_p$ so that $x \otimes \alpha$ represents a via (3.3) and (3.4). Then $x^p \otimes \alpha^p$ represents a^p. But, by Theorem C, $x^p \equiv 0 \pmod p$. Hence $a^p = 0$ as required.

In order to apply Theorem C and its corollary in any given case it is necessary to verify the arithmetical hypothesis, namely that p^{m_i} does not divide d_i. In general this can be quite complicated. In the special case required for Theorem B however, where we have $m_i = mi$, the arithmetic can be dealt with as we shall now show. The following lemma for odd primes is the appropriate generalization of the lemma for $p = 2$ already used in § 1:

LEMMA (3.5). *Let p be an odd prime, m a positive integer not dividing $p-1$ and $1 \leqslant i \leqslant p$. Then, for a suitable integer k, the p-primary factor of $\prod (k^{mi}-k^{mj})$ is at most p^{m-1}, where the product is taken over all integers j with $1 < j < p$ and $j \neq i$.*

Proof. We begin by recalling that the multiplicative group of units of the ring Z_{p^f} (G_f say) is cyclic of order $p^{f-1}(p-1)$. Let k be an integer whose residue class mod p^2 generates G_2. Then it follows that the residue class of k mod p^f generates G_f. Thus we have

$$k^n \equiv 1 \bmod p^f \iff n \equiv 0 \bmod p^{f-1}(p-1),$$

and this holds for all f. With this choice of k we shall compute the p-primary factor p^e of $\prod (k^{mi}-k^{mj})$. Suppose that $(m, p-1) = h$, so that

$$m = ah, \qquad p-1 = bh, \qquad a > 1,$$

and let the p-primary factor of m (or equivalently of a) be p^f. Then we find that

$$e = (f+1)\left(\left[\frac{i-1}{b}\right]+\left[\frac{p-i}{b}\right]\right),$$

where, as usual, $[x]$ denotes the integral part of x. Thus

$$e \leqslant \frac{(f+1)(p-1)}{b} = h(f+1) \leqslant hp^f \leqslant m.$$

Moreover, equality cannot hold in all places since $f+1 = p^f$ implies $f = 0$ and therefore $hp^f = h = m/a < m$. Hence $e < m$ as required.

We are now ready to prove Theorem B. First we observe that replacing X by X^{2pm+1} affects neither the hypotheses nor the conclusion of the theorem. Thus we may suppose the Betti numbers $B_{2k}(X)$ are zero except for

$$k = m, \quad 2m, \quad \ldots, \quad pm.$$

Moreover we may assume that X is connected. But (3.5) implies that p^m does not divide $d_i(m, 2m, \ldots, pm)$ for $1 \leqslant i \leqslant p$ and so the hypotheses of Theorem C are certainly fulfilled. Theorem B then follows from the corollary to Theorem C.

Remark. Of course the corollary to Theorem C applied with $p = 2$ leads to a suitable strengthening of Theorem A along the lines of Theorem B. We leave this to the reader.

4. The mod p Hopf invariant

In this section p denotes an odd prime. Let $f\colon S^{2mp} \to S^{2m+1}$ be any map and let

$$X_f = S^{2m+1} \cup_f e^{2mp+1}$$

be the associated 2-cell complex. The mod p Hopf invariant is defined to be the Steenrod operation

$$P^m\colon H^{2m+1}(X_f; Z_p) \to H^{2mp+1}(X_f; Z_p). \tag{4.1}$$

We propose to prove

THEOREM D. *The* mod p *Hopf invariant is zero for* $m > 1$.

Proof. We begin by recalling that the loop space ΩS^{2m+1} has the following cohomology:

$$\begin{aligned} H^r(\Omega S^{2m+1}; Z) &\cong Z \quad \text{if } r \equiv 0 \bmod 2m, \\ &= 0 \quad \text{otherwise.} \end{aligned}$$

This is an elementary consequence of the Serre spectral sequence. Suppose now that $f\colon S^{2mp} \to S^{2m+1}$ is any map and let

$$g\colon S^{2mp-1} \to \Omega S^{2m+1}$$

be its 'adjoint'. We form the space

$$Y_g = \Omega S^{2m+1} \cup_g e^{2mp}.$$

If m does not divide $p-1$, this satisfies all the conditions of Theorem B except that it is not a finite complex. However, we can approximate Y_g by a finite complex up to any dimension and so the conclusion of Theorem B holds for Y_g: that is the pth power map

$$H^{2m}(Y_g; Z_p) \to H^{2mp}(Y_g; Z_p)$$

is zero. Hence suspending once

$$P^m\colon H^{2m+1}(SY_g; Z_p) \to H^{2mp+1}(SY_g; Z_p) \tag{4.2}$$

is also zero. But, by definition of g, f is homotopic to the composition

$$S^{2mp} \xrightarrow{Sg} S\Omega S^{2m+1} \xrightarrow{e} S^{2m+1},$$

where e is the 'evaluation' map

$$e(t, w) = w(t),$$

and so e extends to a map

$$e'\colon SY_g \to X_f.$$

Since e induces an isomorphism on H^{2m+1}, it follows that e' induces a monomorphism in cohomology, and so the vanishing of (4.2) implies the vanishing of (4.1). This proves the theorem for values of m not dividing $p-1$. However, these exceptional values are easily dealt with by a method due to Adem. In fact, for $1 < m < p-1$, we have

$$P^m = \frac{1}{m} P^1 P^{m-1}$$

[(5) § 24], and so P^m is zero on a 2-cell complex.

REFERENCES

1. J. F. Adams, 'On the non-existence of elements of Hopf invariant one', *Ann. of Math.* 72 (1960) 20–104.
2. —— 'Vector fields on spheres', ibid. 75 (1962) 603–32.
3. —— 'On Chern characters and the structure of the unitary group', *Proc. Cambridge Phil. Soc.* 57 (1961) 189–99.
4. —— 'On the groups $J(X)$ IV', to appear in *Topology*.
5. J. Adem, 'The relations on Steenrod powers of cohomology classes', *Algebraic Geometry and Topology*, Princeton (1957).
6. M. F. Atiyah, 'Power operations in K-theory', to appear.
7. —— and F. Hirzebruch, 'Vector bundles and homogeneous spaces', *Proc. of Symposia on Pure Mathematics* 3, *Differential Geometry*, American Math. Soc. 1961, 7–38.
8. E. Dyer, 'Chern characters of certain complexes', *Math. Z.* 80 (1963) 363–73.
9. A. Liulevicius, 'The factorisation of cyclic reduced powers by secondary cohomology operations', *Memoirs of the American Math. Soc.* No. 42 (1962).
10. N. Shimada and T. Yamanoshita, 'On triviality of the mod p Hopf invariant', *Japanese J. of Math.* 31 (1961) 1–25.

Geometric Dimension of Bundles over RP^n

By

J. F. ADAMS

First I must recall the definition of the groups $K(X)$. Let X be, for example, a finite CW-complex. We can consider vector-bundles ξ over X; more precisely, we consider bundles of vector-spaces over Λ, where Λ is either R, the field of real numbers, or C, the field of complex numbers, or else H, the skew field of quaternions. We divide the vector-bundles ξ into isomorphism classes $[\xi]$. We define the group $K_\Lambda(X)$ by generators and relations; more precisely, we define $K_\Lambda(X)=F/R$, where F is the free abelian group generated by the classes $[\xi]$, and R is the subgroup generated by all elements of the form

$$[\xi]+[\eta]-[\xi\oplus\eta].$$

(Here $\xi\oplus\eta$ is the Whitney sum of ξ and η). If we wish to classify vector-bundles over X, it is natural to consider first the stable classification — that is, to compute the group $K_\Lambda(X)$.

If X is a point P, then a vector-bundle over X is just a vector-space, and is specified up to isomorphism by its dimension. This gives $K_\Lambda(P)\cong Z$.

Next suppose that X is a connected space with a base-point. Then $K_\Lambda(X)$ contains $K_\Lambda(P)$ as a direct summand; we write $\tilde{K}_\Lambda(X)$ for the complementary direct summand. For our present purposes it is generally convenient to regard $\tilde{K}_\Lambda(X)$ as a quotient of $K_\Lambda(X)$ (namely the quotient $K_\Lambda(X)/Z$, where the subgroup Z is generated by the trivial line bundle.)

Let x be an element in $\tilde{K}_\Lambda(X)$. We say that x has geometric dimension $\leq d$ if x is the class in $\tilde{K}_\Lambda(X)$ of some bundle ξ with fibres of dimension d.

The concept of geometric dimension is of obvious use in the theory

4

of immersions. Suppose given a map $f: M \to N$, where M and N are compact smooth manifolds of dimensions m, $m+d$, with tangent bundles τ_M, τ_N. If f is homotopic to an immersion, let ν be the normal bundle for this immersion; then in $K_R(M)$ or $\tilde{K}_R(M)$ we have the equation

$$[\tau_M] \oplus [\nu] = [f^* \tau_N],$$

and so $f^*[\tau_N] - [\tau_M]$ has geometric dimension $\leq d$ (over R).

One of the most classical examples on which to test our methods in this area is the real projective space RP^n. The structure of $\tilde{K}_\Lambda(RP^n)$ is known; it is a cyclic group of order 2^f, where f depends on Λ and n. If $\Lambda = R$, the generator of $\tilde{K}_R(RP^n)$ is the class of the canonical real line bundle over RP^n. From now on the symbol ξ or ξ_R will be reserved for this line bundle. We also write ξ_C for the complexification of ξ_R; this is the generator for $\tilde{K}_C(RP^n)$. Similarly for ξ_H. Again, we write d_Λ for the trivial bundle with fibres of dimension d over Λ. Thus we have $\xi_C = 1_C \xi_R$, $\xi_H = 1_H \xi_R$.

It is found that even if you begin by being interested in the geometric dimension of one element or a few elements in $\tilde{K}_R(RP^n)$, it is better to consider the whole problem and enquire after the geometric dimension of all the elements in $\tilde{K}_R(RP^n)$.

A few results are very elementary.

Proposition 1. *$m\xi$ has geometric dimension $\leq m$.*

This is immediate, because the Whitney sum of m copies of the canonical line bundle is a vector-bundle with fibres of dimension m.

Proposition 2. *Every element of $\tilde{K}_R(RP^n)$ has geometric dimension $\leq n$.*

In fact, an element of $\tilde{K}_R(RP^n)$ can be represented by a map $f: RP^n \to BO$, and there is no obstruction to compressing such a map into $BO(n)$.

Similarly, every element of $\tilde{K}_C(RP^{2n+1})$ and $\tilde{K}_H(RP^{4n+3})$ has geometric dimension $\leq n$ over C or H.

The obvious way to proceed is now to continue this process. That is, one considers maps $f: RP^n \to BO(n)$ and calculates the obstructions to compressing them into $BO(d)$. This method indeed has been very successful (especially in the metastable range), but it does become somewhat technical. In this lecture I want to adopt a very simple-minded approach and ask which elements of $\tilde{K}_R(RP^n)$ have geometric dimension $\leq d$ for certain very low values of d. We start from the following elementary facts.

Proposition 3. *If $n \geq 1$, the only $O(1)$-bundles over RP^n are* 1 *and ξ.*

This is immediate, because $BO(1)$ is an Eilenberg-MacLane space of type $(Z_2, 1)$, so the bundles are classified by $H^1(RP^n; Z_2)$, which is Z_2 for $n \geq 1$.

Proposition 4. *If $n \geq 3$, the only $O(2)$-bundles over RP^n are* 2, $1 \oplus \xi$ *and* 2ξ.

In fact, because $BSO(2)$ is an Eilenberg-MacLane space of type $(Z, 2)$, the orientable bundles are classified by $H^2(RP^n; Z)$, which is Z_2 for $n \geq 2$; we find two orientable bundles, namely 2 and 2ξ. There clearly are non-orientable bundles, for example $1 \oplus \xi$; and two such differ by an element of $H^2(RP^n; Z)$, where the fundamental group $\pi_1(RP^n) = Z_2$ acts non-trivially on the coefficient group Z. This cohomology group with twisted coefficients is 0 for $n \geq 3$.

For $n = 2$, every element of $\tilde{K}_R(RP^2)$ is represented by an $O(2)$-bundle, according to Proposition 2. The only element not already considered is 3ξ, which is represented by the tangent bundle of RP^2.

It is clear that we obtain these results because we know completely the homotopy groups of $O(1)$ and $O(2)$. Since we do not know completely the homotopy groups of any other group $O(n)$, one might imagine that these results would come to a rapid halt. It turns out, however, that we know just enough about a few more groups.

Proposition 5. *If n is sufficiently large ($n \geq 13$ suffices) then the only elements of geometric dimension $\leqq 3$ in $\tilde{K}_R(RP^n)$ are the elements $m\xi$ with $0 \leqq m \leqq 3$.*

A proof will be given below. In this proof, the clumsy condition $n \geq 13$ enters in getting from *Spin*-bundles to *SO*-bundles. This suggests that we should state the result for the group $S^3 = Spin(3)$, regard-

6

ing it as the most tractable group beyond $S^0 = O(1)$ and $S^1 = SO(2)$. Now of course the group S^3 appears in three contexts: as $Spin(3)$, as $SU(2)$ and as $Sp(1)$. A Spin(3)-bundle gives an element of $\tilde{K}_R(RP^n)$, an $SU(2)$-bundle gives an element of $\tilde{K}_C(RP^n)$ and an $Sp(1)$-bundle gives an element of $\tilde{K}_H(RP^n)$.

Theorem 6. (*i*) *The only element of* $\tilde{K}_R(RP^n)$ *which can be represented by a* $Spin(3)$*-bundle is* 0.

(*ii*) *The elements of* $\tilde{K}_C(RP^n)$ *which can be represented by* $SU(2)$*-bundles are precisely the elements* x *and* $(2\xi_C)+x$, *where* x *runs over the kernel of*

$$\tilde{K}_C(RP^n) \longrightarrow \tilde{K}_R(RP^n),$$

that is, the subgroup

$$\begin{array}{lll} 0 & \text{if } n \equiv 1, 2, 3, 4 \text{ or } 5 & \text{mod } 8 \\ Z_2 & \text{if } n \equiv 6, 7 \text{ or } 0 & \text{mod } 8. \end{array}$$

(*iii*) *The elements of* $\tilde{K}_H(RP^n)$ *which can be represented by* $Sp(1)$*-bundles are precisely the elements* y *and* $\xi_H + y$, *where* y *runs over the kernel of*

$$\tilde{K}_H(RP^n) \longrightarrow \tilde{K}_R(RP^n),$$

that is, the subgroup

$$\begin{array}{lll} 0 & \text{if } n \equiv 2, 3 \text{ or } 4 & \text{mod } 8 \\ Z_2 & \text{if } n \equiv 1 \text{ or } 5 & \text{mod } 8 \quad (n > 1) \\ Z_4 & \text{if } n \equiv 6, 7 \text{ or } 0 & \text{mod } 8. \end{array}$$

It is tempting to paraphrase parts (ii) and (iii) by saying that the groups $\tilde{K}_C(RP^n)$ and $\tilde{K}_H(RP^n)$ contain elements of unexpectedly low geometric dimension; but perhaps it would not be prudent to insult topologists by suggesting that their expectations take into account only Propositions 1 and 2. Indeed, I believe the theorem above is already known to the experts.

We can obtain comparable results for a few other Lie groups, although no doubt they rapidly cease to be best possible. The next case should be $Spin(4)$. We have $Spin(4) \cong Sp(1) \times Sp(1)$, and $Sp(1) \times Sp(1)$-bundles determine elements in $\tilde{K}_H(RP^n) \oplus \tilde{K}_H(RP^n)$; but the result for them follows immediately from Theorem 6 (iii).

Theorem 7. *The only elements of* $\tilde{K}_R(RP^n)$ *which can be represented by* $Spin(4)$*-bundles are* 0 *and* 4ξ.

When we come to consider $Spin(5)$-bundles, it is convenient to separete the cases $m\xi$ with $m \equiv 0 \bmod 8$ and $m\xi$ with $m \equiv 4 \bmod 8$. Since we may assume $n \geq 4$, these cases can be distinguished by the Stiefel-Whitney class w_4.

Theorem 8. *If* $n \geq 13$ *the only element of* $\tilde{K}_R(RP^n)$ *which can be represented by a* $Spin(5)$*-bundle with* $w_4 \neq 0$ *is* 4ξ.

The result is equally true for $n \leqq 8$; for $9 \leqq n \leqq 11$ we obtain also -12ξ, and for $n=12$ we obtain also -60ξ.

Theorem 9. *Assuming* $n \geq 4$, *the elements of* $\tilde{K}_R(RP^n)$ *which can be represented by* $Spin(5)$*-bundles with* $w_4 = 0$ *lie at most in the following subgroup of* $\tilde{K}_R(RP^n)$:

$$Z_2 \text{ if } n \equiv 6, 7 \text{ or } 0 \mod 8$$

$$Z_4 \text{ if } n \equiv 1, 2, 3, 4 \text{ or } 5 \mod 8.$$

For small values of n this theorem can be improved, but some non-zero elements can actually occur. It is tempting to suppose that for arbitrarily large n there are $Spin(5)$-bundles with $w_4 = 0$ over RP^n which are not stably trivial. Certainly the corresponding statement is true for $Spin(6)$.

We now begin work on the proofs. First we note that for k even, the operation $\psi^k[1]$ can be defined as a function

$$\psi^k\colon K_H(X) \longrightarrow K_R(X).$$

Lemma 10. *The function*

$$\psi^2\colon \tilde{K}_H(RP^n) \longrightarrow \tilde{K}(RP^n)$$

8

is zero.

Proof. $\psi^2(\xi_H - 1_H) = \psi^2(1_H(\xi_R - 1_R))$

$$= (\psi^2 1_H)(\psi^2 \xi_R - \psi^2 1_R)$$

$$= (\psi^2 1_H)\cdot 0$$

$$= 0.$$

To prove the negative results we use representation-theory, and we now give our conventions on characters. We take the maximal torus in $U(n)$ to consist of the diagonal matrices

$$\begin{bmatrix} z_1 & 0 & \ldots & 0 \\ 0 & z_2 & \ldots & 0 \\ \cdot & \cdot & & \\ \vdots & \cdot & & \\ 0 & 0 & \ldots & z_n \end{bmatrix}$$

with $|z_j| = 1$ for each j. From this, we obtain maximal tori in the other classical groups by the usual injections

$$U(n) \longrightarrow Sp(n)$$

$$U(n) \longrightarrow SO(2n) \longrightarrow SO(2n+1).$$

We take the maximal torus in $Spin(m)$ to be the double cover of that in $SO(m)$.

We can now prove Theorem 6 (i). Let Δ: $Spin(3) \to Sp(1)$ be the usual isomorphism; then its character is given by

$$\chi(\Delta) = z_1^{1/2} + z_1^{-1/2}.$$

Let λ^1: $Spin(3) \to SO(3)$ be the covering map; then its character is given by

$$\chi(\lambda^1) = z_1 + 1 + z_1^{-1}.$$

We have

$$\chi(\psi^2 \Delta) = z_1 + z_1^{-1} = \chi(\lambda^1 - 1),$$

and therefore

$$\psi^2 \Delta = \lambda^1 - 1.$$

We conclude that if η is a $Spin(3)$-bundle, the associated $SO(3)$-bundle is given by

$$\begin{aligned}\lambda^1 \eta &= 1 + \psi^2(\Delta\eta) \\ &= 3 \text{ by Lemma 10}.\end{aligned}$$

This proves Theorem 6 (i).

We now introduce the analogues in K-theory of the symplectic Pontryagin classes. More precisely, we introduce a virtual representation δ_k of $Sp(n)$, whose character is the kth elementary symmetric function of the variables $z_j - 2 + z_j^{-1}$ $(1 \leqq j \leqq n)$. We may define the total Pontryagin class by

$$\delta(\xi) = \sum_{k \geq 0} \delta_n t^k,$$

where t is a formal indeterminate.

Lemma 11. *(i) δ_k is real for k even, symplectic for k odd.*
(ii) $\delta(\xi \oplus \eta) = (\delta\xi)(\delta\eta)$.
(iii) δ_k is zero on $Sp(n)$ for $n < k$.

The proof is routine. It follows that δ_k defines a function

$$\delta_k \colon K_H(X) \longrightarrow K_R(X) \qquad (k \text{ even})$$

$$\text{or} \quad \delta_k \colon K_H(X) \longrightarrow K_H(X) \qquad (k \text{ odd}).$$

Since $\delta(1_H) = 1_R$, these functions factor through the projection $K_H(X) \to \tilde{K}_H(X)$.

We introduce the following lemma to help in evaluating the functions δ_k in RP^n, and also for another purpose.

Lemma 12.

$$1_H \cdot 1_H = 4_R, \; 1_H \cdot \xi_H = 4\xi_R, \; \xi_H \cdot \xi_H = 4_R$$

$$(\xi_H - 1_H)^2 = -2^3(\xi_R - 1_R)$$

$$(\xi_H - 1_H)^3 = +2^4(\xi_H - 1_H)$$

$$(\xi_H - 1_H)^4 = -2^7(\xi_R - 1_R).$$

The proof is trivial if one recalls that $\xi_H = 1_H \cdot \xi_R$.

We turn to Theorem 6 (iii). Take an element $m\xi_H$ of $\tilde{K}_H(RP^n)$. If $m\xi_H$ can be represented by an $Sp(1)$-bundle, then $\delta_2(m\xi_H)=0$. But

$$\delta_2(m\xi_H) = \frac{m(m-1)}{1.2}(\xi_H - 1_H)^2$$

$$= -4m(m-1)(\xi_R - 1_R).$$

Let the order of $\tilde{K}_R(RP^n)$ be 2^r; we conclude that

$$m \equiv 0 \quad \text{or} \quad 1 \qquad \bmod 2^{r-2}.$$

Now $m\xi_H$ in $\tilde{K}_H(RP^n)$ maps to $4m\xi_R$ in $\tilde{K}_R(RP^n)$; we conclude that

$$m\xi_H = y \quad \text{or} \quad \xi_H + y,$$

where y is an element in the kernel of

$$\tilde{K}_H(RP^n) \longrightarrow \tilde{K}_R(RP^n).$$

This proves that an element of $\tilde{K}_H(RP^n)$ which can be represented by an $Sp(1)$-bundle is as described in Theorem 6 (iii). The corresponding half of Theorem 6 (ii) follows immediately, because an $SU(2)$-bundle comes from an $Sp(1)$-bundle via the isomorphism $Sp(1) \to SU(2)$.

Proof of Theorem 7. Let η be a $Spin(4)$-bundle over RP^n. Let $\Delta^+, \Delta^-: Spin(4) \to Sp(1)$ be the two spinor representations, with characters

$$\chi\Delta^+ = z_1^{1/2} z_2^{1/2} + z_1^{-1/2} z_2^{-1/2}$$

$$\chi\Delta^- = z_1^{1/2} z_2^{-1/2} + z_1^{-1/2} z_2^{1/2}.$$

Then $\Delta^+\eta$ and $\Delta^-\eta$ are $Sp(1)$-bundles over RP^n; by Theorem 6 (iii) we have in $K_H(RP^n)$ the equations

11

$$\Delta^+\eta = 1_H + y^+ \quad or \quad \xi_H + y^+$$

$$\Delta^-\eta = 1_H + y^- \quad or \quad \xi_H + y^-$$

where y^+, y^- are as in Theorem 6 (iii). Let λ^1: $Spin(4) \to SO(4)$ be the covering map; then an easy calculation with characters shows that

$$\Delta^+ \cdot \Delta^- = \lambda^1 .$$

So the $SO(4)$-bundle associated to η is given by

$$\lambda^1\eta = (\Delta^+\eta)(\Delta^-\eta) .$$

Substituting the values of $\Delta^+\eta$ and $\Delta^-\eta$, we find four possibilities. By Lemma 12, multiplication by 1_H induces the canonical map $\tilde{K}_H(RP^n) \to \tilde{K}_R(RP^n)$; therefore it annihilates y^+ and y^-. Similarly, multiplication by ξ_H induces minus the canonical map $\tilde{K}_H(RP^n) \to \tilde{K}_R(RP^n)$; therefore it also annihilates y^+ and y^-. Adding, $(\xi_H - 1_H)$ annihilates y^+ and y^-. But since y^+ and y^- are both multiples of $\xi_H - 1_H$, we find $y^+ \cdot y^- = 0$. Thus

$$\lambda^1\eta = 1_H \cdot 1_H,\ 1_H \cdot \xi_H \quad \text{or} \quad \xi_H \cdot \xi_H$$

$$= 4_R \quad \text{or} \quad 4\xi_R .$$

This proves Theorem 7.

Proof of Theorems 8, 9. Let η be a $Spin(5)$-bundle over RP^n. Let Δ: $Spin(5) \to Sp(2)$ be the usual isomorphism, with character

$$\chi(\Delta) = z_1^{1/2} z_2^{1/2} + z_1^{1/2} z_2^{-1/2} + z_1^{-1/2} z_2^{1/2} + z_1^{-1/2} z_2^{-1/2}.$$

Then $\Delta\eta$ is an $Sp(2)$-bundle over RP^n; let its class in $\tilde{K}_H(RP^n)$ be $m\xi_H$. Then we have $\delta_3(m\xi_H) = 0$; that is,

$$\frac{m(m-1)(m-2)}{1.2.3}(\xi_H - 1_H)^3 = 0,$$

i.e.

$$m(m-1)(m-2)2^3(\xi_H - 1_H)^3 = 0 .$$

Let the order of $\tilde{K}_H(RP^n)$ be 2^h; we conclude that

$$m \equiv 1 \qquad \bmod 2^{h-3}$$

or else

$$m \equiv 0 \quad or \quad 2 \qquad \bmod 2^{h-4}.$$

Moreover, we have $\delta_4(m\xi_H) = 0$; that is,

$$\frac{m(m-1)(m-2)(m-3)}{1.2.3.4}(\xi_H - 1_H)^4 = 0$$

i.e. $$m(m-1)(m-2)(m-3)2^4(\xi_R - 1_R) = 0.$$

Let the order of $\tilde{K}_R(RP^n)$ be 2^r; we conclude that

$$m \equiv 0,\ 1,\ 2 \quad or \quad 3 \qquad \bmod 2^{r-5}.$$

Let λ^1: $Spin(5) \to SO(5)$ be the covering map. By an easy calculation with characters, we see that

$$\lambda^2 \Delta = \lambda^1 + 1.$$

By the proof of Lemma 11, or by an easy direct calculation with characters, we have on $Sp(2)$ the formula

$$\lambda^2 = \delta_2 + \rho - 2,$$

where ρ: $Sp(2) \to SO(8)$ is the canonical embedding. For an element

$$\Delta\eta = m(\xi_H - 1_H) + 2_H \in K_H(RP^n)$$

this gives

$$\lambda^2 \Delta\eta = \frac{m(m-1)}{1.2}(-8)(\xi_R - 1_R) + 4m(\xi_R - 1_R) + 6_R,$$

that is,

$$\lambda^1 \eta = m(m-2)4(\xi_R - 1_R) + 5.$$

We can now separate cases. By the last formula, $w_4 \lambda^1 \eta$ is 0 or not according as m is even or odd. Suppose first that m is odd, say

$$m = 1 + 2^{h-3} q.$$

Then

$$\lambda^1\eta-5=(1+2^{h-3}q)(1-2^{h-3}q)4(\xi_R-1_R)$$
$$=(4-2^{2h-4}q^2)(\xi_R-1_R).$$

We conclude that η lies in the stable class 4ξ provided that $2^h-4\geq r$, which is true for $n\geq 13$. This proves Theorem 8.

Suppose secondly that m is even. We have already proved that

$$m(m-1)(m-2)(m-3)2^4(\xi_R-1_R)=0 \quad \text{in} \quad \tilde{K}_R(RP^n);$$

this now implies that

$$m(m-2)2^4(\xi_R-1_R)=0 \quad \text{in} \quad \tilde{K}_R(RP^n),$$

i.e.

$$4(\lambda^1\eta-5)=0.$$

This shows that the stable class of $\lambda^1\eta$ lies at most in the Z_4 subgroup of $\tilde{K}_R(RP^n)$.

Similarly, the equation

$$m(m-1)(m-2)2^3(\xi_H-1_H)=0$$

now implies that

$$m(m-2)2^3(\xi_H-1_H)=0,$$

that is, $2(\lambda^1\eta-5)$ maps to zero in $\tilde{K}_H(RP^n)$. If $n\equiv 6, 7$ or $0 \bmod 8$, then

$$\tilde{K}_R(RP^n)\longrightarrow \tilde{K}_H(RP^n)$$

is iso, and we see that the stable class of $\lambda^1\eta$ lies at most in the Z_2 subgroup of $\tilde{K}_R(RP^n)$. This proves Theorem 9.

Proof of Proposition 5. Suppose the element $m\xi$ in $\tilde{K}_R(RP^n)$ has geometric dimension ≤ 3. If $m\equiv 0 \bmod 4$ then $m\xi$ is representable by a *Spin*(3)-bundle, and Theorem 6 shows that $m\xi=0$. Suppose then that $m\equiv 2 \bmod 4$. Since we are assuming $n\geq 6$, the Stiefel-Whitney class w_6 shows that $m\equiv 2 \bmod 8$. Thus $(m+2)\xi$ is representable by a *Spin*(5)-bundle with $w_4\neq 0$, and Theorem 8 shows that

$$m+2)\xi=4\xi \quad \text{in} \quad \tilde{K}_R(RP^n).$$

Thus $m\xi=2\xi$ in $\tilde{K}_R(RP^n)$.

Finally, suppose we have an $O(3)$-bundle η such that

$$\eta=m(\xi-1)+3 \quad \text{in} \quad K_R(RP^n)$$

with m odd. Then $\xi\eta$ is an $SO(3)$-bundle in the class

$$(3-m)(\xi-1)+3$$

with $(3-m)$ even. By what has just been proved, $\xi\eta$ lies in the stable class 0 or 2ξ; thus $m\xi=\xi$ or 3ξ in $\tilde{K}_R(RP^n)$. This proves Proposition 5.

To prove the positive half of Theorem 6 requires some homotopy-theory. First we make a reduction of the problem, using the following lemma.

Lemma 13. *There is a map*

$$\phi: RP^n \longrightarrow (RP^n/RP^3)\vee(RP^n/RP^{n-4})$$

of type (1, 1).

Proof. The relative homotopy groups of the product $(RP^n/RP^3)\times(RP^n/RP^{n-4})$ modulo the wedge are zero in dimensions $\leq n$.

I now claim that it will be sufficient to obtain maps

$$g: RP^n/RP^{n-4} \longrightarrow BSp(1)$$

which represent the elements y named in Theorem 6 (iii). For suppose we can do this. The element ξ_H can certainly be represented by a map

$$h: RP^n/RP^3 \longrightarrow BSp(1);$$

we can form the map

$$RP^n \xrightarrow{\phi} (RP^n/RP^3)\vee(RP^n/RP^{n-4}) \xrightarrow{(h,g)} BSp(1),$$

and it represents $\xi_H\oplus y$. Composing with the map $BSp(1)\to BSU(2)$, we obtain the elements named in Theorem 6 (ii).

We intend to construct the maps g by exploiting "periodicity" maps.

Lemma 14. *For $n \geq 5$ there is a "periodicity" map*

$$f: S^{n+8} \cup_2 e^{n+9} \longrightarrow S^n \cup_2 e^{n+1}$$

which induces an isomorphism of $\tilde{K}_H^*$.

We note that if such a map f exists for some n, then its suspension Sf serves in dimension $n+1$. For $n \leq 3$ no such map f exists, for if it did, the composite

$$S^{n+8} \xrightarrow{i} S^{n+8} \cup_2 e^{n+9} \xrightarrow{f} S^n \cup_2 e^{n+1} \xrightarrow{j} S^{n+1}$$

would provide a map α with $e_C(\alpha)=1/2$ [2], which is impossible. For $n=4$ trial calculations suggest that the map f does exist; but this requires care.

Proof. The corresponding result in stable homotopy is proved in [2, p. 68]; the same proof works unstably. For instead of taking α to be an element in the stable 7-stem, we take α to be the nonzero element τ'' in the 2-component of $\pi_{12}(S^5)$. (We use the notation of Toda [3].) Then we have $\alpha \cdot (2\iota_{12})=0$, and also

$$(2\iota_5) \cdot \alpha = \alpha + \alpha + [\iota_5, \iota_5](H\alpha) = \nu_5 \eta_8 \nu_9 (4\iota_{12}) = 0 .$$

Finally we have

$$\{2\iota_5, \alpha, 2\iota_{12}\} = \lambda \varepsilon_5$$

for some $\lambda \in Z_2$. But since $\{2\iota_5, \alpha, 2\iota_{12}\}$ is stably zero and ε_5 is stably non-zero, we must have $\lambda=0$. Now the method of [2] applies.

The result is now obvious for for $n \equiv 6 \bmod 8$. In fact, the element of order 4 in $\tilde{K}_H(RP^6)$ can be represented by a map

$$g: RP^6/RP^4 = S^5 \cup_2 e^6 \longrightarrow BSp(1)$$

(see Proposition 2.) Composing with f we obtain a map

$$gf: RP^{14}/RP^{12} = S^{13} \cup_2 e^{14} \longrightarrow BSp(1)$$

which serves, and so on by induction for $n=8k+6$, $k=0, 1, 2, \ldots$.

The result now follows for $n=8k+5$ by restriction, and for $n=8k+7$ since

$$RP^{8k+7}/RP^{8k+4}=(S^{8k+5}\cup{}_2e^{k+6})\vee S^{8k+7}.$$

Next we consider the case $n=8k+8$.
Here we have

$$RP^{8k+8}/RP^{8k+5}=(S^{8k+6}\vee S^{8k+7})\cup e^{8k+8},$$

where the components of the attaching map are $(\eta, 2)$. If we attach a further cell e^{8k+7} by a map with components (2, 0), we do not change $\tilde{K}_H$ or $\tilde{K}_R$, but we obtain a complex X_k equivalent to $(S^{8k+5}\cup{}_2e^{8k+6})\cup(S^1\cup{}_2e^2)$. Such a complex admits a "periodicity" map $X_{k+1}\rightarrow X_k$ for $k\geq 0$ (take the product of the map of Lemma 14 and the identity map of $S^1\cup{}_2e^2$.) Next let y be any element of the Z_4 subgroup of $\tilde{K}_H(RP^8)$; I claim it can be represented by a map $X_0\rightarrow BSp(1)$. This can easily be proved by direct homotopy-theory, or more elegantly by the following remarks.

(i) By choice of y we have $\delta_2 y=0$.

(ii) The element y is the image of a unique element $y'\in\tilde{K}_H(X_0)$ which also satisfies $\delta_2 y'=0$.

(iii) For complexes X of dimension ≤ 10, the condition $\delta_2 y'=0$ is necessary and sufficient for an element $y'\in\tilde{K}_H(X)$ to be represented by a map $X\rightarrow BSp(1)$.

Now we form the composites

$$RP^{16}/RP^{13}\longrightarrow X_1\longrightarrow X_0\longrightarrow BSp(1),$$

etc., and the previous methods apply.

Finally we turn to the case $n=8k+9$. The previous case provides us with a map

$$f\colon RP^{8k+8}/RP^{8k+5}\longrightarrow BSp(1)$$

representing an element of order 4 in $\tilde{K}_H(RP^{8k+8})$. Doubling it, we obtain a map

$$f' : RP^{8k+8}/RP^{8k+5} \longrightarrow BSp(1)$$

representing the element of order 2 in $\tilde{K}_H(RP^{8k+8})$. By construction f factors through X_k and therefore the restriction of f' to RP^{8k+6}/RP^{8k+5} is null homotopic; so f' factors through RP^{8k+8}/RP^{8k+6}. I claim this map can be extended over RP^{8k+9}/RP^{8k+6}. In this complex the top cell e^{8k+9} is attached to the sphere S^{8k+7} by a map in the class η. But by construction, the restriction of f' to S^{8k+7} is twice the restriction of f; therefore its composition with η is null homotopic, and the extension is possible. This deals with the case $n=8k+9$, and completes the proof of Theorem 6.

References

[1] Adams, J. F., Vector Fields on Spheres, *Annals of Math.* **75** (1962), pp. 603–632.

[2] Adams, J. F., On the groups J(X)–IV, *Topology* **5** (1966), pp. 21–71.

[3] Toda, H., Composition methods in homotopy groups of spheres, *Annals of Math. Studies* (49), Princeton U. Press 1962.

LECTURES ON GENERALISED COHOMOLOGY*

by

J. F. Adams

LECTURE 1. THE UNIVERSAL COEFFICIENT THEOREM AND THE KUNNETH THEOREM

It is an established practice to take old theorems about ordinary homology, and generalise them so as to obtain theorems about generalised homology theories. For example, this works very well for duality theorems about manifolds. We may ask the following question. Take all those theorems about ordinary homology which are standard results in everyday use. Which are the ones which still lack a fully satisfactory generalisation to generalised homology theories? I want to devote this lecture to such problems.

As my candidates for theorems which need generalising, I offer you the universal coefficient theorem and the Künneth theorem. I will first try to formulate the conclusions which these theorems should have in the generalised case. I will then make some comments on these formulations, and discuss a certain number of cases in which they are known to be true. I will then comment on the connection between one form of the universal coefficient theorem and the "Adams

---

* Note. These lectures are not arranged in the order in which they were originally given.

spectral sequence". After that I will give some proof under suitable assumptions. Finally I will show that certain results of Conner and Floyd [14] can be related to the universal coefficient theorem.

In discussing the universal coefficient theorem and the Künneth theorem, we will write E_* and F_* for generalised homology theories and E^*, F^* for generalised cohomology theories. In order to avoid tedious notation for relative groups, we will suppose that they are "reduced" theories, defined on some category of spaces with base-point. Thus we can replace the pair X, X' by the space with base-point X/X'. In particular, the coefficient groups for E_* are the groups $E_*(S^0)$, and similarly for the other theories.

The universal coefficient theorem should address itself to the following problems.

(1) Given $E_*(X)$, calculate $F_*(X)$.

(2) Given $E_*(X)$, calculate $F^*(X)$.

(3) Given $E^*(X)$, calculate $F^*(X)$.

(4) Given $E^*(X)$, calculate $F_*(X)$.

The last two problems correspond to the "upside-down universal coefficient theorems" in ordinary homology.

It will surely be necessary to assume some relation between E_* (or E^*) and F_* (or F^*). To begin with, we must suppose given enough products. For example, we need products in order to give sense to the Tor and Ext functors which occur in our statements. We postpone all further discussion of data; the first step is to formulate the conclusions which our generalised theorems ought to assert. We suggest the following.

(UCT1)

Suppose given product maps

$$\mu: E_*(X) \otimes E_*(S^0) \longrightarrow E_*(X)$$

$$\nu: E_*(X) \otimes F_*(S^0) \longrightarrow F_*(X)$$

satisfying suitable axioms. Then there is a spectral sequence

$$\mathrm{Tor}^{E_*(S^0)}_{p,*}(E_*(X), F_*(S^0)) \underset{p}{\Longrightarrow} F_*(X) .$$

The edge-homomorphism

$$E_*(X) \otimes_{E_*(S^0)} F_*(S^0) \longrightarrow F_*(X)$$

is induced by ν .

(UCT2)

Suppose given product maps

$$\mu: E_*(S^0) \otimes E_*(X) \longrightarrow E_*(X)$$

$$\nu: E_*(X) \otimes F^*(X) \longrightarrow F^*(S^0)$$

satisfying suitable axioms. Then there is a spectral sequence

$$\mathrm{Ext}^{p,*}_{E_*(S^0)}(E_*(X), F^*(S^0)) \underset{p}{\Longrightarrow} F^*(X) .$$

The edge-homomorphism

$$F^*(X) \longrightarrow \mathrm{Hom}^*_{E_*(S^0)}(E_*(X), F^*(S^0))$$

is induced by ν .

<u>(UCT3)</u>

Suppose given product maps

$$\mu: E^*(X) \otimes E^*(S^0) \longrightarrow E^*(X)$$

$$\nu: E^*(X) \otimes F^*(S^0) \longrightarrow F^*(X)$$

satisfying suitable axioms. Then there is a spectral sequence

$$\mathrm{Tor}^{E^*(S^0)}_{p,*}(E^*(X), F^*(S^0)) \underset{p}{\Longrightarrow} F^*(X) .$$

The edge-homomorphism

$$E^*(X) \otimes_{E^*(S^0)} F^*(S^0) \longrightarrow F^*(X)$$

is induced by ν .

<u>(UCT4)</u>

Suppose given product maps

$$\mu: E^*(S^0) \otimes E^*(X) \longrightarrow E^*(X)$$

$$\nu: E^*(X) \otimes F_*(X) \longrightarrow F_*(S^0)$$

satisfying suitable axioms. Then there is a spectral sequence

$$\mathrm{Ext}^{p,*}_{E^*(S^0)}(E^*(X), F_*(S^0)) \underset{p}{\Longrightarrow} F_*(X) .$$

The edge-homomorphism

$$F_*(X) \longrightarrow \mathrm{Hom}^*_{E^*(S^0)}(E^*(X), F_*(S^0))$$

is induced by ν .

<u>Note 1</u>. We should spell out some of the axioms on the product maps. We will obviously assume that the product maps have the correct behavior with respect to induced homomorphisms and with respect to suspension. We will assume that the map μ, for $X = S^0$, makes $E_*(S^0)$ (in cases 1 and 2) or $E^*(S^0)$ (in cases 3 and 4) into a graded ring with unit. We will assume that the map μ makes $E_*(X)$ (in cases 1 and 2) or $E^*(X)$ (in cases 3 and 4) into a graded module over $E_*(S^0)$ or $E^*(S^0)$. This module is a left module in cases 2 and 4, a right module in cases 1 and 3. We will assume that the map ν, for $X = S^0$, makes $F_*(S^0)$ (in cases 1 and 4) or $F^*(S^0)$ (in cases 2 and 3) into a graded module over $E_*(S^0)$ or $E^*(S^0)$. This module is a left module in all four cases. This is sufficient to give sense to the Tor and Ext functors in the statements. Again, in cases 1 and 3 we will assume that the product maps

$$\nu: E_*(X) \otimes F_*(S^0) \longrightarrow F_*(X)$$

$$\nu: E^*(X) \otimes F^*(S^0) \longrightarrow F^*(X)$$

factor through $E_*(X) \otimes_{E_*(S^0)} F_*(S^0)$ and $E^*(X) \otimes_{E^*(S^0)} F^*(S^0)$ respectively. In cases 2 and 4 we convert the maps ν into maps

$$F^*(X) \longrightarrow \mathrm{Hom}^*(E_*(X), F^*(S^0))$$

$$F_*(X) \longrightarrow \mathrm{Hom}^*(E^*(X), F_*(S^0))$$

and assume that these actually map into $\mathrm{Hom}^*_{E_*(S^0)}(E_*(X), F^*(S^0))$ and $\mathrm{Hom}^*_{E^*(S^0)}(E^*(X), F_*(S^0))$ respectively. All these four conditions may be viewed as associativity conditions on our products. They give sense to the statements about the edge-homomorphisms.

<u>Note 2</u>. The case of representable functors is particularly important. In this case we suppose given a ring-spectrum E and a spectrum F which is a left module-spectrum over the ring-spectrum E. We take E_* and E^* to be the functors determined by E, as in [31]; we take F_* and F^* to be the functors determined by F. In this case we obtain all the products required for the statements UCT 1-4. For example, in cases 2 and 4 the products ν are Kronecker products. All these products satisfy all the assumptions mentioned in Note 1.

As examples of ring-spectra E, we have MU, and the BU spectrum, and the sphere spectrum S. We also have examples of module-spectra. Any spectrum is a module-spectrum over S; and BU is a module-spectrum over MU, this being the case explored by Conner and Floyd [14].

<u>Note 3</u>. As remarked above, we have yet to discuss the data which might suffice to prove these statements, or the lines of proof which might establish them. The assumptions in Note 1 are intended simply to give meaning to the statements.

<u>Note 4</u>. By assuming extra data, we might expect

to make all these spectral sequences into spectral sequences of modules over $E_*(S^0)$ or $E^*(S^0)$. The extra data would be modelled on the case in which we start from a ring-spectrum E which is commutative, and a module-spectrum F over E. For example, we would take the ring $E_*(S^0)$ or $E^*(S^0)$ to be anticommutative. We spare ourselves the details. If the basic results are proved in any reasonable way, it should not be hard to add such trimmings.

The Künneth theorem (for reduced functors) should address itself to the problem of computing E_* and E^* for the smash-product $X \wedge Y$ in terms of corresponding groups of X and Y. (This corresponds to computing an unreduced theory on $X \times Y$.) We may obtain four statements by substituting in UCT 1 and 4 the functor $F_*(X) = E_*(X \wedge Y)$, and in UCT 2 and 3 the functor $F^*(X) = E^*(X \wedge Y)$. We obtain the following statements.

<u>(KT1)</u>

Suppose given an external product

$$\nu: E_*(X) \otimes E_*(Y) \longrightarrow E_*(X \wedge Y)$$

satisfying suitable axioms. Then there is a spectral sequence

$$\mathrm{Tor}^{E_*(S^0)}_{p,*}(E_*(X),\ E_*(Y)) \underset{p}{\Longrightarrow} E_*(X \wedge Y) .$$

The edge-homomorphism

$$E_*(X) \otimes_{E_*(S^0)} E_*(Y) \longrightarrow E_*(X \wedge Y)$$

is induced by ν .

(KT2)

Suppose given a product

$$\mu : E_*(S^0) \otimes E_*(X) \longrightarrow E_*(X)$$

and a slant product

$$\nu : E_*(X) \otimes E^*(X \wedge Y) \longrightarrow E^*(Y)$$

satisfying suitable axioms. Then there is a spectral sequence

$$\mathrm{Ext}^{p,*}_{E_*(S^0)}(E_*(X), E^*(Y)) \underset{p}{\Longrightarrow} E^*(X \wedge Y) .$$

The edge-homomorphism

$$E^*(X \wedge Y) \longrightarrow \mathrm{Hom}^*_{E_*(S^0)}(E_*(X), E^*(Y))$$

is induced by ν .

(KT3)

Suppose given an external product

$$\nu : E^*(X) \otimes E^*(Y) \longrightarrow E^*(X \wedge Y)$$

satisfying suitable axioms. Then there is a spectral sequence

$$\mathrm{Tor}^{E^*(S^0)}_{p,*}(E^*(X), E^*(Y)) \underset{p}{\Longrightarrow} E^*(X \wedge Y) .$$

The edge-homomorphism

$$E^*(X) \otimes_{E^*(S^0)} E^*(Y) \longrightarrow E^*(X \wedge Y)$$

is induced by ν .

(KT4)

Suppose given a product map

$$\mu: E^*(S^0) \otimes E^*(X) \longrightarrow E^*(X)$$

and a slant product

$$\nu: E^*(X) \otimes E_*(X \wedge Y) \longrightarrow E_*(Y)$$

satisfying suitable axioms. Then there is a spectral sequence

$$\mathrm{Ext}^{p,*}_{E^*(S^0)}(E^*(X), E_*(Y)) \underset{p}{\Longrightarrow} E_*(X \wedge Y) .$$

The edge-homomorphism

$$E_*(X \wedge Y) \longrightarrow \mathrm{Hom}^*_{E^*(S^0)}(E^*(X), E_*(Y))$$

is induced by ν .

Note 5. In KT 1 and 3 it is unnecessary to suppose given the product μ, as it can be obtained by specialising the product ν to the case $Y = S^0$.

Note 6. As each part of the "Künneth theorem" is obtained by transcribing the corresponding part of the "universal coefficient theorem", Notes 1, 3 and 4 above can also be transcribed. Note 1 yields the formal properties of our products μ and ν which we should assume in order to give sense to the statements.

Note 7. The case of representable functors is particularly important. In this case we suppose given a ring-spectrum E. We take E_* and E^* to be the functors determined by E, as in [31]. We then have four classical products - two external products and two slant products [31]. These products satisfy all the formal properties needed to give sense to our statements - see Note 6.

This provides some justification for stating the Künneth theorem in four parts. In fact, we have four products; from each product product we can construct an associated "edge-homomorphism"; the corresponding spectral sequence (if it applies) shows whether or not this homomorphism is an isomorphism.

Note 8. Since each part of the Künneth theorem is obtained by specialising the corresponding part of the universal coefficient theorem, the latter will presumably imply the former, once we get the data settled. (Of course, if we wished to stay inside ordinary homology we could not use this argument.) It should therefore be enough to discuss the universal coefficient theorem.

Note 9. It is almost certain that UCT 3 and UCT 4 will require some finiteness condition, because such a condition is needed for the "upside-down universal coefficient theorems" in ordinary homology. If X is a finite complex, then we can deduce UCT 3 from UCT 1 by S-duality. Let DX be the Spanier-Whitehead dual of X. Suppose given E^*, F^* as in UCT 3. Then we can define theories E_*, F_* on finite complexes by setting

$$E_*(X) = E^*(DX), \quad F_*(X) = F^*(DX) \;;$$

we extend to infinite complexes and spectra by direct limits. We obtain product maps

$$E_*(X) \otimes E_*(S^0) \longrightarrow E_*(X)$$

$$E_*(X) \otimes F_*(S^0) \longrightarrow F_*(X)$$

as required for UCT 1. Applying UCT 1 to DX, we obtain UCT 3 for X.

Similar remarks would apply to deduce UCT 4 from UCT 2, except that the definition

$$F^*(X) = F_*(DX)$$

will only define F^* on finite complexes. At this point we do not know whether it will suffice for UCT 2 to have F^* defined on so small a category. It therefore seems best to begin from a ring-spectrum E and a module-spectrum F. In this case F^* will be defined on a sufficiently large category. We have isomorphisms

$$E_*(DX) \cong E^*(X)$$

$$F^*(DX) \cong F_*(X)$$

and these can be taken to throw the usual products

$$E_*(S^0) \otimes E_*(DX) \longrightarrow E_*(DX)$$

$$E_*(DX) \otimes F^*(DX) \longrightarrow F^*(S^0)$$

onto the usual products

$$E^*(S^0) \otimes E^*(X) \longrightarrow E^*(X)$$

$$E^*(X) \otimes F_*(X) \longrightarrow F_*(S^0) \;.$$

Applying UCT 2 to DX, we obtain UCT 4 for any finite complex X.

Of course, this method of deducing UCT 4 from UCT 2 only gives UCT 4 for representable functors. It is therefore necessary to note that UCT 4 for representable functors implies KT 4 for representable functors. Suppose we start from a ring-spectrum E. Then the functor

$$F_*(X) = E_*(X \wedge Y)$$

is representable; the representing spectrum is given by $F = E \wedge Y$. This spectrum can be made into a (left) module-spectrum over E in the obvious way; this results in a product

$$E^*(X) \otimes F_*(X) \longrightarrow F_*(S^0)$$

which coincides with the usual slant-product

$$E^*(X) \otimes E_*(X \wedge Y) \longrightarrow E_*(Y) .$$

If X is a finite complex, and we apply UCT 4 to X (with this E and F), we obtain KT 4 for X.

The result of this discussion is that to obtain all eight results, under suitable conditions, it should be enough to discuss UCT 1 and UCT 2.

<u>Note 10</u>. Our treatment leads to KT 3 with a finiteness assumption on X but none on Y. Since KT 3 is symmetrical between X and Y, it would be equally reasonable to make a finiteness assumption on Y but none on X. Some finiteness assumption is almost certainly necessary, because

it is so far the corresponding Künneth theorem in ordinary cohomology.

Our treatment leads to KT 4 with a finiteness assumption on X but none on Y. Some finiteness assumption on X is almost certainly necessary, for the usual reason. A finiteness assumption on Y is very likely to be irrelevant. For example, suppose that $E^*(X)$ has a resolution by finitely-generated projectives over $E^*(S^0)$; e.g. this is so if $E = MU$ and X is a finite complex (see Lecture 5). Then $\mathrm{Ext}^{p,*}_{E^*(S^0)}(E^*(X), E_*(Y))$ passes to direct limits as we vary Y; and KT 4 for this X and general Y follows from the case in which Y is a finite complex.

It is now time to discuss some cases in which the statements we have formulated are known to be true.

Note 11. Certain special cases of the statements are classical theorems about ordinary homology.

Note 12. Suppose that $F_*(S^0)$ is flat over $E_*(S^0)$. Then UCT 1 asserts that the edge-homomorphism

$$\varepsilon: E_*(X) \otimes_{E_*(S^0)} F_*(S^0) \longrightarrow F_*(X)$$

is an isomorphism. This is certainly true when X is a finite complex, because as we vary X, ε is a natural transformation between homology functors which is iso for $X = S^0$. If we assume that E_* and F_* pass to direct limits as we vary X, then the same result holds when X is a CW-complex or a spectrum.

Since KT 1 is symmetrical between X and Y, it follows that KT 1 is true if either $E_*(X)$ or $E_*(Y)$ is flat.

Similar remarks apply to UCT 2 if $F^*(S^0)$ is injective, although this case hardly ever arises. One has to approach the case of infinite complexes X by discussing the case of infinite wedge-sums, as in [21].

The same approach does not immediatly prove UCT 1 under the assumption that $E_*(X)$ is flat, because we cannot vary F arbitrarily without losing the products we need. (See Note 14 below.) However, UCT 1 and UCT 2 are trivially true if X is a wedge-sum of spheres; we will use this later.

Note 13. If E is the sphere-spectrum S then any spectrum is a module over S. In this case all the results are true and easy to prove. This will appear as a special case in Note 15 below.

Note 14. Next I have to recall that in the definition of a ring-spectrum, one is allowed various homotopies; for example, the product is supposed to be homotopy-associative. If we do not wish to allow any homotopies, we speak of a strict ring-spectrum. The sphere S is a strict ring-spectrum; otherwise it is usually laborious to show that a given spectrum is a strict ring-spectrum. It has been proved by E. Dyer and D. Kahn (to appear) that if E is a strict ring-spectrum, then KT 1 holds. Their argument also shows that if E is a strict ring-spectrum and F is a strict module-spectrum over E, then UCT 1 holds. The method amounts to constructing an E-free resolution of F; compare the last paragraph of Note 12 above.

This is at least a general theorem. It is likely that one could weaken the conditions on the spectra slightly, by analogy with the case of "A_∞ H-spaces" [28]. Unfortunately, the method does not seem to prove any of the theorems involving Ext; this would require a different sort of resolution.

<u>Note 15</u>. If E is the BU-spectrum and X, Y are finite complexes then KT 3 is a result of Atiyah [6]. (Of course in this case $\mathrm{Tor}_p = 0$ for $p > 1$.) By combining the idea of Atiyah's proof with S-duality, one can obtain a proof of UCT 1 and UCT 2 (and hence of all the rest) for various spectra for which the method happens to work. The spectra E to which the method applies include BO, BU, MO, MU, MSp, S and the Eilenberg-MacLane spectrum $K(Z_p)$.

This method is already known to E. Dyer, and perhaps to many other workers in the field. Since giving the original lecture I have heard that L. Smith has applied the method (a) to consider UCT 1 for the case $E = MU$, $F = K(Z)$ and (b) to consider KT 1 for the case $E = MU$; I am grateful to him for sending me a preprint.

This method is very practical when it works. It definitely doesn't work for $E = K(Z)$. Thus it fails to generalise the classical theorems for ordinary homology.

Note 16. Atiyah [6, footnote on p. 245] has indicated an example in which the edge-homomorphism is not monomorphic; and presumably further such examples can be found. They do not contradict our thesis, because they presumably give examples in which the differentials of the relevant spectral sequence are non-zero.

Next I want to comment on the connection between UCT 2 and the "Adams spectral sequence" [1,2,15]. For this I need some standard ideas from homological algebra, and I give them now in order to avoid interrupting the discussion later.

Let A be an algebra over a ground ring R, and let M be an R-module. Then $A \otimes_R M$ may be made into an A-module by giving it the obvious structure maps; and we have

$$\mathrm{Hom}_A(A \otimes_R M, N) \cong \mathrm{Hom}_R(M,N) .$$

(Hence the same thing is true for Ext .) $A \otimes_R M$ is called an "extended" module. Similarly, let C be a coalgebra over a ring R, and let M be an R-module. Then $C \otimes_R M$ may be made into a C-comodule by giving it the obvious structure maps; and we have

$$\mathrm{Hom}_C(L, C \otimes_R M) \cong \mathrm{Hom}_R(L, M) \ .$$

(Hence the same thing is true for Ext .) $C \otimes_R M$ is called an extended comodule.

In the applications everything will be graded. Also C will be a bimodule over R and the two actions of R on C will be quite distinct; but this does not affect the truth of the clichés presented above.

Let $[X,Y]_*$ be the set of stable homotopy classes of maps from X to Y. I shall argue in Lecture 2 that the most plausible generalisation of the "Adams spectral sequence" would give the following statement.

<u>(ASS)</u>

Under suitable assumptions, there is a spectral sequence

$$\mathrm{Ext}^{p,*}_{E_*(E)}(E_*(X), E_*(Y)) \underset{p}{\Longrightarrow} [X,Y]_* \ .$$

The edge-homomorphism

$$[X,Y]_* \longrightarrow \mathrm{Hom}^*_{E_*(E)}(E_*(X), E_*(Y))$$

assigns to each map f its induced homomorphism $f_*: E_*(X) \longrightarrow E_*(Y)$.

Here E is (as usual) a ring-spectrum. The functors Hom and Ext are defined by considering $E_*(X)$ and

$E_*(Y)$ as comodules with respect to the coalgebra $E_*(E)$. We use $E_*(S^0)$ as the ground ring for our comodules etc. The necessary details are given in Lecture 3.

This result refers to $[X,Y]_*$ for a general Y. If we assume that Y is F, a left module-spectrum over E, then $[X,Y]_*$ becomes $F^*(X)$, and we may hope that this extra data will simplify the computation of the E_2 term. We will now make this more precise. In Lecture 3 we will define a product map

$$m\colon E_*(E) \otimes_{E_*(S^0)} F_*(S^0) \longrightarrow E_*(F) .$$

This map is not one of those we have so far considered, but it is related to the map ν of UCT 1 by the following commutative diagram.

$$\begin{array}{ccc} E_*(E) \otimes_{E_*(S^0)} F_*(S^0) & \xrightarrow{m} & E_*(F) \\ \downarrow{\scriptstyle c\otimes 1} & & \downarrow{\scriptstyle \tau_*} \\ E_*(E) \otimes_{E_*(S^0)} F_*(S^0) & \xrightarrow{\nu} & F_*(E) \end{array}$$

Here τ_* is the isomorphism induced by the switch map $\tau\colon E \wedge F \longrightarrow F \wedge E$, and similarly for c. In Lecture 3 we shall assume that the relevant action of $E_*(S^0)$ on $E_*(E)$ makes $E_*(E)$ into a flat module. So if UCT 1 applies to ν, it will show that ν is an isomorphism, and hence m is an isomorphism. In any case, for each E and F we can check once for all whether this is so. If it is, then the results of Lecture 3 show that $E_*(F)$ is an extended comodule; that is, the isomorphism m throws the diagonal $\psi \otimes 1$ for $E_*(E) \otimes_{E_*(S^0)} F_*(S^0)$ onto the diagonal ψ for $E_*(F)$. In this case we have

$$\mathrm{Ext}^{p,*}_{E_*(E)}(E_*(X),\ E_*(F)) \cong \mathrm{Ext}^{p,*}_{E_*(S^0)}(E_*(X),\ F_*(S^0)) \ .$$

Since $F_*(S^0) \cong F^*(S^0)$ (as modules over $E_*(S^0)$), the statement ASS specialises to UCT 2. (Checking reveals that the edge-homomorphism behaves correctly.)

Since $F_*(X)$ admits an interpretation in terms of stable homotopy, one may ask whether UCT 1 can be related to ASS. Further thought reveals that this is unlikely, as the spectral sequence of UCT 1 involves a filtration starting from 0 and increasing indefinitely, while ASS involves a filtration starting from the whole group $[X,Y]_*$ and decreasing indefinitely. In particular, the edge-homomorphisms run in opposite directions.

I can now explain one motivation for interest in UCT 2. I would like to see further results of the general form of ASS; compare Novikov [23, 24]. It seems that UCT 2 is a special case which sufficiently exhibits many of the difficulties. I would therefore like to see new proofs of UCT 2, as general as possible, in the hope that they may generalise to proofs of ASS.

I will now turn to give further details of the

method mentioned in Note 15. For this purpose I will assume once for all that in what follows the functors E_* and F_* or F^* satisfy Milnor's additivity axiom on wedge-sums [21]. The first step is to deal with a special case which is very restrictive, but important for the applications.

Let X be a CW-complex or a connected spectrum. We assume that the spectral sequence

$$H_*(X;E_*(S^0)) \Longrightarrow E_*(X)$$

is trivial, that is, its differentials are zero. We observe that this spectral sequence is a spectral sequence of modules over $E_*(S^0)$; in the case of UCT 1 it is a spectral sequence of right modules, and in the case of UCT 2 it is a spectral sequence of left modules. The module structure of the E^2 term $H_*(X;E_*(S^0))$ is the obvious one. We assume that for each p, $H_p(X;E_*(S^0))$ is projective as a module over $E_*(S^0)$ (on the left or right as the case may be). Note that for this purpose it is not necessary to assume that $H_p(X)$ is free; for example, if $E_0(S^0)$ is a (commutative) principal ideal ring it will be sufficient if $H_p(X;E_0(S^0))$ is free. Then we conclude:

Proposition 17

With these assumptions, $E_*(X)$ is projective and X satisfies UCT 1 or UCT 2 (as the case may be). That is, the map

$$E_*(X) \otimes_{E_*(S^0)} F_*(S^0) \longrightarrow F_*(X)$$

or

$$F^*(X) \longrightarrow \mathrm{Hom}^*_{E_*(S^0)}(E_*(X), F^*(S^0))$$

is iso.

<u>Proof</u>. Let $E^r_{p,q}(0)$, $E^r_{p,q}(1)$ and $E_r^{p,q}(2)$ be the spectral sequences

$$H_*(X;E_*(S^0)) \Longrightarrow E_*(X)$$

$$H_*(X;F_*(S^0)) \Longrightarrow F_*(X)$$

$$H^*(X;F^*(S^0)) \Longrightarrow F^*(X)$$

It follows immediately from the assumptions on the spectral sequence $E^*_{**}(0)$ that $E_*(X)$ is projective.

The products ν yield homomorphisms

$$E^r_{p,*}(0) \otimes_{E_*(S^0)} F_*(S^0) \longrightarrow E^r_{p,*}(1)$$

$$E_r^{p,*}(2) \longrightarrow \mathrm{Hom}^*_{E_*(S^0)}(E^r_{p,*}(0), F^*(S^0))$$

as the case may be. These homomorphisms send $d^r \otimes 1$ into d^r, or d_r into $(d^r)^*$, as the case may be. (These assertions need detailed proof from the definitions of the spectral sequences, but it can be done using only formal properties of the products ν and the fact that $\otimes$ is right exact while Hom is left exact.) Because of the assumption that the spectral sequence $E^*_{**}(0)$ is trivial (which is essential here), the groups

$E^r_{p,*}(0) \otimes_{E_*(S^0)} F_*(S^0)$ (for $r \geq 2$), equipped with the boundaries $d^r \otimes 1$, form a (trivial) spectral sequence $E^r_{p,q}(3)$. Similarly, the groups $\mathrm{Hom}^*_{E_*(S^0)}(E^r_{p,*}(0), F^*(S^0))$, equipped with the boundaries $(d^r)^*$, form a (trivial) spectral sequence $E_r^{p,q}(4)$. We now have a map of spectral sequences

$$E^r_{p,q}(3) \longrightarrow E^r_{p,q}(1)$$

or

$$E_r^{p,q}(2) \longrightarrow E_r^{p,q}(4)$$

as the case may be. For $r = 2$ it becomes the obvious map

$$H_p(X; E_*(S^0)) \otimes_{E_*(S^0)} F_*(S^0) \longrightarrow H_p(X; F_*(S^0))$$

or

$$H^p(X; F^*(S^0)) \longrightarrow \mathrm{Hom}^*_{E_*(S^0)}(H_p(X; E_*(S^0)), F^*(S^0))$$

as the case may be. But since we are assuming that $H_p(X; E_*(S^0))$ is projective over $E_*(S^0)$ for each p, a theorem on ordinary homology shows that for $r = 2$ the map is iso. Therefore it is iso for all finite r, and the spectral sequence $E^r_{p,q}(1)$ or $E_r^{p,q}(2)$ is trivial.

We next deduce that the map

$$E^\infty_{p,*}(0) \otimes_{E_*(S^0)} F_*(S^0) \longrightarrow E^\infty_{p,*}(1)$$

or

$$E_\infty^{p,*}(2) \longrightarrow \mathrm{Hom}^*_{E_*(S^0)}(E^\infty_{p,*}(0), F^*(S^0))$$

is iso. (If X is not finite-dimensional, this needs

properties of E_* and F_* or F^* with respect to limits, but these follow from the axiom on wedge-sums.)

Let us now introduce notation for the filtration subgroups or quotient groups, as the case may be; say

$$G_{p,*}(0) = \mathrm{Im}(E_*(X^p) \longrightarrow E_*(X))$$

$$G_{p,*}(1) = \mathrm{Im}(F_*(X^p) \longrightarrow F_*(X))$$

$$G^{p,*}(2) = \mathrm{Coim}(F^*(X) \longrightarrow F^*(X^p)) \ .$$

The product $\vee$ yields us homomorphisms

$$G_{p,*}(0) \otimes_{E_*(S^0)} F_*(S^0) \longrightarrow G_{p,*}(1)$$

$$G^{p,*}(2) \longrightarrow \mathrm{Hom}^*_{E_*(S^0)}(G_{p,*}(0), F^*(S^0))$$

as the case may be. (Again, the verification uses only formal properties of the products $\vee$ and the fact that $\otimes$ is right exact while Hom is left exact.) Consider the following commutative diagrams.

$$\begin{array}{ccccccccc}
0 \longrightarrow & G_{p-1,*}(0) \otimes F_*(S^0) & \longrightarrow & G_{p,*}(0) \otimes F_*(S^0) & \longrightarrow & E^\infty_{p,*}(0) \otimes F_*(S^0) & \longrightarrow 0 \\
& \downarrow & & \downarrow & & \downarrow & \\
0 \longrightarrow & G_{p-1,*}(1) & \longrightarrow & G_{p,*}(1) & \longrightarrow & E^\infty_{p,*}(1) & \longrightarrow 0
\end{array}$$

$$
\begin{array}{ccccccc}
0 & \longrightarrow & E_\infty^{p,*}(2) & \longrightarrow & G^{p,*}(2) & \longrightarrow & \\
 & & \downarrow & & \downarrow & & \\
0 & \longrightarrow & \mathrm{Hom}^*(E^\infty_{p*}(0), F_*(S^0)) & \longrightarrow & \mathrm{Hom}^*(G_{p*}(0), F^*(S^0)) & \longrightarrow &
\end{array}
$$

$$
\begin{array}{ccccc}
\longrightarrow & G^{p-1,*}(2) & \longrightarrow & 0 \\
 & \downarrow & & \\
\longrightarrow & \mathrm{Hom}^*(G_{p-1*}(0), F^*(S^0)) & \longrightarrow & 0
\end{array}
$$

Here all the $\otimes$'s and Hom's are taken over $E_*(S^0)$. The first and last rows are exact because $E^\infty_{p,*}(0)$ is projective. An easy induction over p, using the short five lemma, now shows that

$$G_{p,*}(0) \otimes_{E_*(S^0)} F_*(S^0) \longrightarrow G_{p,*}(1)$$

or

$$G^{p,*}(2) \longrightarrow \mathrm{Hom}^*_{E_*(S^0)}(G_{p,*}(0), F^*(S^0))$$

is iso.

In the case of UCT 1, we now pass to direct limits and see that

$$E_*(X) \otimes_{E_*(S^0)} F_*(S^0) \longrightarrow F_*(X)$$

is iso. In the case of UCT 2, we first observe that the spectral sequence $E_r^{p,q}(2)$ satisfies the Mittag-Leffler condition for spectral sequences, and therefore

$$F^*(X) = \varprojlim_p G^{p,*}(2) \ .$$

Because

$$G_{p,*}(0) = G_{p-1,*}(0) \oplus E^\infty_{p,*}(0)$$

and

$$E_*(X) = \varinjlim_p G_{p,*}(0)$$

we have

$$\mathrm{Hom}^*_{E_*(S^0)}(E_*(X), F^*(S^0)) = \varprojlim_p \mathrm{Hom}^*_{E_*(S^0)}(G_{p,*}(0), F^*(S^0)) \ .$$

We can thus pass to inverse limits and see that

$$F^*(X) \longrightarrow \mathrm{Hom}^*_{E_*(S^0)}(E_*(X), F^*(S^0))$$

is iso. This proves Proposition 17.

We next need two further lemmas. For this purpose we assume that we can work in a suitable category in which we can do stable homotopy theory [7, 8, 25]. We assume that the theories E_* and F_* or F^* are defined on this category, and that E_* is represented by an object E in this category. The next two lemmas are stated for E, but they also apply to any other object (such as F, if we have an F.) We assume that E is the direct limit of a given system of finite CW-complexes E_α.

<u>Lemma 18</u>

For any object X and any class $e \in E_p(X)$ there is an E_α and a class $f \in E_p(S^p \wedge DE_\alpha)$ and a map

$g\colon S^p \wedge DE_\alpha \longrightarrow X$ such that $e = g_* f$.

Proof. Take a class $e \in E_p(X)$. Then there is a finite subcomplex $X' \subset X$ and a class $e' \in E_p(X')$ such that $i_* e' = e$. We may interpret e' as a class in $E^{-p}(DX')$; so e' may be represented by a map $h\colon DX' \longrightarrow S^{-p}E$. Since DX' is a finite complex and E is the direct limit of the E_α, we can factor h in the form

$$DX' \xrightarrow{k} S^{-p} \wedge E_\alpha \longrightarrow S^{-p} \wedge E .$$

That is, there is a class f in $E^{-p}(S^{-p} \wedge E_\alpha)$ such that $k^* f = e'$. Dualising back, f may be interpreted as a class in $E_p(S^p \wedge DE_\alpha)$, and we obtain a map

$$Dk\colon S^p \wedge DE_\alpha \longrightarrow X'$$

such that $(Dk)_* f = e'$. We have only to take

$$g = i(Dk)\colon S^p \wedge DE_\alpha \longrightarrow X .$$

This proves Lemma 18.

Lemma 19

For any object X there exists an object of the form

$$W = \bigvee_\beta S^{p(\beta)} \wedge DE_{\alpha(\beta)}$$

and a map $g\colon W \longrightarrow X$ such that

$$g_*\colon E_*(W) \longrightarrow E_*(X)$$

is epi.

The construction is immediate from Lemma 18, by allowing the class e in Lemma 18 to run over a set of generators for $E_*(X)$.

We now introduce the sort of resolution we need. By a "resolution of X with respect to E_*" we shall mean a diagram of the following form, with the properties listed below.

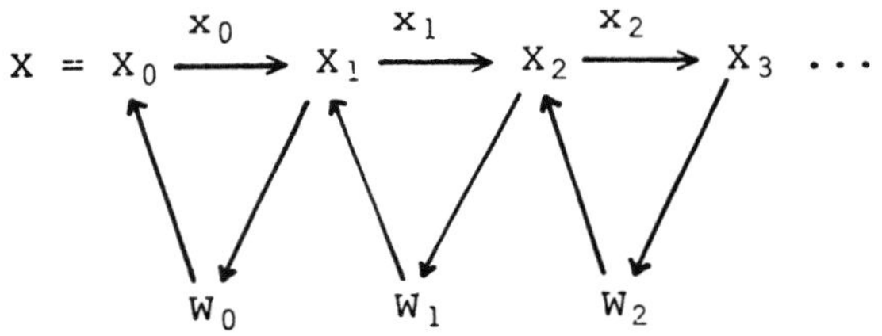

(i) The triangles

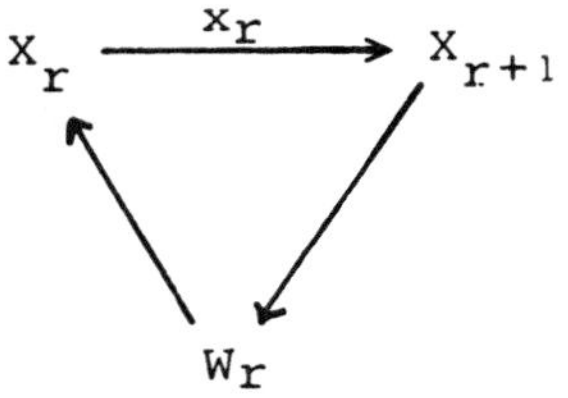

are exact (cofibre) triangles.

(ii) For each r,

$$(x_r)_*: E_*(X_r) \longrightarrow E_*(X_{r+1})$$

is zero.

(iii) For each r, $E_*(W_r)$ is projective over $E_*(S^0)$.

(iv) For each r, W_r satisfies UCT 1 or UCT 2, i.e. the map

$$E_*(X) \otimes_{E_*(S^0)} F_*(S^0) \longrightarrow F_*(X)$$

or

$$F^*(X) \longrightarrow \mathrm{Hom}^*_{E_*(S^0)}(E_*(X), F^*(S^0))$$

is iso.

In order to prove the existence of such resolutions, we introduce the following hypothesis.

Assumption 20

E is the direct limit of finite CW-complexes E_α for which

(i) $E_*(DE_\alpha)$ is projective over $E_*(S^0)$, and

(ii) DE_α satisfies UCT 1 or UCT 2, as the case may be, for the theory F_* or F^*.

In theory we can check this assumption for given E and F. In practice we usually prove it using Proposition 17, which requires strong hypotheses on DE_α but none on F. In practice E is a ring-spectrum, so the use of Proposition 17 involves checking the following two conditions.

(i) The spectral sequence

$$H^*(E_\alpha; E^*(S^0)) \Longrightarrow E^*(E_\alpha)$$

is trivial, and

(ii) For each p, $H^p(E_\alpha; E^*(S^0))$ is projective as a module over $E^*(S^0)$.

Examples.

(i) $E = S$, the sphere spectrum. Take $E_\alpha = S^n$; the conditions are trivially satisfied, and of course Assumption 20 is very easily verified directly.

(ii) $E = K(Z_p)$. The conditions of Proposition 17 are satisfied by any X. It is sufficient to let E_α run over any system of finite complexes whose limit is $K(Z_p)$.

(iii) $E = MO$. It is well known that

$$MO \simeq \bigvee_i S^{n(i)} K(Z_2) \simeq \prod_i S^{n(i)} K(Z_2) .$$

The conditions of Proposition 17 are satisfied by any X. It is sufficient to let E_α run over any system of finite complexes whose limit is MO.

(iv) $E = MU$. We have $H^p(MU; MU^q(S^0)) = 0$ unless p and q are even. Therefore the spectral sequence

$$H^*(MU; MU^*(S^0)) \Longrightarrow MU^*(MU)$$

is trivial. Again, $H^p(MU; MU^*(S^0))$ is free over $MU^*(S^0)$. It is sufficient to let E_α run over a system of finite complexes which approximate MU in the sense that

$$i_*: H_p(E_\alpha) \longrightarrow H_p(MU)$$

is iso for $p \le n$, while $H_p(E_\alpha) = 0$ for $p > n$.

(v) $E = MSp$. A simple adaptation of the method of S. P. Novikov [23, 24] from the unitary to the symplectic case shows that the spectral sequence

$$H^*(MSp; MSp^*(S^0)) \Longrightarrow MSp^*(MSp)$$

is trivial. Again, $H^p(MSp;MSp^*(S^0))$ is free over $MSp^*(S^0)$. The rest of the argument is as in (iv).

(vi) $E = \underline{BU}$. Let us recall that in the spectrum $\underline{BU}$ every even term is the space BU. We have $H^p(BU;\underline{BU}^q(S^0)) = 0$ unless p and q are even. Therefore the spectral sequence

$$H^*(BU;\underline{BU}^*(S^0)) \Longrightarrow \underline{BU}^*(BU)$$

is trivial. Again, $H^p(BU;\underline{BU}^*(S^0))$ is free over $\underline{BU}^*(S^0)$. It is sufficient to let E_α run over a system of finite complexes which approximate as in (iv) to the different spaces BU of the spectrum $\underline{BU}$.

(vii) $E = \underline{BO}$. Let us recall that in the spectrum $\underline{BO}$ every eighth term is the space BSp. I claim that the spectral sequence

$$H^*(BSp;\underline{BO}^*(S^0)) \Longrightarrow \underline{BO}^*(BSp)$$

is trivial. In fact, for each class $h \in H^{8p}(BSp(m))$ we can construct a real representation of Sp(m) whose Chern character begins with h; for each class $h \in H^{8p+4}(BSp(m))$ we can construct a symplectic representation of Sp(m) whose Chern character begins with h. The rest of the argument is as for (vi).

(viii) Cobordism and K-theory with coefficients. The reader will find further examples in Lecture 4.

Assumption 20 allows us to use the method of

Atiyah [6].

The next lemma will construct the resolutions we require; but we state it in a more general form, so that it will also allow us to compare resolutions. We suppose given a diagram of the following form.

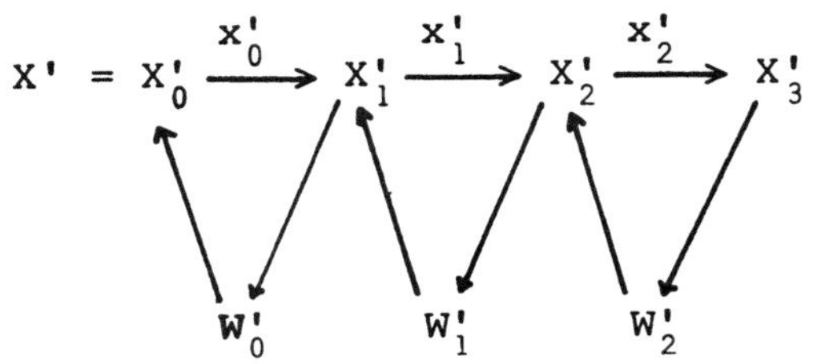

Here the triangles are supposed to be exact (cofibre) triangles, and

$$(x'_r)_*: E_*(X'_r) \longrightarrow E_*(X'_{r+1})$$

is zero for each r. We also suppose given a map $f: X \longrightarrow X'$.

<u>Lemma 21</u>

Under these conditions we can construct a resolution of X with respect to E_* which admits a map over f, in the sense that we can construct the following diagram so that the prisms are maps of exact (cofibre) triangles.

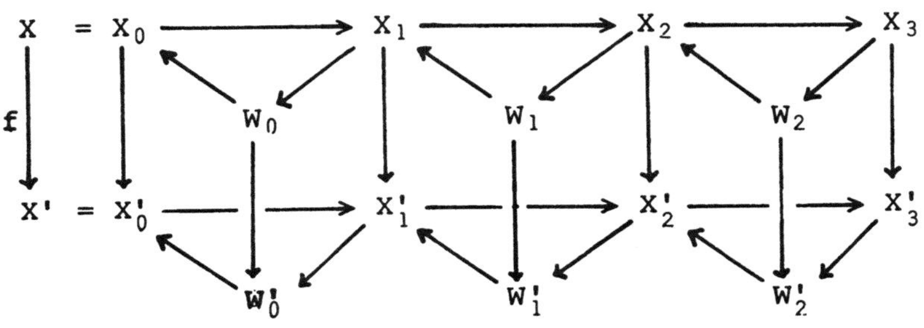

In order to construct a resolution of X with respect to E_*, we need only apply Lemma 21 to the case in which all the objects X'_r and W'_r are trivial.

Proof of Lemma 21. As an inductive hypothesis, suppose the diagram constructed up to the following map.

$$\begin{array}{c} X_r \\ \Big\downarrow{\scriptstyle f_r} \\ X'_r \end{array}$$

Form the following cofibre triangle.

$$\begin{array}{ccc} X_r & \xrightarrow{x'_r f_r} & X'_{r+1} \\ & \nwarrow \quad \swarrow & \\ & Z & \end{array}$$

Then we have the following commutative square.

$$\begin{array}{ccc} X_r & \longleftarrow & Z \\ {\scriptstyle f_r}\Big\downarrow & & \Big\downarrow \\ X'_r & \longleftarrow & W'_r \end{array}$$

Since $(x'_r f_r)_* = 0$, $E_*(Z) \longrightarrow E_*(X_r)$ is epi. By Lemma 19 we can construct a map $W_r \longrightarrow Z$ such that W_r has the form

$$W_r = \bigvee_\beta S^{p(\beta)} \wedge DE_{\alpha(\beta)}$$

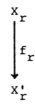

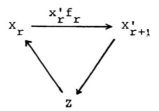

and $E_*(W_r) \longrightarrow E_*(Z)$ is epi. We now have the following commutative square.

$$\begin{array}{ccc} X_r & \longleftarrow & W_r \\ {\scriptstyle f_r}\downarrow & & \downarrow \\ X'_r & \longleftarrow & W'_r \end{array}$$

Here $E_*(W_r) \longrightarrow E_*(X_r)$ is epi. Form the following cofibre triangle.

$$\begin{array}{ccc} X_r & \xrightarrow{x_r} & X_{r+1} \\ & \nwarrow \quad \swarrow & \\ & W_r & \end{array}$$

This triangle can be mapped in the required way, and we have $(x_r)_* = 0$. This completes the induction.

We have constructed a resolution, because W_r inherits the property that $E_*(W_r)$ is projective from its summands $S^p \wedge DE_\alpha$, and similarly for UCT 1, UCT 2 (see Assumption 20). This proves Lemma 21.

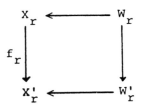

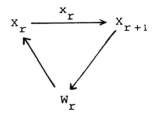

We will now construct the spectral sequences of UCT 1 and UCT 2, using Lemma 21 and the assumption that E_* and F_* or F^* are defined on a sufficiently large category in which we can do stable homotopy theory. Take a resolution of X with respect to E_*, as provided by Lemma 21. By applying the functor F_* or F^*, we obtain a spectral sequence. Now the sequence

$$0 \leftarrow E_*(X) \leftarrow E_*(W_0) \leftarrow E_*(W_1) \leftarrow E_*(W_2) \leftarrow \cdots$$

is a resolution of $E_*(X)$ by projective modules over $E_*(S^0)$. Since the W_r satisfy UCT 1 or UCT 2, the E^1-term of the spectral sequence is obtained by taking this projective resolution and applying $\otimes_{E_*(S^0)} F_*(S^0)$ or $\mathrm{Hom}_{E_*(S^0)}(\quad, F^*(S^0))$. Therefore the E^2-term is the required Tor or Ext.

We have to show that the spectral sequence is independent of the choice of resolution. Suppose given two resolutions, as follows.

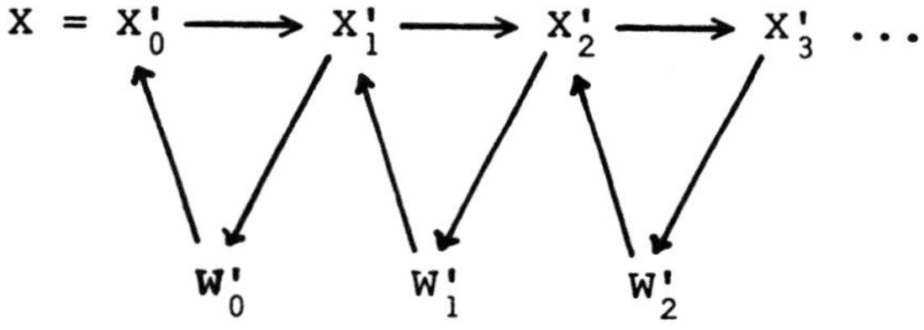

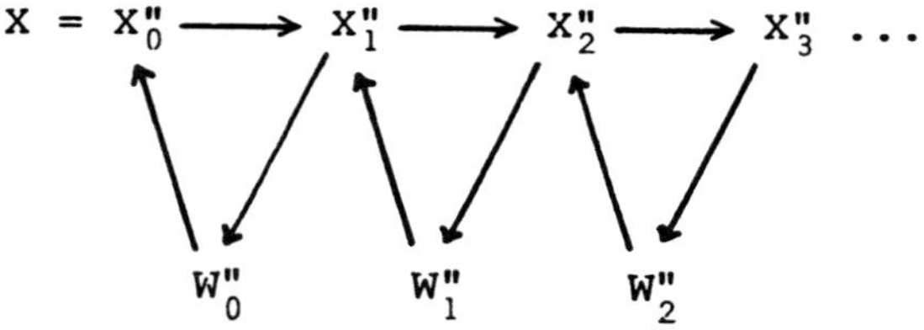

Then we can form the following diagram.

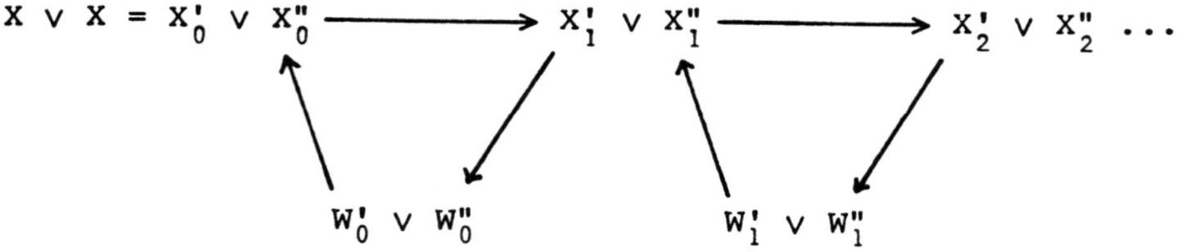

We can now apply Lemma 21 to the map $X \longrightarrow X \vee X$ of type (1,1). We obtain a third resolution and a third spectral sequence which admits comparison maps to or from the first two spectral sequences. ("To" for F_*, "from" for F^*.) But both these comparison maps are iso for $r = 2$ by the comparison theorem of homological algebra; therefore they are iso for all finite r.

It remains to discuss the convergence of these spectral sequences. Given a resolution of X, we can construct a direct limit X_∞ of the objects X_r (by forming a "telescope" or iterated mapping-cylinder). The object X_∞ has the property that

$$E_*(X_\infty) = \varinjlim_r E_*(X_r) = 0 .$$

In the case of UCT 1, for example, the spectral sequence converges in a perfectly satisfactory manner to $F_*(X_\infty, X_0)$. We therefore face the following question.

Problem 22

When can we assert that $E_*(X) = 0$ implies $F_*(X) = 0$ or $F^*(X) = 0$?

This is of course a special case of UCT 1 or UCT 2. When the answer is affirmative, we have (for example) $F_*(X_\infty) = 0$, $F_*(X_\infty, X_0) \cong F_*(X)$ and the spectral sequence of UCT 1 converges in a satisfactory way to $F_*(X)$.

Unfortunately the present state of our knowledge

on Problem 22 appears to be far from satisfactory*. Of course we know special cases; for example, if $E = S$, then $S_*(X) = 0$ implies that X is contractible, and so $F_*(X) = 0$, $F^*(X) = 0$. Again, if $E_*(X) = 0$, then as we vary Y, $E_*(X \wedge Y)$ is a homology functor of Y with zero coefficient groups, therefore zero. Thus the spectral sequence of KT 1 always converges.

At this point we pause to show that our spectral sequences can behave well even in cases which are known to be somewhat pathological.

Example 23. We consider UCT 2 for the case in which X is $K(Z)$, while E and F are the spectrum $\underline{BU}$. We can compute the ordinary homology of the spectrum $\underline{BU}$ by considering that of the space BU and passing to a direct limit; we find

$$H_n(\underline{BU}) = \begin{cases} Q & \text{if } n \text{ is even} \\ 0 & \text{if } n \text{ is odd .} \end{cases}$$

By George Whitehead's remark [31], this is equivalent to

$$\underline{BU}_n(K(Z)) = \begin{cases} Q & \text{if } n \text{ is even} \\ 0 & \text{if } n \text{ is odd .} \end{cases}$$

Now owing to the favourable structure of the ring $\underline{BU}_*(S^0)$, the computation of Ext over this ring reduces to computing Ext over Z. We find

---

* Note added in proof. A satisfactory answer to Problem 22 is now available.

$$\mathrm{Ext}^{p,q}_{\underline{BU}_*(S^0)}(\underline{BU}_*(K(Z)),\underline{BU}^*(S^0))$$

$$= \begin{cases} \mathrm{Ext}_Z(Q,Z) & \text{if } p = 1 \text{ and } q \text{ is even} \\ 0 & \text{otherwise.} \end{cases}$$

This agrees with the result of Hodgkin and Anderson [5, 17].

We will now make some comments on the situation whose exploration was pioneered by Conner and Floyd [14]. We assume that we have representing objects E and F, that E satisfies Assumption 20 and that F satisfies the following hypothesis.

<u>Assumption 24</u>

F is the direct limit of finite CW-complexes F_α for which

(i) $E_*(DF_\alpha)$ is projective over $E_*(S^0)$, and

(ii) DF_α satisfies UCT 1 for the theory F_*.

(Compare Assumption 20.) In practice we generally verify this assumption by using Proposition 17, as for Assumption 20.

<u>Examples</u>.

(i) $E = MU$, $F = \underline{BU}$. In the spectrum $\underline{BU}$ every even term is the space BU. For the space BU we have $H^p(BU;MU^q(S^0)) = 0$ unless p and q are even. Therefore the spectral sequence

$$H^*(BU;MU^*(S^0)) \Longrightarrow MU^*(BU)$$

is trivial. Again, $H^p(BU;MU^*(S^0))$ is free over $MU^*(S^0)$. As in Example (vi) on Assumption 20, it is sufficient to let F_α run over a system of finite complexes which approximate to the different spaces BU of the spectrum $\underline{BU}$ in the sense that

$$i_*: H_p(F_\alpha) \longrightarrow H_p(BU)$$

is iso for $p \leq n$, while $H_p(F_\alpha) = 0$ for $p > n$.

(ii) $E = MSp$, $F = \underline{BO}$. In the spectrum $\underline{BO}$ every eighth term is the space BSp. It follows from the work of Conner and Floyd [14] that the spectral sequence

$$H^*(BSp;MSp^*(S^0)) \Longrightarrow MSp^*(BSp)$$

is trivial. Again, $H^p(BSp;MSp^*(S^0))$ is free over $MSp^*(S^0)$. The rest of the argument is as in (i).

With these assumptions (especially 20 and 24) we have the following results for any X.

<u>Proposition 25</u>

We have

$$\mathrm{Tor}^{p*}_{E_*(S^0)}(E_*(X),F_*(S^0)) = 0 \quad \text{for} \quad p > 0 .$$

The spectral sequence of UCT 1 collapses, and its edge-homomorphism

$$E_*(X) \otimes_{E_*(S^0)} F_*(S^0) \longrightarrow F_*(X)$$

is iso.

Compare Conner and Floyd [14, pp. 60, 63]; but these authors state their theorem with the variance of UCT 3, and use finiteness assumptions.

<u>Proof</u>. It follows from Lemma 19 that given any object X, there exists an object W of the form

$$W = \bigvee_{\beta} S^{p(\beta)} \wedge DE_{\alpha(\beta)} \vee \bigvee_{\gamma} S^{p(\gamma)} \wedge DF_{\alpha(\gamma)}$$

and a map $g: W \longrightarrow X$ such that both

$$g_*: E_*(W) \longrightarrow E_*(X)$$

and

$$g_*: F_*(W) \longrightarrow F_*(X)$$

are epi. Arguing as in Lemma 21, we can now construct a resolution of X with respect to E_* which has the following extra properties.

(i) The objects W_r have the form

$$W_r = \bigvee_{\beta} S^{p(\beta)} \wedge DE_{\alpha(\beta)} \vee \bigvee_{\gamma} S^{p(\gamma)} \wedge DF_{\alpha(\gamma)} .$$

(ii) Not only the homomorphisms

$$(x_r)_*: E_*(X_r) \longrightarrow E_*(X_{r+1})$$

but also the homomorphisms

$$(x_r)_*: F_*(X_r) \longrightarrow F_*(X_{r+1})$$

are zero for all r.

Then the sequence

$$0 \longleftarrow E_*(X) \longleftarrow E_*(W_0) \longleftarrow E_*(W_1) \longleftarrow E_*(W_2) \ \ldots$$

is a resolution of $E_*(X)$ by projectives over $E_*(S^0)$. Consider the following diagram.

$$\begin{array}{ccccc}
E_*(W_0) \otimes_{E_*(S^0)} F_*(S^0) & \longleftarrow & E_*(W_1) \otimes_{E_*(S^0)} F_*(S^0) & \longleftarrow & E_*(W_2) \otimes_{E_*(S^0)} F_*(S^0) \ldots \\
\downarrow \nu_0 & & \downarrow \nu_1 & & \downarrow \nu_2 \\
F_*(W_0) & \longleftarrow & F_*(W_1) & \longleftarrow & F_*(W_2) \ \ldots
\end{array}$$

The homomorphisms ν_r are iso. The lower row is exact by construction. Therefore the upper row is exact, and

$$\mathrm{Tor}^{p,*}_{E_*(S^0)}(E_*(X), F^*(S^0)) = 0 \quad \text{for} \quad p > 0 .$$

We can now consider the following diagram.

$$\begin{array}{ccccccc}
0 & \longleftarrow & E_*(X) \otimes_{E_*(S^0)} F_*(S^0) & \longleftarrow & E_*(W_0) \otimes_{E_*(S^0)} F_*(S^0) & \longleftarrow & \\
& & \downarrow \nu & & \downarrow \nu_0 & & \\
0 & \longleftarrow & F_*(X) & \longleftarrow & F_*(W_0) & \longleftarrow &
\end{array}$$

$$\longleftarrow E_*(W_1) \otimes_{E_*(S^0)} F_*(S^0)$$

$$\Big\downarrow \nu_1$$

$$\longleftarrow F_*(W_1)$$

The upper row is exact because $\otimes$ is right exact, and the lower row is exact by construction. The maps ν_0 and ν_1 are iso. Therefore ν is iso. This completes the proof of Proposition 25.

Since we now know what happens to UCT 1 in this situation, it is natural to ask what happens to UCT 2. For this we need slightly more data. We suppose given two ring-spectra E, F and a map $i: E \longrightarrow F$ of ring-spectra. (For example, $E = MU$ and $F = \underline{BU}$, or $E = MSp$ and $F = \underline{BO}$.) We suppose given also a spectrum G which is a module-spectrum over F, and therefore a module-spectrum over E via i. (For example, $G = F$.) (It would presumably be sufficient to suppose given enough products in homology and cohomology, but let us spare ourselves the details.) We suppose that the pair of theories (E,G) satisfies Lemma 21, so that we can construct a spectral sequence for computing G_* or G^* from E_* as in UCT 1 or UCT 2; we also suppose that the pair of theories (F,G) satisfies Lemma 21, so that we can construct a spectral sequence for computing G_* or G^* from F_* as in UCT 1 or UCT 2.

<u>Proposition 26</u>

(i) The spectral sequence for computing G_* from E_* coincides with the spectral sequence for computing G_* from F_*.

(ii) The spectral sequence for computing G^* from E_* coincides with the spectral sequence for computing G^* from F_*.

<u>Note</u>. By specialising Proposition 26(i) to the case $G = F$, we obtain a result agreeing with Proposition 25; for of course the spectral sequence for computing F_* from F_* collapses.

Proposition 26 will follow almost immediately from the following lemma.

<u>Lemma 27</u>

(i) If $E_*(W)$ is projective over $E_*(S^0)$, then $F_*(W)$ is projective over $F_*(S^0)$.

(ii) If

$$E_*(W) \otimes_{E_*(S^0)} G_*(S^0) \longrightarrow G_*(W)$$

is iso, then

$$F_*(W) \otimes_{F_*(S^0)} G_*(S^0) \longrightarrow G_*(W)$$

is iso.

(iii) If

$$G^*(W) \longrightarrow \mathrm{Hom}^*_{E_*(S^0)}(E_*(W), G^*(S^0))$$

is iso, then

$$G^*(W) \longrightarrow \mathrm{Hom}^*_{F_*(S^0)}(F_*(W), G^*(S^0)) \quad \text{is iso.}$$

<u>Proof</u>.

(i) $F_*(W) \cong E_*(W) \otimes_{E_*(S^0)} F_*(S^0)$, by Proposition 25. So if $E_*(W)$ is projective over $E_*(S^0)$, $F_*(W)$ is projective over $F_*(S^0)$.

(ii) Consider the following commutative diagram.

$$\begin{array}{ccc}
E_*(W) \otimes_{E_*(S^0)} F_*(S^0) \otimes_{F_*(S^0)} G_*(S^0) & \xrightarrow{1 \otimes \nu} & E_*(W) \otimes_{E_*(S^0)} G_*(S^0) \\
\downarrow{\scriptstyle \nu\otimes 1} & & \downarrow{\scriptstyle \nu} \\
F_*(W) \otimes_{F_*(S^0)} G_*(S^0) & \xrightarrow{\nu} & G_*(W)
\end{array}$$

The left-hand column is iso by Proposition 25, the right-hand column is iso by assumption, and the top row is trivially iso. Therefore the bottom row is iso.

(iii) Consider the following commutative diagram.

$$\begin{array}{ccc}
G^*(W) & \xrightarrow{\nu} & \mathrm{Hom}_{F_*(S^0)}(F_*(W), G^*(S^0)) \\
\Big\downarrow{\scriptstyle \nu} & & \Big\downarrow{\scriptstyle \nu^*} \\
 & & \mathrm{Hom}_{F_*(S^0)}\left(F_*(S^0) \otimes_{E_*(S^0)} E_*(W),\ G^*(S^0)\right) \\
 & & \Vert \\
\mathrm{Hom}_{E_*(S^0)}(E_*(W), G^*(S^0)) & \xrightarrow{\nu_*} & \mathrm{Hom}_{E_*(S^0)}\left(E_*(W),\ \mathrm{Hom}_{F_*(S^0)}(F_*(S^0),\ G^*(S^0))\right)
\end{array}$$

The result follows as in part (ii).

Proof of Proposition 26. Take any resolution of X over E_*, say the following.

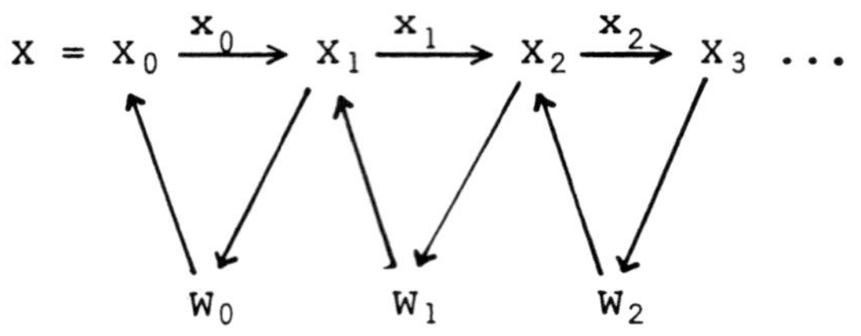

Here the objects W_r are supposed to satisfy UCT 1 or UCT 2 with respect to the functors E_* and G_* or G^*. We will show that it qualifies as a resolution of X over F_*. In fact, since

$$(x_r)_*: E_*(X_r) \longrightarrow E_*(X_{r+1})$$

is zero, the homomorphism

$$(x_r)_*: F_*(X_r) \longrightarrow F_*(X_{r+1})$$

is zero by Proposition 25. The remaining statements which need to be checked are provided by Lemma 27. Proposition 26 follows immediately.

Example. For any X we have

$$\mathrm{Ext}^{p,*}_{MU_*(S^0)}(MU_*(X), \underline{BU}^*(S^0)) = 0 \quad \text{for} \quad p > 1 .$$

This follows immediately from Proposition 26, since the result is trivial for

$$\mathrm{Ext}^{p,*}_{\underline{BU}_*(S^0)}(\underline{BU}_*(X),\underline{BU}^*(S^0))^{\dagger}.$$

The following result is required for use in Lecture 3.

Lemma 28

If $E = \underline{BO}$, $\underline{BU}$, MO, MU, MSp, S or $K(Z_p)$ then $E_*(E)$ is flat as a module over $E_*(S^0)$.

Proof. The cases E = MO, S and $K(Z_p)$ are trivial. In the cases E = MU, MSp we can apply the spectral sequence

$$H_*(E;E_*(S^0)) \Longrightarrow E_*(E)$$

to show that $E_*(E)$ is projective over $E_*(S^0)$; in the case E = MSp this involves remarking that the spectral sequence is trivial, by duality with the spectral sequence

$$H^*(MSp;MSp^*(S^0)) \Longrightarrow MSp^*(MSp)$$

which is known to be trivial (see Assumption 20, Example (v)). In the cases $E = \underline{BU}$, $\underline{BO}$ we apply this argument to the spaces BU, BSp to show that the modules $\underline{BU}_*(BU)$, $\underline{BO}_*(BSp)$ are projective (compare Assumption 20, examples (vi), (vii)). We then remark that a direct limit of projective modules is flat. This proves Lemma 28.

---

† Note added in proof. I have been asked to say explicitly at this point that UCT2 gives the following exact sequence.

$$0 \to \mathrm{Ext}^{1*}_{MU_*(S^0)}\Big(MU_*(X),\underline{BU}^*(S^0)\Big) \to \underline{BU}^*(X) \to \mathrm{Hom}^*_{MU_*(S^0)}\Big(MU_*(X),\underline{BU}^*(S^0)\Big) \to 0$$

LECTURE 2. THE ADAMS SPECTRAL SEQUENCE

In this lecture I want to discuss the prospects of setting up an "Adams spectral sequence" [1, 2, 15] using a generalised homology or cohomology theory. Everything is to be taken as provisional, or as work in progress, and no proofs will be given.

I shall assume that we can work in some stable category, like those supplied by Boardman [7, 8] and Puppe [25]. I shall also suppose that we are given a homology or cohomology functor to use in our constructions. I will suppose that this functor takes values in an abelian category. As long as we are talking generalities, we can then suppose that the functor is covariant; because if it is contravariant, we can replace the abelian category by its opposite. We will write E_* for this homology functor.

I suggest that we now adopt a construction reminiscent of those constructions for Ext which avoid using projectives and injectives. More precisely, I suggest we proceed as follows. Suppose given two objects X, Y in our stable category. Consider diagrams of the following form.

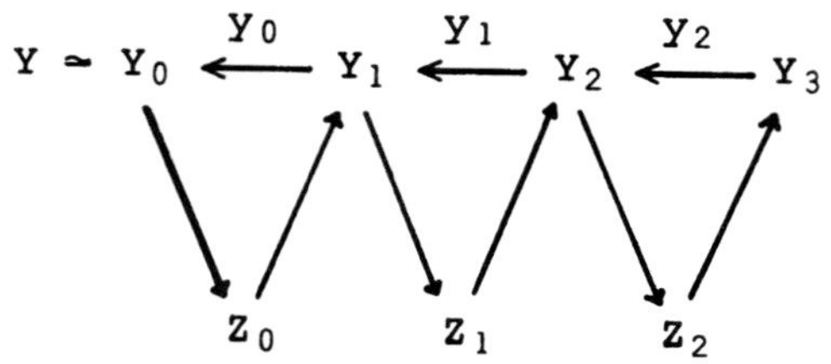

Here the notation $Y \simeq Y_0$ means a homotopy equivalence; and the triangles are supposed to be exact (cofibre) triangles in our stable category. We restrict attention to the diagrams such that

$$E_*(y_r) = 0 : E_*(Y_{r+1}) \longrightarrow E_*(Y_r)$$

for each $r \geq 0$; this is the crucial condition. In this case the sequence

$$0 \longrightarrow E_*(Y) \longrightarrow E_*(Z_0) \longrightarrow E_*(Z_1) \longrightarrow E_*(Z_2) \longrightarrow \cdots$$

is exact. We call such diagrams "filtrations" of Y. If we wish, we can suppose without loss of generality that each y_r is an inclusion map (replace Y_0 by a "telescope").

By mapping X into such a filtration of Y we get a spectral sequence; but this is not yet the spectral sequence we seek. However, we can take all possible filtrations of Y and consider them as the objects of a directed category (in the sense of Grothendieck). (Since I am omitting proofs, I will omit certain details as to how this is done, although they were given in the original lecture.) From each filtration we get a spectral sequence, and we can now take the direct limit of all these spectral sequences; this is the spectral sequence I suggest. Let us call it $SS(X,Y;E_*)$.

I will also omit some arguments in favour of this definition, although they were given in the original lecture.

At this level one should already be able to set up

some formal properties of the spectral sequence. For example, suppose that we have a functor T from one abelian category to another, and that both E_* and TE_* are homology functors. (For examples, see Lecture 1, Proposition 25, or Lecture 4.) Then there clearly is a homomorphism

$$SS(X,Y;E_*) \longrightarrow SS(X,Y,TE_*) ,$$

because every diagram which qualifies as a filtration for E_* also qualifies as a filtration for TE_*. (Compare Lecture 1, Proposition 26.) If E_* and F_* are homology functors which mutually determine each other in this way, then

$$SS(X,Y;E_*) \cong SS(X,Y;F_*) .$$

(For examples, see Lecture 4.)

We can now raise the following question. Suppose that X and Y are finite complexes, and that we consider only filtrations in which each Y_r is equivalent to a finite complex. Do these yield in the limit the same spectral sequence as if we did not restrict the filtrations? This is probably true if the homology theory E_* has sufficiently strong finiteness properties.

We can now consider the behaviour of our constructions under S-duality. Do we have

$$SS(X,Y,E_*) \cong SS(DY,DX,E_*D) \ ?$$

(Note that E_*D is a cohomology theory defined on finite complexes.) This problem leads one to consider also a "dual" approach to the construction.

We consider diagrams of the following form.

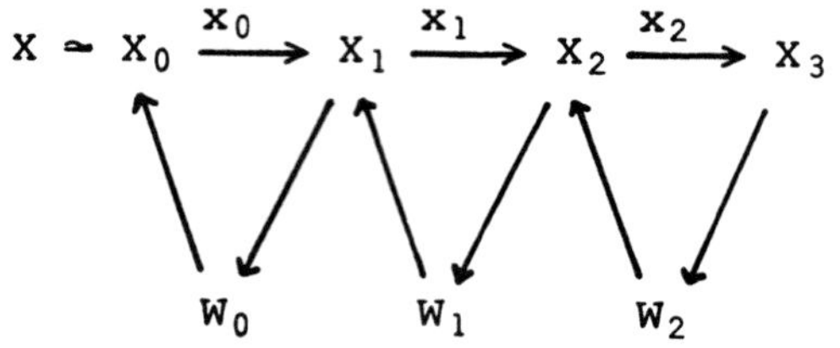

As above, the notation $X \simeq X_0$ means a homotopy equivalence, and the triangles are supposed to be exact (cofibre) triangles in our stable category. We restrict attention to the diagrams such that

$$E_*(x_r) = 0 : E_*(X_r) \longrightarrow E_*(X_{r+1})$$

for each $r \geq 0$. In this case the sequence

$$0 \longleftarrow E_*(X) \longleftarrow E_*(W_0) \longleftarrow E_*(W_1) \longleftarrow E_*(W_2) \longleftarrow \cdots$$

is exact. We call such diagrams "filtrations" of X. If we wish, we can suppose without loss of generality that each x_r is an inclusion map (replace X_0 by a "telescope").

By mapping such a filtration of X into Y we get a spectral sequence. The suggestion would be to vary the filtration (inversely) and take a direct limit of the resulting spectral sequences. Does this give the same spectral sequence as before?

Evidently the situation is like that in homological algebra; there we can define $\mathrm{Ext}^*(L,M)$ by resolving L, or by resolving M, and we want to show that the result is the same. The proof there, as we know, is to resolve both of

them, and show that that gives the same result as resolving either one. Similarly here; one should consider a filtration of X, and also a filtration of Y, and one should try to get a spectral sequence by mapping one to the other. Then one should take a double direct limit, and show that this gives the same spectral sequence as one obtains by filtering either X or Y alone. I haven't tried to write down any details about this.

If one can attain this sort of manipulative ability, one ought to be able to set up various formal properties of the spectral sequences without further assumptions on E_*. For example, there should be a pairing

$$SS(Y,Z;E_*) \otimes SS(X,Y;E_*) \longrightarrow SS(X,Z;E_*)$$

which on the E_∞ level is given by composition.

The next step would be to compute the E_2 term of our spectral sequence. We are supposing that E_* takes values in an abelian category, so we can define Ext by classifying long exact sequences. It is reasonable to hope that we can define a homomorphism from the E_2 term to $\mathrm{Ext}^{**}(E_*(X),E_*(Y))$. The question would be, when can we prove that this homomorphism is an isomorphism? For this purpose one obviously needs to choose the right category, so as to obtain the right Ext groups. More precisely, we need to arrange a very close correspondence between the algebra and the geometry, so that there is some algebraic situation which

gives us a legitimate calculation of the Ext groups and which can be realised geometrically.

At this point all suggestions for proceeding assume that our functor is represented by a spectrum E.

(i) The original formulation asks us to work in cohomology, and consider $E^*(X)$, $E^*(Y)$ as modules over the ring $E^*(E)$ of cohomology operations [1, 2, 23, 24]. This approach has various disadvantages.

(a) In the generalised case $E^*(E)$ is a topologised ring, and $E^*(X)$, $E^*(Y)$ are topologised modules over the topologised ring $E^*(E)$. We have to take account of the topology [24]. Topologised modules usually fail to form an abelian category, owing to the existence of maps $f: L \longrightarrow M$ which are isomorphisms of the module structure, and continuous, but such that f^{-1} is not continuous.

(b) We cannot assert that $E^q(E) = 0$ for $q < 0$; we may have non-zero cohomology operations which lower dimension by any prescribed amount, as well as ones which raise it. Similar remarks apply to our modules. Both (a) and (b) mean that our constructions and calculations lose a certain element of finiteness which is present in the classical case.

(c) By means of examples (which I will now omit, although they were given in the original lecture) we see that even in the classical case of ordinary cohomology with Z_p coefficients, approach (i) only works under finiteness assumptions on Y. In the generalised case, we may see this as follows.

We wish to consider filtrations of Y in which each object Z_r is "free"; in particular, $E^*(Z_r)$ should be "free" in some sense applicable to topologised modules, and we should have

$$[X,Z_r]_* \cong \mathrm{Hom}_{E^*(E)}(E^*(Z_r),E^*(X)) .$$

Since we wish to know about maps from X to Z_r and from Z_r to E, this means in practice that we must stick to the case in which Z_r is both a sum and a product of suspensions S^nE of E. And again, this means in practice that we must stick to the case in which E is connected and Z_r is a countable sum,

$$Z_r = \bigvee_{i=1}^{\infty} S^{n(i)}E ,$$

in which $n(i) \longrightarrow \infty$ as $i \longrightarrow \infty$. In other words, we are compelled to prove or assume that $E^*(Y)$ admits a resolution by "free" topologised graded modules which have only a finite number of "generators" in dimensions less than n (for each n). Although Novikov [24] arranges his work somewhat differently, it is essentially for this purpose that he relies on finiteness properties of E^* which are true in the case $E = MU$ (see Lecture 5). The corresponding properties are unknown for $E = MSp$, and definitely false for $E = S$,

although the Adams spectral sequence works for these spectra in some cases at least.

It may be seen from the examples that trouble (c) arises from a double dualisation. The spectral sequence is covariant in Y, but by taking $E^*(Y)$ we are taking a contravariant functor of Y, and then by taking $\mathrm{Ext}^{**}_{E^*(E)}(E^*(Y), E^*(X))$ we are taking a contravariant functor of $E^*(Y)$. This leads to the next approach.

(ii) The next approach would ask us to follow Cartan and Douady [15], and work in homology, considering $E_*(X)$ and $E_*(Y)$ as modules over the ring $E^*(E)$. In the classical case $E = K(Z_p)$ this works quite well. This is partly owing to the fact that $E_*(E)$ is then an injective module over the ring $E^*(E)$; but this fails to generalise to cases in which $E_*(S^0)$ is not a field. In general the ring $E^*(E)$ retains its previous disadvantages, and this approach suffers from being a compromise or half-way house between (i) and (iii). The way ahead appears to lie in a more whole-hearted acceptance of the idea that homology is better than cohomology.

(iii) My final suggestion is that we should work wholly in homology, and consider $E_*(X)$, $E_*(Y)$ as comodules with respect to the coalgebra $E_*(E)$. We use $E_*(S^0)$ as the ground ring for our comodules etc. The necessary details

are given in Lecture 3. Of course, we need some data for this; in fact, we need to assume that E is a ring-spectrum and $E_*(E)$ is flat over $E_*(S^0)$. This is true for the spectra mentioned in Lecture 1, Lemma 28. Everything now works much better. The comodules $E_*(X)$, $E_*(Y)$ and the coalgebra $E_*(E)$ are discrete; in typical cases we have $E_q(X) = 0$ for sufficiently large negative q, and $E_q(E) = 0$ for $q < 0$. The comodules form an abelian category. Our constructions and calculations regain that element of finiteness which we lost before.

In order to compute $\mathrm{Ext}^{**}_{E_*(E)}(E_*(X), E_*(Y))$, it is sufficient to take a resolution of $E_*(X)$ by comodules which are projective over $E_*(S^0)$, and a resolution of $E_*(Y)$ by extended comodules; the latter play the part of "relative injectives". Both sorts of resolution can be constructed geometrically. For the first, we require a filtration of X such that $E_*(W_r)$ is projective over $E_*(S^0)$ for each r. Such a filtration can be constructed by Lemma 21 of Lecture 1. Moreover, we see that such filtrations are cofinal in the set of all filtrations of X. For the second, we require a filtration of Y such that $E_*(Z_r)$ is an extended comodule for each r. Such a filtration can be constructed in the following way. Let the structure maps of the ring spectrum E be $\mu: E \wedge E \longrightarrow E$ and $i: S^0 \longrightarrow E$. Suppose we have constructed Y_r; the induction starts with $Y_0 = Y$. Take

$Z_r = E \wedge Y_r$, and form the map

$$Y_r \simeq S^0 \wedge Y_r \xrightarrow{i \wedge 1} E \wedge Y_r = Z_r .$$

Then $E_*(Y_r) \longrightarrow E_*(Z_r)$ is mono, since it is defined to be $\pi_*(E \wedge Y_r) \longrightarrow \pi_*(E \wedge E \wedge Y_r)$, and this has a one-sided inverse induced by $E \wedge E \wedge Y_r \xrightarrow{\mu \wedge 1} E \wedge Y_r$. The comodule $E_*(Z_r)$ is extended, by the results of Lecture 3. Form the following cofibre triangle.

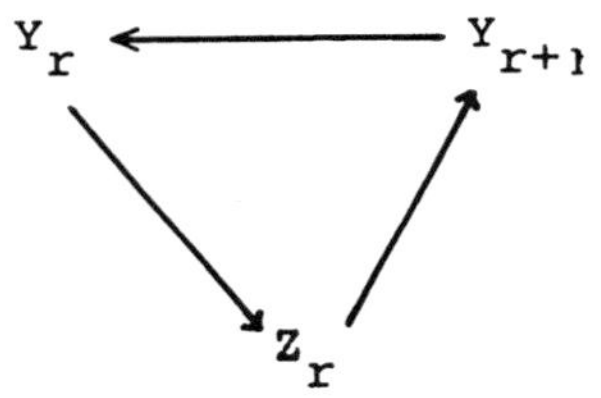

Then $E_*(Y_{r+1}) \longrightarrow E_*(Y_r)$ must be zero. This completes the induction. By adding a few details, we see that such filtrations are cofinal in the set of all filtrations of Y.

We may say that at the present time approach (iii) seems to be promising.

The final step, of course, would be to discuss the convergence of the spectral sequence. I would like to defer this question.

LECTURE 3 HOPF ALGEBRA AND COMODULE STRUCTURE

In the classical case of ordinary cohomology with coefficients Z_p, the mod p Steenrod algebra A^* is a Hopf algebra, and it acts on the left on the cohomology of any space, so that we have an action map $A^* \otimes H^* \longrightarrow H^*$. If we dualise by applying $\mathrm{Hom}_{Z_p}(\ ,Z_p)$, we see that the dual A_* of the Steenrod algebra is also a Hopf algebra; and if the homology H_* of a space is locally finitely generated, we have a coaction map $H_* \longrightarrow A_* \otimes H_*$. (The finiteness condition is actually unnecessary, but we do not need to spend time on that here.)

It is the object of this lecture to see how the material mentioned above generalises to the case of a generalised homology theory. We will begin by stating our assumptions; then we will list the structure maps we propose to introduce, and list their principal formal properties. Next we will give the definitions of the structure maps, and comment on the proofs of the formal properties. Then we give two propositions which relate A_* to A^* in the generalised case. Finally, we use these two propositions to show that if we specialise to the classical case of ordinary cohomology with Z_p coefficients, all our structure maps specialise to those classically considered.

It will be convenient to write as if we are working in a stable category in which we have smash-products with the usual properties; but if the reader objects to this, our statements can be "demythologised" by known methods. We shall suppose given a ring-spectrum E, so that we are given a product map $\mu: E \wedge E \longrightarrow E$ and a unit map $i: S^0 \longrightarrow E$. These are supposed to have the usual properties; that is, the following diagrams are homotopy-commutative.

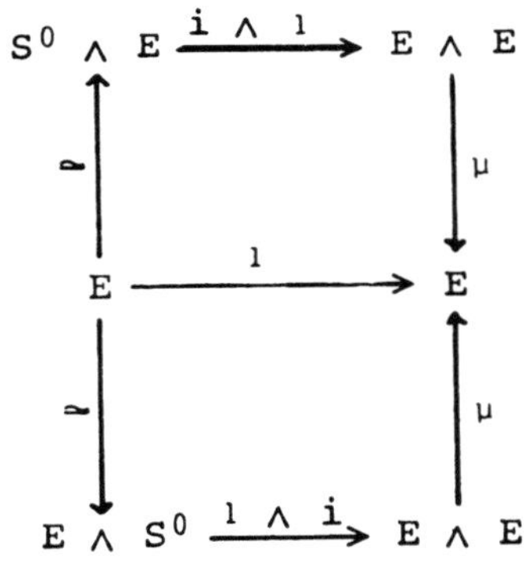

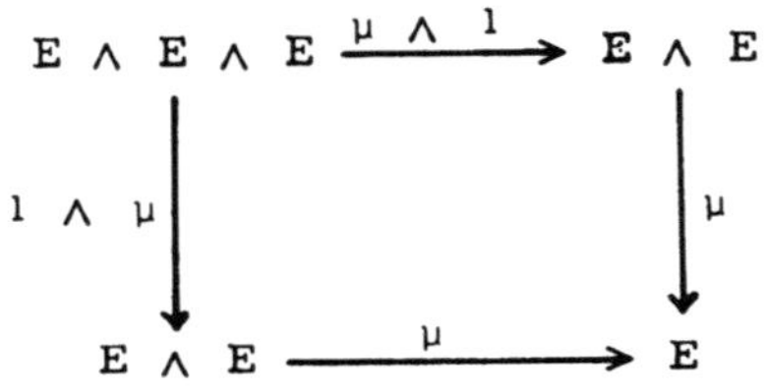

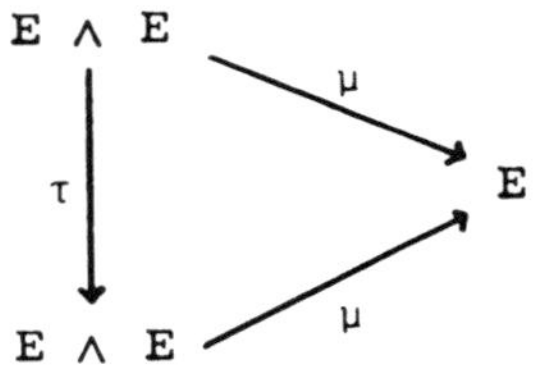

Here τ is the usual switch map.

We recall that the homology groups of a spectrum X with coefficients in E are given by

$$E_n(X) = [S^n, E \wedge X] = \pi_n(E \wedge X) .$$

- 58 -

The classical case is given by taking E to be the Eilenberg-MacLane spectrum $K(Z_p)$. The analogue of A_* in the generalised case is therefore $E_*(E) = \pi_*(E \wedge E)$, the homology of E with coefficients in E. The analogue of Z_p is $E_*(S^0) = \pi_*(E)$. Since E is a ring-spectrum, we have various products. More precisely, suppose given a pairing $\mu: E \wedge F \longrightarrow G$ of spectra. Then we shall have to consider three products, which appear in the following commutative diagram.

$$\begin{array}{ccc}
\pi_*(E \wedge X) \otimes \pi_*(F \wedge Y) & \xrightarrow{\nu} & \pi_*(G \wedge X \wedge Y) \\
\uparrow{\scriptstyle \tau_* \otimes 1} & & \uparrow{\scriptstyle (\tau \wedge 1)_*} \\
\pi_*(X \wedge E) \otimes \pi_*(F \wedge Y) & \xrightarrow{m} & \pi_*(X \wedge G \wedge Y) \\
\downarrow{\scriptstyle 1 \otimes \tau_*} & & \downarrow{\scriptstyle (1 \wedge \tau)_*} \\
\pi_*(X \wedge E) \otimes \pi_*(Y \wedge F) & \xrightarrow{\nu'} & \pi_*(X \wedge Y \wedge G)
\end{array}$$

Here the product ν is the usual external homology product, as used (for example) in Lecture 1, Note 7. The product ν' is a back-to-front version of ν. The product m is defined as follows. Suppose given maps

$$f: S^p \longrightarrow X \wedge E, \qquad g: S^q \longrightarrow F \wedge Y .$$

Then $m(f \otimes g)$ is the following composite.

$$S^p \wedge S^q \xrightarrow{f \wedge g} X \wedge E \wedge F \wedge Y \xrightarrow{1 \wedge \mu \wedge 1} X \wedge G \wedge Y .$$

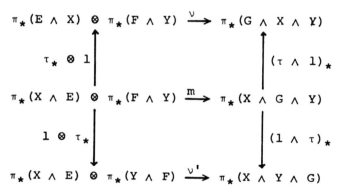

Since it is important for us in this lecture to keep factors in their correct order, we will use m as our basic product. By taking $X = S^0$ or $Y = S^0$, we obtain the following special cases.

$$m: \pi_p(E) \otimes \pi_q(F \wedge Y) \longrightarrow \pi_{p+q}(G \wedge Y)$$

$$m: \pi_p(X \wedge E) \otimes \pi_q(F) \longrightarrow \pi_{p+q}(X \wedge G)$$

$$m: \pi_p(E) \otimes \pi_q(F) \longrightarrow \pi_{p+q}(G) .$$

In particular, $\pi_*(E)$ is an anticommutative ring with unit. For any Y, $\pi_*(E \wedge Y)$ is a left module over $\pi_*(E)$; the product map

$$m: \pi_*(E) \otimes \pi_*(E \wedge Y) \longrightarrow \pi_*(E \wedge Y)$$

is the usual one, and coincides with the map μ considered in UCT 2 (see Lecture 1, Note 2). For any X, $\pi_*(X \wedge E)$ is a right module over $\pi_*(E)$. The product

$$m: \pi_*(X \wedge E) \otimes \pi_*(E \wedge Y) \longrightarrow \pi_*(X \wedge E \wedge Y)$$

factors to give a map

$$\pi_*(X \wedge E) \otimes_{\pi_*(E)} \pi_*(E \wedge Y) \longrightarrow \pi_*(X \wedge E \wedge Y),$$

which we also call m.

We have product maps

$$m: \pi_*(E) \otimes \pi_*(E \wedge E) \longrightarrow \pi_*(E \wedge E)$$

$$m: \pi_*(E \wedge E) \otimes \pi_*(E) \longrightarrow \pi_*(E \wedge E),$$

and thus $\pi_*(E \wedge E)$ becomes a bimodule over $\pi_*(E)$. It should be noted that the two actions of $\pi_*(E)$ on $\pi_*(E \wedge E)$

are in general quite distinct; this is the main difference between the generalised case and the classical case, in which we have only one action of Z_p on A_*. The presence of these two actions means that the generalised case demands a little more care than the classical case.

We now assume that $\pi_*(E \wedge E)$ is flat as a right module over $\pi_*(E)$ (using the right action). By using the switch map

$$\tau: E \wedge E \longrightarrow E \wedge E$$

to interchange the two factors, we check that it is equivalent to assume that $\pi_*(E \wedge E)$ is flat as a left module over $\pi_*(E)$ (using the left action). This hypothesis is somewhat restrictive, but it is satisfied in many important cases, notably the cases

$$E = \underline{BO}, \underline{BU}, MO, MU, MSp, S \text{ and } K(Z_p)$$

(see Lecture 1, Lemma 28).

With this hypothesis, we will see that $\pi_*(E \wedge E)$ is a Hopf algebra in a fully satisfactory sense, and that for any spectrum X, $\pi_*(E \wedge X)$ is a comodule over the coalgebra $\pi_*(E \wedge E)$. We will now make this more precise by listing the structure maps we shall introduce, and giving their principal properties.

The structure maps comprise a product map

$$\phi: \pi_*(E \wedge E) \otimes \pi_*(E \wedge E) \longrightarrow \pi_*(E \wedge E),$$

two "unit" maps

$$\eta_L : \pi_*(E) \longrightarrow \pi_*(E \wedge E)$$

$$\eta_R : \pi_*(E) \longrightarrow \pi_*(E \wedge E)$$

a counit map

$$\varepsilon : \pi_*(E \wedge E) \longrightarrow \pi_*(E)$$

a canonical anti-automorphism

$$c : \pi_*(E \wedge E) \longrightarrow \pi_*(E \wedge E)$$

a diagonal map

$$\psi = \psi_E : \pi_*(E \wedge E) \longrightarrow \pi_*(E \wedge E) \otimes_{\pi_*(E)} \pi_*(E \wedge E)$$

and for each spectrum X, a coaction map

$$\psi = \psi_X : \pi_*(E \wedge X) \longrightarrow \pi_*(E \wedge E) \otimes_{\pi_*(E)} \pi_*(E \wedge X) .$$

(The diagonal map ψ_E is obtained by specialising the coaction map ψ_X to the case $X = E$.)

It is important to note that in the tensor-product $\pi_*(E \wedge E) \otimes_{\pi_*(E)} \pi_*(E \wedge X)$, the action of $\pi_*(E)$ on the left-hand factor $\pi_*(E \wedge E)$ is the right action. (The action of $\pi_*(E)$ on the right-hand factor $\pi_*(E \wedge X)$ is the usual left action.) This is exactly what we need to use the tensor-product notation in a systematic way.

The tensor-product $\pi_*(E \wedge E) \otimes_{\pi_*(E)} \pi_*(E \wedge X)$ can be considered as a left module over $\pi_*(E)$, by using the left action of $\pi_*(E)$ on $\pi_*(E \wedge E)$; that is,

$$\lambda(e \otimes x) = (\lambda e) \otimes x$$

$(\lambda \in \pi_*(E),\ e \in \pi_*(E \wedge E),\ x \in \pi_*(E \wedge X))$.
The coaction map ψ_X is a map of left modules over $\pi_*(E)$.

In particular, the previous two paragraphs apply to the case $X = E$. Here the tensor-product $\pi_*(E \wedge E) \otimes_{\pi_*(E)} \pi_*(E \wedge E)$ can also be considered as a right module over $\pi_*(E)$, by using the right action of $\pi_*(E)$ on the right-hand factor. The diagonal map ψ_E is a map of bimodules over $\pi_*(E)$.

The behaviour of the other structure maps with respect to the actions of $\pi_*(E)$ will emerge from the properties given below. The tensor-product on which the product map ϕ is defined can be taken over the integers.

The principal properties of these structure maps are as follows. The product map ϕ is associative, anticommutative and has a unit element 1. The maps η_L, η_R, ε and c are homomorphisms of graded rings with unit. The left action of $\pi_*(E)$ on $\pi_*(E \wedge E)$ is given by

$$\lambda e = \phi((\eta_L \lambda) \otimes e) \qquad (\lambda \in \pi_*(E),\ e \in \pi_*(E \wedge E)) .$$

Similarly, the right action of $\pi_*(E)$ on $\pi_*(E \wedge E)$ is given by

$$e\lambda = \phi(e \otimes (\eta_R \lambda)) \qquad (e \in \pi_*(E \wedge E),\ \lambda \in \pi_*(E)) .$$

We have

$$\varepsilon\eta_L = 1,\ \varepsilon\eta_R = 1,\ c\eta_L = \eta_R,\ c\eta_R = \eta_L,$$

$$\varepsilon c = \varepsilon,\ c^2 = 1 .$$

These properties determine the behaviour of ϕ, η_L, η_R, ε and c with respect to the actions of $\pi_*(E)$. In particular, ε is a map of bimodules.

The coaction map is natural for maps of X. The coaction map is associative, in the sense that the following diagram is commutative.

$$\begin{array}{ccc}
\pi_*(E\wedge X) & \xrightarrow{\psi_X} & \pi_*(E\wedge E) \otimes_{\pi_*(E)} \pi_*(E\wedge X) \\
\downarrow \psi_X & & \downarrow 1 \otimes \psi_X \\
\pi_*(E\wedge E) \otimes_{\pi_*(E)} \pi_*(E\wedge X) & \xrightarrow{\psi_E \otimes 1} & \pi_*(E\wedge E) \otimes_{\pi_*(E)} \pi_*(E\wedge E) \otimes_{\pi_*(E)} \pi_*(E\wedge X)
\end{array}$$

(Note that $1 \otimes \psi_X$ is defined because ψ_X is a map of left modules over $\pi_*(E)$, and $\psi_E \otimes 1$ **is** defined because ψ_E is a map of right modules over $\pi_*(E)$.) In particular, we can specialise this diagram to the case $X = E$, and we see that the diagonal map is associative.

The behaviour of the diagonal with respect to the product is given by the following commutative diagram.

$$\begin{array}{ccc}
\pi_*(E\wedge E) \otimes \pi_*(E\wedge E) & \xrightarrow{\phi} & \pi_*(E\wedge E) \\
\downarrow \psi_E \otimes \psi_E & & \downarrow \psi_E \\
[\pi_*(E\wedge E) \otimes_{\pi_*(E)} \pi_*(E\wedge E)] \otimes [\pi_*(E\wedge E) \otimes_{\pi_*(E)} \pi_*(E\wedge E)] & \xrightarrow{\phi} & \pi_*(E\wedge E) \otimes_{\pi_*(E)} \pi_*(E\wedge E)
\end{array}$$

Here the map Φ is defined by

$$\Phi(e \otimes f \otimes g \otimes h) = (-1)^{pq}\phi(e \otimes g) \otimes \phi(f \otimes h)$$

where $f \in \pi_p(E \wedge E)$, $g \in \pi_q(E \wedge E)$. It has to be verified that this formula does give a well-defined map of the product of tensor products over $\pi_*(E)$, but this can be done using the facts stated above.

The behaviour of the diagonal map on the unit is given by $\psi_E(1) = 1 \otimes 1$. It follows that we have

$$\psi_E \eta_L \lambda = (\eta_L \lambda) \otimes 1 \ , \ \psi_E \eta_R \lambda = 1 \otimes (\eta_R \lambda) \qquad (\lambda \in \pi_*(E)) \ .$$

The behaviour of the diagonal map with respect to the counit is given by the following commutative diagram.

$$\begin{array}{ccc}
\pi_*(E \wedge X) & \xrightarrow{\ \psi_X\ } & \pi_*(E \wedge E) \otimes_{\pi_*(E)} \pi_*(E \wedge X) \\
\Big\downarrow 1 & & \Big\downarrow \varepsilon \otimes 1 \\
\pi_*(E \wedge X) & \xleftarrow{\ \cong\ } & \pi_*(E) \otimes_{\pi_*(E)} \pi_*(E \wedge X)
\end{array}$$

Here the bottom arrow is given by the usual left action of $\pi_*(E)$ on $\pi_*(E \wedge X)$. The map $\varepsilon \otimes 1$ is defined because ε is a map of right modules over $\pi_*(E)$. Similarly, we have the following commutative diagram.

$$\begin{array}{ccc}
\pi_*(E \wedge E) & \xrightarrow{\ \psi_E\ } & \pi_*(E \wedge E) \otimes_{\pi_*(E)} \pi_*(E \wedge E) \\
\Big\downarrow 1 & & \Big\downarrow 1 \otimes \varepsilon \\
\pi_*(E \wedge E) & \xleftarrow{\ \cong\ } & \pi_*(E \wedge E) \otimes_{\pi_*(E)} \pi_*(E)
\end{array}$$

Here the bottom arrow is given by the right action of $\pi_*(E)$ on $\pi_*(E \wedge E)$. The map $1 \otimes \varepsilon$ is defined because ε is a map of left modules over $\pi_*(E)$.

The behaviour of the diagonal with respect to the canonical anti-automorphism c is given by the following commutative diagram.

$$\begin{array}{ccc}
\pi_*(E \wedge E) & \xrightarrow{\psi_E} & \pi_*(E \wedge E) \otimes_{\pi_*(E)} \pi_*(E \wedge E) \\
\downarrow c & & \downarrow C \\
\pi_*(E \wedge E) & \xrightarrow{\psi_E} & \pi_*(E \wedge E) \otimes_{\pi_*(E)} \pi_*(E \wedge E)
\end{array}$$

Here the map C is defined by

$$C(e \otimes f) = (-1)^{pq} cf \otimes ce$$

$(e \in \pi_p(E \wedge E),\ f \in \pi_q(E \wedge E))$.

It has to be verified that this formula does give a well-defined map of the tensor product over $\pi_*(E)$, but this can be done using the facts stated above.

The following commutative diagrams express that property of the canonical anti-automorphism which in the classical case is taken as its definition.

$$\begin{array}{ccc}
\pi_*(E \wedge E) & \xrightarrow{\varepsilon} & \pi_*(E) \\
\downarrow \psi_E & & \downarrow \eta_L \\
\pi_*(E \wedge E) \otimes_{\pi_*(E)} \pi_*(E \wedge E) & \xrightarrow{\phi(1 \otimes c)} & \pi_*(E \wedge E)
\end{array}$$

$$\begin{array}{ccc}
\pi_*(E \wedge E) & \xrightarrow{\quad \varepsilon \quad} & \pi_*(E) \\
\Big\downarrow{\scriptstyle \psi_E} & & \Big\downarrow{\scriptstyle \eta_R} \\
\pi_*(E \wedge E) \otimes_{\pi_*(E)} \pi_*(E \wedge E) & \xrightarrow{\phi(c\otimes 1)} & \pi_*(E \wedge E)
\end{array}$$

It has to be verified that $\phi(1\otimes c)$ and $\phi(c\otimes 1)$ do give well-defined maps of the tensor product over $\pi_*(E)$, but this can be done using the facts stated above.

This completes the list of properties of our structure maps. We also require one further formal property in order to show that certain comodules $E_*(X)$ are extended (see Lectures 1, 2). Let F be a left module-spectrum over the ring-spectrum E; for example, we might have $F = E \wedge Y$. Then the following diagram is commutative.

$$\begin{array}{ccc}
\pi_*(E \wedge E) \otimes_{\pi_*(E)} \pi_*(F) & \xrightarrow{\quad m \quad} & \pi_*(E \wedge F) \\
\Big\downarrow{\scriptstyle \psi_E \otimes 1} & & \Big\downarrow{\scriptstyle \psi_F} \\
\pi_*(E\wedge E) \otimes_{\pi_*(E)} \pi_*(E\wedge E) \otimes_{\pi_*(E)} \pi_*(F) & \xrightarrow{1\otimes m} & \pi_*(E\wedge E) \otimes_{\pi_*(E)} \pi_*(E\wedge F)
\end{array}$$

The map $1 \otimes m$ is defined because m is a map of left modules over $\pi_*(E)$.

We now give the definition of our structure maps. The product ϕ is given by either way of chasing round the following commutative square.

$$\begin{array}{ccc}
\pi_*(E \wedge E) \otimes \pi_*(E \wedge E) & \xrightarrow{\nu'} & \pi_*(E \wedge E \wedge E) \\
\downarrow \nu & & \downarrow (\mu \wedge 1)_* \\
\pi_*(E \wedge E \wedge E) & \xrightarrow{(1 \wedge \mu)_*} & \pi_*(E \wedge E)
\end{array}$$

(For ν and ν', see the discussion of products at the beginning of this lecture.) In other words, suppose given

$$f: S^p \longrightarrow E \wedge E, \ g: S^q \longrightarrow E \wedge E;$$

then $\phi(f \otimes g)$ is the following composite.

$$S^p \wedge S^q \xrightarrow{f \wedge g} E \wedge E \wedge E \wedge E \xrightarrow{1 \wedge \tau \wedge 1} E \wedge E \wedge E \wedge E \xrightarrow{\mu \wedge \mu} E \wedge E.$$

We have maps

$$E \simeq E \wedge S^0 \xrightarrow{1 \wedge i} E \wedge E$$

$$E \simeq S^0 \wedge E \xrightarrow{i \wedge 1} E \wedge E$$

which map E into $E \wedge E$ as the left and right factors. We define η_L and η_R to be the corresponding induced homomorphisms. We define ε and c to be the homomorphisms induced by

$$\mu: E \wedge E \longrightarrow E$$

and

$$\tau: E \wedge E \longrightarrow E \wedge E .$$

It only remains to define ψ_X.

Lemma 1

If $\pi_*(X \wedge E)$ is flat as a right module over $\pi_*(E)$, then $m: \pi_*(X \wedge E) \otimes_{\pi_*(E)} \pi_*(E \wedge Y) \longrightarrow \pi_*(X \wedge E \wedge Y)$ is iso.

Proof. This is essentially the trivial case of KT 1 (see Lecture 1, Note 12). The map m is a natural transformation between homology functors of Y which is iso for $Y = S^0$; therefore it is iso for any finite complex Y. Pass to direct limits.

We now define

$$h: \pi_*(X \wedge Y) \longrightarrow \pi_*(X \wedge E \wedge Y)$$

to be the homomorphism induced by

$$X \wedge Y \simeq X \wedge S^0 \wedge Y \xrightarrow{1 \wedge i \wedge 1} X \wedge E \wedge Y .$$

The map h is essentially the Hurewicz homomorphism in E-homology.

If $\pi_*(X \wedge E)$ is flat, we can consider the following composite.

$$\pi_*(X \wedge Y) \xrightarrow{h} \pi_*(X \wedge E \wedge Y) \xrightarrow{m^{-1}} \pi_*(X \wedge E) \otimes_{\pi_*(E)} \pi_*(E \wedge Y) .$$

We define $\psi = m^{-1}h$. In particular, since we are assuming that $\pi_*(E \wedge E)$ is flat, we can specialise to the case $X = E$; we take the resulting map ψ for our coaction map ψ_Y. This completes the definition of the structure maps.

The proofs of all the formal properties are by

diagram-chasing. In proving any property of ψ_X, of course we have to make our diagram up out of two subdiagrams, one for h and one for m. For example, in proving that the coaction map is associative, we first prove two more elementary results; ψ_X is natural for maps of X, and $\psi_F m = (1 \otimes m)(\psi_E \otimes 1)$ (which is the diagram required to prove that $E_*(F)$ is an extended comodule). We now set up the following diagram.

$$\begin{array}{ccc}
\pi_*(E \wedge X) & \xrightarrow{\psi_X} & \pi_*(E \wedge E) \otimes_{\pi_*(E)} \pi_*(E \wedge X) \\
\downarrow h & & \downarrow 1 \otimes h \\
\pi_*(E \wedge E \wedge X) & \xrightarrow{\psi_{E \wedge X}} & \pi_*(E \wedge E) \otimes_{\pi_*(E)} \pi_*(E \wedge E \wedge X) \\
\uparrow m & & \uparrow 1 \otimes m \\
\pi_*(E \wedge E) \otimes_{\pi_*(E)} \pi_*(E \wedge X) & \xrightarrow{\psi_E \otimes 1} & \pi_*(E \wedge E) \otimes_{\pi_*(E)} \pi_*(E \wedge E) \otimes_{\pi_*(E)} \pi_*(E \wedge X)
\end{array}$$

Here the top square is commutative because h is induced by a map

$$X \simeq S^0 \wedge X \xrightarrow{i \wedge 1} E \wedge X ,$$

and ψ_X is natural for maps of X. Similarly, the bottom square is commutative by the second result mentioned, taking $F = E \wedge X$. This gives the required result. The two subsidiary results are proved in the same way.

In proving the behaviour of the diagonal with respect to the product, it is convenient to prove a slightly more general result first. Suppose that $\pi_*(A \wedge E)$, $\pi_*(B \wedge E)$ and $\pi_*(A \wedge B \wedge E)$ are all flat; then the following diagram is commutative.

$$\begin{array}{ccc}
\pi_*(A\wedge X) \otimes \pi_*(B\wedge Y) & \longrightarrow & \pi_*(A\wedge B\wedge X\wedge Y) \\
\downarrow \psi\otimes\psi & & \downarrow \psi \\
[\pi_*(A\wedge E) \otimes_{\pi_*(E)} \pi_*(E\wedge X)] \otimes [\pi_*(B\wedge E) \otimes_{\pi_*(E)} \pi_*(E\wedge Y)] & \longrightarrow & \pi_*(A\wedge B\wedge E) \otimes_{\pi_*(E)} \pi_*(E\wedge X\wedge Y)
\end{array}$$

Here the upper horizontal map is the obvious product, and the lower horizontal map sends $e \otimes f \otimes g \otimes h$ into $(-1)^{pq}\nu'(e \otimes g) \otimes \nu(f \otimes h)$ (see the discussion of products at the beginning of this lecture). This diagram is proved commutative in the same way as before - separate h and m. Next observe that since the functor $\pi_*(E \wedge E) \otimes_{\pi_*(E)}$ preserves exactness, applying it twice preserves exactness; that is, the right module

$$\pi_*(E \wedge E \wedge E) \cong \pi_*(E \wedge E) \otimes_{\pi_*(E)} \pi_*(E \wedge E)$$

is flat. So we may specialise to the case $A = B = E$. Now apply naturality to the map

$$A \wedge B = E \wedge E \xrightarrow{\mu} E ;$$

we see that the following diagram is commutative.

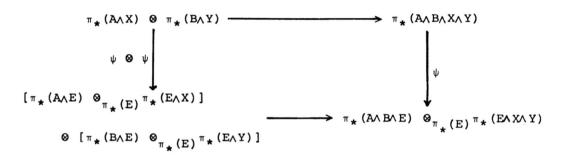

$$\begin{array}{ccc}
\pi_*(E\wedge X) \otimes \pi_*(E\wedge Y) & \xrightarrow{\quad\nu\quad} & \pi_*(E\wedge X\wedge Y) \\
\Big\downarrow{\scriptstyle \psi_X \otimes \psi_Y} & & \Big\downarrow{\scriptstyle \psi_{X \wedge Y}} \\
[\pi_*(E\wedge E) \otimes_{\pi_*(E)} \pi_*(E\wedge X)] & & \\
\otimes\ [\pi_*(E\wedge E) \otimes_{\pi_*(E)} \pi_*(E\wedge Y)] & \longrightarrow & \pi_*(E\wedge E) \otimes_{\pi_*(E)} \pi_*(E\wedge X\wedge Y)
\end{array}$$

Here the lower horizontal map sends $e \otimes f \otimes g \otimes h$ into $(-1)^{pq}\phi(e \otimes g) \otimes \nu(f \otimes h)$. This diagram gives the behaviour of the coaction map with respect to the external homology product. Finally we specialise to the case $X = Y = E$ and apply naturality to the map

$$X \wedge Y = E \wedge E \xrightarrow{\mu} E .$$

We obtain the required commutative diagram.

The proof of the remaining formal properties does not call for any special comment.

We now turn to further formulae, involving cohomology, which will help to show that our definitions specialise correctly to the classical case. We recall that the cohomology groups of a spectrum X with coefficients in E are given by

$$E^{-n}(X) = [S^n \wedge X, E] .$$

We have a Kronecker product

$$E^{-p}(X) \otimes E_q(X) \longrightarrow \pi_{p+q}(E)$$

defined as follows. Suppose given maps

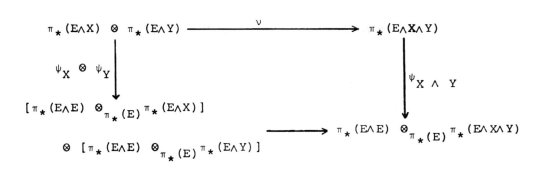

$$f: S^p \wedge X \longrightarrow E, \quad g: S^q \longrightarrow E \wedge X .$$

Then $\langle f,g \rangle$ is the following composite.

$$S^p \wedge S^q \xrightarrow{1\wedge g} S^p \wedge E \wedge X \xrightarrow{1\wedge \tau} S^p \wedge X \wedge E \xrightarrow{f\wedge 1} E \wedge E \xrightarrow{\mu} E .$$

In particular, we have the cohomology groups $E^*(E)$. Since these are defined in terms of maps from E to E (up to suspension), they act on the left on the homology and cohomology groups $E_*(X)$ and $E^*(X)$. The precise definitions are as follows. Suppose given maps

$$a: S^p \wedge E \longrightarrow E, \quad f: S^q \longrightarrow E \wedge X, \quad g: S^r \wedge X \longrightarrow E .$$

Then af is

$$S^p \wedge S^q \xrightarrow{1\wedge f} S^p \wedge E \wedge X \xrightarrow{a\wedge 1} E \wedge X ,$$

and ag is

$$S^p \wedge S^r \wedge X \xrightarrow{1\wedge g} S^p \wedge E \xrightarrow{a} E .$$

In this way $E^*(E)$ becomes a ring with unit, and $E_*(X)$, $E^*(X)$ become left modules over this ring.

We will show that the action of $E^*(E)$ on $E_*(X)$ is determined by the coaction map ψ_X. Suppose $a \in E^*(E)$, $x \in E_*(X)$ and $\psi_X x = \sum_i e_i \otimes x_i$, where $e_i \in E_*(E)$, $x_i \in E_*(X)$. Then we have:

<u>Proposition 2</u>

$$ax = \sum_i \langle a, ce_i \rangle x_i .$$

To prove this proposition, we set up the following diagram.

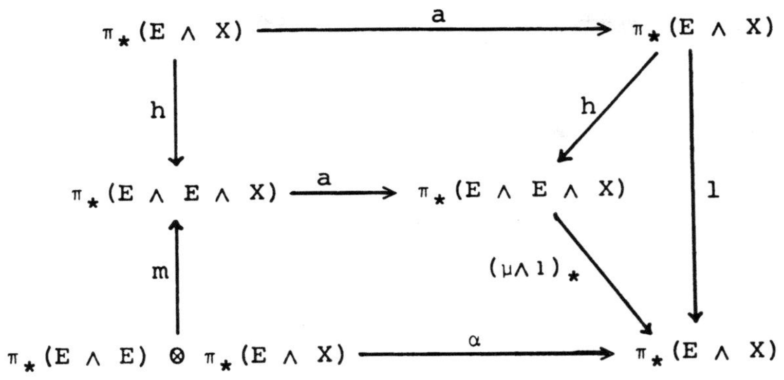

Here α is defined by

$$\alpha(e \otimes x) = \langle a, ce\rangle x \ .$$

It is easy to show that the diagram is commutative. This proves Proposition 2.

In the case when an element $z \in E^*(X)$ is determined by the values of $\langle z, x\rangle$ for all $x \in E_*(X)$, it is reasonable to ask for a calculation of the action of $E^*(E)$ on $E^*(X)$ in terms of ψ_X. There is a choice of formulae which answer this question; here I will give one which seems neater than that which I actually gave in Seattle. Suppose $a \in E^*(E)$, $y \in E^p(X)$, $x \in E_*(X)$ and $\psi_X x = \sum_i e_i \otimes x_i$, where $e_i \in E_{q(i)}(E)$, $x_i \in E_*(X)$. Then we have:

<u>Proposition 3</u>

$$\langle ay, x\rangle = \sum_i (-1)^{pq(i)} \langle a, e_i \langle y, x_i\rangle\rangle .$$

The formula on the right makes sense, because e_i lies in $\pi_*(E\wedge E)$, and $\langle y, x_i\rangle$ lies in $\pi_*(E)$, which acts on the right on $\pi_*(E\wedge E)$.

To prove the proposition, we first define

$$y_*: \pi_*(F \wedge X) \longrightarrow \pi_*(F \wedge E)$$

(for any F) as follows. Suppose given $y: S^p \wedge X \longrightarrow E$ and $f: S^r \longrightarrow F \wedge X$; let $y_* f$ be the composite

$$S^p \wedge S^r \xrightarrow{1 \wedge f} S^p \wedge F \wedge X \xrightarrow{\tau \wedge 1} F \wedge S^p \wedge X \xrightarrow{1 \wedge y} F \wedge E .$$

Then we easily check that

$$\langle ay, x\rangle = \langle a, y_* x\rangle .$$

We now set up the following diagram.

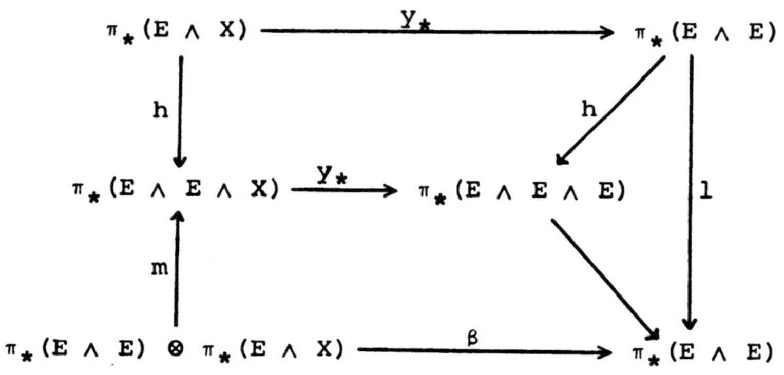

Here β is defined by

$$\beta(e \otimes x) = (-1)^{pq}\, e\langle y, x\rangle$$

for $e \in E_q(E)$. It is easy to show that this diagram is commutative. This shows that

$$y_* x = \sum_i (-1)^{pq(i)} e_i \langle y, x_i\rangle ,$$

and proves Proposition 3.

We will now discuss the way in which our constructions

specialise to the case $E = K(Z_p)$. It is sufficiently clear from the definitions that ϕ, η_L, η_R and ε specialise to their classical counterparts ϕ, η, η and ε. The right action of $\pi_*(E) = Z_p$ on $\pi_*(E \wedge E) = A_*$ coincides with the left action, because the unit acts as a unit on either side, and so the result follows for integer multiples of the unit. It follows that in Proposition 3 we can bring the factor $\langle y, x_i \rangle$ to the left of e_i; and after that we can bring it outside the Kronecker product, so as to obtain the following formula.

$$\langle ay, x \rangle = \sum_i (-1)^{pq(i)} \langle a, e_i \rangle \langle y, x_i \rangle .$$

It follows that ψ_X is indeed the dual of the action map $A^* \otimes H^* \longrightarrow H^*$, and (specialising to the case $X = E$) that ψ_E is the dual of the composition map $A^* \otimes A^* \longrightarrow A^*$. Thus ψ_E and ψ_X specialise to their classical counterparts.

Since we have seen that

$$\phi(1 \otimes c)\psi_E = \eta_L \varepsilon$$

and

$$\phi(c \otimes 1)\psi_E = \eta_R \varepsilon ,$$

it now follows that c specialises to its classical counterpart.

It remains only to point out one difference between the classical case and the generalised case. In the generalised case we have introduced a left action of $E^*(E)$ on

$E_*(X)$. This does not specialise to the action of A^* on H_* which is usually considered in the classical case, since the latter is a right action, defined by

$$\langle y, xa\rangle = (-1)^{(p+q)r}\langle ay, x\rangle$$

$(y \in H^p,\ x \in H_q,\ a \in A^r)$.

The connection between the two actions may be read off from Proposition 2 and 3. We have

$$xa = (-1)^{qr}(ca)x \qquad (x \in H_q,\ a \in A^r)\ .$$

Thus the left and right actions differ by the canonical anti-automorphism, as one might expect.

LECTURE 4 SPLITTING GENERALISED COHOMOLOGY THEORIES WITH COEFFICIENTS

S. P. Novikov [23, 24] has emphasised the importance of the generalised cohomology theory provided by complex cobordism. This is a representable functor; if we take it "reduced", we have

$$MU^n(X) = [X, S^n MU] .$$

It has been proved by Brown and Peterson [10] that if one neglects all the primes except one prime p, then the MU-spectrum splits as a sum or product:

$$MU \underset{p}{\simeq} \bigvee_i S^{n(i)} BP(p) \simeq \prod_i S^{n(i)} BP(p) .$$

Here $BP(p)$ means the Brown-Peterson spectrum. The sum coincides with the product since $BP(p)$ is connected and $n(i) \longrightarrow \infty$ as $i \longrightarrow \infty$. The business of neglecting all primes except one may be formalised conveniently by introducing coefficients. Let Q_p be the ring of rational numbers a/b with b prime to p. Then we can form $MU^*(X; Q_p)$, and we have

$$MU^n(X; Q_p) \cong \prod_i L^{n + n(i)}(X) ,$$

where

$$L^m(X) = [X, S^m L]$$

and L is a suitable version of the Brown-Peterson spectrum.

This situation has been considered by S. P. Novikov

[24]. Potentially it is very profitable. The cohomology theory L^* is just as powerful as $MU^*(\ ;Q_p)$; for example, it gives rise to the same "Adams spectral sequence" (see Lecture 2). However, the groups $L^*(X)$ are much smaller than the groups $MU^*(X;Q_p)$; similarly for the coefficient groups $L^*(S^0)$, the ring of operations $L^*(L)$ and the Hopf algebra $L_*(L)$ (see Lecture 3). For all these reasons, calculations with L should be smaller and easier than calculations with MU.

Unfortunately, these benefits have not yet been fully realised in practice. The reason is that the splittings given by Brown and Peterson, and by Novikov, are not canonical; they involve large elements of choice. It is doubtless because of this that these authors have not yet given such helpful and illuminating formulae for the structure of $L^*(L)$, etc., as are available for the structure of $MU^*(MU)$, etc.

I therefore propose the following thesis. When we split a cohomology theory into summands, we should try to do so in a canonical way, issuing in helpful and enlightening formulae. To secure these ends I would even be willing to split the theory into summands larger than the irreducible ones. The method which I propose is to take a suitable ring of cohomology operations, say A, and construct in it canonical idempotents, say e. Then whenever A acts on a module, say H, H will split as the direct sum $eH \oplus (1-e)H$.

I will first show how this thesis applies to K-theory. Not only is the case of K-theory somewhat easier, but for technical reasons it is useful as a tool in attacking cobordism. For K-theory I shall give a treatment which seems tolerably complete and satisfactory (Lemma 1 to Lemma 9 below). I will then turn to cobordism (Lemma 10 to Theorem 19 below). Here the theory is somewhat less complete, but it is sufficient to show the existence of canonical summands in cobordism with suitable coefficients.

Let R be a subring of the rationals. Let $K^*(X;R)$ be ordinary, complex K-theory, with coefficients in R. We write K for K^0; then $K(X;R)$ is a representable functor; we write BUR for the representing space. We require some information on $K^*(BUR;R)$. All that is really needed is that its Lim^1 subgroup [21] is zero; but our method will prove more. It is for this purpose that we introduce the first few lemmas.

Let d be a positive integer, and let $f: BU \longrightarrow BU$ be the map obtained by taking the identity map of the space BU and adding it to itself d times, using the H-space structure of BU.

<u>Lemma 1</u>

If d is invertible in R then

$$f_*: H_*(BU;R) \longrightarrow H_*(BU;R)$$

and

$$f^*: H^*(BU;R) \longrightarrow H^*(BU;R)$$

are isomorphisms.

Proof. We will prove that f^* is epi. Suppose, as an inductive hypothesis, that the image of f^* contains the Chern classes $c_1, c_2, \ldots, c_{n-1}$. Then it contains all decomposable elements in $H^{2n}(BU;R)$. For any primitive element $p_n \in H^{2n}(BU;R)$ we have $f^*p_n = dp_n$. But we can find such a p_n which is a non-zero multiple of c_n mod decomposable elements. Therefore $f^*c_n = dc_n$ mod decomposables. Since d is invertible in R, c_n lies in the image of f^*. This completes the induction and proves that f^* is epi; by duality, f_* is mono.

A precisely dual argument shows that f_* is epi and f^* is mono. Indeed, the preceding paragraph was written so as to dualise correctly. One needs some minimal knowledge of $H_*(BU;R)$ as a ring under the Pontryagin product, and the fact that f is an H-map, so that f_* is a homomorphism of rings. This proves Lemma 1.

Next, let R_1, R_2 be two subrings of the rationals. We have an obvious map $i: BU \longrightarrow BUR_1$.

Lemma 2

If $R_1 \subset R_2$, the maps

$$i_*: H_*(BU;R_2) \longrightarrow H_*(BUR_1;R_2)$$

$$i^*: H^*(BUR_1;R_2) \longrightarrow H^*(BU;R_2)$$

$$(i\times i)_*: H_*(BU\times BU;\ R_2) \longrightarrow H_*(BUR_1\times BUR_1;\ R_2)$$

$$(i\times i)^*: H^*(BUR_1\times BUR_1;\ R_2) \longrightarrow H^*(BU\times BU;\ R_2)$$

are isomorphisms.

Proof. If $R_1 = Z$ the result is trivial, so we may assume $R_1 \neq Z$. We now construct a model for BUR_1. Consider the positive integers invertible in R_1 and arrange them in a sequence $d_1, d_2, d_3, \ldots$. For each d_n we have a map $f_n: BU \longrightarrow BU$, as in Lemma 1. Take the maps

$$BU \xrightarrow{f_1} BU \xrightarrow{f_2} BU \longrightarrow \cdots \longrightarrow BU \xrightarrow{f_n} BU \longrightarrow \cdots$$

and form a "telescope" or iterated mapping-cylinder; this gives a construction for BUR_1. The map $i: BU \longrightarrow BUR_1$ is the injection of the first copy of BU. We have

$$H_*(BUR_1;R_2) = \varinjlim(H_*(BU;R_2), f_{n*}) .$$

Now the result about i_* follows from Lemma 1. The result about $(i \times i)_*$ follows from the Künneth theorem. The results about i^* and $(i \times i)^*$ follow from the universal coefficient theorem. This proves Lemma 2.

Lemma 3

Suppose $R_1 \subset R_2$. Then the maps

$$i_*: K_*(BU;R_2) \longrightarrow K_*(BUR_1;R_2)$$

$$i^*: K^*(BUR_1;R_2) \longrightarrow K^*(BU;R_2)$$

$$(i\times i)^*: K^*(BUR_1\times BUR_1;\ R_2) \longrightarrow K^*(BU\times BU;\ R_2)$$

are isomorphisms. The maps i^* and $(i \times i)^*$ are also homeomorphisms with respect to the filtration topology.

Proof. Let P be a point. Consider the usual spectral sequence

$$H_*(X;K_*(P;R_2)) \Longrightarrow K_*(X;R_2) .$$

By Lemma 2, the map $i: BU \longrightarrow BUR_1$ induces an isomorphism between the spectral sequences for $X = BU$ and for $X = BUR_1$. This proves the result about i_*. The proof for i^* and $(i \times i)^*$ is similar, using the spectral sequence

$$H^*(X;K^*(P;R_2)) \Longrightarrow K^*(X;R_2) .$$

The space BUR_1 is an H-space; let $\mu: BUR_1 \times BUR_1 \longrightarrow BUR_1$ be the product map, and let $\pi,\tilde{\omega}: BUR_1 \times BUR_1 \longrightarrow BUR_1$ be the projections onto the two factors. We retain the assumption that $R_1 \subset R_2$, and consider the set of primitive elements in $\tilde{K}(BUR_1;R_2)$, that is, the set of elements a such that $\mu^*a = \pi^*a + \tilde{\omega}^*a$. This set may be identified with the set of cohomology operations

$$a: \tilde{K}(X;R_1) \longrightarrow \tilde{K}(X;R_2)$$

which are defined for all connected X, natural, and additive in the sense that

$$a(x + y) = a(x) + a(y) .$$

(If an operation is additive, it follows that it is R_1-linear.) Such operations need not be stable.

This set is to be topologised as a subset of $\tilde{K}(BUR_1;R_2)$; in other words, an operation $\underline{a}$ is close to zero if it vanishes in all CW-complexes of dimension n.

According to Lemma 3 above, the set of operations $\underline{a}$ to be considered is essentially independent of R_1, so long as $R_1 \subset R_2$. (This fact would be trivial if we were dealing only with finite CW-complexes X, since then we have $\tilde{K}(X;R_1) \cong \tilde{K}(X) \otimes R_1$, $\tilde{K}(X;R_2) \cong \tilde{K}(X) \otimes R_2$.) We therefore write $\tilde{A}(R_2)$ for the set of operations introduced above, and regard it primarily as the ring of cohomology operations on $\tilde{K}(X;R_2)$.

We define

$$A(R) = R + \tilde{A}(R) .$$

By making the first summand R act in the obvious way on $K(P;R)$, the set $A(R)$ may be identified with the set of cohomology operations

$$a: K(X;R) \longrightarrow K(X;R)$$

which are defined for all X, natural, and additive (hence R-linear).

<u>Lemma 4</u>

If $R_1 \subset R_2$, we have a monomorphism

$$\iota: A(R_1) \longrightarrow A(R_2)$$

such that for each $a \in A(R_1)$ and each X the following diagram is commutative.

$$\begin{array}{ccc} K(X;R_1) & \xrightarrow{i_*} & K(X;R_2) \\ \Big\downarrow a & & \Big\downarrow \iota a \\ K(X;R_1) & \xrightarrow{i_*} & K(X;R_2) \end{array}$$

This follows from the preceding discussion together with the fact that

$$i_*: \tilde{K}(BU;R_1) \longrightarrow \tilde{K}(BU;R_2)$$

is monomorphic.

Because of this lemma, it will be sufficient to construct idempotents in $A(Q)$ and then prove that they are defined over some suitable subring of the rationals. But over Q the idempotents are obvious. The Chern character allows us to identify $K(X;Q)$ with the product

$$\prod_n H^{2n}(X;Q) .$$

Let us define e_n to be projection on the nth factor:

$$e_n(h^0,h^2,\ldots,h^{2n-2},h^{2n},h^{2n+2},\ldots) = (0,0,\ldots,0,h^{2n},0,\ldots) .$$

Then e_n is an idempotent in $A(Q)$.

I now choose a positive integer d, and seek to construct a "fake K-theory" with one non-zero coefficient group every $2d$ dimensions. The required idempotents are obvious. Take a residue class of integers mod d, say

$\alpha \in Z_d$, and define

$$E_\alpha = \sum_{n\in\alpha} e_n \in A(Q) .$$

This sum is convergent in the topology which $A(Q)$ has. If we use the Chern character to identify $K(X;Q)$ with $\prod_n H^{2n}(X;Q)$, as above, then we have

$$E_\alpha(h^0,h^2,h^4,\ldots) = (k^0,k^2,k^4,\ldots)$$

where

$$k^{2n} = \begin{cases} h^{2n} & \text{if } n \in \alpha \\ 0 & \text{if } n \notin \alpha \end{cases} .$$

<u>Theorem 5</u>

E_α lies in $A(R)$, where $R = R(d)$ is the ring of rationals a/b such that b contains no prime p with $p \equiv 1 \bmod d$.

For example, if $d = 2$, R is the ring of fractions $a/2^f$.

For the proof, we need to work with a representation of $A(R)$. Let η be the canonical line bundle over CP^∞; then $K(CP^\infty;R)$ is the ring of formal power-series $R[[\zeta]]$, where $\zeta = \eta - 1$. We define an (R-linear) homomorphism

$$\theta: A(R) \longrightarrow K(CP^\infty;R)$$

by

$$\theta(a) = a(\eta) .$$

Lemma 6

θ is an isomorphism.

Proof. First we show that θ is mono. Let $a \in A(R)$ be such that $a(\eta) = 0$. Then by naturality $a(1) = 0$, so $a \in \tilde{A}(R)$. Let ξ be the universal $U(n)$-bundle over $BU(n)$; then $\xi - n$ is the universal element in $\tilde{K}(BU(n))$. Since a is additive, the splitting principle shows that $a(\xi - n) = 0$ in $\tilde{K}(BU(n);R)$. Let $i: BU \longrightarrow BUR$ be as above. We have

$$\tilde{K}(BU;R) = \varprojlim_{n} \tilde{K}(BU(n);R) \quad ;$$

it follows that $a(i) = 0$ in $\tilde{K}(BU;R)$. By Lemma 3 we have $a = 0$ in $\tilde{K}(BUR;R)$.

Next we show that θ is an epimorphism. For each n we can find an integral linear combination a_n of the operations ψ^k [3, 4] such that

$$a_n \eta = \zeta^n \quad ;$$

more precisely,

$$a_n = \sum_{0 \le k \le n} (-1)^{n-k} \frac{n!}{k!n-k!} \psi^k \quad .$$

For any sequence of elements $r(n) \in R$, the sum

$$\sum_{n=1}^{\infty} r(n) a_n$$

is convergent in the filtration topology on $\tilde{K}(BU;R)$ and defines a primitive element $\tilde{a}$ of $\tilde{K}(BU;R)$, that is, an element $\tilde{a} \in \tilde{A}(R)$. It remains only to take

$$a = r(0)\Psi^0 + \tilde{a} \in A(R) .$$

We have

$$a(\eta) = \sum_{n=0}^{\infty} a(n)\zeta^n .$$

This proves Lemma 6.

We observe that the isomorphism θ of Lemma 6 becomes a homeomorphism if we give $K(CP^\infty;R)$ the filtration topology. The filtration topology coincides with the usual topology on $R[[\zeta]]$: a power-series is close to zero if its first n coefficients vanish.

The isomorphism θ of Lemma 6 throws the monomorphism ι of Lemma 4 onto the obvious inclusion map

$$R_1[[\zeta]] \subset R_2[[\zeta]] .$$

We now return to the proof of Theorem 5. Let $x \in H^2(CP^\infty;Z)$ be the generator, so that

$$\mathrm{ch}\ \eta = \sum_n \frac{x^n}{n!} .$$

Consider the power-series

$$\log(1 + \zeta) = \zeta - \frac{\zeta^2}{2} + \frac{\zeta^3}{3} - \frac{\zeta^4}{4} \ldots .$$

Since ch commutes with sums, products and limits, we have

$$\begin{aligned} \mathrm{ch}\log(1 + \zeta) &= \log \mathrm{ch}(1 + \zeta) \\ &= \log \exp x \\ &= x . \end{aligned}$$

Now we have

$$\mathrm{ch}\; e_n\eta = \frac{x^n}{n!} = \mathrm{ch}\left(\frac{(\log(1+\zeta))^n}{n!}\right).$$

Therefore

$$e_n\eta = \frac{(\log(1+\zeta))^n}{n!}.$$

We now make a formal manipulation in $Q[t][[\zeta]]$, the ring of formal power-series in ζ with coefficients which are polynomials in t. Namely:

$$\sum_n t^n e_n\eta = \sum_n \frac{t^n(\log(1+\zeta))^n}{n!}$$

$$= \exp(t \log(1+\zeta))$$

$$= (1+\zeta)^t$$

$$= 1 + t\zeta + \frac{t(t-1)}{1.2}\zeta^2 + \ldots .$$

This is true as a formal identity in the ring cited.

Now consider $E_\alpha\eta = \sum_{n\in\alpha} e_n\eta = \sum_r \frac{a_r}{b_r}\zeta^n$, say. We wish to show that the coefficients $\frac{a_r}{b_r}$ lie in $R = R(d)$. Take any prime p such that $p \equiv 1 \bmod d$; we wish to show that a_r/b_r is a p-adic integer. Since d divides $p - 1$, I can find in the p-adic integers a primitive d^{th} root of 1, say ω. Set $\rho = \omega^m$, where the integer m is fixed for the moment. Then $\rho^d = 1$ and ρ^α makes sense. We have

$$\sum_\alpha \rho^\alpha E_\alpha \eta = \sum_n \rho^n e_n \eta$$
$$= 1 + \rho\zeta + \frac{\rho(\rho-1)}{1.2}\zeta^2 + \cdots$$
$$= c_m(\zeta), \quad \text{say.}$$

Here the binomial coefficient

$$b(t) = \frac{t(t-1)\ldots(t-r+1)}{1.2 \quad \ldots \quad r}$$

maps Z to Z and is continuous in the p-adic topology; therefore it maps p-adic integers to p-adic integers. So $c_m(\zeta)$ is a formal power-series in ζ with coefficients which are p-adic integers. Take $m = 1,2,\ldots,d$; we obtain d equations for the d unknowns $E_\alpha \eta$. The solution is

$$E_\alpha \eta = d^{-1} \sum_{1 \le m \le d} \omega^{-m\alpha} c_m(\zeta) .$$

Since d^{-1} is a p-adic integer, this is a formal power-series whose coefficients are p-adic integers. This proves Theorem 5.

The properties of the elements $E_\alpha \in A(R)$ are as follows.

Theorem 7

(i) $E_\alpha^2 = E_\alpha$

(ii) $E_\alpha E_\beta = 0$ if $\alpha \neq \beta$

(iii) $\sum_\alpha E_\alpha = 1$

(iv) For any x, y in $K(X;R)$

we have a "Cartan formula"

$$E_\alpha(xy) = \sum_{\beta+\gamma=\alpha} (E_\beta x)(E_\gamma y) \ .$$

<u>Proof</u>. By Lemma 4, $\iota: A(R) \longrightarrow A(Q)$ is a monomorphism. So parts (i), (ii) and (iii) follow from the corresponding equations in $A(Q)$, which are obvious. We turn to part (iv). The result is trivial when either x or y lies in $K(P;R)$, so it is sufficient to prove it when x and y lie in $\tilde{K}(X;R)$. It is sufficient to prove it for external products. Let both x and y be the universal elements in $\tilde{K}(BU)$; then the result holds in $\tilde{K}(BU \times BU;\ Q)$, by an obvious calculation using the Chern character. Since

$$\tilde{K}(BU \times BU;\ R) \longrightarrow \tilde{K}(BU \times BU;\ Q)$$

is monomorphic, the result holds in $\tilde{K}(BU \times BU;\ R)$. Let x and y be the universal element in $\tilde{K}(BUR;R)$; then the result holds in $\tilde{K}(BUR \times BUR;\ R)$ by Lemma 3. The case in which x and y are general follows by naturality. This proves part (iv) and completes the proof of Theorem 7.

Theorems 5 and 7 lead immediately to the results on the splitting of $K^*(X;R)$ (and indeed of $K_*(X:R)$, if required). As above, we are supposing given a positive integer d; $R = R(d)$ is as in Theorem 5, and α runs over Z_d.

Corollary 8

(i) We have a natural direct sum splitting

$$K(X;R) \cong \sum_{\alpha} K_{\alpha}(X),$$

where

$$K_{\alpha}(X) = E_{\alpha}K(X;R) .$$

(ii) $K_{\alpha}(X)$ is a representable functor.

(iii) If $x \in K_{\beta}(X)$ and $y \in K_{\gamma}(X)$, then $xy \in K_{\beta+\gamma}(X)$.

(iv) We have

$$\tilde{K}_{\alpha}(S^n) = \begin{cases} R & \text{if } \tfrac{1}{2}n \in \alpha \\ 0 & \text{otherwise} . \end{cases}$$

(v) Define

$$\phi: \tilde{K}_{\alpha}(X) \longrightarrow \tilde{K}_{\alpha+1}(S^2 \wedge X)$$

by taking the external product with a generator of $\tilde{K}_1(S^2)$. Then ϕ is an isomorphism.

Proof. Part (i) follows from Theorem 7 parts (i), (ii), (iii). For part (ii), observe that a direct summand of an exact sequence is an exact sequence, and that we have no trouble about verifying the axiom about disjoint unions (for K_{α}) or wedge-sums (for $\tilde{K}_{\alpha}$). Part (iii) follows from Theorem 7 part (iv). For part (iv), make the obvious calculation in $\tilde{K}(S^{2m};Q) \cong H^{2m}(S^{2m};Q)$. For part (v), let the representing space for $\tilde{K}_{\alpha}$ be BUR_{α}; convert the homomorphism ϕ into a map $BUR_{\alpha} \longrightarrow \Omega^2 BUR_{\alpha+1}$, and check as in part (iv) that this map induces an isomorphism of homotopy groups.

It follows from part (v), iterated d times, that the representable functor $K_\alpha(X)$ is periodic with period $2d$, in the same sense that standard K-theory is periodic with period 2. We therefore have no difficulty extending it to a graded cohomology theory $K_\alpha^*(X)$. Alternatively, we can first take the spectrum

$$BUR_\alpha, \; BUR_{\alpha+1}, \; BUR_{\alpha+2}, \; \ldots$$

and then take the resulting cohomology theory.

It follows from part (iii) that for $\alpha = 0$ the theory K_0 has products.

Let BUR_α be the representing space for $\tilde{K}_\alpha$, as above; then we have

$$BUR \simeq \prod_\alpha BUR_\alpha \; .$$

It is easy to obtain the rational cohomology of the factors BUR_α by inspecting their homotopy groups. In fact, $H^*(BUR_\alpha;Q)$ is a polynomial algebra on generators of dimension $2n$, where n runs over the positive integers in the residue class α.

Before moving on to cobordism, we need one more result. Given d, we have a map

$$E_0 : BU \longrightarrow BUR$$

where $R = R(d)$. Let us define $E_0^!$ so that the following diagram is commutative.

$$\begin{array}{ccc} H^*(BUR;Q) & \xrightarrow{\quad E_0^* \quad} & H^*(BU;Q) \\ & {\searrow}^{\cong} \qquad\qquad {\nearrow}_{E_0^!} & \\ & H^*(BU;Q) & \end{array}$$

We remark that in what follows, $H^*(X;Q)$ really arises as $E^0(X)$, where E is the spectrum

$$\prod_{-\infty<n<+\infty} K(Q,2n) \ .$$

Thus H^* should be interpreted as a direct product of groups H^p, while H_* should be interpreted as a direct sum of groups H_p. Let

$$\text{todd} \in H^*(BU;Q)$$

be the characteristic class which has the following properties.

(i) $\text{todd}(\xi_1 \oplus \xi_2) = (\text{todd}\ \xi_1)(\text{todd}\ \xi_2)$.

(ii) If η is the canonical line bundle over CP^∞ and $x \in H^2(CP^\infty)$ is the generator (so that $\text{ch}\ \eta = e^x$) then

$$\text{todd}\ \eta = \frac{e^x - 1}{x} \ .$$

Then we have the following result.

<u>Lemma 9</u>

There is a characteristic class

$$\tau \in K(BU;R)$$

such that

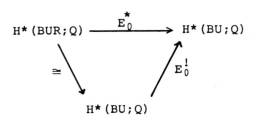

$$\frac{E_0^! \mathrm{todd}}{\mathrm{todd}} = \mathrm{ch}\ \tau .$$

Here $R = R(d)$ is as in Theorem 5. The motivation for this result is best seen from the proof of Theorem 14.

<u>Proof</u>. Let todd' be the class in $H^*(BUR;Q)$ which maps to todd in $H^*(BU;Q)$. Then we easily see that

$$\mathrm{todd}'(\xi_1 \oplus \xi_2) = (\mathrm{todd}'\xi_1)(\mathrm{todd}'\xi_2)$$

for ξ_1, ξ_2 in $K(X;R)$. We also have

$$(E_0^! \mathrm{todd})\xi = \mathrm{todd}' E_0 \xi$$

for ξ in $K(X)$. It is now easy to see that

$$\left(\frac{E_0^! \mathrm{todd}}{\mathrm{todd}}\right)(\xi_1 \oplus \xi_2) = \left(\frac{E_0^! \mathrm{todd}}{\mathrm{todd}}\xi_1\right)\left(\frac{E_0^! \mathrm{todd}}{\mathrm{todd}}\xi_2\right) .$$

Now $\frac{E_0^! \mathrm{todd}}{\mathrm{todd}}$ is certainly equal to $\mathrm{ch}\ \tau$ for some τ of augmentation 1 in $K(BU;Q)$. Using the last formula, we find that

$$\mathrm{ch}\ \tau(\xi_1 \oplus \xi_2) = (\mathrm{ch}\ \tau(\xi_1))(\mathrm{ch}\ \tau(\xi_2)) .$$

Therefore

$$\tau(\xi_1 \oplus \xi_2) = \tau(\xi_1) \cdot \tau(\xi_2)$$

in $K(X;Q)$ for any X. We wish to show that $\tau \in K(BU;R)$. For this purpose it is now sufficient to consider $\tau(\eta)$, where η is the canonical line bundle over CP^∞; if $\tau(\eta)$ lies in $K(CP^\infty;R)$ then the splitting principle shows that τ lies in $K(BU;R)$.

Next let S be some ring containing the rationals. Let $G(CP^n;S)$ be the multiplicative group of elements of augmentation 1 in $H^*(CP^n;S)$. Then we can define a homomorphism

$$\text{todd}: K(CP^n;S) \longrightarrow G(CP^n;S)$$

by

$$\text{todd}(\xi \otimes s) = (\text{todd}\,\xi)^s$$

for $\xi \in K(CP^n)$. Here $(1+x)^s$ is defined by the usual binomial series

$$(1+x)^s = 1 + sx + \frac{s(s-1)}{1.2}x^2 + \dots ;$$

in this case the series is finite. On $K(CP^n;R)$ the homomorphism agrees with todd'. Passing to inverse limits, we obtain a homomorphism

$$\text{todd}: K(CP^\infty;S) \longrightarrow G(CP^\infty;S) .$$

(Here $G(CP^\infty;S)$ is the multiplicative group of elements of augmentation 1 in $H^*(CP^\infty;S)$.) On $K(CP^\infty;R)$ this homomorphism agrees with todd'.

Take an indeterminate t and take $S = Q[t]$. Consider

$$\text{todd}(1+\zeta)^t = \text{todd}(1 + t\zeta + \frac{t(t-1)}{1.2}\zeta^2 + \dots) .$$

This is an element of $G(CP^\infty;Q[t])$, that is, it is a formal power-series in x with coefficients which are polynomials in t; say

$$\text{todd}(1+\zeta)^t = 1 + p_1(t) + p_2(t)x^2 + \dots .$$

But for any integer n, $(1+\zeta)^n$ is a line bundle, and we have

$$\mathrm{todd}(1 + \zeta)^n = \frac{e^{nx}-1}{nx}$$

$$= 1 + \frac{nx}{2!} + \frac{n^2x^2}{3!} + \dots .$$

So for integer values of t we have

$$p_r(n) = \frac{n^r}{(r+1)!} \quad ;$$

thus

$$p_r(t) = \frac{t^r}{(r+1)!}$$

and

$$\mathrm{todd}(1 + \zeta)^t = \frac{e^{tx}-1}{tx} \quad .$$

Consider now $(\tau(\eta))^d$. A priori this is a power-series in ζ with rational coefficients. I claim that these coefficients actually lie in R. To prove this, choose a prime p such that $p \equiv 1 \mod d$; we wish to prove that the coefficients of $(\tau(\eta))^d$ are p-adic integers. We work over the p-adic integers, and manipulate as follows.

$$\mathrm{ch}(\tau(\eta))^d = \mathrm{ch}\ \tau(d\eta)$$

$$= \left(\frac{E_0^! \mathrm{todd}}{\mathrm{todd}}\right)(d\eta)$$

$$= \frac{\mathrm{todd}' E_0 d\eta}{\mathrm{todd}\ d\eta}$$

$$= \frac{\mathrm{todd}\ dE_0\eta}{\mathrm{todd}\ d\eta}$$

Now the basic remark in the proof of Theorem 5 is that

$$dE_0\eta = \sum_\rho (1+\zeta)^\rho$$

where ρ runs over $\rho_n = \omega^m$ for $1 \le m \le d$, and ω is a primitive d^{th} root of unity as in Theorem 5. Thus we have

$$\mathrm{ch}(\tau(\eta))^d = \prod_\rho \frac{\mathrm{todd}(1+\zeta)^\rho}{\mathrm{todd}(1+\zeta)}$$

$$= \prod_\rho \frac{e^{\rho x}-1}{\rho(e^x-1)} \qquad \text{(by the remarks above)}$$

$$= \mathrm{ch} \prod_\rho \frac{(1+\zeta)^\rho - 1}{\rho\zeta}$$

Thus we have

$$(\tau(\eta))^d = \prod_\rho \frac{(1+\zeta)^\rho - 1}{\rho\zeta} \quad .$$

But for each ρ the coefficients of the power series

$$\frac{(1+\zeta)^\rho - 1}{\zeta}$$

are p-adic integers; and the denominator

$$\prod_\rho \rho = (-1)^{d-1}$$

is invertible. Therefore the coefficients in the power-series $(\tau(\eta))^d$ are p-adic integers. This proves that these coefficients lie in R, as claimed.

Finally, since d is invertible in R, we deduce that the coefficients of $\tau(\eta)$ lie in R. This proves Lemma 9.

We now turn to cobordism.

Let R be a subring of the rationals. Let $MU^*(X;R)$ be complex cobordism with coefficients in R. This is a representable functor; we write MUR for the representing spectrum. We require the same information as before.

Lemma 10

If $R_1 \subset R_2$, the maps

$$i_*: H_*(MU;R_2) \longrightarrow H_*(MUR_1;R_2)$$

$$i^*: H^*(MUR_1;R_2) \longrightarrow H^*(MU;R_2)$$

$$(i\wedge i)_*: H_*(MU\wedge MU;\ R_2) \longrightarrow H_*(MUR_1\wedge MUR_1;\ R_2)$$

$$(i\wedge i)^*: H^*(MUR_1\wedge MUR_1;\ R_2) \longrightarrow H^*(MU\wedge MU;\ R_2)$$

are iso.

Proof. Let Y be a Moore spectrum with

$$\pi_n(Y) = 0 \quad \text{for} \quad n < 0\ ,$$

$$H_n(Y) = \begin{cases} R & \text{for} \quad n = 0 \\ 0 & \text{for} \quad n \neq 0\ . \end{cases}$$

Then we may take $MU \wedge Y$ as a construction for MUR. This leads immediately to the result.

Lemma 11

Suppose $R_1 \subset R_2$. Then the maps

$$i_*: K_*(MU;R_2) \longrightarrow K_*(MUR_1;R_2)$$

$$i^*: MU^*(MUR_1;R_2) \longrightarrow MU^*(MU;R_2)$$

$$(i\wedge i)^*: MU^*(MUR_1\wedge MUR_1;\ R_2) \longrightarrow MU^*(MU\wedge MU;\ R_2)$$

are iso. The maps i^* and $(i \wedge i)^*$ are also homeomorphisms with respect to the filtration topology.

The proof is the same as for Lemma 3.

We now consider the set $MU^0(MUR_1;R_2)$. This set may be identified with the set of cohomology operations

$$b: MU^n(X;R_1) \longrightarrow MU^n(X;R_2)$$

which are defined for all X and n, natural, and stable (therefore additive and R_1-linear). This set is topologised by the filtration topology. According to Lemma 11, the set to be considered is essentially independent of R_1, so long as $R_1 \subset R_2$. We therefore write $B(R_2)$ for the set of operations just introduced, and regard it primarily as the ring of stable cohomology operations of degree zero on $MU^*(X;R_2)$.

Lemma 12

If $R_1 \subset R_2$, then we have a monomorphism

$$\iota: B(R_1) \longrightarrow B(R_2)$$

such that for each $b \in B(R_1)$, each X and each n the the following diagram is commutative.

$$\begin{array}{ccc} MU^n(X;R_1) & \xrightarrow{i_*} & MU^n(X;R_2) \\ \downarrow{b} & & \downarrow{\iota b} \\ MU^n(X;R_1) & \xrightarrow{i_*} & MU^n(X;R_2) \end{array}$$

This follows from the preceding discussion, together with the fact that

$$i_*: MU^0(MU;R_1) \longrightarrow MU^0(MU;R_2)$$

is monomorphic. (Compare Lemma 4.)

Because of this lemma, it will be sufficient to construct an idempotent in $B(Q)$. But over Q, stable homotopy theory becomes trivial. We will give the next construction in slightly greater generality than is needed now, for use later. Let $f: X \longrightarrow MUQ$ be a map. Then we define $f_!$ so that the following diagram is commutative.

$$\begin{array}{ccc} H_*(X;Q) & \xrightarrow{f_*} & H_*(MUQ;Q) \\ & {\scriptstyle f_!}\searrow \quad \nearrow{\scriptstyle \cong} & \\ & H_*(MU;Q) & \end{array}$$

Of course we can make a similar definition with H_* replaced by K_*, or with MU, MUQ replaced by BU, BUQ.

Now we define

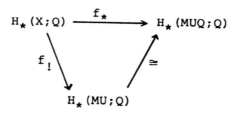

$$\theta: MU^0(X;Q) \longrightarrow \mathrm{Hom}_Q(H_*(X;Q), H_*(MU;Q))$$

as follows. If $f: X \longrightarrow MUQ$ is a map, then $\theta(f) = f_!$.

<u>Lemma 13</u>

θ is an isomorphism.

If we assign the obvious topology to the Hom group, then θ becomes a homeomorphism. If $X = MU$, then θ carries composition in $B(Q)$ into composition in the Hom group.

This lemma is a known consequence of Serre's C-theory [27].

I now choose a positive integer d, and seek to construct a "fake cobordism theory" whose coefficient groups are periodic with one multiplicative generator every $2d$ dimensions. Let $E_0 \in A(Q)$ be as above. Then we define $\varepsilon \in B(Q)$ to be the element such that the following diagram is commutative.

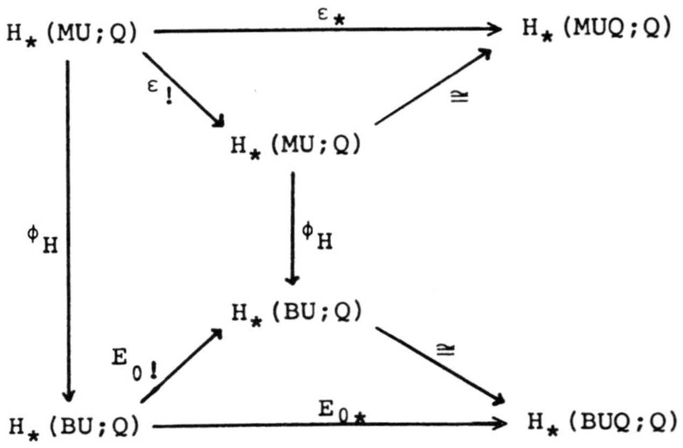

(Here ϕ_H is the Thom isomorphism in homology.) It is clear that ε is idempotent; indeed ε is the most obvious idempotent in sight.

Theorem 14

ε lies in $B(R)$, where $R = R(d)$ is the ring of rationals a/b such that b contains no prime p with $p \equiv 1 \mod d$, as in Theorem 5.

The proof will require two intermediate results.

Lemma 15

A map $f\colon S^p \longrightarrow MUQ$ factors through MUR if and only if

$$f_!\colon K_*(S^p;Q) \longrightarrow K_*(MU;Q)$$

maps $K_*(S^p;R)$ into $K_*(MU;R)$.

This is the theorem of Stong and Hattori [16, 29]. Note that if S^p is regarded as a space rather than as a spectrum, then $K_*(S^p)$ must be taken reduced.

Lemma 16

Let X be a connected spectrum such that $H_r(X)$ is free for all r. Then a map $f\colon X \longrightarrow MUQ$ factors through MUR if and only if

$$f_!\colon K_*(X;Q) \longrightarrow K_*(MU;Q)$$

maps $K_*(X;R)$ into $K_*(MU;R)$.

<u>Proof</u>. It is trivial that if f factors, then $f_!$ maps $K_*(X;R)$ into $K_*(MU;R)$. We wish to prove the converse. First assume that X is finite-dimensional, say $(n-1)$-connected and $(n+d)$-dimensional. We proceed by induction over d. The result is true if X is a wedge of spheres, by Lemma 15. We may now assume we have a cofibering

$$A \xrightarrow{i} X \xrightarrow{j} B$$

with the following properties.

(i) For $r \le m$ we have

$$i_*: H_r(A) \cong H_r(X), \quad H_r(B) = 0 .$$

(ii) For $r \le m$ we have

$$H_r(A) = 0, \ j_*: H_r(X) \cong H_r(B) .$$

(iii) The result holds for A and B.

Now suppose given a map $f: X \longrightarrow MUQ$ such that $f_!$ maps $K_*(X;R)$ into $K_*(MU;R)$. Then $fi: A \longrightarrow MUQ$ maps $K_*(A;R)$ into $K_*(MU;R)$. By (iii), we have the following commutative diagram.

$$\begin{array}{ccc} A & \xrightarrow{i} & X \\ {\scriptstyle g}\downarrow & & \downarrow{\scriptstyle f} \\ MUR & \xrightarrow{i'} & MUQ \end{array}$$

Now the spectral sequence

$$H^*(X;MUR^*(S^0)) \Longrightarrow MUR^*(X)$$

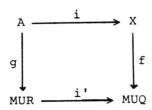

is trivial (since the differentials are zero mod torsion and the groups are torsion-free). We deduce that

$$i^*: MUR^*(X) \longrightarrow MUR^*(A)$$

is epi. So g extends over X; say we have $h: X \longrightarrow MUR$ such that $hi = g$. Then we have

$$f = i'h + kj$$

for some $k: B \longrightarrow MUQ$. Then evidently $(kj)_!$ maps $K_*(X;R)$ into $K_*(MU;R)$. Now the spectral sequence

$$H_*(X;K_*(S^0;R)) \Longrightarrow K_*(X;R)$$

is trivial (since the differentials are zero mod torsion and the groups are torsion-free). We deduce that

$$j_*: K_*(X;R) \longrightarrow K_*(B;R)$$

is epi. Therefore $k_!$ maps $K_*(B;R)$ into $K_*(MU;R)$. By (iii), k factors through MUR. Therefore f factors through MUR. This completes the induction and proves the result when X is finite-dimensional.

We now tackle the case of a general X. Approximate X by X^n such that $i_*: H_r(X^n) \longrightarrow H_r(X)$ is iso for $r \le n$ and $H_r(X^n) = 0$ for $r > n$. We have

$$MUR^*(X) = \varprojlim_n MUR^*(X^n)$$

$$MUQ^*(X) = \varprojlim_n MUQ^*(X^n)$$

(since the usual spectral sequences satisfy the Mittag-Leffler condition). Take a map $f: X \longrightarrow MUQ$ such that $f_!$ maps $K_*(X;R)$ into $K_*(MU;R)$. Then the composite

$$X^n \xrightarrow{i_n} X \xrightarrow{f} MUQ$$

is such that $(fi_n)_!$ maps $K_*(X^n;R)$ into $K_*(MU;R)$. Hence fi_n factors through an element $g_n \in MUR^0(X^n)$. Since $MUR^*(X^n) \longrightarrow MUQ^*(X^n)$ is mono, the elements g_n define an element of $\varprojlim_n MUR^*(X^n)$ and thus give a factorisation of f. This proves Lemma 16.

<u>Proof of Theorem 14</u>. Let $\varepsilon: MU \longrightarrow MUQ$ be as above. We aim to apply Lemma 16 to ε. We equip ourselves with various formal remarks.

(i) The following diagram is not commutative.

$$\begin{array}{ccc} K_*(MU;Q) & \xrightarrow{ch} & H_*(MU;Q) \\ \downarrow \phi_K & & \downarrow \phi_H \\ K_*(BU;Q) & \xrightarrow{ch} & H_*(BU;Q) \end{array}$$

In fact, for a suitable choice of ϕ_K we have

$$ch\ \phi_K z = \text{todd} \cdot \phi_H\ chz \ .$$

(Here the product of a cohomology class and a homology class is taken in the sense of the cap product. The reader who prefers to work entirely in cohomology may write out an argument dual to the one which follows, to verify that ε satisfies the analogue of Lemma 16 for K^*.)

(ii) The following diagrams are commutative.

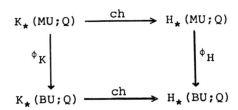

$$\begin{array}{ccc} K_*(MU;Q) & \xrightarrow{\varepsilon_!} & K_*(MU;Q) \\ {\scriptstyle ch}\downarrow & & \downarrow{\scriptstyle ch} \\ H_*(MU;Q) & \xrightarrow{\varepsilon_!} & H_*(MU;Q) \end{array}$$

$$\begin{array}{ccc} K_*(BU;Q) & \xrightarrow{E_{0!}} & K_*(BU;Q) \\ {\scriptstyle ch}\downarrow & & \downarrow{\scriptstyle ch} \\ H_*(BU;Q) & \xrightarrow{E_{0!}} & H_*(BU;Q) \end{array}$$

(iii) If $u \in H^*(BU;Q)$, $v \in H_*(BU;Q)$ we have

$$E_{0!}((E_0^!u) \cdot v) = u \cdot (E_{0!}v) .$$

Now we wish to check that $\varepsilon\colon MU \longrightarrow MUQ$ satisfies the conditions of Lemma 16. So take any element x in $K_*(MU;R)$; we wish to check that $\varepsilon_! x$ lies in $K_*(MU;R)$. Since ϕ_K is iso, it is sufficient to prove that $\phi_K\varepsilon_! x$ lies in

$$\phi_K K_*(MU;R) = K_*(BU;R) .$$

But we have

$$\begin{aligned} ch\phi_K\varepsilon_! x &= \text{todd} \cdot \phi_H ch\varepsilon_! x \\ &= \text{todd} \cdot \phi_H \varepsilon_! ch\ x \\ &= \text{todd} \cdot E_{0!}\phi_H ch\ x \\ &\qquad\qquad (\text{by definition of } \varepsilon) \\ &= E_{0!}(E_0^! \text{todd} \cdot \phi_H ch\ x) \end{aligned}$$

$$= E_{0!}\left(\frac{E_0^! \text{todd}}{\text{todd}} \cdot \text{ch}\phi_K x\right)$$

$$= E_{0!}(\text{ch}\ \tau \cdot \text{ch}\ \phi_K x)$$

(where τ is as in Lemma 9)

$$= E_{0!}\text{ch}(\tau \cdot \phi_K x)$$

$$= \text{ch}E_{0!}(\tau \cdot \phi_K x) .$$

Since ch is iso, we have

$$\phi_K \varepsilon_! x = E_{0!}(\tau \cdot \phi_K x) .$$

But $\tau \in K^*(BU;R)$ and $\phi_K x \in K_*(BU;R)$, so $\tau \cdot \phi_K x \in K_*(BU;R)$. Again, we have $E_0 \colon BU \longrightarrow BUR$, so $E_{0!}$ maps $K_*(BU;R)$ into $K_*(BU;R)$. Thus $\phi_K \varepsilon_! x$ lies in $K_*(BU;R)$ and $\varepsilon_! x$ lies in $K_*(MU;R)$. Therefore ε satisfies the conditions of Lemma 16, and $\varepsilon \in B(R)$. This proves Theorem 14.

The properties of the element $\varepsilon \in B(R)$ are as follows.

<u>Theorem 17</u>

(i) $\varepsilon^2 = \varepsilon$ in $B(R)$.

(ii) For any x, y in $MU^*(X;R)$ we have

$$\varepsilon(xy) = (\varepsilon x)(\varepsilon y) .$$

<u>Proof</u>. Since $B(R) \longrightarrow B(Q)$ is mono, part (i)

follows trivially from the corresponding equation in $B(Q)$.

To prove part (ii), we have to compare the following composites.

$$MU \wedge MU \xrightarrow{\mu} MU \xrightarrow{\varepsilon} MUQ$$

$$MU \wedge MU \xrightarrow{\varepsilon \wedge \varepsilon} MUQ \wedge MUQ \xrightarrow{\mu} MUQ \ .$$

We have to compare $(\varepsilon\mu)_!$ with $(\mu(\varepsilon \wedge \varepsilon))_!$. If we compose with the map

$$H_*(MU;Q) \otimes H_*(MU;Q) \longrightarrow H_*(MU \wedge MU;Q) \ ,$$

we obtain the two ways of chasing round the following commutative diagram.

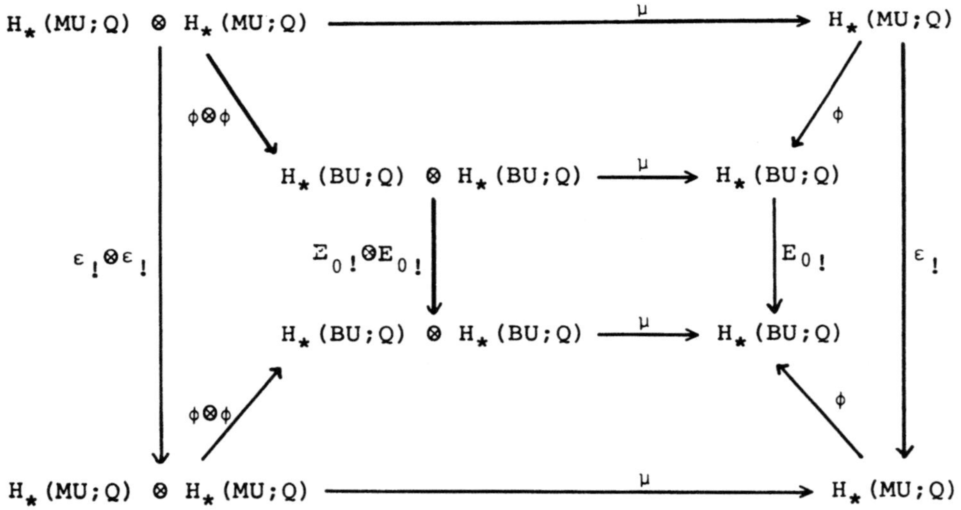

Here the commutativity of the central square arises from the fact that E_0 is additive; that is, the following square is commutative.

$$\begin{array}{ccc} BU \times BU & \xrightarrow{\mu} & BU \\ \Big\downarrow{\scriptstyle E_0 \times E_0} & & \Big\downarrow{\scriptstyle E_0} \\ BUQ \times BUQ & \xrightarrow{\mu} & BUQ \end{array}$$

(Here μ is the product map in BU which represents addition in $\tilde{K}$.) This proves that

$$(\varepsilon\mu)_! = (\mu(\varepsilon \wedge \varepsilon))_! ,$$

and (using Lemma 13) that the following square is homotopy-commutative.

$$\begin{array}{ccc} MU \wedge MU & \xrightarrow{\mu} & MU \\ \Big\downarrow{\scriptstyle \varepsilon \wedge \varepsilon} & & \Big\downarrow{\scriptstyle \varepsilon} \\ MUQ \wedge MUQ & \xrightarrow{\mu} & MUQ \end{array}$$

In other words, we have the formula

$$\varepsilon(xy) = (\varepsilon x)(\varepsilon y)$$

for the external product, when x and y are both the generator in $MU^*(MU)$ and the equality takes place in $MU^*(MU \wedge MU;\ Q)$. Since

$$MU^*(MU \wedge MU;\ Q) \longrightarrow MU^*(MU \wedge MU;\ R)$$

is mono, the equality holds in $MU^*(MU \wedge MU;\ R)$. Since

$$MU^*(MUR \wedge MUR;\ R) \longrightarrow MU^*(MU \wedge MU;\ R)$$

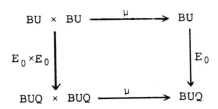

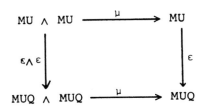

is iso, the equality holds in $MU^*(MUR \wedge MUR;\ R)$ when x and y are both the generator in $MU^*(MUR;R)$. Therefore it always holds. This proves Theorem 17.

S. P. Novikov [24] has shown that multiplicative cohomology operations on MU^* are characterised by their values on the generator $\omega \in MU^2(CP^\infty)$. It might perhaps be of interest to examine $\varepsilon\omega$, and to see if this provides an alternative approach to ε.

We now define $MU_0^*(X) = \varepsilon MU^*(X;R)$.

<u>Corollary 18</u>

$MU_0^*(X)$ is a cohomology theory with products, and is a representable functor.

The proof that $MU_0^*(X)$ is a representable functor is exactly as for Corollary 8, using Theorem 17 (i). The fact that $MU_0^*(X)$ has products is immediate from Theorem 17 (ii).

We write MUR_0 for the representing spectrum for $MU_0^*(X)$. In order to lend credibility to the idea that MUR_0 is an acceptable "Thom complex" corresponding to the space BUR_0, we remark that the following diagram factors to give a unique "Thom isomorphism" ϕ_0.

$$\begin{array}{ccccc} H_*(MUR_0;R) & \xrightarrow{i_*} & H_*(MUR;R) & \xleftarrow{\cong} & H_*(MU;R) \\ \Big\downarrow{\scriptstyle \phi_0} & & & & \Big\downarrow{\scriptstyle \phi} \\ H_*(BUR_0;R) & \xrightarrow{i_*} & H_*(BUR;R) & \xleftarrow{\cong} & H_*(BU;R) \end{array}$$

This follows immediately from the definition of ε.

Theorem 19

(i) The coefficient ring $\pi_*(MUR_0)$ is a polynomial ring over R with generators in dimensions $2d$, $4d$, $6d$, ...

(ii) $MU^*(X;R)$ is a direct product of theories isomorphic to $MU_0^*(X)$.

Note. In part (ii) the splitting is not asserted to be canonical, but the injection of $MU_0^*(X)$ and the projection onto $MU_0^*(X)$ are of course canonical; this is sufficient for the applications.

Proof. For any connected algebra A, let $Q(A)$ be its indecomposable quotient. Then

$$\varepsilon: \pi_*(MUR) \longrightarrow \pi_*(MUR)$$

induces

$$Q(\varepsilon): Q(\pi_*(MUR)) \longrightarrow Q(\pi_*(MUR))$$

with $Q(\varepsilon) \cdot Q(\varepsilon) = Q(\varepsilon)$. We have

$$Q(\mathrm{Im}\varepsilon) \cong \mathrm{Im}(Q\varepsilon) .$$

Now $Q(\pi_*(MUR))$ is R-free with generators x_1, x_2, x_3,... in dimensions 2, 4, 6,... [20, 30]. For each x_n we have either $Q(\varepsilon)x_n = x_n$ or $Q(\varepsilon)x_n = 0$. We may thus choose a homogeneous R-base for $\mathrm{Im}Q(\varepsilon)$ and extend it to a homogeneous R-base for $Q(\pi_*(MUR))$. Lift the basis elements in $\mathrm{Im}Q(\varepsilon)$ to elements g_i in $\mathrm{Im}\varepsilon$, and lift the remaining basis elements in any way to elements h_j. Then $\pi_*(MUR)$ is the polynomial algebra generated by the g_i and h_j, and $\mathrm{Im}\varepsilon$ is precisely the subalgebra generated by the g_i. But this subalgebra is polynomial. It remains only to find the dimensions of the generators.

We have

$$\pi_*(MUR_0) \otimes Q \cong H_*(MUR_0;Q) \cong H_*(BUR_0;Q)$$

(by the remarks above). But as remarked above, $H^*(BUR_0;Q)$ is a polynomial algebra with generators in dimension 2d, 4d, 6d,... . Now part (i) follows by counting dimensions over Q.

The preceding proof actually shows that $\pi_*(MUR)$ is free as a module over $\pi_*(MUR_0)$. Choose a $\pi_*(MUR_0)$-free base for $\pi_*(MUR)$ (beginning with the unit element 1) and represent the basis elements by maps

$$f_j\colon S^{n(j)} \longrightarrow MUR .$$

We now consider the map

$$g\colon \bigvee_j S^{n(j)} \wedge MUR_0 \longrightarrow MUR$$

which on the jth factor is given by

$$S^{n(j)} \wedge MUR_0 \xrightarrow{f_j \wedge i} MUR \wedge MUR \xrightarrow{\mu} MUR .$$

It is clear that g induces an isomorphism of homotopy groups. Since MUR_0 is connected and $n(j) \longrightarrow \infty$ as $j \longrightarrow \infty$ the infinite wedge-sum is also a product. Therefore

$$[X,MUR] \cong \prod_j [X, S^{n(j)} \wedge MUR_0] .$$

This proves part (ii).

We have now accomplished our object of splitting $MU^*(X;R)$ into a direct sum of similar functors. I believe that the functors MU_0^* and K_0^*, together with the spectrum MUR_0 and the space BUR_0, are of some interest. I would like to give further results to prove that MUR_0 is related to BUR_0 as MU is to BU; for lack of time in writing up these notes I offer the following in the disguise of an exercise.

<u>Exercise 20</u>

Show that Proposition 25 of Lecture 1 (the Conner-Floyd theorem) applies to the case $E = MUR_0$, $F = \underline{BUR}_0$.

<u>Hints</u>.

(a) $H^p(MUR_0;R) = 0$ unless $p \equiv 0 \mod 2d$. Therefore $K_\alpha(MUR_0) = 0$ for $\alpha \neq 0$, and $K_0(MUR_0)$ is the whole of $K(MUR_0;R)$. Take the orientation class u in $K(MU)$, map it into $K(MU;R)$, lift it into $K(MUR;R)$ and restrict it to $K(MUR_0;R)$; the result must lie in $K_0(MUR_0)$. This gives the necessary orientation class.

(b) In checking Assumption 20 and 24 of Lecture 1, exercise care in approximating MUR_0 and BUR_0 by finite complexes.

LECTURE 5. FINITENESS THEOREMS

In this lecture I want to give an exposition of certain finiteness theorems in algebra which seem useful in algebraic topology. These results are slight generalisations of known results on coherent rings; one may find the latter in Bourbaki [9, pp. 62-63]. I became interested in the subject in the course of reproving certain results of S. P. Novikov [24]. Independently, Joel M. Cohen became interested in similar results for a different topological application. I am most grateful to Cohen for sending me preprints of his two papers [12, 13]. (So far as I know these papers have not yet appeared.)

The following results 1-5 will serve as illustrations of the sort of topological application which I have in mind.

Theorem 1 (S. P. Novikov)

If X is a finite CW-complex, then $MU^*(X)$ is finitely-generated as a module over the coefficient ring $MU^*(S^0)$.

The methods I will give also yield the following result, which is slightly stronger.

Theorem 2

Let X be a finite CW-complex. Then $MU^*(X)$, considered as a module over the coefficient ring $MU^*(S^0)$, admits a resolution of finite length

$$0 \longrightarrow C_n \longrightarrow C_{n-1} \longrightarrow \cdots \longrightarrow C_1 \longrightarrow C_0 \longrightarrow MU^*(X) \longrightarrow 0$$

by finitely-generated free modules.

Since giving the original lecture I have heard that this result is also known to P. E. Conner and L. Smith; it may also be known to other workers in the field. I am grateful to L. Smith for sending me a preprint.

I will not quote the results of Cohen verbatim, but will reword them to suit the present lecture. I will use the words "almost all" to mean "with a finite number of exceptions".

Theorem 3 (J. M. Cohen)

Let X be a spectrum whose stable homotopy groups $\pi_r(X)$ are finitely generated, and are zero for almost all r. Then $H^*(X;Z_p)$ is finitely-presented as a module over the mod p Steenrod algebra A.

This result can be used to show that under mild restrictions, a space Y (as distinct from a spectrum) must have infinitely many non-zero stable homotopy groups. Even better for this purpose is the variant which follows next. We will say that an abelian group G is p-trivial if $p: G \longrightarrow G$ is iso. Spelling this out, it asks that the torsion subgroup of G should contain no elements of order p, and that the torsion-free quotient of G should be divisible by p.

Theorem 4 (J. M. Cohen)

Let X be a connected spectrum whose stable homotopy groups $\pi_r(X)$ are p-trivial for almost all r. Then the A-module $H^*(X; Z_p)$ can be presented by generators in only finitely many dimensions and relations in only finitely many dimensions.

In particular, of course, the theorem applies if $\pi_r(X) = 0$ for almost all r. The difference between this case and Theorem 3 is that if the groups $\pi_r(X)$ are not finitely-generated, then $H^*(X; Z_p)$ may need infinitely many generators in some dimensions.

Corollary 5 (J. M. Cohen)

Let Y be a space such that $\tilde{H}^*(Y; Z_p) \neq 0$. Then there are infinitely many values of r such that the stable homotopy group $\pi_r^S(Y)$ is not p-trivial (and therefore non-zero).

This answers a question of Serre [26, p.219].

To prove these results, we will present a slight axiomatisation of Bourbaki's results. We will first set up our assumptions, definitions and general theory. From Corollary 12 onwards we turn to the topological applications, and sketch the proof of the results given above. Topologists looking for motivation might perhaps turn to the passage beginning immediately after Example 14.

We suppose given a graded ring R with unit. The word "module" will mean a graded left R-module, unless otherwise specified. We suppose given a class $\underline{C}$ of projective modules. The class $\underline{C}$ is supposed to satisfy two axioms*.

(i) If $P \cong Q$ and $P \in \underline{C}$, then $Q \in \underline{C}$.

(ii) If $P \in \underline{C}$ and $Q \in \underline{C}$, then $P \oplus Q \in \underline{C}$.

Examples.

(i) We define $\underline{F}$ to be the class of finitely-generated free modules.

(ii) We define $\underline{D}$ to be the class of free modules with generators in only a finite number of dimensions.

(iii) We define $\underline{E}$ to be the class of free modules such that for each n there are only a finite number of generators in dimensions $\leq n$.

(iv) We define $\underline{O}$ to be the class containing only the zero module.

In what follows the symbols $\underline{F}$ and $\underline{D}$ will always have the meanings just given to them. In proving Theorems 1, 2 and 3 we take $\underline{C} = \underline{F}$; in proving Theorem 4 and Corollary 5 we take $\underline{C} = \underline{D}$. The axiomatisation simply saves us from giving the same proof twice over.

Definition 6

An R-module M is of $\underline{C}$-type n if it has a projective resolution

---

* Note added in proof. It should also be assumed that $0 \in \underline{C}$.

$$0 \longleftarrow M \longleftarrow C_0 \longleftarrow C_1 \longleftarrow \cdots \longleftarrow C_r \longleftarrow \cdots$$

such that $C_r \in \underline{C}$ for $0 \leq r \leq n$. (Compare Bourbaki p. 60, exercise 6.)

<u>Examples</u>.

(i) All modules are of $\underline{C}$-type -1.

(ii) A module is of $\underline{C}$-type ∞ if and only if it has a projective resolution by modules in $\underline{C}$.

(iii) A module of $\underline{F}$-type 0 is a finitely-generated module.

(iv) A module of $\underline{F}$-type 1 is a finitely-presented module.

(v) A module of $\underline{D}$-type 0 is one which can be generated by generators in only finitely many dimensions.

(vi) A module of $\underline{D}$-type 1 is one which can be presented using generators in only finitely many dimensions and relations in only finitely many dimensions.

Thus, the conclusion of Theorem 1 states that $MU^*(X)$ is of $\underline{F}$-type 0. The conclusion of Theorem 3 states that $H^*(X;Z_p)$ is of $\underline{F}$-type 1. The conclusion of Theorem 4 states that $H^*(X;Z_p)$ is of $\underline{D}$-type 1.

We could also say that M is <u>of $\underline{C}$-cotype n</u> if it has a projective resolution such that $C_r \in \underline{C}$ for $r > n$ With $\underline{C} = \underline{0}$, for example, we would be discussing homological dimension. It would perhaps be interesting to see if known results about homological dimension generalize to cotype (perhaps in the presence of extra assumptions on $\underline{C}$). In particular, is the analogue of Lemma 7 (iii) below true for cotype*? We will not pursue this further here.

If we do not need to emphasise $\underline{C}$, we will write "type" for "$\underline{C}$-type". The basic property of Definition 6 is as follows.

Lemma 7

Suppose given an exact sequence

$$0 \longrightarrow L \xrightarrow{i} M \xrightarrow{j} N \longrightarrow 0$$

of R-modules.

(i) If L is of type $(n - 1)$ and M is of type n , then N is of type n .

(ii) If L is of type n and N is of type n , then M is of type n .

(iii) If M is of type n and N is of type $(n + 1)$, then L is of type n .

(Compare Bourbaki p.60, exercise 6 a, c, d. For the most significant special case see Bourbaki p. 37, Lemma 9.)

Proof. We begin with part (ii). Given resolutions

$$0 \longleftarrow L \longleftarrow C'_0 \longleftarrow C'_1 \longleftarrow \cdots \longleftarrow C'_r \longleftarrow \cdots$$

$$0 \longleftarrow N \longleftarrow C''_0 \longleftarrow C''_1 \longleftarrow \cdots \longleftarrow C''_r \longleftarrow \cdots$$

of L and N , one knows how to construct a resolution of M in which $C_r = C'_r \oplus C''_r$; see [11, p.80]. If $C'_r \in \underline{C}$ for $r \leq n$, and $C''_r \in \underline{C}$ for $r \leq n$, then $C_r \in \underline{C}$ for

* Note added in proof. An affirmative answer to this problem has been obtained by Mrs. S. Cormack.

$r \leq n$. This proves part (ii).

We proceed similarly for part (i). Suppose that we are given resolutions

$$0 \longleftarrow L \longleftarrow C'_0 \longleftarrow C'_1 \longleftarrow \cdots \longleftarrow C'_r \longleftarrow \cdots$$

$$0 \longleftarrow M \longleftarrow C_0 \longleftarrow C_1 \longleftarrow \cdots \longleftarrow C_r \longleftarrow \cdots$$

of L and M . By constructing a chain map over $i\colon L \longrightarrow M$ and forming its mapping cylinder, we can construct a resolution for N in which $C''_0 = C_0$ and $C''_r = C_r \oplus C'_{r-1}$ for $r \geq 1$. If $C_r \in \underline{C}$ for $r \leq n$ and $C'_r \in \underline{C}$ for $r \leq n - 1$, then $C''_r \in \underline{C}$ for $r \leq n$. This proves part (i).

To prove part (iii), we begin by considering the special case in which M is projective. Since N is of type $(n + 1)$, we have an exact sequence

$$0 \longrightarrow K \longrightarrow F \longrightarrow N \longrightarrow 0$$

with $F \in \underline{C}$ and K of type n . Compare this with the exact sequence

$$0 \longrightarrow L \longrightarrow M \longrightarrow N \longrightarrow 0 \quad .$$

By Schanuel's Lemma [18, p.101] we have

$$L \oplus F \cong M \oplus K \quad .$$

So we have an exact sequence

$$0 \longrightarrow F \longrightarrow M \oplus K \longrightarrow L \longrightarrow 0 \quad .$$

Here F is of type ∞ and $M \oplus K$ is of type n by part (ii). Therefore L is of type n by part (i).

We now turn to the general case. Since M is of type n , and the result is empty for $n = -1$, we may suppose given an exact sequence

$$0 \longrightarrow K \longrightarrow F \xrightarrow{q} M \longrightarrow 0$$

with $F \in \underline{C}$ and K of type $(n - 1)$. Let P be the kernel of the composite $jq: F \longrightarrow N$; then P has type n by the special case already considered. We can construct the following diagram.

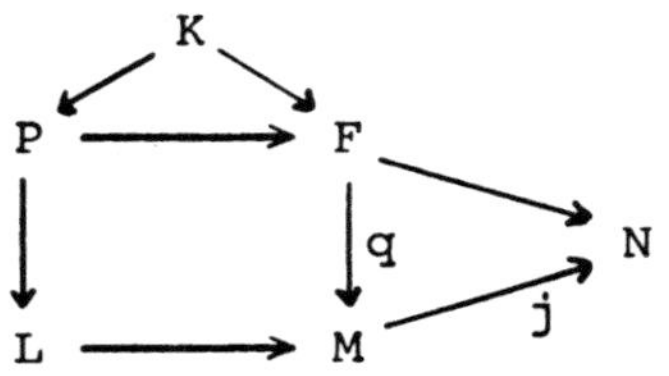

The sequence

$$0 \longrightarrow K \longrightarrow P \longrightarrow L \longrightarrow 0$$

is exact. Here K has type $(n - 1)$ and P has type n, so L has type n by part (i). This completes the proof of Lemma 7.

For technical reasons we need the following corollary.

<u>Corollary 8</u>

Suppose given an exact sequence

$$0 \longrightarrow K \longrightarrow C_0 \longrightarrow C_1 \longrightarrow \cdots \longrightarrow C_{n-1} \xrightarrow{d} C_n \longrightarrow M \longrightarrow 0$$

in which C_r is of type r. Then M is of type n.

<u>Proof</u>. The result is true for $n = 0$, by 7(i). As an inductive hypothesis, suppose it true for $(n - 1)$. Then $d(C_{n-1})$ is of type $(n - 1)$, and we have the following exact sequence.

$$0 \longrightarrow d(C_{n-1}) \longrightarrow C_n \longrightarrow M \longrightarrow 0 .$$

So M is of type n by 7 (i). This completes the induction and proves Corollary 8.

The next question which we consider arises as follows. The "Noetherian" case is essentially that in which all modules of $\underline{F}$-type 0 are of $\underline{F}$-type ∞. The "coherent" case is essentially that in which all modules of $\underline{F}$-type 1 are of $\underline{F}$-type ∞. (See Bourbaki, p. 61 exercise 7a and p. 63 exercise 12d, or below). Although it is not necessary for the applications, it seems worth describing a hierarchy of more subtle cases; the nth case is that in which all modules of type n are of type ∞.

Theorem 9

Suppose given $\underline{C}$ and $n \geq 0$. Then the following conditions are all equivalent.

(i) If $C \in \underline{C}$ and P is a submodule of C of type $(n-1)$, then P is of type n.

(ii) If M is of type n and P is a submodule of M of type $(n-1)$, then P is of type n.

(iii) Suppose given an exact sequence

$$0 \longrightarrow K \longrightarrow C_n \longrightarrow C_{n-1} \longrightarrow \cdots \longrightarrow C_1 \longrightarrow C_0 \longrightarrow M \longrightarrow 0$$

in which C_r is of type n for each r. Then K is of type n.

(iv) Suppose given an exact sequence

$$C_n \longrightarrow C_{n-1} \longrightarrow \dots \longrightarrow C_1 \longrightarrow C_0 \longrightarrow M \longrightarrow 0$$

in which $C_r \in \underline{C}$ for each r. Then we can extend it to an exact sequence

$$C_{n+1} \longrightarrow C_n \longrightarrow C_{n-1} \longrightarrow \dots \longrightarrow C_1 \longrightarrow C_0 \longrightarrow M \longrightarrow 0$$

in which $C_{n+1} \in \underline{C}$.

(v) Every module of type n is of type ∞.

We note that in conditions (iii) and (iv) the module M at the right-hand end of the sequence is included only to avoid making an exception of the case $n = 0$. If $n \geq 1$, we can suppose given the sequence

$$C_n \longrightarrow C_{n-1} \longrightarrow \dots \longrightarrow C_1 \xrightarrow{d} C_0$$

and define $M = C_0/dC_1$.

<u>Proof of Theorem 9</u>. First we prove that (i) implies (ii). Suppose that M is of type n. Then by definition, we can find a sequence

$$0 \longrightarrow K \longrightarrow C_0 \xrightarrow{j} M \longrightarrow 0$$

with $C_0 \in \underline{C}$ and K of type $(n-1)$. Let P be a submodule of M of type $(n-1)$; then we have the following exact sequence.

$$0 \longrightarrow K \longrightarrow j^{-1}P \longrightarrow P \longrightarrow 0 .$$

Since K and P are of type $(n-1)$, $j^{-1}P$ is of type $(n-1)$ by 7 (ii). Since $j^{-1}P$ is a submodule of C_0 and $C_0 \in \underline{C}$,

$j^{-1}P$ is of type n by 9 (i), which we are assuming. Hence P is of type n by 7 (i). This proves (ii).

We prove that (ii) implies (iii). Suppose given an exact sequence

$$0 \longrightarrow K \longrightarrow C_n \xrightarrow{d} C_{n-1} \longrightarrow \dots \longrightarrow C_1 \xrightarrow{d} C_0 \xrightarrow{\varepsilon} M \longrightarrow 0$$

in which C_r is of type n for each r. Let $Z_r \subset C_r$ be the submodule

$$\mathrm{Im}(d\colon C_{r+1} \longrightarrow C_r) = \mathrm{Ker}(d\colon C_r \longrightarrow C_{r-1}) ,$$

with the obvious interpretation for $r = 0,n$. Then by Corollary 8 (or trivially if $n = 0$) Z_0 is of type $(n-1)$. Since Z_0 is a submodule of C_0 and we are assuming 9 (ii), Z_0 is of type n. Assume as an inductive hypothesis that Z_{r-1} is of type n. We have the following exact sequence.

$$0 \longrightarrow Z_r \longrightarrow C_r \longrightarrow Z_{r-1} \longrightarrow 0 .$$

So Z_r is of type $(n-1)$ by 7 (iii). Since Z_r is a submodule of C_r and we are assuming 9 (ii), Z_r is of type n. This completes the induction. The induction proves that $K = Z_n$ is of type n. This proves (iii).

We prove that (iii) implies (iv). Suppose given an exact sequence

$$C_n \xrightarrow{d} C_{n-1} \longrightarrow \dots \longrightarrow C_1 \longrightarrow C_0 \longrightarrow M \longrightarrow 0$$

in which $C_r \in \underline{C}$ for each r. Then certainly C_r is of type n. Let Z_n be as in the proof of (iii); then by (iii), Z_n is of type $n \geq 0$. Thus we can find an epimorphism

$$C_{n+1} \longrightarrow Z_n$$

with $C_{n+1} \in \underline{C}$. This proves (iv).

We prove that (iv) implies (v). Suppose given a module M of type n. By definition, we have an exact sequence

$$C_n \longrightarrow C_{n-1} \longrightarrow \ldots \longrightarrow C_1 \longrightarrow C_0 \longrightarrow M \longrightarrow 0$$

in which $C_r \in \underline{C}$ for each r. By (iv) we can extend it to an exact sequence

$$C_{n+1} \longrightarrow C_n \longrightarrow \ldots \longrightarrow C_1 \longrightarrow C_0 \longrightarrow M \longrightarrow 0$$

in which $C_{n+1} \in \underline{C}$. Now (iv) applies again to the sequence

$$C_{n+1} \longrightarrow C_n \longrightarrow \ldots \longrightarrow C_2 \longrightarrow C_1 \longrightarrow Z_0 \longrightarrow 0 .$$

Continue by induction. The induction constructs a resolution of M by modules C_r in $\underline{C}$ and shows that M is of type ∞. This proves (v).

We prove that (v) implies (i). Suppose given $C \in \underline{C}$ and a submodule $P \subset C$ of type $(n-1)$. Then we have an exact sequence

$$0 \longrightarrow P \longrightarrow C \longrightarrow C/P \longrightarrow 0 .$$

Here C/P is of type n (by 7 (i) or direct from the definition). By 9 (v), which we are assuming, C/P is of type $(n+1)$. Therefore P is of type n by 7 (iii). This proves (i). We have completed the proof of Theorem 9.

It now seems reasonable to make the following definition.

Definition 10

The ring R is $(n,\underline{C})$-coherent if the equivalent conditions stated in Theorem 9 are satisfied.

It is clear from 9 (v) that if R is $(n,\underline{C})$-coherent, it is $(m,\underline{C})$-coherent for $m \geq n$.

Examples.

(i) The ring R is $(0,\underline{F})$-coherent if and only if it is (left) Noetherian.

(ii) We say that R is finite-dimensional if it has non-zero components in only finitely many dimensions, so that $R = \sum_{-N}^{N} R_n$. Such a ring is $(0,\underline{D})$-coherent; the proof is trivial.

(iii) Coherence, as defined in Bourbaki, is $(1,\underline{F})$-coherence. More precisely, condition 9 (i) says in this case that every submodule P of $\underline{F}$-type 0 in C is of $\underline{F}$-type 1. This coincides with Bourbaki's condition "C is pseudo-coherent". If

$$0 \longrightarrow C' \longrightarrow C \longrightarrow C'' \longrightarrow 0$$

is exact and C', C" satisfy this condition, then so does C. (This follows easily from Lemma 7; see Bourbaki p. 62 exercise 11a). So in order to check the condition for every C in $\underline{F}$, it is sufficient to check it for C = R (compare Bourbaki p. 63 exercise 12a). This proves the equivalence of our definition with Bourbaki's.

We will now prove that for $n \geq 1$ the property of n-coherence passes to suitable direct limits. We suppose given a (graded) ring R containing subrings R^{α}, and make the following assumptions. First, we assume that $\underline{C}$ is either $\underline{F}$ or $\underline{D}$, and we divide cases accordingly. If $\underline{C} = \underline{F}$, we assume that for any finite set of elements $r_1, r_2, \ldots, r_n$ in R we can find an α such that $r_1, r_2, \ldots, r_n$ lie in R^{α}. If $\underline{C} = \underline{D}$, we assume that for any finite set of dimensions $n, m, \ldots, p$ we can find an α such that $R_n, R_m, \ldots, R_p$ are contained in R^{α}. This assumption ensures that the R^{α} approximate sufficiently closely to R, in a sense depending on $\underline{C}$. Secondly, we assume that, for each α, R is free as a right module over R^{α}. With these assumptions we have:

Theorem 11

(i) For $0 < n < \infty$, the R-modules of type n are precisely those of the form $R \otimes_{R^{\alpha}} M^{\alpha}$, where R^{α} runs over the subrings and M^{α} runs over the R^{α}-modules of type n.

(ii) If $n > 0$ and R^{α} is $(n,\underline{C})$-coherent for each α then R is $(n,\underline{C})$-coherent.

(Compare Bourbaki p. 63 exercise 12e. A check through the proof below shows that for $n = 1$ we need only assume that R is flat, rather than free, as a right module over R^{α}.)

Proof. For part (i), we begin by showing that the R-modules of the form $R \otimes_{R^\alpha} M^\alpha$ are of type n. For suppose that M^α is of type n; then there is an exact sequence of R^α-modules

$$C_n^\alpha \longrightarrow C_{n-1}^\alpha \longrightarrow \ldots \longrightarrow C_1^\alpha \longrightarrow C_0^\alpha \longrightarrow M^\alpha \longrightarrow 0$$

with C_t^α in $\underline{F}$ or $\underline{D}$ as the case may be. The functor $R \otimes_{R^\alpha}$ preserves exactness, so we have the following exact sequence.

$$R \otimes_{R^\alpha} C_n^\alpha \longrightarrow \ldots \longrightarrow R \otimes_{R^\alpha} C_0^\alpha \longrightarrow R \otimes_{R^\alpha} M^\alpha \longrightarrow 0 .$$

Here the modules $R \otimes_{R^\alpha} C_t^\alpha$ are free and lie in $\underline{F}$ or $\underline{D}$ as the case may be. Thus $R \otimes_{R^\alpha} M^\alpha$ is of type n.

To prove the converse, suppose given an R-module M of type n, where $0 < n < \infty$. Then we have an exact sequence of R-modules

$$C_n \xrightarrow{d} C_{n-1} \longrightarrow \ldots \longrightarrow C_1 \xrightarrow{d} C_0 \longrightarrow M \longrightarrow 0$$

with $C_t \in \underline{C}$ for each t. Choose R-free bases in each C_t; then each map d can be represented by a matrix r_{ij}. If $\underline{C} = \underline{F}$, there are only a finite number of elements r_{ij} in all. If $\underline{C} = \underline{D}$, the elements r_{ij} lie in only a finite number of dimensions. In either case, we can find an α such that all the elements r_{ij} lie in R^α. Let C_t^α be the free R^α-module generated by the R-free base of C_t. Then the maps d restrict to give

$$C_n^\alpha \xrightarrow{d^\alpha} C_{n-1}^\alpha \longrightarrow \ldots \longrightarrow C_1^\alpha \xrightarrow{d^\alpha} C_0^\alpha .$$

The original sequence

$$C_n \xrightarrow{d} C_{n-1} \longrightarrow \cdots \longrightarrow C_1 \xrightarrow{d} C_0$$

is, up to isomorphism,

$$R \otimes_{R^\alpha} C_n^\alpha \xrightarrow{1 \otimes d^\alpha} R \otimes_{R^\alpha} C_{n-1}^\alpha \longrightarrow \cdots \longrightarrow R \otimes_{R^\alpha} C_1^\alpha \xrightarrow{1 \otimes d_\alpha} R \otimes_{R^\alpha} C_0^\alpha \ .$$

Since R is free as a right module over R^α, this sequence (as a sequence of groups) is isomorphic to a direct sum of copies of the sequence

$$C_n^\alpha \xrightarrow{d^\alpha} C_{n-1}^\alpha \longrightarrow \cdots \longrightarrow C_1^\alpha \xrightarrow{d^\alpha} C_0^\alpha \ .$$

Since the original sequence was exact, the sequence

$$C_n^\alpha \xrightarrow{d^\alpha} C_{n-1}^\alpha \longrightarrow \cdots \longrightarrow C_1^\alpha \xrightarrow{d^\alpha} C_0^\alpha$$

must be exact. We can define $M^\alpha = C_0^\alpha / dC_1^\alpha$, and M^α is an R^α-module of type n, since C_t^α lies in $\underline{F}$ or $\underline{D}$ as the case may be. Since $R \otimes_{R^\alpha}$ preserves exactness, the sequence

$$R \otimes_{R^\alpha} C_1^\alpha \xrightarrow{1 \otimes d^\alpha} R \otimes_{R^\alpha} C_0^\alpha \longrightarrow R \otimes_{R^\alpha} M^\alpha \longrightarrow 0$$

is exact, and we have

$$M \cong R \otimes_{R^\alpha} M^\alpha \ .$$

This proves part (i).

To prove part (ii), we assume that R^α is n-coherent for each α. Let M be an R-module of type n. By part (i) M has the form $M \cong R \otimes_{R^\alpha} M^\alpha$ with M^α of type n. By 9 (v) for R^α, M^α is of type $(n+1)$. By part (i), M

is of the type $(n+1)$. We have shown that each R-module of type n is of type $(n+1)$. By the proof that 9 (v) implies 9 (i), this is sufficient to show that R is $(n,\underline{C})$-coherent.

Corollary 12

The ring $MU^*(S^0)$ is $(1,\underline{F})$-coherent but not Noetherian.

In fact, $MU^*(S^0)$ is a polynomial ring $Z[x_1, x_2, \ldots, x_n, \ldots]$ on a countable set of generators [20, 30]. Each finite subset of the generators generates a Noetherian subring, and we take these subrings for the R^α in Theorem 11. (Compare Bourbaki p. 63 exercise 12f.)

Corollary 13

The Steenrod algebra A is both $(1,F)$-coherent and $(1,D)$-coherent, but neither Noetherian nor finite-dimensional.

In fact, any finite subset of A, and any finite-dimensional part $\sum_{r=0}^{N} A_r$ of A, is contained in a Hopf subalgebra which is finite [19], and therefore both $(0,\underline{F})$-coherent and $(0,\underline{D})$-coherent. We take such subalgebras for the R^α in Theorem 11; the whole algebra is free over R^α since R^α is a Hopf subalgebra [22].

Example 14

The stable homotopy groups of spheres form (under

composition) a graded ring which is neither $(1,\underline{F})$-coherent nor $(1,\underline{D})$-coherent.

We may now summarise our guiding philosophy. The most classical finiteness theorems in algebra concern finitely-generated modules over a Noetherian ring. In our applications, however, we have to use rings which are not Noetherian. The Noetherian condition gives us finiteness results on submodules. But in algebraic topology and in homological algebra we can do without information about general submodules, provided that we have information about kernels. (I mean, of course, kernels of maps from one "good" module to another.) In other words, we can use the following result.

Corollary 15

Suppose that R is $(1,\underline{C})$-coherent, that L and M are modules of $\underline{C}$-type 1 and that $f: L \longrightarrow M$ is an R-map. Then $\operatorname{Ker} f$ is of $\underline{C}$-type 1.

This follows immediately from Theorem 9 (iii).

Corollary 16

Suppose that R is $(1,\underline{C})$-coherent, and that

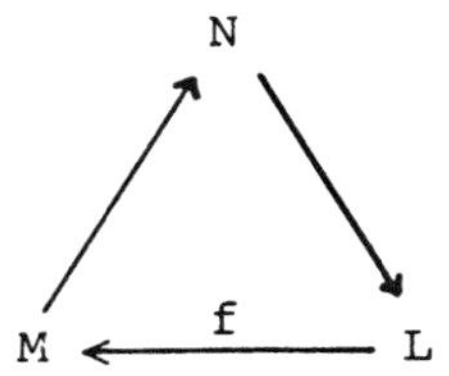

is an exact triangle of R-modules in which L and M are of $\underline{C}$-type 1. Then N is of $\underline{C}$-type 1.

<u>Proof</u>. Coker f is of type 1 by Corollary 8 and Ker f is of type 1 by Corollary 15. Thus N is of type 1 by Lemma 7 (ii).

For the next proposition we assume that the class $\underline{C}$ contains any free module on one generator. This is true, of course, for $\underline{C} = \underline{F}$ and $\underline{C} = \underline{D}$. We assume that E^* is a (reduced) generalised cohomology theory with products, and that the coefficient ring $E^*(S^0)$ is $(1,\underline{C})$-coherent.

<u>Proposition 17</u>

If X is a finite CW-complex, then $E^*(X)$ is a module of $\underline{C}$-type ∞ over $E^*(S^0)$.

<u>Proof</u>. The result is true if $X = S^n$, for $E^*(S^n)$ is a free module over $E^*(S^0)$ on one generator. This serves to start an induction over the number of cells in X. If X is not a sphere, we can find a cofibering

$$A \longrightarrow X \longrightarrow B$$

in which A and B have fewer cells than X. (For example, take A to be any proper subcomplex of X.) As our inductive hypothesis, we suppose that $E^*(A)$ and $E^*(B)$ are of type 1. The cofibering gives the following exact triangle of modules over $E^*(S^0)$.

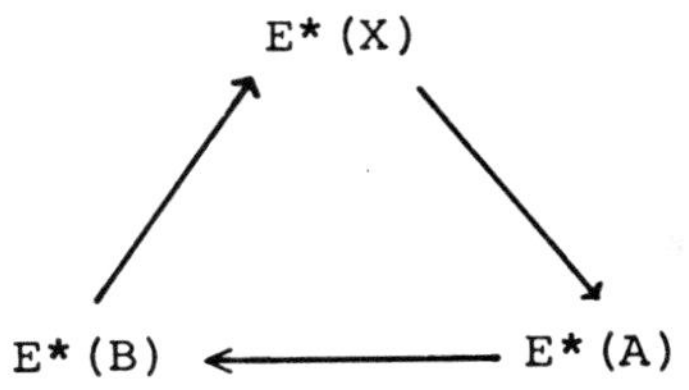

By Corollary 16, $E^*(X)$ is of type 1. This completes the induction. Of course, by Theorem 9 (v) a module of type 1 is of type ∞. This proves Proposition 17.

It is clear that Theorem 1 follows immediately from Corollary 12 and Proposition 17.

To prove Theorem 2, one uses Theorem 11 to reduce the problem to the study of a module M^α over a polynomial ring R^α on finitely many generators (see Corollary 12). For M^α we know the existence of a resolution of the sort required; take such a resolution and apply $R \otimes_{R^\alpha}$, as in the proof of Theorem 11.

We will sketch the proof of Theorem 4. Let G be an abelian group which is p-trivial, and let $K(G)$ be the corresponding Eilenberg-MacLane spectrum. Then $H^*(K(G);Z_p) = 0$, for $p: G \longrightarrow G$ must induce an isomorphism p_* of $H_*(K(G);Z_p)$, but $p_* = 0$. Next let X be a connected spectrum such that $\pi_r(X)$ is p-trivial for each r; then again we have $H^*(X;Z_p) = 0$. It follows that the general case of Theorem 4 can be deduced, without changing the module $H^*(X;Z_p)$, from the special case in which $\pi_r(X)$ is zero for almost all r.

Next let F be a free abelian group; one can show that $H^*(K(F);Z_p)$ is of D-type 1. This allows us to deduce the same result for a general Eilenberg-MacLane spectrum $K(G)$; we consider a fibering

$$K(F_1) \longrightarrow K(F_2) \longrightarrow K(G)$$

and apply Corollary 16 to the resulting exact triangle of cohomology modules.

Now we can prove the result for a spectrum X with just n non-zero homotopy groups. This is done by induction over n, as for Proposition 17, but applying Corollary 16 to the exact triangle of cohomology modules arising from a suitable fibering. This completes the proof.

The proof of Theorem 3 can now safely be left to the reader.

To deduce Corollary 5, we suppose given a space Y which contradicts Corollary 5, so that $\tilde{H}^*(Y;Z_p) \neq 0$ and Theorem 4 applies to the corresponding spectrum. Let y be a non-zero class of lowest dimension in $\tilde{H}^*(Y;Z_p)$; then

$$P^{p^f} y = 0$$

for all sufficiently large f; this makes it extremely plausible that $\tilde{H}^*(Y;Z_p)$ cannot have a presentation with relations in only finitely many dimensions, and this can indeed be proved. This contradicts Theorem 4 and proves Corollary 5.

REFERENCES

[1] Adams, J. F., "Une Relation entre Groupes d'Homotopie et Groups de Cohomologie", *Comptes Rendues de l'Acad. des Sci., Paris*, 245; 24-26 (1957).

[2] __________, "On the Structure and Applications of the Steenrod Algebra", *Comm. Math. Helv.*, 32; 180-214 (1958).

[3] __________, "Vector Fields on Spheres", *Bull. Amer. Math. Soc.*, 68; 39-41 (1962).

[4] __________, "Vector Fields on Spheres", *Annals of Math.*, 75; 603-632 (1962).

[5] Anderson, D. W., and Hodgkin, L., "The K-theory of Eilenberg-MacLane Complexes", (preprint, to appear).

[6] Atiyah, M. F., "Vector Bundles and the Künneth Formula", *Topology*, 1; 245-248 (1962).

[7] Boardman, J. M., Thesis, Cambridge, 1964.

[8] __________, "Stable Homotopy Theory", mimeographed notes, University of Warwick, 1965 onward.

[9] Bourbaki, N., "Algebre Commutative", Chapters 1 and 2 of *Éléments de Mathématique*, 27, Hermann, (1961). (Act. Sci. et Ind. 1290)

[10] Brown, E. H., and Peterson, F. P., "A Spectrum Whose Z_p Cohomology is the Algebra of Reduced p^{th} Powers", *Topology*, 5; 149-154 (1966).

[11] Cartan, H., and Eilenberg, S., *Homological Algebra*, Princeton University Press (Princeton Mathematical Series, no. 19) (1956).

[12] Cohen, J. M., "Coherent Graded Rings", (preprint, to appear).

[13] __________, "The Non-existence of Spaces of Finite Stable Homotopy Type", (preprint, to appear).

[14] Conner, P. E., and Floyd, E. E., *The Relation of Cobordism to K-theories*, (Lecture Notes in Mathematics, no. 28) Springer, (1966).

[15] Douady, A., Seminaire H. Cartan 11 (1958/59), exposés 18, 19.

[16] Hattori, A., "Integral Characteristic Numbers for Weakly Almost Complex Manifolds", *Topology*, 5; 259-280 (1966).

[17] Hodgkin, L., "K-theory of Eilenberg-MacLane Complexes I, II", (preprints).

[18] MacLane, S., *Homology*, (Grundlehren der Math. #114) Springer, (1963).

[19] Milnor, J., "The Steenrod Algebra and Its Dual", *Annals of Math*, 67; 150-171 (1958).

[20] __________, "On the Cobordism Ring Ω^* and a Complex Analogue", *Amer. Jour. Math.*, 82; 505-521 (1960).

[21] __________, "On Axiomatic Homology Theory", *Pacific Jour. Math.*, 12; 337-341 (1962).

[22] __________, and Moore, J. C., "On the structure of Hopf Algebras", *Annals of Math.*, 81; 211-264 (1965).

[23] Novikov, S. P., "Rings of Operations and Spectral Sequences", *Doklady Akademii Nauk S.S.S.R.*, 172; 33-36 (1967).

[24] __________, *Izvestija Akademii Nauk S.S.S.R., Serija Matematiceskaja*, 31; 855-951 (1967).

[25] Puppe, D., "Stabile Homotopietheorie I", *Math. Annalen*, 169; 243-274 (1967).

[26] Serre, J.-P., "Cohomologie Modulo 2 des Complexes d'Eilenberg-MacLane", *Comm. Math Helv.*, 27; 198-232 (1953).

[27] __________, "Groupes d'Homotopie et Classes de Groupes Abéliens", *Annals of Math.*, 58; 258-294 (1953).

[28] Stasheff, J. D., "Homotopy Associativity of H-spaces I", *Trans. Amer. Math. Soc.*, 108; 275-292 (1963).

[29] Stong, R. E., "Relations Among Characteristic Numbers I", *Topology*, 4; 267-281 (1965).

[30] Thom, R., "Travaux de Milnor sur le Cobordisme", *Séminaire Bourbaki*, 11, no. 180 (1958/59).

[31] Whitehead, G. W., "Generalized Homology Theories", *Trans. Amer. Math. Soc.*, 102; 227-283 (1962).

ALGEBRAIC TOPOLOGY IN THE LAST DECADE

J. F. ADAMS

1. **Introduction.** I am grateful to the American Mathematical Society for offering me this rather challenging assignment of reporting on the progress of algebraic topology in the past decade. I should tell you how I will interpret this rather large subject. I will say little or nothing about the applications of algebraic topology. There will be little or nothing about the Atiyah-Singer Index Theorem. There will be little or nothing about the recent dramatic progress in the topology of manifolds stemming from the work of Kirby and Siebenmann. I know that such pieces of work deserve to be recorded in letters of gold; but on the one hand I expect that we will hear all about them at the International Congress of Mathematicians, Nice 1970; and on the other hand, I do not know them well enough to do them justice.

I will be concerned mainly with progress inside algebraic topology; and here I will be more concerned with methods which are likely to be useful in the future than with particular results. I hope that everything I say will come into the class of things which can usefully be remembered by those who have the care of graduate students.

Since algebraic topologists need to apply algebra, I will also mention certain pieces of algebra which may be useful to topologists. If I had been reporting on the progress of the previous decade, 1950–1960, I would surely have had to say something about homological algebra. Similarly for the last decade. So I will begin this lecture on the boundary between algebra and algebraic topology. In my second lecture I will take the theme of generalized homology and cohomology theories. In my third lecture I will discuss various more miscellaneous topics.

What, then, are the pieces of algebra which we have learnt in the past decade, which topologists should remember?

AMS 1970 *subject classifications.* Primary 18F25, 55-02, 55B20, 55E50; Secondary 18G10, 55B45, 55C05, 55E05, 55E10, 55G20, 55H05, 55H15, 55H20.

2. **Algebraic K-theory.** Homological algebra starts from the regrettable fact that not all modules are projective. Similarly, algebraic K-theory starts from the regrettable fact that not all projective modules are free.

This subject began in the previous decade, with a most germinal seminar by J.-P. Serre [1]. In this seminar, Serre exploits an analogy between vector bundles and projective modules. Let X be a compact Hausdorff space. Let $A = R(X)$ be the ring of continuous real-valued functions $f: R \to X$. Let ξ be a real vector-bundle over X. Let $\Gamma(\xi)$ be the set of continuous sections of ξ; then $\Gamma(\xi)$ is a module over A. Moreover, it is a finitely generated projective module; for we can find a complementary bundle η so that $\xi \oplus \eta$ is trivial of dimension n, say $\xi \oplus \eta \cong n$; then we have

$$\Gamma(\xi) \oplus \Gamma(\eta) \cong \Gamma(n) \cong A^n.$$

This construction sets up a (1-1) correspondence between isomorphism classes of vector bundles over X and isomorphism classes of finitely generated projective modules over A. The proofs here were carefully written up by R. G. Swan [2].

By means of this analogy, we can use our knowledge of vector bundles to suggest results, proofs and constructions for projective modules.

For example, suppose given any ring A. We consider functions f which assign to each finitely generated projective A-module P a value $f(P)$. Each function f is supposed to map into some abelian group, and to satisfy the following axioms:

(i) If $P \cong Q$, then $f(P) = f(Q)$.

(ii) $f(P \oplus Q) = f(P) + f(Q)$.

Among such functions f there is one which is universal; the group in which it takes its values is written $K_0(A)$ and called the projective class group of A. If we impose the further axiom

(iii) $f(A) = 0$,

we obtain the reduced group $\tilde{K}_0(A)$.

If we substitute $A = R(X)$, then $K_0(A)$ becomes the group $K^0(X)$ of Grothendieck, Atiyah, and Hirzebruch. However, for other applications we take smaller rings A, and here we rely on the algebraists to compute the group $K_0(A)$ for us. Similar remarks apply to the group $K_1(A)$, when I get so far. For an account of the theorems which allow you to compute these groups, see H. Bass [3].

As an application, I want to cite the work of C. T. C. Wall on finiteness obstructions; see C. T. C. Wall [4], [5].

Let X be a CW-complex, and let A be the integral group ring of its fundamental group, $A = Z(\pi_1(X))$. Let $\tilde{X}$ be the universal cover of X, and let us consider the chain groups $C_*(\tilde{X})$. (It does not matter much what sort of chain groups we use; let us use cellular chains.) Then $C_*(\tilde{X})$ is a chain complex of A-modules. Let us suppose, to begin with, that we can replace $C_*(\tilde{X})$ by another chain complex C', chain-equivalent over A to $C_*(\tilde{X})$, with the following properties.

(i) Each component C'_n is a finitely generated projective module over A.

(ii) The groups $H_n(C')$ and $H^n(C'; M)$ (where M is any A-module) are zero for n sufficiently large, say $n \geqq \nu$.

I will not discuss how to find such a C'; but if we cannot find such a C', then it is not even plausible to suppose that X is equivalent to a finite complex.

Suppose, then, that we have such a C'. By a further reduction, we can obtain a chain complex C chain equivalent over A to C' with the following properties.

(i) Each component C_n is a finitely generated projective module over A.

(ii) $C_n = 0$ for n sufficiently large, say $n \geqq \nu$.

This is very easy; you just take the chain complex C' and chop it off at a suitable point. Observe, however, that even if all the modules C'_n are free (which is the usual case), the last nonzero module C_n need not be free; we can only say that it is projective.

We can now form Wall's invariant

$$\chi(C) = \sum_i (-1)^i [C_i] \in \tilde{K}_0(A).$$

This depends only on the chain-equivalence class of C, and so it depends only on the homotopy type of X.

If X is homotopy-equivalent to a finite complex K then we can take $C = C_*(\tilde{K})$; $C_n(\tilde{K})$ is free, so $\chi(C) = 0$. On the other hand, there exist examples in which $\chi(C) \neq 0$. In this way Wall was able to answer an outstanding problem raised by J. H. C. Whitehead: if X is dominated by a finite CW-complex, does it follow that X is homotopy-equivalent to some finite complex? The answer is "no".

This work probably went along with Wall's work on surgery for non-simply-connected manifolds M. In this case also A is the group ring $Z(\pi_1(M))$. Wall's surgery obstructions take their values in a group G which can be considered as a Grothendieck group. To construct G, one considers projective A-modules P equipped with extra structure maps, such as a bilinear pairing $\varphi: P \times P \to A$. Until the appearance of Wall's book [6], the reference is C. T. C. Wall [7].

Another sort of module with extra structure is a finitely generated projective module P provided with an automorphism $f: P \to P$. If we think a bit about the properties of the determinant in the classical case, we see that we should consider functions d which assign to each pair (P, f) an element $d(f)$ in some multiplicative abelian group, and satisfy the following axioms.

(i) If

$$P \xrightarrow{f} P \xrightarrow{g} P,$$

then

$$d(gf) = d(g)d(f).$$

(ii) If

$$\begin{array}{ccccccccc} 0 & \longrightarrow & P & \xrightarrow{i} & Q & \xrightarrow{j} & R & \longrightarrow & 0 \\ & & \downarrow{\scriptstyle f} & & \downarrow{\scriptstyle g} & & \downarrow{\scriptstyle h} & & \\ 0 & \longrightarrow & P & \xrightarrow{i} & Q & \xrightarrow{j} & R & \longrightarrow & 0 \end{array}$$

(where the row is exact), then

$$d(g) = d(f)d(h).$$

(The classical case is that

$$\det \begin{bmatrix} B & C \\ 0 & D \end{bmatrix} = \det(B) \cdot \det(D).)$$

Among such functions d there is one which is universal; the group in which it takes its values is written $K_1(A)$.

Suppose that we are given an acyclic chain complex

$$0 \leftarrow C_0 \leftarrow C_1 \leftarrow \cdots \leftarrow C_n \leftarrow 0$$

in which the C_i are finitely generated free A-modules with given bases, and the boundary maps are A-maps. (Data a little weaker than "given bases" would actually suffice.) Then it turns out that we can associate to such an acyclic chain complex an invariant in $K_1(A)$. This way of looking at things has led to a much clearer understanding of Reidermeister-Franz torsion, and of J. H. C. Whitehead's theory of "simple homotopy type". For a good exposition, see J. Milnor [**8**].

The groups $K_i(A)$ for $i \geqq 2$ are so far more interesting to number theorists than to topologists; but for the latest material, see R. G. Swan [**9**].

3. **Derived functors of the inverse-limit functor.** The next piece of algebra which I think deserves comment concerns the derived functors of the inverse-limit functor. Let I be a directed set of indices α; then we can consider inverse systems of abelian groups $\mathbf{G} = \{G_\alpha, g_{\alpha\beta}\}$. These inverse systems form the objects of a category. The inverse limit is representable in this category; for let $\mathbf{Z}$ be the inverse system with $Z_\alpha = Z$, $z_{\alpha\beta} = 1$; then

$$\text{Hom}\,(\mathbf{Z}, \mathbf{G}) \cong \varprojlim_\alpha \mathbf{G}.$$

Moreover, this category is one in which we can do homological algebra; in particular, there are enough injectives. We can therefore form the functors

$$\lim{}^i\, \mathbf{G} = \text{Ext}^i\,(\mathbf{Z}, \mathbf{G}).$$

We have $\lim^0 = \varprojlim_\alpha$.

The use of these functors goes back to J. Milnor [**10**]. Let H^* be a generalized cohomology theory satisfying the wedge axiom, which says that the canonical map

$$\tilde{H}^*\Big(\bigvee_\alpha X_\alpha\Big) \to \prod_\alpha \tilde{H}^*(X_\alpha)$$

is an isomorphism. (One can use H^* instead of $\tilde{H}^*$ if one uses the disjoint union instead of the wedge.) Suppose given an increasing sequence of CW-pairs (X_n, A_n), and set

$$X = \bigcup_n X_n, \qquad A = \bigcup_n A_n.$$

Then Milnor shows that we have an exact sequence

$$0 \to \lim{}^1 H^{q-1}(X_n; A_n) \to H^q(X; A) \to \lim{}^0 H^q(X_n; A_n) \to 0.$$

In particular, substituting X for X_n and X_n for A_n, we have

$$\lim{}^1 H^*(X; X_n) = 0, \qquad \lim{}^0 H^*(X; X_n) = 0.$$

In this situation, when I is countable, we have $\lim^i \mathbf{G} = 0$ for $i \geqq 2$.

The careful use of these derived functors enables one to overcome many of the difficulties which arise in the use of inverse limits. For example, suppose that we wish to construct a map $\mu: BU \wedge BU \to BU$ corresponding to the tensor product of virtual bundles of virtual dimension zero. Let X_n be an increasing sequence of finite CW-complexes whose union is BU. We easily construct maps $\mu_n: X_n \wedge X_n \to BU$ and verify that they give an element of

$$\lim{}^0 [X_n \wedge X_n, BU].$$

Since

$$[BU \wedge BU, BU] \to \lim{}^0 [X_n \wedge X_n, BU]$$

is an epimorphism, we see that there is a map $\mu: BU \wedge BU \to BU$ which restricts to each μ_n (up to homotopy). Now suppose we wish to check the associativity of μ, that is, to check that $\mu(\mu \wedge 1) \sim \mu(1 \wedge \mu): BU \wedge BU \wedge BU \to BU$. We easily check that $\mu(\mu \wedge 1)$ and $\mu(1 \wedge \mu)$ have the same image in

$$\lim{}^0 [X_n \wedge X_n \wedge X_n, BU].$$

In order to check that $\mu(\mu \wedge 1)$ and $\mu(1 \wedge \mu)$ are homotopic, we simply check that

$$\lim{}^1 [X_n \wedge X_n \wedge X_n, U] = 0.$$

This "$\lim^1 = 0$" argument, which is commonplace today, was unknown before 1960; I attribute it to John Milnor.

For some applications it is important to know how inverse limits work in spectral sequences. For the first examples, see D. W. Anderson's thesis **[11]** and M. F. Atiyah **[12]**. Suppose, for example, that we take a generalized cohomology theory H satisfying the wedge axiom and a CW-complex X containing an increasing sequence of subcomplexes $\emptyset = X_{-1} \subset X_0 \subset X_1 \subset \cdots \subset X_n \subset \cdots \subset X$. Suppose also that

$$\lim{}^0 H^*(X; X_n) = 0, \qquad \lim{}^1 H^*(X; X_n) = 0.$$

(For example, we might have $X = \bigcup_n X_n$.) Applying H^*, we obtain a half-plane spectral sequence with

$$E_1^{p,q} = H^{p+q}(X_p; X_{p-1}).$$

In what sense does this spectral sequence converge? We may be interested in three conditions.

(i) Observe that $E_{r+1}^{p,q} \to E_r^{p,q}$ is monomorphic for $r > p$. So we can ask that the map

$$E_\infty^{p,q} \to \lim_r^0 E_r^{p,q}$$

should be an isomorphism.

(ii) Similarly, we can ask that

$$\lim_r^1 E_r^{p,q} = 0.$$

(iii) Let $F^{p,q}$ be the filtration quotients of $H^{p+q}(X)$, so that we have exact sequences

$$0 \to E_\infty^{p,q} \to F^{p,q} \to F^{p-1,q+1} \to 0$$

and $F^{-1,q} = 0$. We can ask that the map

$$H^n(X) \to \lim_p^0 F^{p,n-p}$$

should be an isomorphism.

THEOREM 3.1. *Condition* (ii) *is equivalent to* (i) *plus* (iii).

In practice we verify condition (ii); it is equivalent to check that $E_{p+1}^{p,q}$ is complete for the topology defined by the subgroups $E_r^{p,q}$, $r > p$. (Here we use words so that "complete" does not imply "Hausdorff"; it means that each Cauchy sequence has a limit, perhaps not unique.) We then use the theorem to deduce that conditions (i) and (iii) hold.

This result is stronger than previous results which rely on the Mittag-Leffler condition for spectral sequences. I have taken most of it from a seminar by J. M. Boardman, but it may also be included in the work of Eilenberg and Moore [**13**].

We can also generalize the result of Milnor. For convenience I consider the absolute case. Let X be any CW-complex which is the union of a directed set of subcomplexes X_α. Then we have a spectral sequence

$$\lim_\alpha^p H^q(X_\alpha) \underset{p}{\Rightarrow} H^{p+q}(X).$$

And if you want to know in what sense this spectral sequence is convergent, it is convergent in the sense that Theorem 3.1 holds.

It has been claimed by Jensen that there is an example of an inverse system **G** of abelian groups such that $\lim^i \mathbf{G} \neq 0$ for all $i > 0$, but I have not seen the proof. The reference is C. U. Jensen [**14**].

4. **Coalgebraic structures.** The next piece of progress which I think deserves comment is the increasing use of coalgebraic structures. Suppose I have in sight some ground ring R. Then an algebra A over R is an R-module provided with an R-linear product map $\varphi: A \otimes_R A \to A$. Of course φ has to satisfy suitable axioms. Similarly, an A-module M is an R-module provided with an R-linear product map $\varphi': A \otimes_R M \to M$. Again, φ' has to satisfy suitable axioms.

Dually, a coalgebra C over R is an R-module provided with an R-linear coproduct map $\psi: C \to C \otimes_R C$. Of course ψ has to satisfy suitable axioms. Similarly, a C-comodule N is an R-module provided with an R-linear map $\psi': N \to C \otimes_R M$. Again, ψ' has to satisfy suitable axioms.

These definitions, of course, were current long before 1960. I was writing about the cobar construction in 1956, and coalgebras were not new then. But I do claim that in the past decade people have become increasingly used to thinking in coalgebraic terms. Probably topologists were much influenced by the work of Milnor and Moore [**15**].

However, I want to illustrate the point from the homology of fibre spaces. At the end of the last decade it appeared that if you wanted to do homological calculations in fiberings, and if you wanted to work at the chain level, so as to get results more precise than those afforded by the Serre spectral sequence, then the best prospect for the future was some form of twisted tensor product; see E. H. Brown [**17**].

Of course one had the work of Cartan on "constructions", from the previous decade; and it was not long before Eilenberg put matters in the following form. The ordinary theory of Tor and Ext refers to modules over a ring A. However, it does not take much trouble to replace the ring A by a differential graded algebra. Let G be (say) a topological group, and BG its classifying space; then we have

$$H_*(BG; R) \cong \mathrm{Tor}_*^{C_*(G;R)}(R, R), \qquad H^*(BG; R) \cong \mathrm{Ext}^*_{C_*(G;R)}(R, R).$$

It follows that we have spectral sequences

$$\mathrm{Tor}_*^{H_*(G;R)}(R, R) \Rightarrow H_*(BG; R), \qquad \mathrm{Ext}^*_{H_*(G;R)}(R, R) \Rightarrow H^*(BG; R).$$

The canonical joke at this point is that one defines these spectral sequences by filtering according to the number of bars used by Eilenberg and Mac Lane in their bar construction.

These spectral sequences were obtained in a much more geometrical way by Rothenberg and Steenrod [**18**]. Their work remains usable when one wishes to replace ordinary homology by generalized homology.

So far we have been talking about going from G to BG. The converse problem is to go from X to ΩX. The results, at any rate, can be formulated by turning the arrows around; the reference is S. Eilenberg and J. C. Moore [**19**].

First we dualize the definition of the tensor product. Suppose that A is a differential graded algebra, and L, M are differential graded algebras over A, L being a right module and M a left module. Then the definition of $L \otimes_A M$ says that the following sequence is exact:

$$L \otimes A \otimes M \xrightarrow{\varphi\otimes 1 - 1\otimes\varphi} L \otimes M \to L \otimes_A M \to 0.$$

(Here $\otimes$ means $\otimes_R$.) Now let C be a differential graded coalgebra, and let L, M be differential graded comodules over C, L being a right comodule and M a left comodule. We define the cotensor product $\square_C$ so that the following sequence is exact:

$$0 \to L \,\square_C\, M \to L \otimes M \xrightarrow{\psi \otimes 1 - 1 \otimes \psi} L \otimes C \otimes M.$$

The cotensor product has derived functors, and they are written Cotor.

Now suppose given the following diagram.

$$\begin{array}{ccc} E' & \longrightarrow & E \\ {\scriptstyle \pi'}\downarrow & & \downarrow{\scriptstyle \pi} \\ B' & \xrightarrow{\;b\;} & B \end{array}$$

We assume that π is a fibering and that π' is the fibering induced by b. In particular, if B' is a point, then E' will be the fibre F of π. Then $C_*(B; R)$ can be made into a coalgebra, and $C_*(B'; R)$, $C_*(E; R)$ can be made into comodules over this coalgebra. One theorem of Eilenberg and Moore states that under mild restrictions we have

$$H_*(E'; R) \cong \mathrm{Cotor}_*^{C_*(B;R)}(C_*(B'; R), C_*(E; R)).$$

With this viewpoint, the cobar construction is regarded as a canonical resolution for computing Cotor.

Under suitable assumptions, such as the assumption that R is a field, we get a spectral sequence

$$\mathrm{Cotor}_*^{H_*(B;R)}(H_*(B'; R), H_*(E; R)) \Rightarrow H_*(E'; R).$$

(Filter according to the number of cobars used by Coeilenberg, etc.) This is the Eilenberg-Moore spectral sequence, which has already found applications.

5. **Generalized homology and cohomology theories.** A generalized homology or cohomology theory is a functor which satisfies the first six axioms of Eilenberg and Steenrod, but does not necessarily satisfy the seventh axiom, the dimension axiom. As long as we are dealing with finite complexes these axioms suffice. If we have to deal with infinite complexes we have to impose further axioms. In homology, it is appropriate to assume that the obvious map

$$\operatorname{dir\,lim}_{\alpha} H_n(X_\alpha) \to H_n(X)$$

is an isomorphism; here X_α runs are the finite subcomplexes of X. In cohomology we assume the wedge axiom: the obvious map

$$\tilde{H}^n\left(\bigvee_\alpha X_\alpha\right) \to \prod_\alpha \tilde{H}^n(X_\alpha)$$

is an isomorphism.

In the past decade we have seen a lot of progress in this area. It may be that the progress did not come earlier because we did not have convincing examples of such functors; for the beginning of K-theory, see Atiyah and Hirzebruch [**20**]. But by now we have several sorts of K-theory and several sorts of bordism or cobordism, and it is clear that they work; for example, they allow you to solve geometric problems.

The first theorem in this subject is E. H. Brown's Representability Theorem (see E. H. Brown [**21**], [**22**]). Suppose given a contravariant functor H. It should be defined on the category in which the objects are CW-complexes with base-point, and the morphisms are homotopy classes of maps; both maps and homotopies are to preserve the base-point. H should take values in the category of sets. Then Brown's Theorem gives very simple necessary and sufficient conditions in order that we have a natural isomorphism $H(X) \cong [X, Y]$ for some fixed CW-complex Y. The first condition is the wedge axiom: the obvious map $H(\bigvee_\alpha X_\alpha) \to \prod_\alpha H(X_\alpha)$ is an isomorphism of sets. The second condition is the Mayer-Vietoris axiom. Consider the following diagram.

$$\begin{array}{ccc} H(U \cup V) & \longrightarrow & H(U) \\ \downarrow & & \downarrow \\ H(V) & \longrightarrow & H(U \cap V) \end{array}$$

We ask that if $u \in H(U)$, $v \in H(V)$ have the same image in $H(U \cap V)$, then they should be the images of a single class $w \in H(U \cup V)$.

Brown's Theorem, then, says that if a functor $H(X)$ satisfies these two conditions, then it has the form $[X, Y]$ up to natural isomorphism. This is a very useful theorem, because we can use it to construct CW-complexes Y.

(i) For the most trivial example, suppose given any space Z and consider the functor $[X, Z]$; we get a CW-complex Y and a natural isomorphism $[X, Y] \cong [X, Z]$. That is, we get a CW-complex weakly equivalent to Z.

(ii) Take $H(X)$ to be the ordinary homology group $\tilde{H}^n(X; \pi)$. We obtain a complex Y, and we easily check that we have obtained an Eilenberg-Mac Lane complex of type (π, n); for

$$\begin{aligned} \pi_r(Y) = [S^n, Y] &\cong \tilde{H}^n(S^r; \pi) \\ &= \begin{cases} \pi & (n = r), \\ 0 & (n \neq r). \end{cases} \end{aligned}$$

(iii) Suppose given a topological group G. We want to construct the classifying space BG. So we define a functor $H(X)$ by taking the G-bundles over X and classifying them into isomorphism classes. (Actually one has to modify this definition a little to take account of the base-point, but this is just a technicality.) We check that $H(X)$ satisfies the axioms, and we receive a CW-complex Y with the property that G-bundles over X are classified by maps $f: X \to Y$.

It may sometimes happen that one's functor H is defined not on all CW-complexes, but only on some subcategory. For example, suppose we define the Grothendieck-Atiyah-Hirzebruch functor $\tilde{K}(X)$ in terms of vector-bundles, in the classical way. This definition is good only for finite-dimensional complexes X; if we tried to use it for infinite-dimensional complexes, it would not satisfy the wedge axiom. All the same, the theorem of E. H. Brown remains true if the functor H is given only on finite-dimensional complexes. Of course, the complex Y which is constructed will not be finite dimensional, in general.

Similarly, suppose that the functor H is given only on finite CW-complexes. In this case I will assume that the functor H takes values in the category of groups. If so, then E. H. Brown's Theorem is still true. (Again, the complex Y which is constructed will not be a finite complex, in general.)

Suppose now that we have a generalized cohomology theory H. Then we have a sequence of functors $\tilde{H}^n$ satisfying the conditions of Brown's Theorem. So we get a sequence of representing complexes Y_n such that $\tilde{H}^n(X) \cong [X, Y_n]$. Also we have isomorphisms $\tilde{H}^n(X) \cong \tilde{H}^{n+1}(SX)$; these correspond to weak equivalences $Y_n \simeq \Omega Y_{n+1}$; by taking the adjoint we obtain maps $SY_n \rightarrow Y_{n+1}$.

In this way we reach the notion of a *spectrum*, first published by Lima in the previous decade. According to G. W. Whitehead, a spectrum is a sequence of CW-complexes E_n provided with maps $e_n: SE_n \rightarrow E_{n+1}$.

EXAMPLES. (i) The Eilenberg-Mac Lane spectrum, in which $E_n = K(\Pi, n)$.

(ii) The sphere spectrum, in which $E_n = S^n$.

(iii) The MU-spectrum of Thom and Milnor, in which $E_{2n} = MU(n)$.

(iv) The BU-spectrum, in which E_{2n} is the space BU.

Now we come to the work of G. W. Whitehead [**23**]. According to G. W. Whitehead, a spectrum determines not only a generalized cohomology theory, but also a generalized homology theory. The definition is

$$\tilde{E}_n(X) = \lim_{n \to \infty} \pi_{m+n}(E_m \wedge X).$$

For example, taking $E = MU$ we obtain a homology theory called complex bordism. It was already known that bordism theories were important; see Atiyah [**24**].

THEOREM 5.1 *Every generalized homology theory defined on* CW-*complexes can be obtained by G. W. Whitehead's construction from a suitable spectrum E.*

It is convenient to use the same letter for a spectrum and for its associated homology and cohomology theory. For this reason I will use K for the BU-spectrum and KO for the BO-spectrum.

Now I want to go on and point out that it is highly desirable to have a good category of spectra in which we can work and do stable homotopy theory. For one thing, we would like to make the following definition for the generalized cohomology of a spectrum:

$$E^n(X) = [S^{-n}X, E].$$

Here $S^{-n}X$ means a suitable desuspension of X, and $[A, B]$ means homotopy classes of maps from A to B in our category of spectra. Unfortunately we cannot do this if we use G. W. Whitehead's definition of a "map of spectra".

Again, it would be very convenient if we had smash products of spectra in our category. For example, when G. W. Whitehead considers products in generalized cohomology theories, he has to introduce the notion of a pairing of spectra from E and F to G. It would be very convenient if this were simply a map $\mu: E \wedge F \to G$ in our category.

Let me give an example to show how easy life is when you have the right formalism.

PROPOSITION 5.2. *Assume that E is a ring-spectrum, F is a module-spectrum over E, and $E_*(X) = 0$. Then $F^*(X) = 0$.*

This is both a special case of the Universal Coefficient Theorem and a lemma for use in setting up the general case.

PROOF. Take any map $f: X \to F$. We can factor it in the following way.

$$\begin{array}{ccccc} & & E \wedge X & & \\ & \overset{i\wedge 1}{\nearrow} & & \overset{1\wedge f}{\searrow} & \\ S \wedge X & & & & E \wedge F \\ \uparrow{\scriptstyle \simeq} & & & & \downarrow{\scriptstyle \mu} \\ X & & \xrightarrow{\quad f \quad} & & F \end{array}$$

(Here S is the sphere-spectrum, which acts as a unit for the smash-product; and $i: S \to E$ is the unit map for the ring-spectrum E.) We are given $\pi_*(E \wedge X) = 0$, so $E \wedge X$ is contractible and $f \sim 0$. Similarly with X replaced by $S^{-n}X$.

This proof is certainly very easy; but you try doing it without a category in which you can make this argument.

The good category in which to do stable homotopy theory was constructed by Boardman. We all hope that he will publish a readable and definitive account of it; till then, I recommend the exposition by R. Vogt **[25]**.

Let me give a quick description of a category equivalent to Boardman's. I restrict attention to spectra E_n in which each map $e_n: SE_n \to E_{n+1}$ is an isomorphism from the complex SE_n to a subcomplex of E_{n+1}. This is no great restriction. On occasion we neglect the maps e_n and regard S^rE_n as embedded in E_{n+r}. A subspectrum $E' \subset E$ is said to be *cofinal* if for each cell $e \subset E_n$ there exists an r such that $S^re \subset E'_{n+r}$. That is, each cell of E gets into E' after enough suspensions.

The maps which we now take from E to F are sequences $f_n: E'_n \to F_n$, where E'_n is a cofinal subspectrum of E and the following diagram is strictly commutative for each n.

$$\begin{array}{ccc} SE'_n & \longrightarrow & E'_{n+1} \\ {\scriptstyle Sf_n}\downarrow & & \downarrow{\scriptstyle f_{n+1}} \\ SF_n & \longrightarrow & F_{n+1} \end{array}$$

An example will illustrate the reason we take this definition. Take two spectra with

$$E_n = S^{n+3} \vee S^{n+7} \vee S^{n+11} \vee \cdots, \qquad F_n = S^n.$$

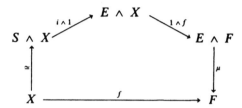

We would like to make a map from E to F whose component from S^{n+4k-1} to S^n is a generator for the image of J in the stable $(4k - 1)$-stem. But there is no single value of n for which all the requisite maps exist as maps into S^n; we have to concede that for the different cells of E the maps come into existence for different values of n. In other words, the maps exist on a cofinal subspectrum of E. The slogan is "cells now—maps later".

The notion of homotopy is now obvious; a homotopy has to be given on a cofinal subspectrum of the spectrum $\{E_n \wedge (I/\emptyset)\}$.

This gives a category which has all the good properties of the category of CW-complexes.

Let me come back to generalized homology and cohomology. It is now more or less accepted that practically everything which one used to do with ordinary homology and cohomology can be carried over to generalized homology and cohomology. We have the great majority of the classical tools, and we can calculate as well as one could reasonably expect. Of course, before you calculate $E_*(X)$ for any other X, you must know the result when X is a point, that is, the coefficient groups $\pi_*(E)$. I should summarize progress in this direction.

For real and complex K-theory the coefficient groups are determined by the Bott periodicity theorem. I should perhaps cite the proof of this theorem by Atiyah and Bott **[26]**. We have also the generalization to the real case by R. Wood **[27]**, and its subsequent further generalization by Atiyah **[28]**, and by M. Karoubi **[29]**.

For the various bordism theories, $\pi_*(MO)$ and $\pi_*(MSO)$ were computed before the decade started. So was $\pi_*(MU)$, although Milnor did not immediately publish his paper **[30]**. A very firm hold on $\pi_*(MU)$ was provided by the Hattori-Stong theorem. See R. Stong **[31]**, Hattori **[32]**. Although this theorem was originally presented as determining the image of $\pi_*(MU)$ in $H_*(MU)$, it seems now to be accepted that it is best stated in terms of $K_*(MU)$. Further enlightenment was provided by the work of Quillen **[33]**. I have tried to give an exposition in my lecture notes, University of Chicago, 1970. On SU-bordism and spin-bordism I refer you to the following sources: Conner and Floyd **[34]**; Anderson, Brown, and Peterson **[35]**, **[36]**; Wall **[37]**. The structure of $\pi_*(MSp)$ is still not known.

Let me now go on to talk about products. The basic products are defined in G. W. Whitehead's paper, cited above; they are four in number.

(i) An external product in cohomology, say

$$u \otimes v \to u \mathbin{\bar{\wedge}} v : E^p(X) \otimes F^q(Y) \to (E \wedge F)^{p+q}(X \wedge Y).$$

Of course, if you have a pairing of spectra, that is, a map of spectra $E \wedge F \to G$, you can apply the resulting induced homomorphism to obtain an answer in $G^{p+q}(X \wedge Y)$.

(ii) An external product in homology,

$$u \otimes v \to u \mathbin{\underline{\wedge}} v : E_p(X) \otimes F_q(Y) \to (E \wedge F)_{p+q}(X \wedge Y).$$

(iii) and (iv). Two slant products

$$u \otimes v \to u/v \; : \; E^p(X \wedge Y) \otimes F_q(Y) \to (E \wedge F)^{p-q}(X),$$
$$u \otimes v \to u\backslash v \; : \; E^p(X) \otimes F_q(X \wedge Y) \to (E \wedge F)_{q-p}(Y).$$

Note. If you wish to make the notation convenient, give a fraction the same variance as its numerator, the opposite variance from its denominator.

The definitions of the products are all easy if one can work in a suitable category.

Apart from obvious naturality properties, the products satisfy two anti-commutative laws and eight associative laws and have a unit. The external products are associative and anticommutative. The remaining six associative laws are easily interpreted as obvious rules for manipulating fractions; for example,

$$u \mathbin{\overline{\wedge}} (v/w) = (u \mathbin{\overline{\wedge}} v)/w.$$

Unfortunately the eight associative laws are not given in G. W. Whitehead's paper; any student of the subject should certainly make a list of them. For the case of ordinary homology and cohomology, six out of the eight can be found scattered through the pages of Spanier's book.

We can now summarize the position about duality in manifolds. Let E be a ring-spectrum, and suppose that M is a manifold for which you have chosen an orientation class in, say, E-cohomology of the tangent bundle. Then the account of duality in Spanier's book goes over word for word; the statements are all true, and the proofs are formally the same once you have the toolkit.

What has to be done for any particular E, then, is to construct good orientations for some useful class of bundles. This can be nontrivial. For example, consider Atiyah, Bott, and Shapiro [**38**]. In this paper they construct a KO-orientation for Spin-bundles. They take particular care that their orientation behaves well on a bundle which happens to be the Whitney sum of two others; in my terminology, they take care to construct a map of ring-spectra

$$M\,\mathrm{Spin} \to KO.$$

One can also orient U-bundles over complex K-theory, and of course G-bundles over MG for any of the usual choices for G (take the identity map $MG \to MG$).

I understand from Wall that Sullivan has shown that if you ignore the prime 2, then PL-bundles can be oriented over KO; but I myself cannot answer for the statement or the proof.

I would now like to cite the theorem of Conner and Floyd [**39**]. I just said that you can orient U-bundles over K. More precisely, we have a map of ring-spectra from MU to K. This leads to a remarkably simple sort of Universal Coefficient Theorem; we have an isomorphism

$$MU_*(X) \otimes_{\pi_*(MU)} \pi_*(K) \xrightarrow{\cong} K_*(X).$$

Similarly for

$$MSp_*(X) \underset{\pi_*(MSp)}{\otimes} \pi_*(KO) \xrightarrow{\cong} KO_*(X).$$

The theory of characteristic classes with values in a generalized cohomology theory is now fairly well understood, at least if you stick to any of the usual sorts of bundle and any of the usual cohomology theories. For example, take $U(n)$ bundles; the Chern classes in ordinary cohomology have analogues in complex K-theory (Grothendieck, Atiyah) and in complex cobordism (see the work by Conner and Floyd cited above).

The theory of formulae of "Riemann-Roch type", involving a multiplicative factor, is also well understood; indeed it goes back to the previous decade. The most elementary situation here is as follows. Take a bundle ξ (over X, say) which has been oriented over two theories E^*, F^*; and suppose given a map of ring-spectra $\alpha : E \to F$. Then we have the following diagram, but it is not commutative.

$$\begin{array}{ccc} \tilde{E}^*(X^\xi) & \xrightarrow{\alpha} & \tilde{F}^*(X^\xi) \\ \Big\uparrow{\scriptstyle \varphi_E} & & \Big\uparrow{\scriptstyle \varphi_F} \\ E^*(X) & \xrightarrow{\alpha} & F^*(X) \end{array}$$

(Here X^ξ means the Thom complex of ξ, and φ_E, φ_F are Thom isomorphisms.) We have

$$\varphi_F^{-1}\alpha\varphi_E x = \alpha x(\varphi_F^{-1}\alpha\varphi_E 1).$$

Here $\varphi_F^{-1}\alpha\varphi_E 1$ is a characteristic class of ξ, which can be evaluated in practical cases. Similarly if we study manifolds and consider duality isomorphisms (instead of φ), homomorphisms induced by maps of manifolds (instead of α) or the "Umkehrungshomomorphismus".

For the more interesting applications of generalized homology and cohomology we require cohomology operations in E-cohomology, or something similar. Here one can obviously consider $E^*(E)$, which is the algebra of stable cohomology operations on E-cohomology. If we take E to be the Eilenberg-Mac Lane spectrum $K(Z_p)$, we get the Steenrod algebra. We can also form $E_*(E)$. If we take E to be the Eilenberg-Mac Lane spectrum $K(Z_p)$, we get the dual of the Steenrod algebra, which was studied by Milnor. I have shown that under mild restrictions on E, $E_*(E)$ is a Hopf algebra in a good sense, and $E_*(X)$ is a comodule over $E_*(E)$ [**40**].

Examples. (i) $E = K$ and KO; real and complex K-theory. $E^*(E)$ is very bad. There are however unstable operations in K-theory. Atiyah has related them to Steenrod's method for constructing operations in ordinary cohomology: see Atiyah [**41**]. In this case $E_*(E)$ is very good. It has been computed by Adams, Harris, and Switzer [**42**].

(ii) $E = MU$. $E_*(E)$ and $E^*(E)$ are strictly dual. They are both very good, and have been computed by Novikov [**43**]. The Steenrod approach was carried out by tom Dieck [**44**].

(iii) $E = BP$, the Brown-Peterson spectrum. $E_*(E)$ and $E^*(E)$ are strictly dual. They are both very good. $E^*(E)$ has been computed by Quillen [33]. The interpretation in terms of $E_*(E)$ is given in my lecture notes, Chicago 1970.

(iv) $E = MSp$. The analogues of Novikov's results on MU are true, but until we know $\pi_*(MSp)$ they are not so much use.

(v) $E = bu$, connective K-theory. Some progress both on $E_*(E)$ and on cohomology operations with suitable coefficients has been made by D. W. Anderson.

I have left myself with little space to mention applications. We have had quite a lot of fun with K-theory; I suspect that the quick profits have probably been reaped, except maybe for some of the fancy forms of K-theory, like the equivariant K-theory of Atiyah and Segal. As for bordism functors like complex bordism, Conner and Floyd have already showed us that it can be very profitable in studying manifolds. I think some of the recent work of Larry Smith, Toda, and others on the spaces $V(n)$ shows that it will probably soon be profitable in homotopy theory.

6. **The groups** $J(X)$. Historically, the subject to be considered in this section arose from two sources. The first was the study of the stable J-homomorphism

$$J : \pi_r(SO) \to \pi_{n+r}(S^n) \qquad (n > r + 1);$$

this is of interest to homotopy-theorists, and also to differential geometers. The second was a method which occurred to Atiyah for proving some results of I. M. James about the problem of vector-fields on spheres. This led to Atiyah's celebrated paper [45].

Let us take a finite CW-complex X, say connected. We can take the real vector-bundles over X, and proceeding in the usual way we obtain the Grothendieck group $KO(X)$. We can also define a quotient group of $KO(X)$ by considering an equivalence relation between bundles which is cruder than isomorphism, namely, fibre homotopy equivalence of the associated sphere bundles. This quotient group is called $J(X)$. Equivalently, $J(X)$ is the image of the following homomorphism:

$$[X, Z \times BO] \to [X, Z \times BG].$$

Here BG is the classifying space for spherical fibrations. This homomorphism is of some interest to differential topologists.

For the usual reason, we have $J(X) \cong Z + \tilde{J}(X)$. One of the central observations in Atiyah's paper was the fact that $\tilde{J}(X)$ is a finite group. If we substitute $X = S^{r+1}$, then $\tilde{J}(S^{r+1})$ is the image of the stable J-homomorphism

$$J : \pi_r(SO) \to \pi_{n+r}(S^n) \qquad (n > r + 1).$$

In order to compute $J(X)$, we must do two things. First, if two sphere-bundles are not fibre homotopy equivalent, we must construct invariants to prove it. The best-known fibre homotopy invariants are the Stiefel-Whitney classes. The reason they are invariant is as follows. Take an $(n - 1)$-sphere bundle E_0 over X, and regard it as the boundary of the associated disc bundle E. We have the following diagram.

$$\begin{array}{ccc} H^n(E, E_0; Z_2) & \xrightarrow{Sq^i} & H^{n+i}(E, E_0; Z_2) \\ \uparrow \varphi & & \uparrow \varphi \\ H^0(X; Z_2) & & H^i(X; Z_2) \end{array}$$

Here φ is the Thom isomorphism, and the Stiefel-Whitney class w_i is given by $w_i = \varphi^{-1}Sq^i\varphi 1$. It is clear that a fibre homotopy equivalence will give a homotopy equivalence from the pair E, E_0 to another such pair, and this will not alter anything. So w_i is a fibre homotopy invariant.

We can copy this argument, replacing $H^*(X; Z_2)$ by $KO^*(X)$ and Sq^i by the operation Ψ^k. However the proof of invariance needs to be changed a little, because a fibre homotopy equivalence $E, E_0 \to E', E'_0$ will generally carry the orientation class $\varphi'1$ not into $\varphi 1$, but into some other class. We have to allow for this in formulating our invariant. In fact, we have to consider the vector $\{\varphi^{-1}\Psi^k\varphi 1\}$ not as an element of $\prod_k KO(X)$, but as an element of a suitable quotient.

We can now define a quotient $J'(X)$ of $KO(X)$ as follows: two bundles ξ and η are to become equal in $J'(X)$ if the fibre homotopy invariant indicated above takes the same value on ξ as on η. It follows immediately that the quotient map $KO(X) \to J'(X)$ factors through $J(X)$; we have the following commutative diagram.

$$\begin{array}{ccc} KO(X) & \longrightarrow & J(X) \\ & \searrow & \downarrow \\ & & J'(X) \end{array}$$

The second thing we have to do to compute $J(X)$ is as follows. If two sphere-bundles are fibre homotopy equivalent, we must give a method for proving it. For this purpose I got interested some time ago in the following formulation.

(6.1) Suppose given a finite complex X, an element $x \in KO(X)$ and an integer k. Then there is an integer n such that $k^n(\Psi^k - 1)x$ maps to zero in $J(X)$.

I proved a few simple special cases of this statement, and explored its relation to other statements. Everybody now seems agreed to call it the Adams conjecture, so I suppose I had better go along with that.

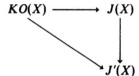

The first piece of progress with it was that Quillen found a line of argument, using algebraic geometry, which would have been a proof if the algebraic geometers had proved one theorem which he needed; unfortunately they had not. The reference is Quillen [**46**]. Sullivan also did a good deal of work on the problem, but I am not too well informed about it. He may well have a proof. Finally Quillen came up with a proof of the Adams conjecture which is now circulating in preprint form. Atiyah has given Quillen's proof in lectures at Oxford. It now seems safe to assume that Quillen's proof is correct, so we can now make use of the Adams conjecture (6.1).

The obvious way to make use of it is to define a group $J''(X)$ so that (6.1) provides us with the following commutative diagram.

$$\begin{array}{ccc} & \nearrow & J''(X) \\ & & \downarrow \\ KO(X) & \longrightarrow & J(X) \\ & & \downarrow \\ & \searrow & J'(X) \end{array}$$

But I had already proved that the quotient map $J''(X) \to J'(X)$ is an isomorphism; see [**47**]. So we have $J''(X) = J(X) = J'(X)$, and we can compute $J(X)$.

For example, the proof of the Adams conjecture should allow one to compute $J(QP^n)$, and so settle the problem of the cross-sections of the symplectic Stiefel fiberings.

As a more classical example, it follows that the image of the stable J-homomorphism

$$J : \pi_r(SO) \to \pi_{n+r}(S^n)$$

has precisely the order suggested by the work of Milnor and Kervaire [**48**], as improved by Atiyah and Hirzebruch [**20**].

I should also mention that if we restrict attention to $X = S^{r+1}$, then there is another proof in the pipeline about the behaviour of the classical stable J-homomorphism. It comes from Mark Mahowald, and I will sketch some of the ideas behind it.

If we want to compute the stable homotopy groups $\pi_*^S(X)$, or at least their p-components, we have a spectral sequence

$$\mathrm{Ext}_{A_*}^{s,t}(Z_p, \tilde{H}_*(X; Z_p)) \underset{s}{\Rightarrow} \pi_{t-s}^S(X).$$

Here Z_p and $\tilde{H}_*(X; Z_p)$ are considered as comodules over the coalgebra A_* dual to the Steenrod algebra. We would like to construct an analogous spectral sequence in which ordinary homology theory $\tilde{H}_*(X; Z_p)$ is replaced by some generalized homology theory. So we take a ring-spectrum E, and form cofibre sequences by induction:

$$X = X_0 \simeq S \wedge X_0 \xrightarrow{i \wedge 1} E \wedge X_0 \longrightarrow X_1,$$
$$X_1 \simeq S \wedge X_1 \xrightarrow{i \wedge 1} E \wedge X_1 \longrightarrow X_2, \quad \text{etc.}$$

Mapping the sphere-spectrum S into these X_{i-1} we obtain a spectral sequence. It is easy to prove that it converges to $\pi_*^S(X)$ if we assume suitable hypotheses on

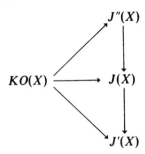

E; for example, assume

$$\pi_r(E) = 0 \quad \text{for } r < 0, \qquad H_0(E) = Z, \qquad H_1(E) = 0.$$

If $E_*(E)$ is flat over $\pi_*(E)$ then the E_2 term of the spectral sequence is

$$\mathrm{Ext}^{s,t}_{E_*(E)}(E_*(S), E_*(X));$$

but you do not have to stick to this case if you can compute the E_2 term or the E_1 term some other way.

This construction is so easy that everybody thinks it's his idea. In particular, I think it is my idea. I even think I published it; see "Lectures on generalized cohomology" [**40**].

The case $X = S$, $E = bu$ (connective complex K-theory) has been studied by D. W. Anderson. Mahowald's proof is based on a study of the case $X = S$, $E = bo$ (connective real K-theory). It also uses ideas from his previous paper [**49**], in which he proved the result up to the dimension 2^{12}. I understand that an announcement by Mahowald is to appear in the Bulletin of the American Mathematical Society [**50**].

7. **Stable homotopy theory.** The material on the J-homomorphism leads me on to stable homotopy theory.

Well, now, some topologists do have a certain love-hate relationship with the stable homotopy groups of spheres. By using their preferred methods they have, in the past decade, about doubled the dimensional range in which they are intimate with these creatures. However, the philosophy does seem to be gaining ground that for real interest one should try for theorems which give an infinite amount of information. I would like to refer to one method which seems new and worthwhile. It was first published by Toda for the case of an odd prime p; see [**51**], [**52**]. He has used it to settle some quite subtle questions in stable homotopy theory.

I will describe the construction for the case $p = 2$; the generalization to the case of other primes is fairly obvious. Let X be, say, a finite CW-complex with base-point x_0. Form

$$\frac{X \times X \times S^n}{(x_0 \times X \times S^n) \cup (X \times x_0 \times S^n)}$$

and let Z_2 act on it by

$$(x, y, s) \longrightarrow (y, x, -s).$$

Let the quotient be $Q^n(X)$.

If $X = S^r$ we can work out what $Q^n(X)$ is; it is obtained by suspending a suitable quotient of real projective spaces RP^a/RP^b.

The functor Q^n does not send cofiberings into cofiberings; it is not "linear", but "quadratic". Nevertheless, the deviation from linearity is small, and if X is a small complex like $S^r \cup_f e^t$ we can work out a good deal about the cell-structure of $Q^n(X)$. This leads to relations between elements of homotopy groups, suggests the construction of secondary homotopy operations, and is otherwise useful.

For an application which I think is quite worthwhile, see D. S. Kahn [53].

8. **Unstable homotopy theory.** Here I would like to refer to various points. There has been some work on unstable analogues of the Adams spectral sequence. First we had the work of Massey and Peterson [54]. Their method does impose certain restrictions on X, but these are satisfied if $X = S^n$. Then we had a good deal of work by the MIT school, using CSS methods; see in particular, Bousfield, Curtis, Kan, Quillen, Rector, and Schlesinger [55], and Rector [56].

I would also like to cite the paper by Boardman and Steer [57]. It seems to me that this paper says the last word on the formula which replaces the distributive law for $(f + g)h$.

I would also like to mention the work of Barratt and Mahowald [58]. This is a very clear and satisfactory result, but most people seem to feel that they do not quite know how to fit it into their general picture of homotopy theory.

To end this section, I would like to mention the work of Mahowald [59]. Certainly the results here contain more than a finite amount of information. However, the methods lead me on to the subject of cohomology operations, which is the next topic.

9. **Secondary operations in ordinary cohomology.** There are some questions in homotopy theory which one naturally reduces to questions about secondary operations, either by obstruction theory or by the method of killing homotopy groups. If we dismiss the questions which we can easily answer, we are left with various more subtle questions. For example, one may want to know a Cartan formula for $\Phi(xy)$, including those terms which involve primary operations but no secondary operations. Or one may know that a secondary operation Φ applied in a particularly low dimension gives the same result as some primary operation, and then one wants to know which primary operation it is.

For such questions the most powerful method and the most precise information is contained in the papers of Kristensen. See Kristensen [60], [61], [63]; Kristensen and Madsen [62].

Outsiders should be warned, however, that the method is hard work. Although painless general methods like acyclic models are used whenever possible, the basic method is one which relies on definite cochain operations.

10. **Iterated loop-spaces and infinite loop-spaces.** One of the basic papers here, of course, came in the previous decade: Kudo and Araki [64]. Kudo and Araki set up the basic operations in homology with mod 2 coefficients. Early in this decade we had the corresponding operations in homology with mod p coefficients: Dyer and Lashof [65]. In general, however, the subject was unjustly neglected until later in the decade. Then we had some very successful work on G and BG, the classifying space for spherical fibrations. Some of the references are: A. Tsuchiya [66]; R. J. Milgram [67]; J. Peter May [68]; Ib Madsen [69].

You have heard Peter May explain how a fairly clean way of setting up some of the basic machinery is provided by the work of Boardman and Vogt [**70**]; see also mimeographed notes with the same title, University of Warwick, 1968. The work of Boardman and Vogt, in its turn, may perhaps be made even smoother by forthcoming work of G. Segal.

NOTE ADDED IN PROOF. This subject has been making rapid progress recently. See Part IV of the survey by J. Stasheff in this symposium [**16**].

11. ***H*-spaces.** I will treat this topic in the most sketchy fashion, since I am sure that Stasheff will give you a much better account of it later. I would like to mention two points about H-spaces. First the well-known work of Peter Hilton and Joseph Roitberg [**71**]. J. Hubbuck [**72**] succeeded in classifying finite complexes which are homotopy-commutative H-spaces.

REFERENCES

1. J.-P. Serre, *Modules projectifs et espaces fibrés à fibre vectorielle*, Séminaire P. Dubreil, Dubreil-Jacotin et C. Pisot, 1957/58, Exposé 23, Secrétariat mathématique, Paris, 1958. MR **31** #1277.

2. R. G. Swan, *Vector bundles and projective modules*, Trans. Amer. Math. Soc. **105** (1962), 264–277. MR **26** #785.

3. H. Bass, *Algebraic K-theory*, Benjamin, New York, 1968. MR **40** #2736.

4. C. T. C. Wall, *Finiteness conditions for* CW-*complexes*, Ann. of Math. (2) **81** (1965), 56–69. MR **30** #1515.

5. ———, *Finiteness conditions for* CW-*complexes*. II, Proc. Roy. Soc. Ser. A **295** (1966), 129–139.

6. ———, *Surgery on compact manifolds*, Academic Press, New York, 1970.

7. ———, *Surgery of non-simply-connected manifolds*, Ann. of Math. (2) **84** (1966), 217–276. MR **35** #3692.

8. J. Milnor, *Whitehead torsion*, Bull. Amer. Math. Soc. **72** (1966), 358–426. MR **33** #4922.

9. R. G. Swan, *Non-abelian homological algebra and K-theory*, Proc. Sympos. Pure Math., vol. 17, Amer. Math. Soc., Providence, R.I., 1970.

10. J. Milnor, *On axiomatic homology theory*, Pacific J. Math. **12** (1962), 337–341. MR **28** #2544.

11. D. W. Anderson, *The real K-theory of classifying spaces*, Thesis, Univ. of California, Berkeley, Calif., 1964.

12. M. F. Atiyah, *Characters and cohomology of finite groups*, Inst. Hautes Études Sci. Publ. Math. No. 9 (1961), 23–64. MR **26** #6228.

13. S. Eilenberg and J. C. Moore, *Limits and spectral sequences*, Topology **1** (1962), 1–23.

14. C. U. Jensen, *On the vanishing of* $\varprojlim^{(i)}$, J. Algebra **15** (1970), 151–166.

15. J. W. Milnor and J. C. Moore, *On the structure of Hopf algebras*, Ann. of Math. (2) **81** (1965), 211–264. MR **30** #4259.

16. James D. Stasheff, *H-spaces and classifying spaces: Foundations and recent developments*, Proc. Sympos. Pure Math., vol. 22, Amer. Math. Soc., Providence, R.I., 1971.

17. E. H. Brown, Jr., *Twisted tensor products*. I, Ann. of Math. (2) **69** (1959), 223–246. MR **21** #4423.

18. M. Rothenberg and N. E. Steenrod, *The cohomology of classifying spaces of H-spaces*, Bull. Amer. Math. Soc. **71** (1965), 872–875. MR **34** #8405.

19. S. Eilenberg and J. C. Moore, *Homology and fibrations*. I. *Coalgebras, cotensor product and its derived functors*, Comment. Math. Helv. **40** (1966), 199–236. MR **34** #3579.

20. M. F. Atiyah and F. Hirzebruch, *Riemann-Roch theorem for differentiable manifolds*, Bull. Amer. Math. Soc. **65** (1959), 276–281. MR **22** #989.

21. E. H. Brown, Jr., *Cohomology theories*, Ann. of Math. (2) **75** (1962), 467–484; correction, ibid. (2) **78** (1963), 201. MR **25** #1551; MR **27** #749.

22. ———, *Abstract homotopy theory*, Trans. Amer. Math. Soc. **119** (1965), 79–85. MR **32** #452.

23. G. W. Whitehead, *Homology theories and duality*, Trans. Amer. Math. Soc. **102** (1962), 277–283.

24. M. F. Atiyah, *Bordism and cobordism*, Proc. Cambridge Philos. Soc. **57** (1961), 200–208. MR **23** #A4150.

25. R. Vogt, *Boardman's stable category*, Lecture Note Series, no. 21, Aarhus Universitet, Aarhus, 1970.

26. M. F. Atiyah and R. Bott, *On the periodicity theorem for complex vector bundles*, Acta Math. **112** (1964), 229–247. MR **31** #2727.

27. R. Wood, *Banach algebras and Bott periodicity*, Topology **4** (1965), 371–389. MR **32** #3062.

28. M. F. Atiyah, *K-theory and reality*, Quart. J. Math. Oxford Ser. (2) **17** (1966), 367–386. MR **34** #6756.

29. M. Karoubi, *Algèbres de Clifford et K-theorie*, Ann. Sci. École Norm. Sup. (4) **1** (1968), 161–270. MR **39** #287.

30. J. W. Milnor, *On the cobordism ring Ω^* and a complex analogue*. I, Amer. J. Math. **82** (1960), 505–521. MR **22** #9975.

31. R. E. Stong, *Relations among characteristic numbers*. I, Topology **4** (1965), 267–281. MR **33** #740.

32. A. Hattori, *Integral characteristic numbers for weakly almost complex manifolds*, Topology **5** (1966), 259–280. MR **33** #742.

33. D. Quillen, *On the formal group laws of unoriented and complex cobordism theory*, Bull. Amer. Math. Soc. **75** (1969), 1293–1298. MR **40** #6565.

34. P. E. Conner and E. E. Floyd, *The* SU-*bordism theory*, Bull. Amer. Math. Soc. **70** (1964), 670–675. MR **29** #5253.

35. D. W. Anderson, E. H. Brown, Jr., and F. P. Peterson, SU-*cobordism*, KO-*characteristic numbers, and the Kervaire invariant*, Ann. of Math. (2) **83** (1966), 54–67. MR **32** #6470.

36. ———, *The structure of the spin cobordism ring*, Ann. of Math. (2) **86** (1967), 217–298. MR **36** #2160.

37. C. T. C. Wall, *Addendum to a paper of Conner and Floyd*, Proc. Cambridge Philos. Soc. **62** (1966), 171–175. MR **32** #6472.

38. M. F. Atiyah, R. Bott and A. Shapiro, *Clifford modules*, Topology **3** (1964), suppl. 1, 3–38. MR **29** #5250.

39. P. E. Conner and E. E. Floyd, *The relation of cobordism to K-theories*, Lecture Notes in Math., no. 28, Springer-Verlag, Berlin and New York, 1966. MR **35** #7344.

40. J. F. Adams, *Lectures on generalized cohomology*, Lecture Notes in Math., no. 99, Springer-Verlag, Berlin and New York, 1966.

41. M. F. Atiyah, *Power operations in K-theory*, Quart. J. Math. Oxford Ser. (2) **17** (1966), 165–193. MR **34** #2004.

42. Adams, Harris, and Switzer, *Hopf algebras of cooperations for real and complex K-theory*, Proc. London Math. Soc. (to appear).

43. S. P. Novikov, *The methods of algebraic topology from the viewpoint of cobordism theory*, Izv. Akad. Nauk SSSR Ser. Mat. **31** (1967), 855–951 = Math. USSR Izv. **1** (1967), 827–913. MR **36** #4561.

44. T. tom Dieck, *Steenrod-Operationen in Kobordismen-Theorien*, Math. Z. **107** (1968), 380–401. MR **39** #6302.

45. M. F. Atiyah, *Thom complexes*, Proc. London Math. Soc. (3) **11** (1961), 291–310. MR **24** #A1727.

46. D. G. Quillen, *Some remarks on étale homotopy theory and a conjecture of Adams*, Topology **7** (1968), 111–116. MR **37** #3572.

47. J. F. Adams, *On the groups $J(X)$*. III, Topology **3** (1965), 193–222. MR **33** #6627.

48. J. W. Milnor and M. W. Kervaire, *Bernoulli numbers, homotopy groups, and a theorem of Rohlin*, Proc. Internat. Congress Math. 1958, Cambridge Univ. Press, New York, 1960, pp. 454–458. MR **22** #12531.

49. M. Mahowald, *On the order of the image of J*, Topology **6** (1967), 371–378. MR **35** #3663.

50. ———, *On the order of the image of the J-homomorphism*, Bull. Amer. Math. Soc. **76** (1970), 1310–1313.

51. H. Toda, *An important relation in homotopy groups of spheres*, Proc. Japan Acad. **43** (1967), 839–842. MR **37** #5872.

52. ———, *Extended p-th powers of complexes and applications to homotopy theory*, Proc. Japan Acad. **44** (1968), 198–203. MR **37** #5873.

53. D. S. Kahn, *Cup-i products and the Adams spectral sequence*, Topology **9** (1970), 1–9. MR **40** #6552.

54. W. S. Massey and F. P. Peterson, *The* mod 2 *cohomology structure of certain fibre spaces*, Mem. Amer. Math. Soc. No. 74 (1967). MR **37** #2226.

55. A. K. Bousfield, E. B. Curtis, D. M. Kan, D. G. Quillen, D. L. Rector and J. W. Schlesinger, *The* mod-*p lower central series and the Adams spectral sequence*, Topology **5** (1966), 331–342. MR **33** #8002.

56. D. L. Rector, *An unstable Adams spectral sequence*, Topology **5** (1966), 343–346.

57. J. M. Boardman and B. Steer, *On Hopf invariants*, Comment. Math. Helv. **42** (1967), 180–221. MR **36** #4555.

58. M. G. Barratt and M. E. Mahowald, *The metastable homotopy of $O(n)$*, Bull. Amer. Math. Soc. **70** (1964), 758–760. MR **31** #6229.

59. M. E. Mahowald, *Some Whitehead products in S^n*, Topology **4** (1965), 17–26. MR **31** #2724.

60. L. Kristensen, *On secondary cohomology operations*, Math. Scand. **12** (1963), 57–82. MR **28** #2550.

61. ———, *On a Cartan formula for secondary cohomology operations*, Math. Scand. **16** (1965), 97–115. MR **33** #4926.

62. L. Kristensen and Ib Madsen, *On evaluation of higher order cohomology operations*, Math. Scand. **20** (1967), 114–130. MR **36** #5936.

63. L. Kristensen, *On secondary cohomology operations*. II., Conf. on Algebraic Topology (Univ. of Illinois at Chicago Circle, Chicago, Ill., 1968), Univ. of Illinois at Chicago Circle, Chicago, Ill., 1969, pp. 117–133. MR **40** #3539.

64. T. Kudo and S. Araki, *Topology of H_n-spaces and H-squaring operations*, Mem. Fac. Sci. Kyūsū Univ. Ser. A. **10** (1956), 85–120. MR **19**, 442.

65. E. Dyer and R. K. Lashof, *Homology of iterated loop spaces*, Amer. J. Math. **84** (1962), 35–88. MR **25** #4523.

66. A. Tsuchiya, *Characteristic classes for spherical fiber spaces*, Proc. Japan Acad. **44** (1968), 617–622. MR **40** #2115.

67. R. J. Milgram, *The* mod 2 *spherical characteristic classes*, Ann. of Math. (2) **92** (1970), 238–261.

68. J. Peter May, *Geometry of iterated loop spaces*, in preparation; see also his contribution to this symposium.

69. Ib Madsen, *On the action of the Dyer-Lashof algebra in $H_*(G)$ and $H_*(G/\mathrm{Top})$*, Thesis, Chicago, Ill., 1970.

70. J. M. Boardman and R. M. Vogt, *Homotopy-everything H-spaces*, Bull. Amer. Math. Soc. **74** (1968), 1117–1122. MR **38** #5215. See also: University of Warwick Mimeographed Notes, 1968.

71. P. Hilton and J. Roitberg, *On principal S^3 bundles over spheres*, Ann. of Math. (2) **90** (1969), 91–107. MR **39** #7624.

72. J. Hubbuck, *On homotopy commutative H-spaces*, Topology **8** (1969), 119–126.

University of Manchester and
University of Cambridge, England

CPSIA information can be obtained at www.ICGtesting.com
262462BV00005B/5/P

9 780521 110679